Boat Joinery & Cabinetmaking

SIMPLIFIED

Boat Joinery & Cabinetmaking

SIMPLIFIED

Fred P. Bingham

Illustrated by the Author

INTERNATIONAL MARINE

CAMDEN, MAINE

Formerly published as *Practical Yacht Joinery* in 1983

Published by International Marine

10 9 8 7 6 5 4 3 2

Library of Congress

Bingham, Fred P., 1907–
Boat joinery and cabinetmaking simplified / Fred P. Bingham.
p. cm.
Rev. ed. of: Practical yacht joinery. 1983.
Includes bibliographical references and index.
ISBN 0-87742-354-7
1. Yacht building. 2. Joinery. 3. Cabinetwork. I. Bingham, Fred P., 1907– Practical yacht joinery. II. Title.
VM331.B585 1993
623.8'223—dc20 93-2392
CIP

Questions regarding the content of this book should be addressed to:

International Marine
P.O. Box 220
Camden, ME 04843

Questions regarding the ordering of this book should be addressed to:

TAB Books
A Division of McGraw-Hill, Inc.
Blue Ridge Summit, PA 17294
1-800-233-1128

Boat Joinery and Cabinetmaking Simplified is printed on 60-pound Renew Opaque Vellum, an acid-free paper which contains 50 percent recycled waste paper (preconsumer) and 10 percent postconsumer waste paper.

Printed by R. R. Donnelley, Harrisonburg, VA
Composition by Farrar Associates, White Horse Beach, MA
Design by James Brisson
Production by Janet Robbins
Edited by J.R. Babb, Michael Brosnan, Tom McCarthy

DEDICATION

The original *Practical Yacht Joinery*, published in 1983, was sparked by a casual remark of exasperation.

When my son Bruce was still designing yachts back in 1974, I often peeked over his shoulder. As I was an old-time boatbuilder (then building the hull plug for his famous *Flicka*) I would offer now and then a suggestion such as, "If you designed it this way, it might be easier"—or stronger, or better looking, or less expensive. Often he would make the changes then and there. But after a few months of this, taken with good grace and some warm discussion, he exploded, "If you know so much, why don't you go write a book? To get you out of here I'll do your . . . drawings!" That day I started jotting down ideas, and six years later I was ready for his incomparable illustrations (remember *Sailor's Sketchbook?*), without which I would never have written a word. But by then Bruce had been living aboard and cruising between Long Island, the Bahamas, and Florida for five years. He was up to his spreaders in illustrations, rarely in communication with the world ashore, and at least two years away from starting my drawings. So I brushed up on my World War II training in perspective drawing and, with trepidation (and encouragement from International Marine), I produced about 400 drawings for the following pages.

During those years my wife, Vivian, pushed me when I faltered, put up with my messy desk and boards and our considerable lack of income, and accepted it all with a serene facade. So now I thank Bruce for his instigation, International Marine for its vision, and Vivian for her support. Without any one of these three, nothing would have happened.

CONTENTS

ACKNOWLEDGMENTS

This acknowledgment may be a little different; most are primarily a list of influential people and of books that were useful to the author or recommendations to the learning-seeker. This one is an attempt to boil down into a page or two all the influences that shaped one long lifetime.

I would whither and blow away without water nearby. I learned to sail while I was trying my footsteps. My father guided my woodworking education when I was seven or thereabouts. At thirteen I built a precise 36-inch sailing model that required an outboard for pursuit. I used and enjoyed this woodworking ability at every opportunity throughout my life.

I began filing and classifying pages from *The Rudder* and *Yachting* in the late 1920s, lost my file in 1940, began again and gave the file to my son Bruce in 1974. I made a religion around the Herreshoffs, Nat and L. Francis; compiled fat folders about the great sailyacht designers—John Alden, William Atkin and John, S.S. Crocker, Uffa Fox, Charles G. MacGregor, Olin Stephens, Winthrop Warner, Fenwick Williams, Ralph Winslow, Charles Wittholz, Nelson Zimmer, others too many to name, and especially the great Howard I. Chapelle (I built his 36-foot "three sail bateau"). In 1936, Chap's *Yacht Designing and Planning*, added to my old *Skene's Elements of Yacht Design*, cemented the path of my life. In 1941 came Chapelle's boatbuilder's bible, *Boatbuilding*, and in 1951 *American Small Sailing Craft*. From these and many other works as well as from my studies of those yacht design geniuses, I acquired a modest knowledge of lines, construction, rigs, sails, and beauty. In the meantime, long before those dates, I had built, rigged, and sailed several skiffs and, in 1931, a real sailboat, a Snipe. Lightning struck. I *had* to build boats.

A heavy influence was *the* pioneer in plywood, Charles G. MacGregor, a *complete* naval architect, engineer, author, and iconoclast. He was almost unknown for his large commercial vessels and famous racing and fishing schooners (registered under names of prestigious firms), such as *Gertrude Thibaud*, but famed for countless power and sail yachts under his own banner. In times when plywood was a seven-letter word he designed durable plywood aircraft (*Mosquito Bomber*) in the 1930s, was a consultant to plywood manufacturers and then a regular feature contributor to *The Rudder*. His smallest and best-known design was the 8-foot dinghy, *Sabot*. In the four-year life of my Detroit boat shop, beginning in 1937, I built five of MacGregor's designs: five *Norseman* 15-foot fin-keelers, a 27-foot voyaging sloop, *Threesome*, four *Defender* 23-foot fin-keelers, three plank-on-frame *Shipmite* cruising sloops, and started a 42-foot ketch when the Depression nabbed me in 1940. The 23-foot raised-deck cruisers had (*horrors*!) masts stepped on plywood decks with just one heavy beam under, no column for support. Impossible! But all were successful and long lived. He taught me that there were two or more ways of doing almost everything, and that low-cost does not mean cheap. My book larnin' and modest experience enabled me to work successfully with my team of French, Scottish, Swedish, and German *artists* in wood. And not only to learn from *them*, but to introduce them to this "new" boatbuilding medium, plywood. Early in 1940 at Fisher Boat Works I was exposed to some of the finest workmanship—experimental torpedo boats, yacht-like sub-chasers, and major yacht repairs. And there for six months I was privileged to hide away in the same drafting room with Nelson Zimmer, N.A., a talented yacht designer and meticulous draftsman whose beautiful boats have been featured in *WoodenBoat* magazine, and who now devotes his time to very large yachts and special passenger ships. As for me, the coming war lured me into aircraft production, technical writing, then industrial advertising and other occupations far removed from my love. From then on for twenty-five years I was an amateur boatbuilder.

In 1972, chafing for a woodworking activity, I set up a small shop and built custom cabinets, bookcases, entertainment centers, and simple contemporary furniture of my design; I gained a small income and tons of satisfaction. Many of my cabinetwork methods and tricks were born of trial and error and contingencies of one-man operations. Working alone also bolstered my visits to the Santa Barbara Library. Here are most of the books and other publications that were helpful in my shop and in writing the original *Practical Yacht Joinery*:

New Complete Woodworking Handbook, by Jeannette T. Adams and Emmanuele Stieri

The Home Workshop, by William R. Wellman

Complete Book of Home Workshops, by David K. Manners

How to Build Your Own Workshop Equipment, by Arthur Wakeling

Fun with a Saw, by R. J. de Cristoforo

Complete Book of Woodworking, by Rosario Capotosto

Cabinetmaking, by Paul Haynie

Woodwork Joints, by Charles N. Hayward

How to Work with Tools and Wood, by Robert N. Wood and N.H. Mager

These acknowledgments must include the excellent *Boatbuilding Manual* (revised edition), originally a series by Robert M. Steward in *The Rudder*; L. Francis Herreshoff's *Common Sense of Yacht Design*, also from *The Rudder* (now lamentably out of print.) Today, for the ultimate in boat construction, buy, beg, borrow, or steal *WoodenBoat* magazine; for equal quality on cabinetry, furniture, etc., there is one leader, *Fine Woodworking*, and many others with which I am not familiar but that, nevertheless, have earned popularity. Local libraries are excellent sources of information on woodworking and boatbuilding—the latter in both classical and contemporary designs. Pay them a visit.

INTRODUCTION

In revising and expanding this book, I have not removed or changed substantially anything that won applause from our boat-loving friends. But I *have* created a new section for you builders and would-be builders of woodwork for your homes, offices, shops, and cabins. The job is to show you the close relationship between the two, and that the minute details in the earlier sections are almost entirely applicable to the broader woodworking field. This expanded edition is for you who want to learn *how* to select, use, and maintain your tools with which to create wooden things of beauty, usefulness, and value. Readers of *Practical Yacht Joinery* taught me that tools are tools, that "yacht joinery tools" and procedures are doing similar jobs for them in both environments. One reader wrote, "I used PYJ extensively when fitting out the interior of a small house I've been building over the past ten years or so. Without it, I never would have been able to figure out how to make apparently complex but actually easy-to-make joints for the built-in furniture we used to save space, nor would I have found that poor-man's radial-arm saw, the SLAT."

If you now use tools, techniques, and tips similar to those described in *Practical Yacht Joinery* and this expanded edition, you are qualified to build just about any complex cabinetry as well as the fine, sometimes specialized, joinery for a lovely yacht. If you can accept a new or merely unusual approach, a lot of know-how can be yours in a few hours of pleasant reading.

By definition a yacht is a watercraft designed, built, and fitted for pure pleasure or racing (*pleasure?*). However, it must also embody beauty of form and finish, excellent materials and fittings, and meticulous workmanship—the *same* words that apply to your better cabinets and furniture but not necessarily to your utility storage. But the yacht must also have the strength to butt into a head-sea for days on end, make pleasing progress under all conditions, provide safety and comfort as well as *pride* of ownership—and there the two meet. A sound yacht can be a long-term investment, paying dividends from its first day in the water, and, like a good home, it may appreciate or depreciate in value. But if it is of sloppy construction it may sell for a fraction of the price of its better-built counterpart. So it is with a treasure chest, a home, a bookcase. You will decide whether yours will be a yacht or a clunker, an heirloom or a crate; *you* can control quality more than you realize, even though at this moment you may not be a skilled craftsman. Only by building the *best* way you know how can you earn that pride of accomplishment, the *real* reward. Do you realize that your simple contemporary cabinet, if well done, can become an art object to be willed to your children's children? But first you must learn how to build one the best way.

The last fifteen years or so I have seen an *explosion* of interest in construction of wooden boats and completion of fiberglass hulls or kits, both home-built and professional, and even more so in cabinetmaking and in the hundreds of new tools. Is this growth responsible for wonderful magazines such as *WoodenBoat* and *Fine Woodworking*, or have such distinguished publications sparked this interest and fanned the flames? This atmosphere gives me assurance that this "how-to" book is timely, filling a need not just for you who are less sure of yourselves, but also for you who are at or near the "pro" level. For example, Chapter Five is on building your own bench and floor machines, tools, jigs, and accessories, some from kits, more from scraps in your shop—valuable to carpenters, cabinetmakers, and boatbuilders. If you fit in these high categories, take only the bits you *need* from Chapters One through Six (the selection, uses, and maintenance of hand and power tools); one or two shop tips could be answers to old problems. Chapters Seven through Fifteen are oriented toward boat work, describing alternate ways of construction—cabin soles (floors), bulkheads, laminated beams, decks, cabin structures, hatches, cockpits, rails, spars. But also such homey woodwork as doors, drawers, tables, settees (berths), counters, medicine chests, and vanities, plus the myriad joints that tie all together. Be prepared for some methods to be less than traditional to boatbuilding or cabinetry, but born of one-man small-shop expediency, simplicity, convenience, and especially low-cost.

In Chapter 16 I continue descriptions of work procedures, fabricating and assembly, noting options when appropriate. Except for a cedar chest masquerading as a seaman's chest, all are from our old shop, shown in my non-professional photographs and drawings everyone understands. Jobs range from a "first time" simple bedside telephone stand to a contemporary mahogany hutch. Chapter Seventeen is the *piece de resistance*—a plan for an 11½-foot sloop-rigged dinghy (car-toppable) built on the fast-growing "stitch-and-tape" system for plywood and fiberglass, requiring no building jig or experience. It is a lot more interesting to build and sail than the ubiquitous 8-foot variety so popular through the past forty-five years. I include this hoping that our new cabinetmaking friends will decide to give that boat-filled world a try.

Plans for boats are much more complex than for cabinets. Plans sources differ, too. Consequently, I prescribe a large dose of advice for you who lean to the nautical, but one a bit easier to swallow for you drylanders.

Boat lovers: Don't fall irrevocably in love with a specific design. Face facts: yours may be a multi-thousand dollar job. Can you pay the bills that pile up for one, two, three years? Have you a free or nearly free building location? Can the boat be covered, with tools, machines, and materials safe from weather and thieves? Have you the capabilities in woodworking, fiberglassing, and mechanics with which to bring this dream to a satisfying, divorceless conclusion? It's possible that this book may bolster your courage, so look far ahead, aim high, plan for ten, twenty years down the road. This is a job equal to planning and building a small home; a complete professionally built 30-foot, 11,000-pound cruising sailyacht (at 1993 prices) sells for anywhere from $90,000 to $135,000 ($8.50 to $12 per pound). The cost of your materials may range from 30 percent to 50 percent of that, less if you are a scrounger (spars, hardware, engine, sails, lead, etc.). Maybe your confidence will lean you toward a 30-foot cutter instead of a 25-foot sloop. If your workmanship is more than adequate, your yacht may be worth $45,000 to $60,000; your skill can make a vast difference. And here is news: your yacht with living facilities is mortgageable, just like a second home, so your "sweat equity" might provide a source for funds after the completion—*if* design, materials, and workmanship are sound.

For boat plans, read the magazines, recent and old; notice the designer whose boats please you—racing, daysailing, coastal cruising, bluewater voyaging, racing cruising? Write him, compliment him, ask about "stock" plans, the type and size you think you want. Explain your experiences—sailing, boatbuilding, carpentry, repairing fiberglass, welding. Tell him how many you are likely to have aboard, your family situation, your long-term plans, a hint as to your financial position and anticipated launching date, whether you are working from scratch, a bare hull, or a kit. His fee may range from $250 to possibly $450, and should included patterns (less if you loft it) and consultation. Some designers price proportionate to displacement (the weight of water displaced by a floating object); some furnish many prints, manuals, article reprints; others may not, so *ask*. Check ads in *WoodenBoat*, *Cruising World*, *SAIL*, and other magazines for mail order plans and catalog outfits such as Benford Design Group, Fred Bingham Yacht Design, Clarkcraft, Glen-L Marine, Ken Hankinson Associates, Bruce Roberts, and others. Most of these also offer kits, frame sets, etc.

Mystic Seaport Museum in Mystic Seaport, Connecticut, is a fabulous source for all kinds of boat and yacht designs of the past—from dinghies, peapods, and wherries to famous yachts and the racing fishing schooners of the 1920s to 1940s. Write to: 50 Greenmandville Ave., Mystic, CT 06355, or phone 203-572-0711.

Now a plea: Plans are a trifle, a fraction of one percent of the total, so it is sheer mental incompetence to try something you dreamed up in order to save a pittance and risk many dollars and your own and other's lives. It's OK to digress slightly, squeeze in a locker, a roomier head/vanity, a navigation desk—but a shower plus holding tank (not in the plans)? Watch out! If you require more headroom or less draft, or more berths, or greater power, etc., please buy another design. Never add to or change a hull or deck structure in any way without consulting the designer. Seemingly small interior changes that shift, raise, lower, increase, or decrease weights in the wrong direction can cause a disaster. Bolster your ego in a less dangerous way!

Read the books I mentioned in my acknowledgments. And don't forget "the old hat trick." Put on your hat and go visit marinas, boat yards, sales lots, brokerages. Take along your camera. Talk to boat owners especially, but *never*, never step aboard a boat without permission from the owner. If you treat him with courtesy he might invite you to take a look below, possibly photograph interesting joiner work. Shoot all the details, learn how each was made and put together, from rails, decks, and hatches, to pin rails, winch bases, cockpits, to stemhead treatments, bowsprit pulpits, and spars. Record teak decking, gratings, joints in covering boards and rail caps. Don't trust your memory; photograph a permanent record.

Now for you Drylanders: How do you get started in cabinetry? First, peruse the words above directed at boat lovers; there are parallels. I repeat with emphasis my advice in the Acknowledgments. Read! Read! Read!

Second, don't settle for something you just knocked together out of your noggin! Build to a *plan*, even if it's so basic it proves only that the piece will fit your space, shows dimensions of lumber or plywood, and footage needed. You can make all kinds of mistakes on paper, but none in wood. Profit from the experience of the many fine designers and writers in this field. Public libraries provide shelf after shelf of fine books on woodworking; look for those that do tell you *how* to construct the illustrated projects. Amazingly, many supermarkets carry as many craft manuals as do the lumberyards and hardware stores, mostly paperbacks at very good prices. Some of the larger newspapers run syndicated features on woodworking projects with plans offered for minimal fees.

Don't forget the enormously popular magazines, *Fine Woodworking*, *Fine Homebuilding*, *Popular Mechanics*, *Home Mechanics*, and perhaps a dozen others with "wood" in their titles, all available at better newsstands.

Third, try "the old hat trick" for yourself; put yours on, then go forth to examine the excellent specimens in furniture and unfinished woodwork stores. Perhaps a neighboring cabinetmaker could be talked into letting you look at some of his workmanship.

Fourth, watch "This Old House," the great television series that show the details of carpentry, cabinetmaking, and home modernizing.

Last but far from least, if you're looking for a fascinating and profitable lifetime occupation, or just to improve your present skills, look for an industrial arts course, perhaps a night school.

I hope that *Boat Joinery and Cabinetmaking Simplified* becomes a substitute for experience gained by working at the side of a crotchety old craftsman, picking up some of his secrets, his odd ways, which, in the end, work miracles. I can't offer everything, or be all things to all men, but I emphasize that in addition to learning, the main factors in the woodworker's *art* are patience, perseverance, practice, and experimentation.

PART 1

1 The Woodworker's Basic Hand Tools

The selection, use, and maintenance of woodworking tools and machinery require close scrutiny, good judgment, and careful work habits. Close scrutiny to ensure that you select tools that will stand you in good stead over a long period; good judgment to ensure that you don't go overboard and purchase more tools than you will need; and careful work habits for obvious reasons. The next four chapters will describe the selection, use, and maintenance of tools and machines suitable for yacht joinery and general cabinetry.

Not all are essential; almost any job can be done with hand tools alone. If, on the other hand, you are going to invest a large sum in the yacht or furniture itself, plan also to set aside a substantial kitty for tools.

If you have already built your hull, you probably have much of what is needed to complete the job. If you plan to purchase a hull, however, and the proposed vessel's value would be about $50,000 if professionally built, you probably should allocate as much as $2,500 for tools and equipment. If your yacht will be in the $5,000 class, you might benefit by investing up to $1,000 on tools. Don't be alarmed. If you buy quality machines and tools, you could recoup 75 percent of their cost, perhaps more, if you decide to sell them. Depending on the rate of inflation, you might even make a profit.

As you read along, you'll be able to decide which tools you need and can rent or finance right now, and which to build and add later as work progresses. Any tools you make from my instructions will sell for at least half the price of the average professionally manufactured item. Thus, you stand to save substantially if you build some of your own tools.

If you are looking for information on the traditional tools used by skilled boatbuilding craftsmen in the "good old days," you won't find much here. If you have access to special-purpose tools and learn how to use them, more power to you. By all means pick up any handmade wooden planes you find, but be prepared to pay high prices.

With the hand tools listed in this chapter you'll be able to do almost anything in joinery and cabinetry that you want to do. Throughout history, fine work has been done with far less. If you acquire a modest assortment of power tools, however, many of these hand tools will not be needed. Moreover, don't rush out and buy all these tools at once. When you anticipate the need for a tool you don't own, try to decide whether there will be additional uses for it. If it turns out to be a one-time use only, borrow or rent the needed tool. For example, why buy a circle cutter for a drill press if you can do the job with a keyhole saw, electric sabersaw, or "jig saw"?

Figure 1-1. *The author's tool box. When every slot is full, the set of tools is complete.*

SOME USEFUL BOOKS

For more complete information on tools, I recommend the many fine books on cabinetmaking and workshop procedures available. Several that I found especially valuable are:

The Home Workshop, by William R. Wellman, Van Nostrand-Reinhold, 1953.

The Complete Woodworking Handbook, by J.T. Adams and Emanuele Stieri, Arco, 1969.

The Complete Book of Home Workshops, by David X. Manners, Popular Science Publishing, 1969.

While these may be out of print, but still available in libraries, there are many others probably as good. Read as many as you can handle, as well as the boatbuilding references mentioned in the Introduction.

WHERE TO FIND TOOLS

Quality hand tools are available at local hardware stores, chains such as Sears and Ward's (obtain their fine special tool catalogs), industrial tool and equipment houses, building supply and home-improvement centers, lumberyards, and so on. Wherever you buy, play it safe and buy only well-known name brands. The higher quality is worth the small difference in price, and the guarantees are worth having.

Less obvious sources for tools are sometimes worthwhile. In California there is an institution called the "swap meet," although little is swapped. In other regions they are called flea markets. Some people try to sell junk; very frequently, however, you can pick up useful tools at very good prices. You may be able to bargain. He asks $2, you offer $1. Perhaps you get the tool for $1.75. Here are some of the items and their prices that I found in the 1970s at "swap meets":

Three 36-inch bar clamps, $5 each
Dozens of C-clamps, $.75 to $2
Electric motors, $2 to $15, depending on horsepower
Small Stanley router, $22
A 24-inch jointer plane, $7.50
Antique drill press, $5
A 50-foot steel tape, $1.50

I've even found interesting things at rummage sales. Would you believe a Delta Homecraft jointer (somewhat rusted, but little used, with blades still sharp), minus fence, for $5? A heavy-duty Ward's reciprocating sander for $3? I have heard that you can "steal" tools at pawn shops if you are the first customer on Monday mornings. And don't forget the used-tool store, although I have rarely seen anything much more than 10 or 15 percent under new list price. I make a habit of reading the classified Miscellaneous For Sale and Machinery columns. Garage and yard sales, too, are often productive.

THE BASIC HAND TOOLS

Nail or claw hammer
6-inch try square or combination bevel square
Carpenter's square
12- to 16-foot pocket tape
Plane (9–10 inch), smooth or jack
Crosscut saw, 10 point
Backsaw, 12 inch, 14 point or more
Ripsaw, 5 to 7 point
Two straight screwdrivers
Two Phillips-head screwdrivers
Hand drill or breast drill
Wood bits, ¼ to 1 inch
Brace for wood bits
Set of twist drills, ¼ to ½ inch
Set of small twist drills, 1/16 to ¼ inch
Wood countersink, squared shank
Rat-tail file
Rasp, flat/half-oval combination (called shoemaker's)
Kit of nail sets
Fine oilstone
Brad awl or prick punch
Scribing compass
Portable woodworker's vise
Chisels, ¼ to 1 inch
Hacksaw
Keyhole saw
Marking gauge
Bevel gauge
Six 3- or 4-inch C-clamps
Six 6-inch C-clamps
Grinder

OTHER TOOLS YOU'LL WISH YOU HAD

Slick, a long chisel 1½ inches or wider
Yankee screwdriver (rare)
Spokeshave
Drawknife
Expansion bit
Counterbore bits, ⅜ and ½ inch (for plugs)
Plug cutters, ⅜ and ½ inch
Bench vise
Dado plane or rabbet plane
Hand scraper
Cabinet scraper
Fore plane, 18 inch
Jointer plane, 22–24 inch
Block plane, low angle
Miter box
Three or four pipe clamps or bar clamps over 36 inches

SELECTING HAND TOOLS

Before offering some tips on the use of the basic hand tools, I'll comment on their selection. I have used many tools that were from 25 to 50 years old. I occasionally used a 24-inch wooden jointer plane over 100 years old. My Yankee screwdriver was purchased in 1937. Never reject a tool *simply* because it is old. A plane never wears out, although the blade may become quite short from years of grinding. Blades, of course, are easily replaced. Chisels can last for generations if the handles are renewed. The same may be said of many of the other tools listed.

First, look for a quality name. Finish per se means little. Some of the leading manufacturers produce tools that are not as nicely ground, polished, and/or enameled as their top-of-the-line products. In most cases the materials are the same, the tools perform just as well, and the prices are considerably lower. Always look for defects such as hairline cracks in a used tool. And never buy used tools that show evidence of welding or brazing.

If you are unfamiliar with tools, their features and relative costs, write to manufacturers or pick up catalogs at tool supply houses. A few hours of study will give you much of what you need.

Send for the catalogs of Woodcraft, 313 Montvale Ave., Woburn, MA 01888; Brookstone, 127 Vose Farm Rd., Peterborough, NH 03458; Woodworker's Supply of New Mexico, 5704 Alemeda Place N.E., Albuquerque, NM 87113; and others offering high-quality and many rather hard-to-find tools. An excellent source of information is a carpenter, cabinetmaker, boatbuilder, or almost anyone else in the world of wood construction. He'll tell you the names to look for, such as Black & Decker, Bosch, Chicago Pneumatic, Craftsman, Delta, Estwing, Foley-Belsaw (or Belsaw), Jorgensen, Milwaukee, Plumb, Porter-Cable, Skil, Stanley, Vaughan, Williams & Hussey, and others equally trustworthy, including the relative late-bloomers, such as Freud, Hitachi, Makita, Ryobi, and some I have not yet met.

USING HAND TOOLS

The use of many hand tools may seem quite obvious. Remember, however, that other readers may not have your experience, skill, and dexterity. Note too that people with know-how and real skill are sometimes using and maintaining tools improperly. So please be tolerant, and look for any information from which you can benefit.

Hammers

Your hammer should be a claw hammer, shown in Figure 1-2. The curved claws are designed for pulling nails easily, but not for ripping apart old nailed work. The framing hammer with rather straight claws is usually a 20-ounce tool, used for framing or general construction work and useful for splitting, jamming between boards, and so on. Yours should be 16 ounces for general construction and joinery, and one of about 12 ounces for much of the lighter cabinetry and trim work.

Figure 1-2. *Hammers. Top to bottom: 16-ounce nail or claw hammer, 16-ounce framing hammer, 12-ounce cabinetmaker's hammer.*

Look for a slightly *convex* hammer face. A face ground dead flat will bend finish nails and is more likely to mar your work. Of course, never drive a finish nail flush; you let it protrude 1⁄16 to 1⁄8 inch and then set it below the surface with a nail set (Figure 1-3). No man breathes who has not let his hammer slip off the nail and dent the wood, so make sure you use a hammer with a convex face. Wooden handles (or hafts) are still preferred by many, but always replace a split or loose one immediately. Fit the new handle using any cutting tool—spokeshave, chisel, plane, rasp, and so on—and drive in two steel wedges (available at any hardware store). The steel-shaft type of hammer is satisfactory, but I suggest that it be from a good U.S. manufacturer such as Plumb, Estwing, Stanley, Vaughan, Craftsman, etc. In any event, if the hammer balances well in your hand and feels good, that's it.

A word of warning. Never attempt to pull a nail without a block or pad under the claws to protect the work surface. Use a piece of 1⁄4-inch plywood if the nail head is almost into the wood, then a piece of 3⁄4-inch stuff up close to the nail as it is drawn out. To pull a finishing nail with just the head out, jam the tapered slot of the claw hard against the nail so the inner sharp edges cut into the shank, then insert a piece of 1⁄16-inch metal between hammer and work (Figure 1-4).

If you should get paint, glue, or grease on the face of your hammer, wipe or sand it off or you will surely bend nails you are hammering. If you are driving the serrated (or ring) Anchorfast nails (Figure 1-5) occasionally used in boat work, you have a problem. Unless the *lead* or *pilot hole* is correct,

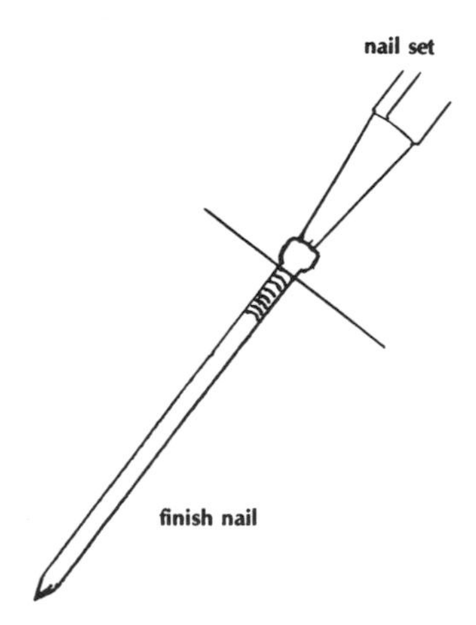

Figure 1-3. *Finish nail and nail set.*

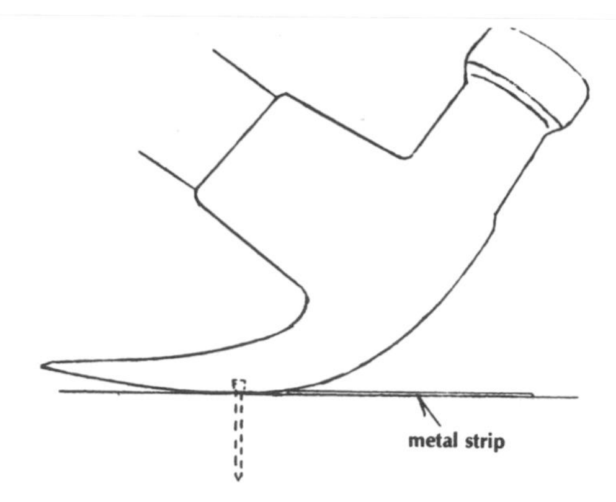

Figure 1-4. *Pulling a finish nail.*

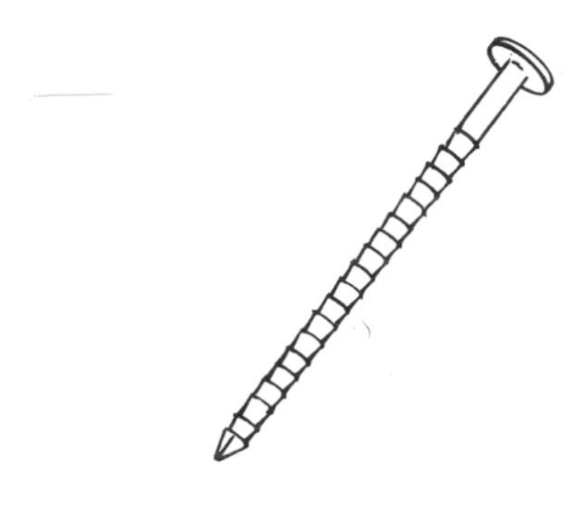

Figure 1-5. *Ringed nail.*

these soft nails, if bronze, bend easily, especially in oak. If you try to draw the nail, it will pull out all the surrounding wood with it or the head will come off. Don't try. Instead, grasp the head with pliers or Vise-Grips and bend the nail back and forth until it breaks off. If part still projects, set it below the wood surface with a large nail set and drive a new nail close by.

Squares

Most good try squares have hardwood or steel heads (Figure 1-6). If one shows signs of having been used as a tack hammer, reject it. You will use it frequently, so keep it handy. Keep it clean and rub a little oil into it occasionally. These suggestions apply to your carpenter's (or framing) square, too (Figure 1-7). If you find a good one, but it is somewhat rusted, sand it with emery cloth after soaking with Liquid Wrench. If you can't read the marks

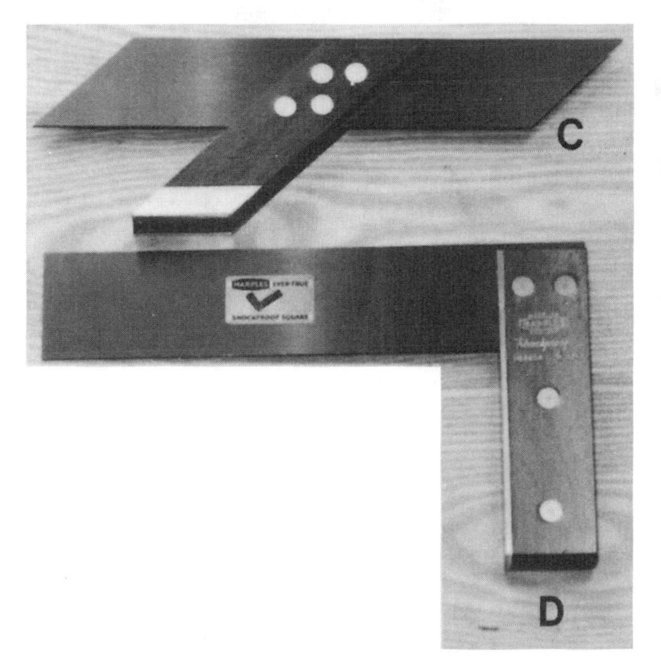

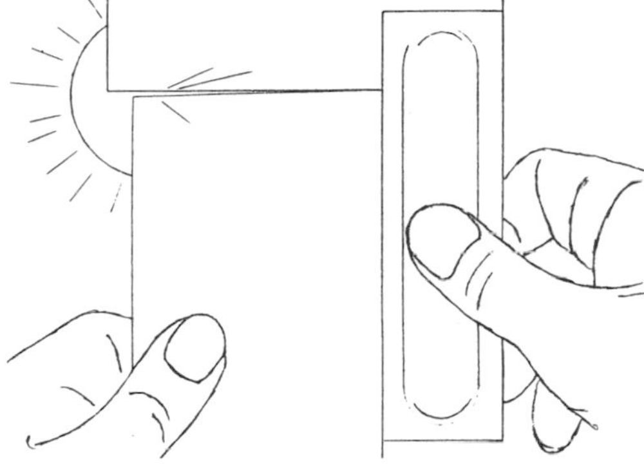

Above: Figure 1-6. *Hardwood try square with steel handle is inferior to those with rosewood handles but takes rough handling.* **Top right: Figure 1-7.** *Carpenter's or framing square is useful for laying out all kinds of angles, checking mitered joints, as a power-saw guide, and so on. The 12-inch sliding-T 45-degree bevel square is always useful.*

Figure 1-8. *Checking accuracy with a square.*

easily (this sometimes happens with new ones, too), wipe a contrasting paint into the indentations, then wipe the surface clean. The carpenter's square is needed for squaring larger parts such as cabinet doors, for drawer assembly, for checking table saw alignment, and for many other jobs. I make frequent use of a steel square as a guide for my power saw when the work is too long or too heavy to cut on the table saw. Simply clamp the square accurately with two small C-clamps and run the sole or base of the saw along the square. For checking, hold the work piece and the square between you and the light, as shown in Figure 1-8.

Levels

I use an old wooden level with brass trim. The new aluminum ones, however, with 45-degree-angle features, adjustable bubbles, and so on, are quite versatile (Figure 1-9). A 24-inch level is usually

Figure 1-9. *The extra 6-inch length of this 30-inch aluminum level is handy. A 48-inch level is desirable. If you use an inexpensive wooden level, check it frequently.*

Top: Figure 1-10. *The little torpedo level is a must for working in close quarters. Laid on a straightedge, it takes the place of a long level.* **Center: Figure 1-11.** *A bevel gauge or square is used constantly in boat work.* **Bottom: Figure 1-12.** *The hand crosscut saw is used frequently, the ripsaw rarely. Ten or 12 points is best for clean work, 8 points for framing.*

adequate for inside work, but you should have a longer one. If you want to use an old one or the cheap wooden type you find on bargain tables, check it against the store's walls, columns, doorways, and floors. It must read the same with either face on the surface. If the bubble is always off to the same side of the mark, don't use the level. The little torpedo-type level (Figure 1-10) made of die-cast zinc or aluminum is a necessity in restricted areas. In a pinch, a 4-inch line level will do for leveling shelves and for work where you can't get in with the 24-incher. Don't, however, count on these short levels for long surfaces such as cabin sole beams, bulkheads, and berths, unless taped to a straightedge.

Speaking of levels, here is an important point. *Do not plan* to deck your hull or cover it over in some way and then complete the interior after launching. Your hull is not yet on its lines, and always in motion so nothing is level or plumb.

Bevel Gauge

The bevel gauge (Figure 1-11) is a simple little tool indispensable for picking up angles in degrees from a protractor, setting saw blades for bevel cuts, marking for repeated angle cuts, and a myriad of other uses. You can make one out of hardwoods. The newer ones have plastic bodies, instead of the traditional hardwoods. They'll still last your lifetime.

Pocket Tape

A 12- or 16-foot tape measure is essential for taking inside measurements. If you are building scaffolds, molds, and spars, however, you must have a 50-foot steel tape.

Saws

Unless you have a hand power saw, you might need four handsaws. The first is a 26-inch crosscut, with 10 points or more to the inch, most used (Figure 1-12); then a ripsaw, usually 5 to 7 points; then a backsaw, 14 points, with a 12- to 14-inch blade; and, finally, a keyhole or compass saw. The crosscut and ripsaw may have either a straight back (stiffer) or a skew back (curved and limber). I think you might saw a straighter line with the straight back, but it's no big thing. The possibility of sawing a lot of plywood suggests the need for a 12-point fine-toothed crosscut saw to reduce splintering. Otherwise, the backsaw is used for fine cuts and in a miter box.

To start a saw cut, place the side of your left thumb above the teeth to guide the saw safely on the waste side of the mark (Figure 1-13). Never saw on the mark so that you lose it. You need some stock to plane off to square up the rough edge. You can split the line when you get to be an expert. It is a good idea to pick up the habit of marking with

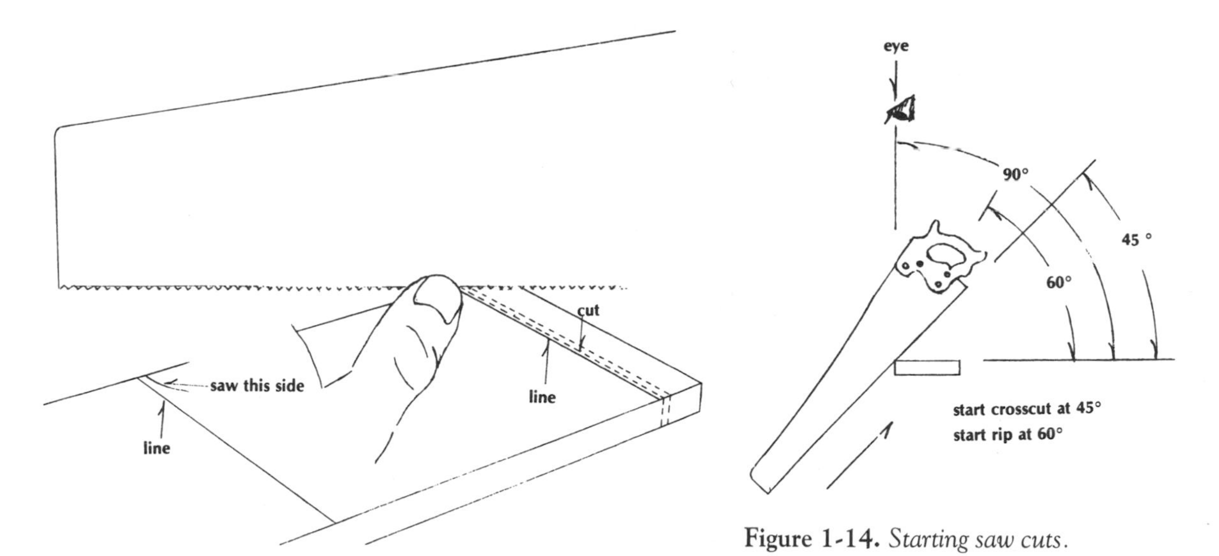

Figure 1-13. *Starting a saw cut.*

Figure 1-14. *Starting saw cuts.*

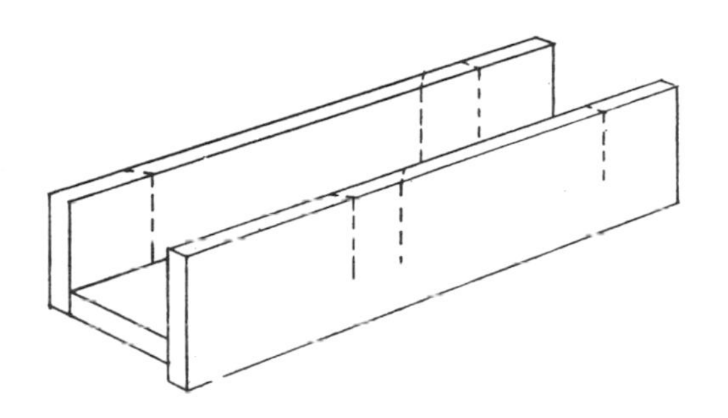

Figure 1-15. *Layout of simple miter box.*

the square, and then making another light mark on the waste side. This will prevent sawing on the wrong side of the mark—a real calamity. And don't think this is an error limited to greenhorns. To get a really clean saw cut, mark with a knife. A score about 1/64 inch deep is plenty to give you a nice clean edge, and it's much more accurate than the sharpest pencil.

Start a cross-grain cut with the saw at about a 45-degree angle above the work, with your head directly over the beginning point (Figure 1-14). Rip at a higher angle, about 60 degrees. Always start with one or two upward strokes, dragging toward you with the end of your thumb still guiding the saw. If your cut wanders from a straight line, twist the handle for the next few strokes to bring it back. Of course, the saw-blade side must be at right angles to the surface of the work. Ease the pressure as you end the cut, to avoid splitting off a chip.

A backsaw is often used in a miter box. In its simplest form, a miter box is an arrangement of three pieces (preferably hardwood or plywood) nailed together to form a U-shaped channel (Figure 1-15). Two 45-degree cuts and one of 90 degrees are carefully laid out and sawed with the backsaw *perfectly* square to the bottom of the box. These cuts must be absolutely true or the miter box will just waste hours of your time. You can buy a quite satisfactory miter box in hardwood, metal, or even plastic at any lumberyard or hardware store (Figure 1-16). More elaborate affairs run from about $45 up to over $150 (Figure 1-17). Most of these provide rapid setting of many angles. Even the wooden cheapie is fine for quick 90-degree cuts in small stuff, so keep one handy.

It's necessary to tack or clamp the miter box to a bench or a plank on sawhorses. When cutting, hold the piece hard against the back of the box, taking extreme care that the cut will be exactly where you want it. Always make a light pencil mark showing the direction of the miter (Figure 1-18); it's easy to become confused. Never set an irregularly shaped molding on the molded back (Figure 1-19); you won't get a true angle. The work

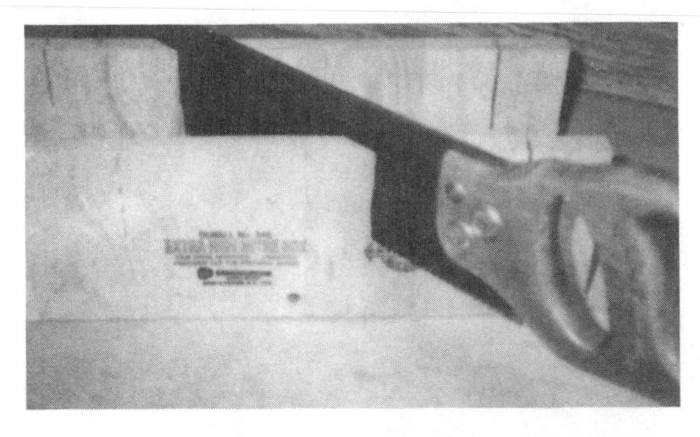

Top: Figure 1-16. *This store-bought miter box has extra depth for large moldings. It is adequate for most work, but not for precise molding joints.* **Above: Figure 1-17.** *A precision miter box, one that cuts within .005 inch, is a necessity for fine handwork. Such boxes take 20-inch and longer blades.*

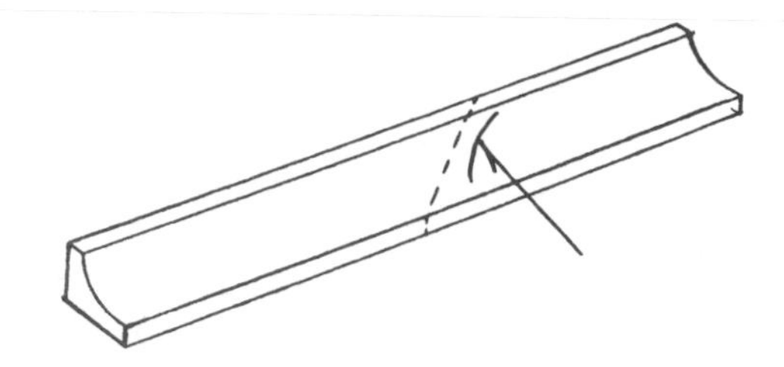

Figure 1-18. *General direction of miter cut.*

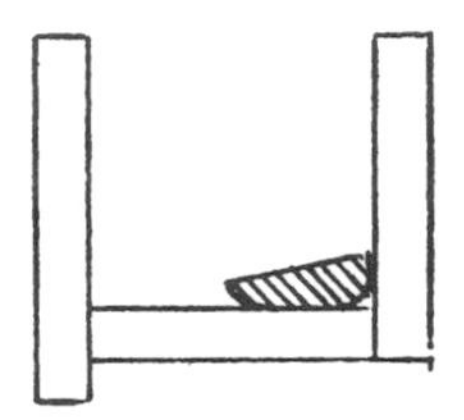

Figure 1-19. *Never lay moldings with irregular side down.*

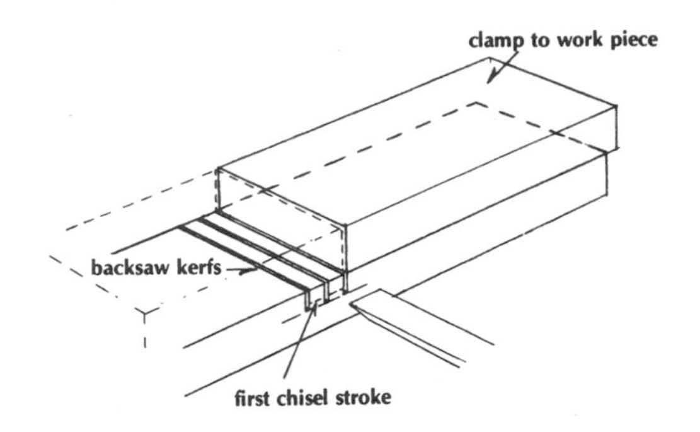

Figure 1-20. *Sawing dado with blocks.*

will tend to creep sideways, so hold it tight or clamp it. Start cutting with a short pull or two, then lower the saw to the horizontal as the cut deepens. If you find that a miter has been cut just a bit too long, clamp another piece that has just been cut off against that end. The saw will cut through almost as if the union were not present.

Tack a piece of ¼-inch plywood to the bottom of the miter box to take the saw cuts, rather than saw up the box itself. Also, provide some kind of support for the ends of long pieces of molding, or you may get splitting at the cut end.

It is possible to dado (cut a groove across the grain) with a backsaw by clamping a block right on the line, placing the blade against this block, then clamping another block against the blade so there is no wobble. Repeat this for the other side of the dado. Watch your measurements carefully!

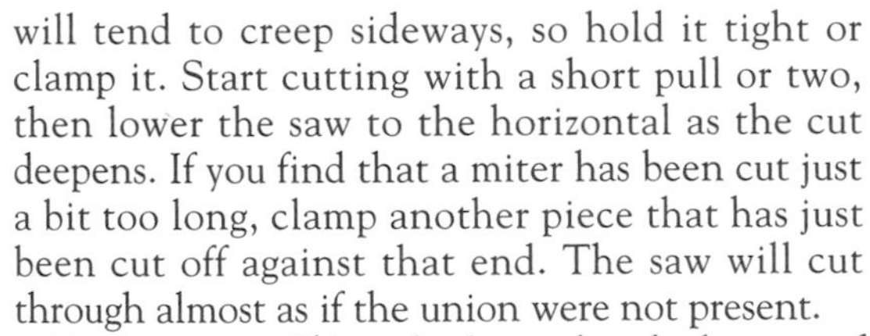

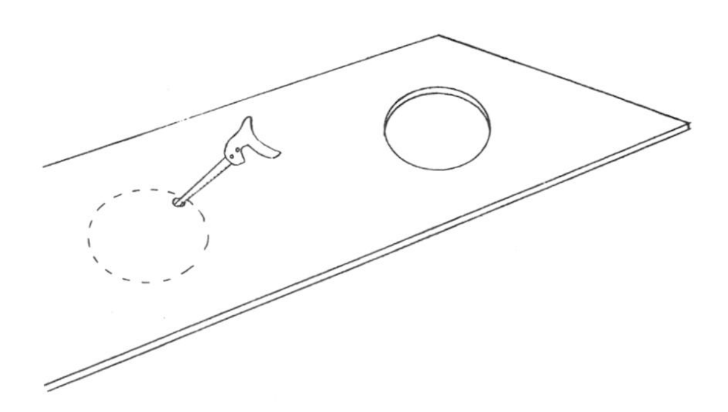

Figure 1-21. *Keyhole sawing operation.*

Figure 1-22. *Planes. Clockwise from top: 24-inch jointer, 15-inch jack plane, 6-inch block plane, 9-inch smooth plane.*

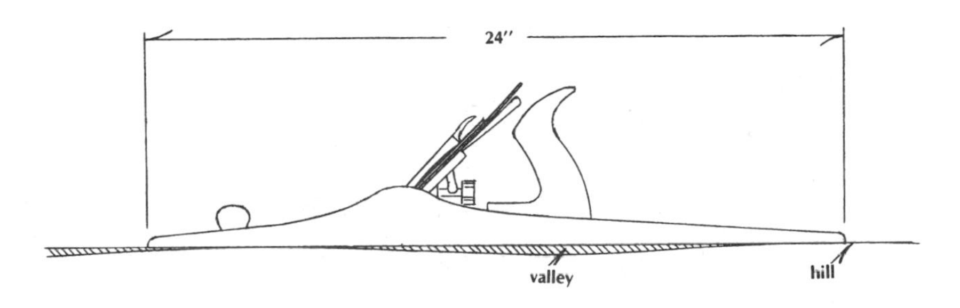

Figure 1-23. *Jointing an edge.*

Lastly, make another cut between these two cuts to make it easier to chisel out the dado (Figure 1-20). Take care, too, that your saw does not rock, which would produce a dado shallower in the middle. More on this later.

A handy device for use with any crosscut saw is a bench hook, an 8- to 12-inch-long 1 by 6 with a cleat screwed and glued across each end on opposite sides. You place one cleat in a vise, or clamp, or tack the assembly to a bench or plank. Then you hold the work piece against the farther cleat while you saw. Best to clamp it at the end of the workbench so you may crosscut properly (Figure 5-30).

By the way, if your handsaw refuses to cut straight, take it to a saw shop. Even new saws sometimes have poor blade settings, or an amateur may have done a bad job. More set on one side will make even the best saw wander. I will say no more about sharpening saws because this is a matter for experts. If you are interested in learning the art, some of the good woodworking texts give instructions.

The other saw you will require is a keyhole saw. This tool has a narrow tapered blade about 12 inches long. A keyhole saw is useful for cutting quite sharp curves of any kind, such as openings for portlights. To make a circular cut, bore a hole about 1 inch in diameter near the mark and start the cut from this point (Figure 1-21). The blade of a keyhole saw is coarse, so the cut is rough. It's best to saw at least 1⁄16 inch from the line so the rough surface can be dressed out with a spokeshave, rasp, chisel, or a coarse curved sanding block. Note that keyhole saw blades are very soft, there being little or no temper. As a result, they bend easily. Just straighten a bent blade by hand and return to the job.

Planes

The plane is your most-used tool (see Figure 1-22). No sawed surface is left rough. At the beginning, you may get by with a smooth plane about 9 or 10 inches in length, especially if your sawing is accurate. Such a plane will remove quite a lot of stock, but its relatively light weight and size is not the best for this task. The next size is the jack plane, about 11 to 15 inches in length, used for taking off material to a line, or split the line with the smooth plane set fine. Check with the try square as the work progresses.

The next largest is the jointer plane, about 24 inches in length. This tool is for long, straight work. It takes out high spots by spanning from one to the next more accurately than a jack plane can, as shown in Figure 1-23. A line of fine wooden (light weight) planes is shown in Figure 1-24.

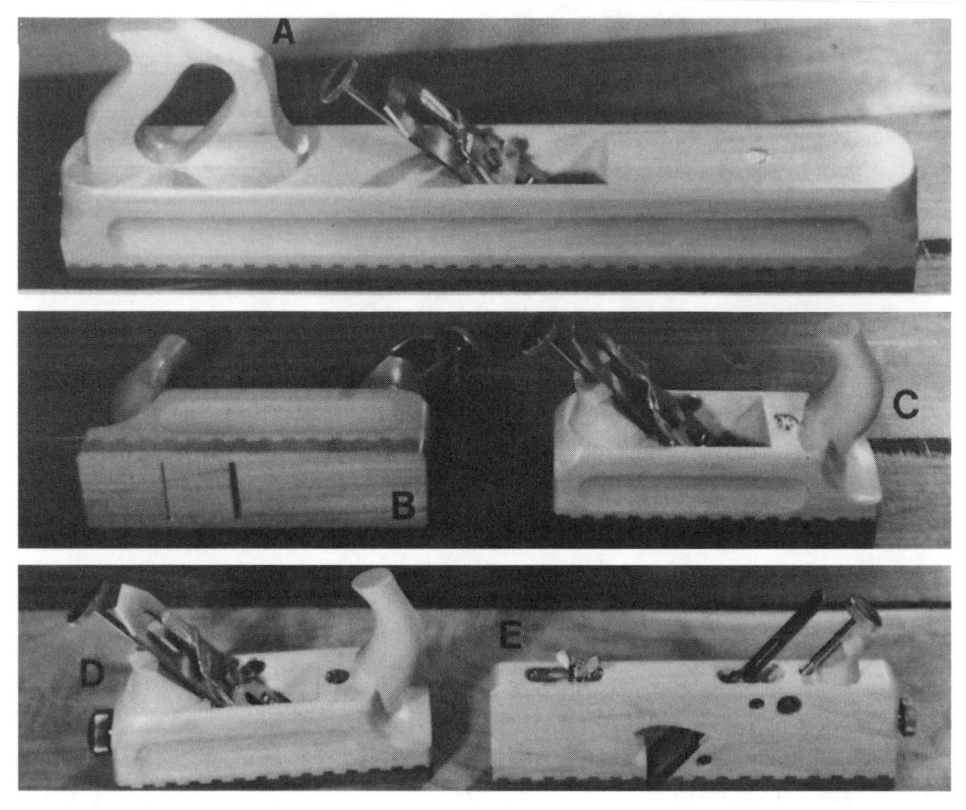

Figure 1-24. *Planes. (A) Primus 25 ⅝-inch jointer plane, 2 ⅜-inch blade. (B) Smooth plane, for fine cuts in wild grain; adjustable throat. (C) General smooth plane for finishing. (D) Jack plane. The low angle permits thick cuts; the workhorse of planes. (E) Shoulder rabbet plane. Adjustable front shoe provides great accuracy. All of these planes are European red beech; soles are lignum vitae. They have been made by Emmerich since 1852. (Courtesy Woodworker's Supply)*

At the other end in size is the block plane, which runs from 5 to 6 inches in length. The blade is set at a low angle desirable for planing across the grain and for plywood edges. This tool is great for all very fine work, such as dressing the surfaces of jointed parts after assembly. It will take off an almost transparent shaving, but must be kept sharp to produce satisfactory results (Figure 1-32).

Another special-purpose plane is the dado plane. (Remember, a dado is a groove cut across the grain, and a groove is a groove cut with the grain.) A dado set is a combination of blades used on a table saw to cut dadoes and grooves. Since you don't have a table saw, you have to make these cuts with a backsaw and a chisel. The dado plane, which is very narrow, is run through the dado or groove to clean up the bottom, remove saw kerfs, and smooth out the cut. In Chapter Five, on tools you can make, I show you how to make a dado plane using a harrow chisel as the iron.

On occasion, you may need a scrub or roughing plane (Figure 1-25). Usually quite narrow, perhaps 1¼ inches wide, the blade is ground to a low arc (Figure 1-26). This plane is used to remove material rapidly, straight or diagonally across the grain (especially on oak). Then, after nearing the line with a jack or jointer to fair the surface, you finish up with a sharp smooth plane. An extra blade for your smooth plane may be ground to an arc for such roughing jobs. The scrub plane is similar to the

Figure 1-25. *An unusual 80-year-old scrub or roughing plane with 1 ¼-inch rounded blade. Used diagonally across the grain, also for hollowing planking. Also used as a rabbet plane with a square blade.*

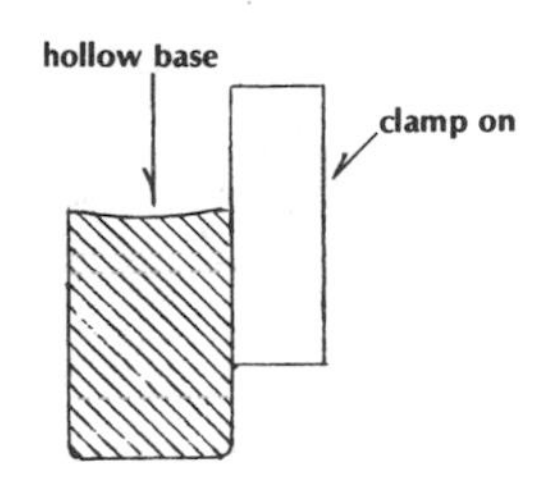

Figure 1-27. *Rail can be hollowed with plane.*

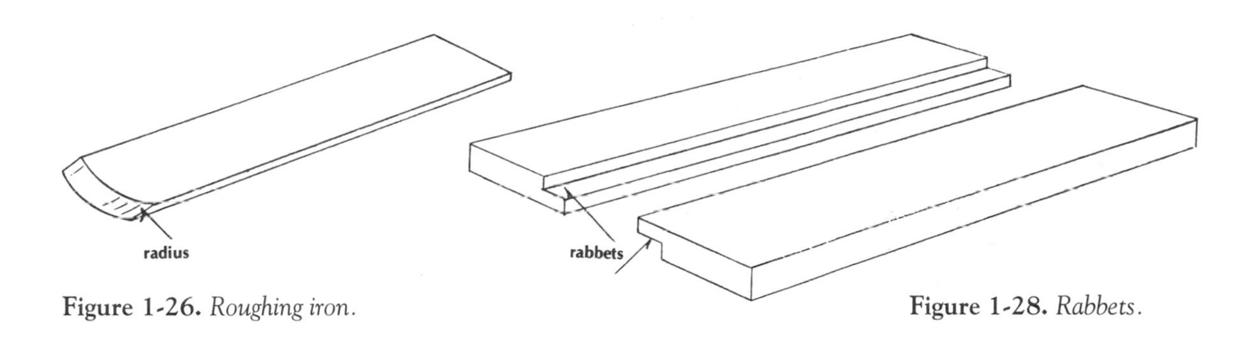

Figure 1-26. *Roughing iron.*

Figure 1-28. *Rabbets.*

planes (usually wooden—see Chapter Five) used for hollowing the backs of planking. Hollowing is good practice when a piece such as a handrail, toerail, hatchrail, and so on, must be bedded down securely, especially fastened to a curved surface. In such cases it is best to clamp a guide piece on the side of the work piece so the hollowing is centered accurately (Figure 1-27).

There will be many rabbet joints in your vessel. As shown in Figure 1-28, a rabbet is a recess running along the edge or end of a piece. Another matching piece fits into the first piece. Simply clamp a stiff fence to guide the plane. Fine rabbet planes, however, come in sets with a guide rail or fence for planing parallel to an edge. The sole is slotted at the right side so the blade cuts to the side of the rabbet. Some planes have an additional seat in the forward end so the blade can plane to ¼ inch of corners. This is called a bullnose. Several rabbet planes are shown in Figure 1-29.

Wooden planes were once the mark of a real craftsman in boatbuilding. They were made of

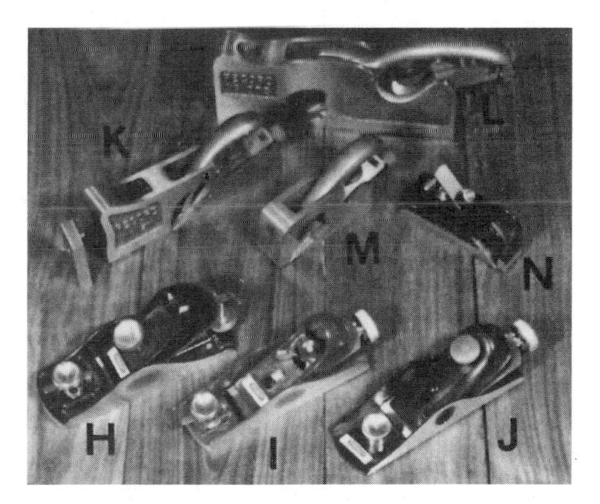

Figure 1-29. *Rabbet planes. (H) Fine 21-degree block plane. (I, J) 12-degree planes. (K, L) Blades on these planes may be moved forward to bullnose position. (M) Four-inch adjustable-nose plane. (N) Plane that makes coarse to fine cuts. (Courtesy Woodworker's Supply)*

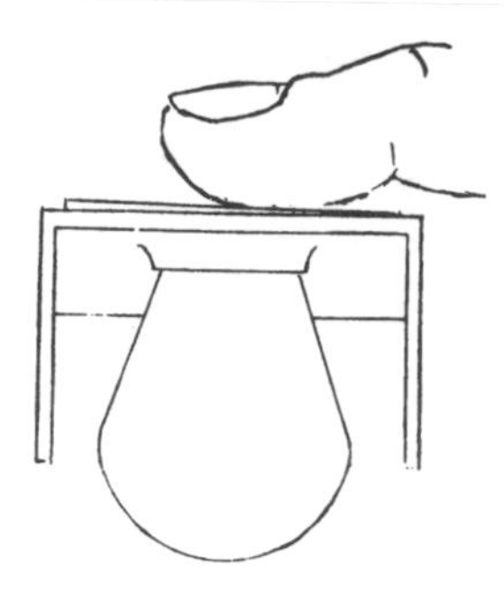

Figure 1-30. *Brush thumb against iron and/or sight toward edge.*

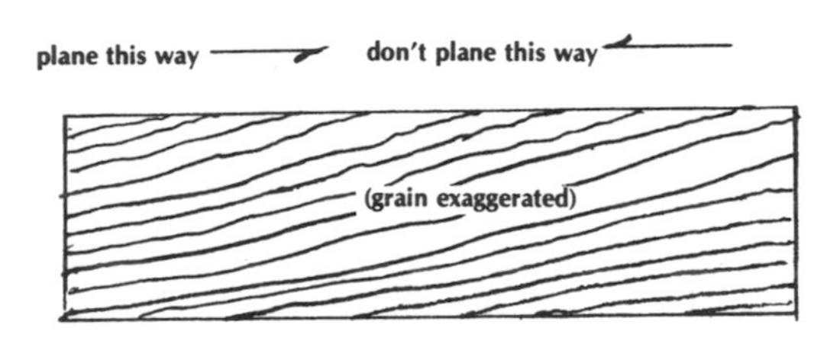

Figure 1-31. *Planing direction.*

lignum vitae, beech, and other extremely hard, wear-resistant woods. These planes were generally much lighter than their cast-iron counterparts and were prized for planing hull planking, dressing spars, and other tedious jobs. Directions for making wooden planes can be found in books on making tools the old way. Or they can be purchased from such firms as Woodcraft Supply, Brookstone, and others. Be prepared for rather steep prices. Many say they are worth it.

Setting a Plane

Always set a plane blade for the minimum cut for the first few strokes to find the direction of the grain. The depth of cut may then be increased, but keep in mind that it is far better to take it off in many fine shavings than to hack it off in a few strokes. Also, the finer cut allows you to check repeatedly for squareness of the edge. There are two ways to find the proper adjustment of the plane blade. One is to sight down the sole toward the cutting edge of the blade. You should be able to see whether a side is high or low and whether the squared blade will cut fine or coarse. The second way is to brush lightly against the cutting edge with the ball of the thumb (never lengthwise on the cutting edge), as shown in Figure 1-30. This second method will come naturally with experience. To me, it is the only way.

To adjust the blade, hold the plane, sole up. Turn the knurled nut as you sight down the sole toward the cutting edge, until the edge appears as a hairline. It probably won't be perfectly parallel to the sole, so adjust by moving the lever right or left. Brush the blade with your thumb: does it drag about the same near both corners? Try the setting on a piece of scrap. Back the blade or advance it until you get a fine shaving. Now examine your work piece for grain direction. Never plane against the grain, as the surface will be very rough and large chips may tear out (Figure 1-31). If the grain is in long waves (reversing direction), you may have to plane both ways by reversing the piece in the vise and finish by block sanding. Remember, you want sharp, fine shavings.

When you are through with a plane, even momentarily, lay it on its side, never on its sole. And when you store a plane, retract the blade so there is no chance of it being nicked or dulled.

The Plane Stroke

Start a stroke with a bit of weight down on the forward knob, then apply equal pressure on knob and handle through the middle of the stroke. At the end, let up pressure on the knob, but do not lift. Your shavings should have a fine feather edge at each end if your strokes are long and you don't rush.

If you need to plane square across the grain at the end of a piece, do so before planing the edges with the grain. If you do not, you will surely split off the far edge (Figure 1-32A). This is not easy with a smooth plane, and it is even more difficult with a jack. The proper tool is the little block plane, generally used with one hand or with only slight pressure and control with the forward hand. The low angle of the blade has a slicing effect and produces a nice finish. To further prevent splintering, plane toward the center or to a safe distance from the far edge, then reverse the piece in the vise and again plane to the center or somewhat beyond, watching the marks and checking for square both ways (Figure 1-32B). Another way, if there is ample stock, is to plane off a little corner of the far edge. Now you can stroke all the way across the end without danger of splitting (Figure 1-32C).

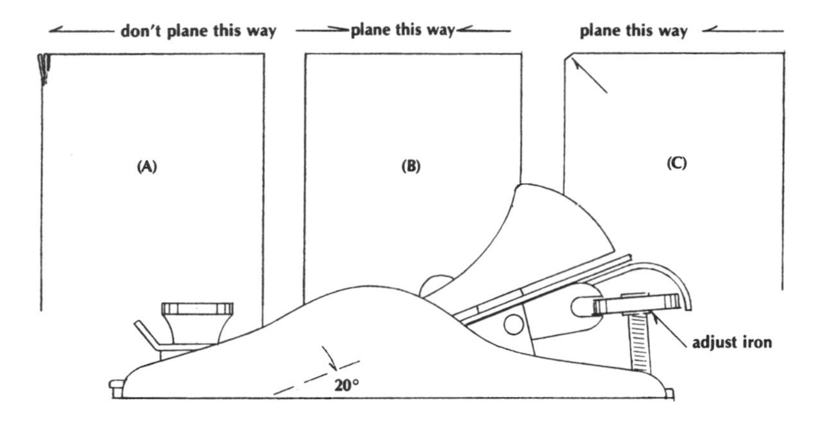

Figure 1-32. *Plane end grain with low-angle block plane.*

Sharpening a Plane Blade (or "Iron")

If you keep in mind that a plane is actually a wide chisel held to a precisely controlled depth of cut and angle, you will better understand its sharpening needs. The blade must be sharpened like a chisel, hollow ground on an electric grinding wheel or your hand grinding wheel. The cutting edge must then be whetted on an oilstone. Your grinding wheel has some kind of tool rest that can be adjusted at all angles to the face of the wheel. To get started, find a chisel or plane blade with the proper bevel of 25 to 30 degrees. Lay this flat on the tool rest and swing the rest up or down until the bevel is a close match to the cutting surface of the grinding wheel (Figure 1-33). Move the tool back slightly, and turn the wheel a few times by hand while you touch the edge against it. You'll see a bright spot on either the cutting edge or the back of the bevel. Adjust the tool rest so the grinding wheel touches at or near the center of the bevel. Now you're ready to grind. Be sure you have a dish or can of water in reach.

With the wheel turning at full speed, slide the blade right and left repeatedly across the tool rest (Figure 1-34). Do not rush, but always keep the blade *moving*. If you stop, the steel will become overheated. Dip the blade in water after each five or six passes across the wheel. If the edge suddenly turns blue or black, plunge it instantly into the water. The burned area (annealed) has lost temper and will not take an edge, being dead soft. It will have to be ground out, with *frequent* cooling. It is

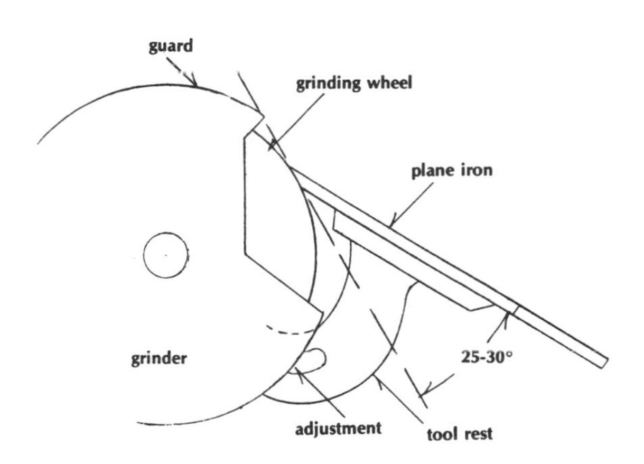

Figure 1-33. *Using the grinding wheel.*

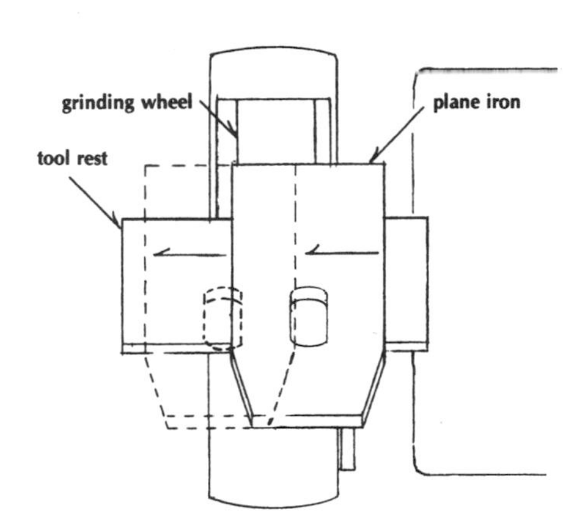

Figure 1-34. *Using the grinding wheel.*

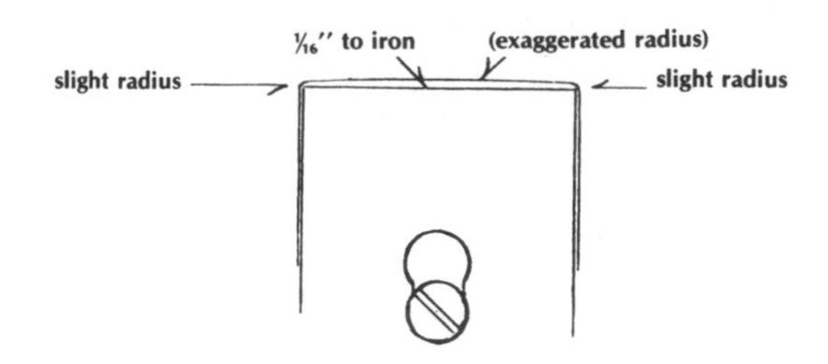

Figure 1-35. *Plane iron refinements.*

good practice to grind a tiny radius on each corner of the cutting edge so there is no sharp corner to mark the work (Figure 1-35). I think most craftsmen grind their plane blades so the edge has a slight, almost invisible curve; you should learn to do this, too.

Whetting or Honing

Now you're ready to whet or hone the blade. Hold the oilstone firmly by tacking a frame around it on the bench, as shown in Figure 1-36, or on a small board. Spread three or four drops of machine oil thinned with kerosene on the fine surface of the stone. Now hold the blade in your best hand at an angle of about 25 to 30 degrees so just the *cutting edge* is in contact. I was taught as a boy to rock the bevel up and down slightly so its back edge also touched the stone *lightly*. Keep the bevel close to 25 degrees. A greater angle produces an edge less likely to chip, but a finer edge cuts like a razor if you take care of it. An angle of less than 25 degrees may produce an edge too fragile for practicality.

The first two fingers of both hands now apply a light pressure while you slide the blade about in a figure-eight or elliptical pattern. After half a dozen strokes, check the cutting edge. It should have a bright strip extending all the way across the blade, from corner to corner; it may not be more than 1/32 inch wide. That's enough, as more simply reduces the life of the bevel between grindings. Do, however, whet your plane blades and chisels frequently. The honing devices shown in Figures 1-37 and 1-38 are excellent, as they control the bevel angle with precision.

At this point, there will be a burr or "wire edge" on the upper side of the blade. Carefully lay it flat on the stone, bevel side up, and slide it around in a figure-eight or circular motion, five or six times. Do not allow the blade to lift; it must remain flat, since lifting it would destroy the cutting edge and it would have to be reground. Give another two or three strokes to the cutting edge, and another on the back. Then slice the sharp edge across the corner of a piece of scrap wood. This removes the last of the burr. Slicing with the grain should now take off a nice sliver without effort. If it doesn't, back to the oilstone! This hollow-ground edge can be whetted many times, but grind it before the bevel becomes flat. Some old-timers strop the edge on a piece of leather tacked to a small paddle (Figure 1-39). Patternmakers use a neat trick. They touch the beveled edge for four or five seconds to a cotton polishing buff on the grinder arbor. And then

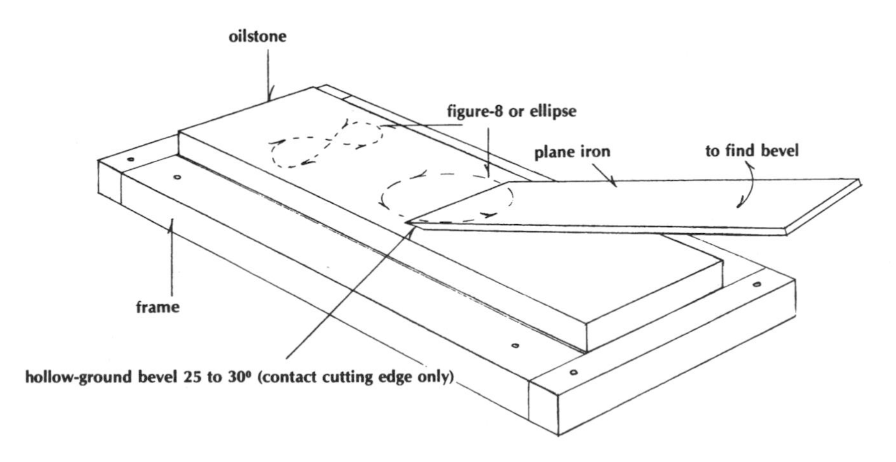

Figure 1-36. *Honing or whetting a plane iron or chisel.*

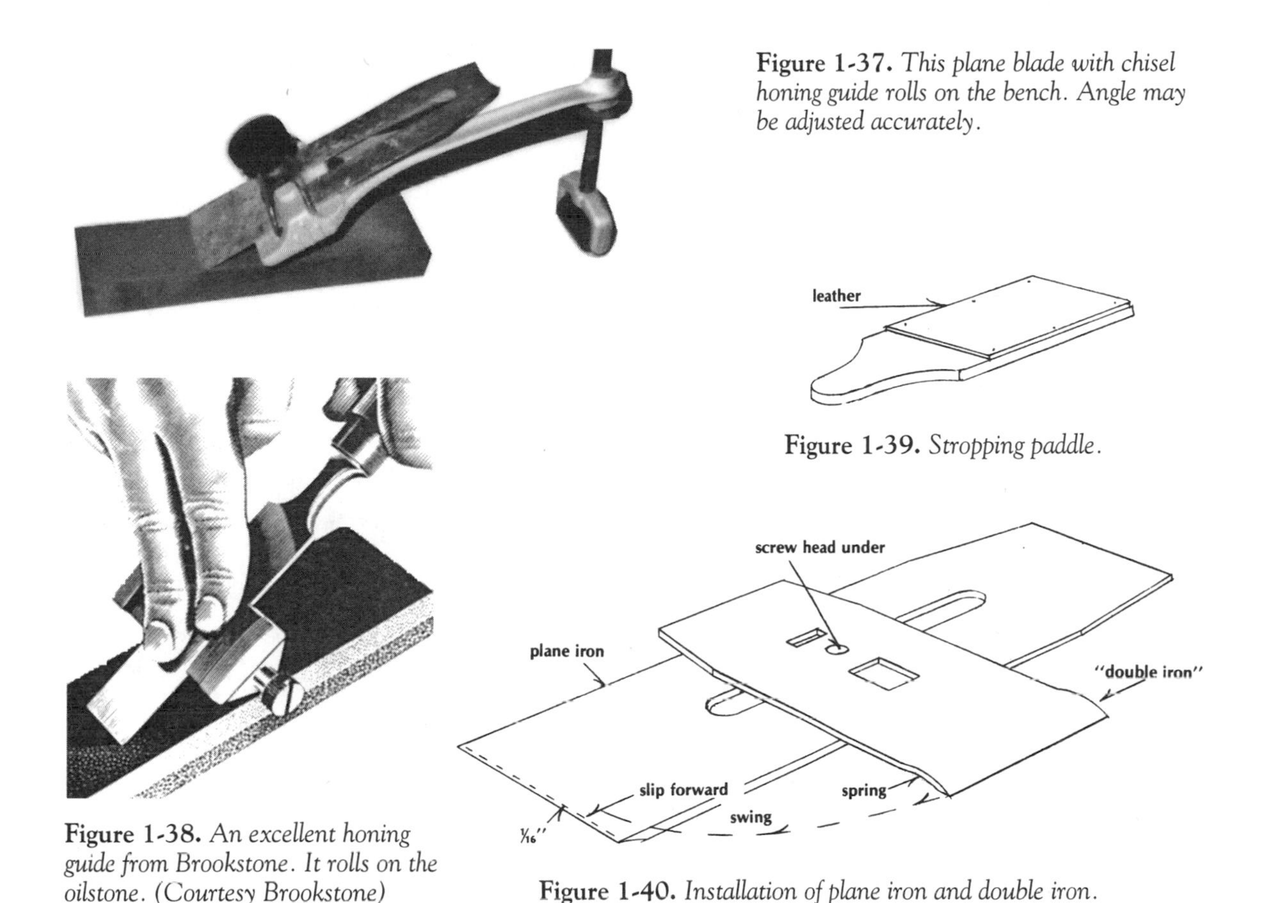

Figure 1-37. *This plane blade with chisel honing guide rolls on the bench. Angle may be adjusted accurately.*

Figure 1-38. *An excellent honing guide from Brookstone. It rolls on the oilstone. (Courtesy Brookstone)*

Figure 1-39. *Stropping paddle.*

Figure 1-40. *Installation of plane iron and double iron.*

briefly touch the back of the blade. The result? A razor-sharp edge.

Assembling the Plane

Assemble the plane by placing the blade crosswise on the "double iron," with the large screw just covering the slot (Figure 1-40). Rotate the blade with care until it is parallel to the double iron, but always extending well beyond the curved spring edge. Don't *ever* slide the blade so that the cutting edge contacts the hooked part of the double iron, as this will ruin your beautiful edge. Now bring the double iron forward slowly until it is about 1/16 inch from the edge of the blade. Tighten the screw. Set this assembly in its seat and place the heavy lever cap on top. Wiggle the assembly slightly to be sure it is properly seated, then lock it with the cam lever. The adjusting screw can be turned in or out so that the cam operates with considerable force. It should, however, be easy to unlock. Do not use a hammer. You might break the plane. Use your thumb instead.

Incidentally, the assembly of a *block* plane is different, as the low angle of the blade requires that it be placed with the bevel up (Figure 1-41). There are several types of adjustment mechanisms. Some are cam levers, others are vertical and horizontal screws. For dressing plywood edges, look for an extremely low blade angle.

Always start your work with the blade retracted, then bring it out until you see fine shavings. Of course, a plane can be used with coarser settings for rapid removal of stock. In my opinion, however, it's best to saw close to the line. Every now and then you will have to plane to a crooked line to get a close fit to an unfair surface. This may be done by using your block plane diagonally or directly across the grain. A rasp or coarse production paper on a block will do the same job.

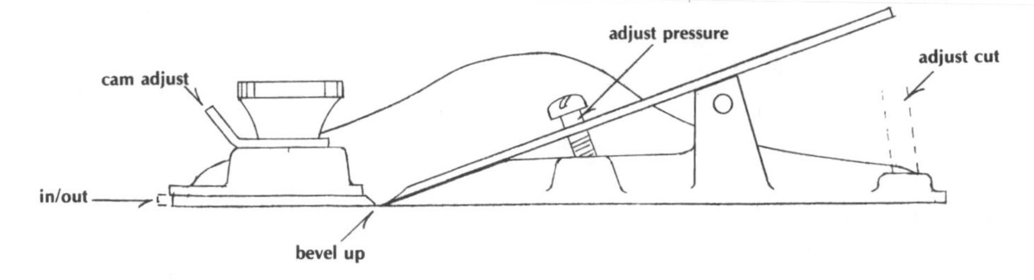

Figure 1-41. *Assembly of a block plane.*

Screwdrivers

You will need both the straight slot type and the Phillips head (cross) style screwdrivers—two sizes of each. Do not use a tip that is too thick to reach the bottom of the slot or wider than the screw head. This will mar your work. Too narrow a tip may jump out of the slot, burr, and chew it up (Figure 1-42). A Phillips-head bit that is too small will spin around in the cross and possibly ruin both bit and screw. In general, the longer the screwdriver, the better. Screw and tool should always be aligned in the same axis, impossible with a stubby. The wide, flat portion just above the tip permits use of a small wrench for maximum torque. Square shank tools also may be turned this way, a big help with large screws. The "English pattern" cabinet screwdriver does not have a wide flat, so you can drive screws into counterbores without tearing up the hole (Figure 1-43). But note that the upper shank is flattened for a wrench.

I have a little steel and plastic screwdriver set with a receptacle in the handle for storing four small octagonal bits—two Phillips, two straight (Figure 1-44). The chuck has a magnet built in the end; thus, the tip of a bit becomes magnetized when seated, so it holds screws while you are starting them in difficult spots. This tool is useful for picking up spilled screws, parts, nails, and so on. Not designed for heavy production, this screwdriver is ideal for small hinges, latches, and other hardware.

Hand Drills

Two types of hand drills are useful for limited drilling. The lightweight egg-beater style of drill (Figure 1-45), is quite adequate for drilling pilot holes in hardwood, or for screws in thin trim where

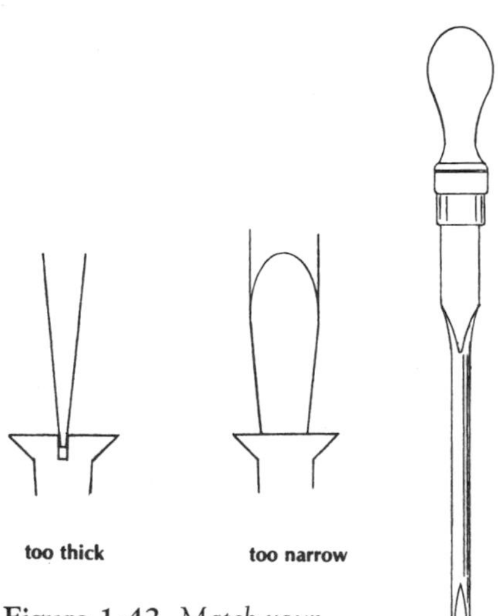

Figure 1-42. *Match your screwdriver to the screw.*

Figure 1-43. *"English pattern" cabinet screwdriver.*

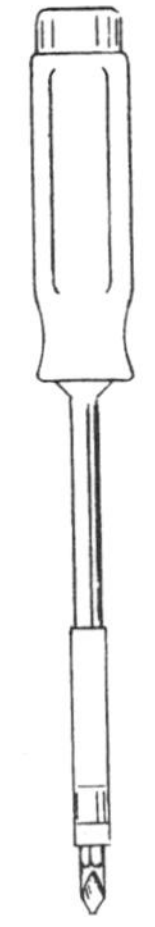

Figure 1-44. *"Four-bit" magnetic screwdriver.*

Figure 1-45. *The eggbeater-style light-duty hand drill is excellent for pilot holes and for starting small screws. Bits are carried in the handle. (Courtesy Brookstone)*

Figure 1-46. *Pressing the chest against this breast drill develops lots of pressure. It is adequate for occasional metalwork, where time is not essential, and fine for wood.*

oval heads are left exposed. Another tool useful for lead holes is the automatic "push drill." A heavier type is the breast drill (Figure 1-46). This tool has a curved plate on the upper end of the shaft that provides a good strong grip, or you can lay your chest against it and really develop power. With a ½-inch chuck (not keyed) and two speeds, I have slowly but successfully drilled ¼-inch holes in steel, bronze, and brass. This tool is handy when I have no power for an electric drill. This drill is adequate for counterboring for plugs and drilling in soft wood to ½-inch diameter. Some old boatbuilders bore all the ⅜- or ½-inch holes with a regular auger bit or Forstner bit, then follow up with a lead bit with the tip ground to a taper. The lead drill in a patent plug counterbore can also be tapered by grinding carefully.

Brace and Bits

The brace holds the squared end of an auger bit (Figure 1-47). A brace will take a square-end screwdriver bit also, for driving large screws. For boring large holes and driving large screws and in restricted spots, use the brace's ratchet. You can buy cheap versions of this tool, but better braces have ball bearings in the head and turning grips, and last a lifetime.

There are several types of auger bit. The most frequently seen is the solid center or straight core with a single twist (Figure 1-48). This type cuts fast and clears chips easily, especially in wood full of gum. The double twist or fluted bit is not as fast, but it makes an accurate and clean hole and is good in softwoods (Figure 1-49). This type of bit is good for plug boring, as it does not chip around the hole,

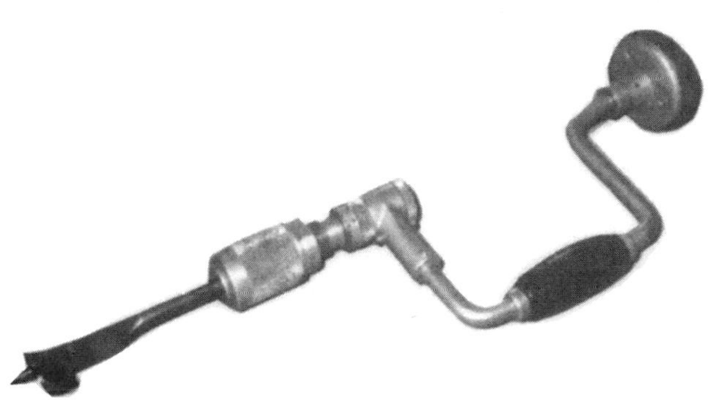

Figure 1-47. *Brace with a powerful chuck. It is ball-bearing throughout.*

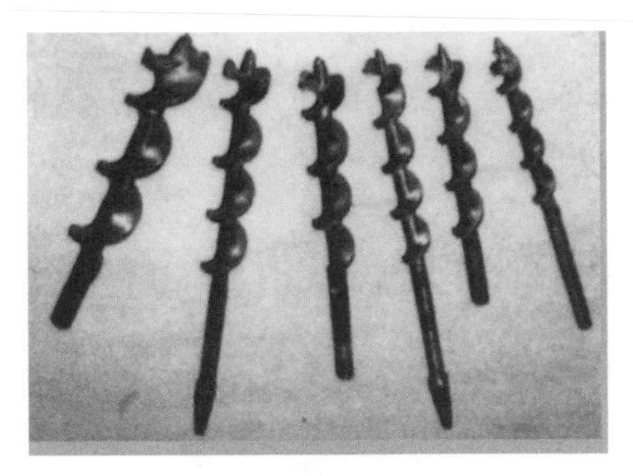

Figure 1-48. *Solid-center bit has double spurs and cutters with a medium-fast screw pitch. Available in sets 1/4 to 1 inch, in 16ths. A must.*

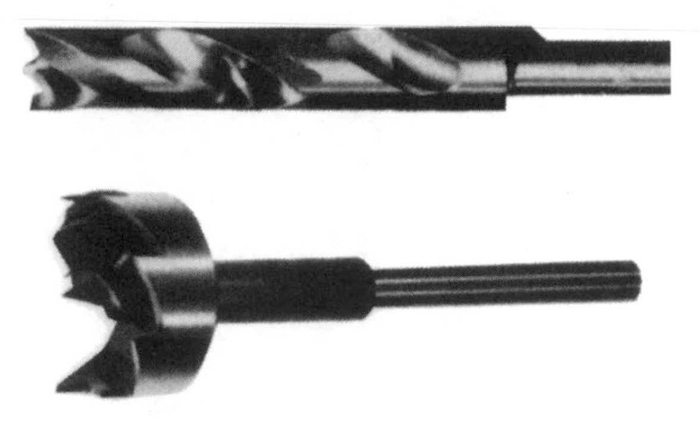

Figure 1-50. Top: *Machine brad point bores cleanly across or along grain.* **Above:** *Multi-spur does difficult jobs better—veneers, angles, overlapping, and so on. (Courtesy Woodworker's Supply)*

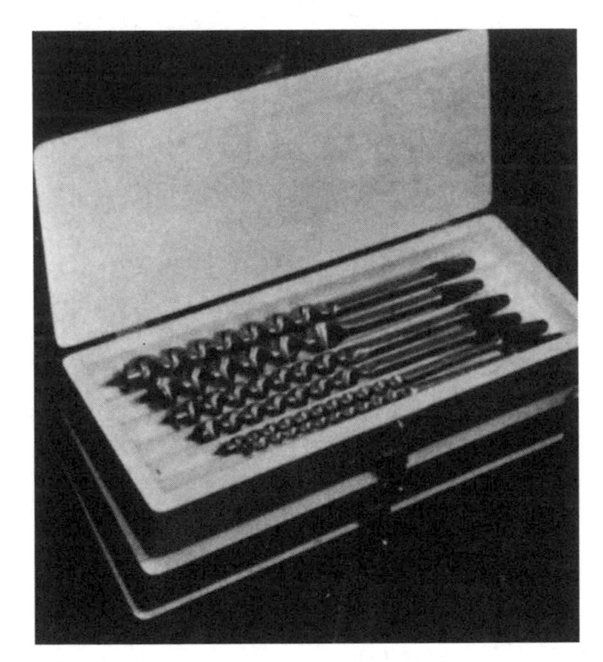

Figure 1-49. *The traditional Jennings double-twist bits clear chips easily. The tapered bodies prevent binding. Just about the best available.*

even at slow hand speeds. Boring end grain with a standard auger bit, on the other hand, is difficult. Best is the Forstner bit, which usually has no screw or lips, so it can be hard to start accurately. Some, however, are provided with a spur or point for centering (Figure 1-50), often used in end grain, where the point can be pressed in slightly, then turned. Such a bit planes out the hole. The Forstner can also be used close to the end of a piece, where an ordinary bit would split the wood.

Bits come in sets of 1/4, 5/16, 3/8, and so on, in 16ths of an inch up to 1 inch, or in various combinations in smaller sets. Of course, you can buy any single bit you may need. The sizes are stamped into the square end. One-quarter inch reads "4," 7/16 inch reads "7," and so on. All the bits described here can be used in a slow-speed breast drill or electric drill by sawing off the squared shank.

Ship's Auger Bits

Eventually you may need a couple of ship's auger bits (Figure 1-51), sometimes called barefoot bits because they have no screw or spur. They are used to bore deep, straight holes for drifts, long bolts, and dowels in such places as cabin trunk sides, in keel and deadwood, centerboard, rudder, or wherever a long fastening is needed. Knots and hard grain have a tendency to deflect an ordinary screw auger bit and make it wander; the barefoot bit, however, has no point to be deflected.

To start, drill a shallow hole with an ordinary bit, then follow up with the ship's auger. Chips, especially if wet, green, or gummy, must be cleared frequently, or the bit will pack solidly and twist off. If you should have this problem, place a clench ring or washer over the bit and saw it off about one diameter above the ring, then peen it over into a rivet. If it breaks off below the surface, fill the hole with hot tar or tallow and plug with wood. Drill another hole nearby, but a little to one side, so it's not in the same grain, to prevent a split from occur-

Figure 1-51. *Ship's auger bits for boring long, straight holes. There is no point to be deflected. Start in shallow hole drilled with a conventional bit. For cabin sides, centerboards, rudders, keels, and deadwood.*

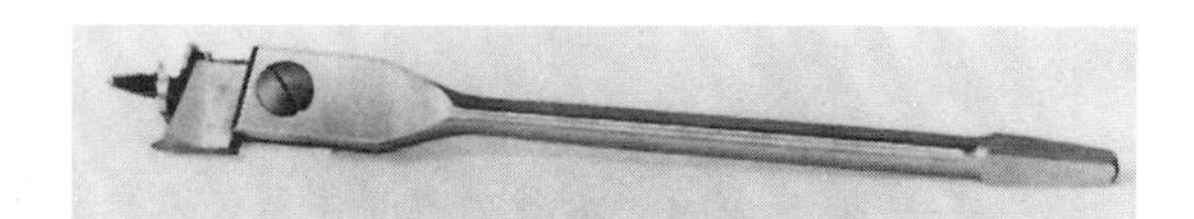

Figure 1-52. *Expansion bits bore up to 4-inch diameters with hand brace. Do not use a drill press or electric drill. Cutter is adjustable for precision. Sharp spur cuts clean circle. (Courtesy Brookstone)*

ring. As Howard Chapelle says in *Boatbuilding*, "A good many boats have an auger bit serving as a drift in their keels."

Expansion Bits

You may have frequent use for an expansion bit (Figure 1-52.) It has an adjustable cutter for boring diameters up to 3 inches. Always use an expansion bit in a hand brace, for they are very dangerous in a drill press or an electric drill. They are prone to grab the wood, and you and the drill suddenly rotate. Do not try to bore through with an expansion bit. When the point of the screw shows on the back side, reverse the piece in the vise and bore from that side. If you don't, you will create a badly chewed-up mess. You may have to chisel out the remaining plug. If, after a few turns, the screw no longer pulls, you can make it bore by rocking it slightly from side to side as you turn the brace. Better yet, when you start to bore, clamp a block of scrap to the work piece so the screw will continue to pull all the way through, advisable when you do any boring with the back side accessible.

Twist Drill Bits

You can use twist drills, or machine bits (Figure 1-53), for much woodwork, but where a clean hole is mandatory, they are not always satisfactory. They are good for bolt and lead holes, of course. High-speed steel is best, although long "electrician's bits" are available in carbon steel. These should be turned at low speed to prevent heating and annealing (softening). High-speed steel takes the heat, holds its edge 10 times longer than carbon steel, but is brittle. Your brace may not be powerful enough to hold the round shanks when drilling deep holes; borrow a ⅜- or ½-inch electric drill. For drilling through deadwood, keels, cabin side, centerboards, and so on, it is common

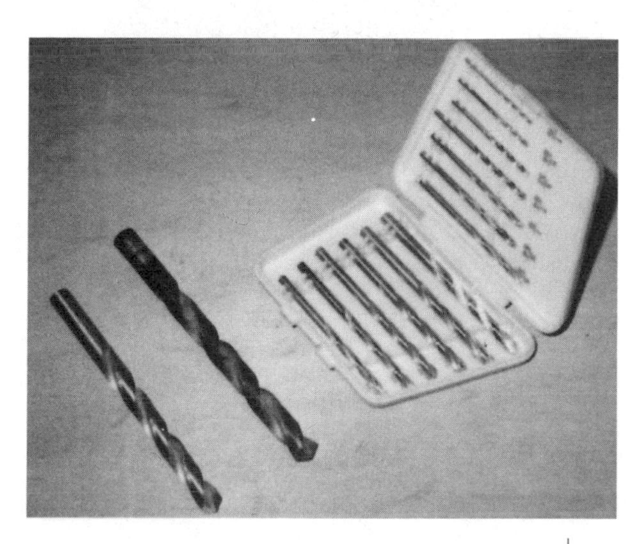

Figure 1-53. *Twist drills ground to about 60-degree angle for wood, 40-degree angle for metal. Weld shank extensions (of high-speed steel) for deep holes.*

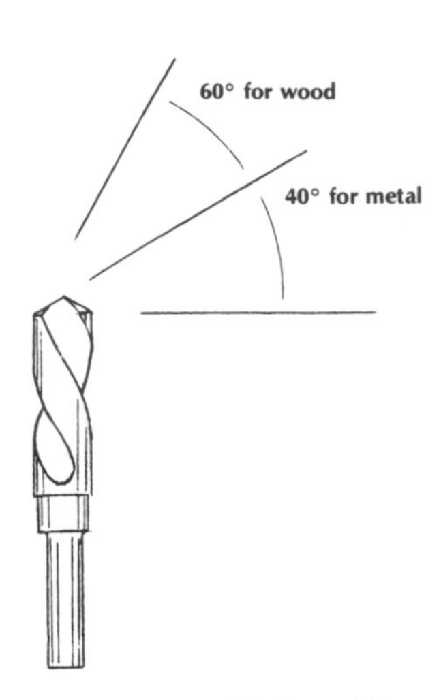

Figure 1-54. *Blacksmith's drill bit.*

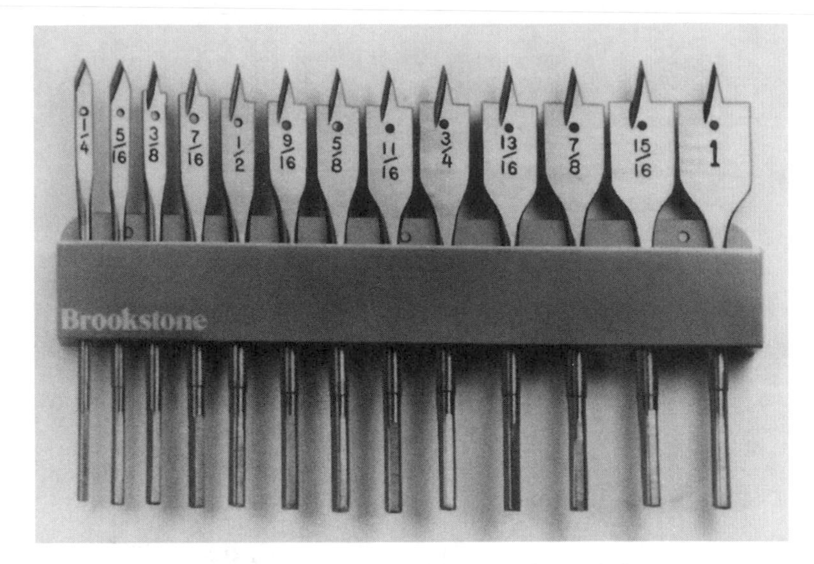

Figure 1-55. *A fine set of spade bits. These are fast and clean-cutting. They are filed sharp easily and are inexpensive. They cut best at fairly high speeds. (Courtesy Brookstone)*

to have the shanks on ordinary twist drills extended by welding or brazing on a foot or so of iron rod of 1/16 inch smaller diameter, to reduce friction. Be sure to back out the chips frequently, or you'll lose a good drill bit.

Occasionally, I have used a low-priced twist drill called a blacksmith's bit. Available in sets of three or four in carbon steel, from 1/4 to 3/4 inch in diameter, some have 1/4-inch shanks for use in a small electric drill, breast drill, or even a brace. These bits are usually somewhat shorter than the average twist drill bit in tool steel. They won't handle much work in iron, so you'll be grinding often (Figure 1-54).

Grinding Twist Drills

Twist drills used in wood are not for use in metal. An angle of 40 degrees is supplied for metalwork. Wood requires about 60 degrees, as shown in Figure 1-54. While grinding drill bits is an art, it is one you can learn easily from the many good books available. If the idea appeals to you, there are many quite reasonable drill-grinding devices you can buy.

Spade Bits

Every hardware store has a very efficient bit called the spade bit (Figure 1-55). These bits employ a kind of scraping-planing action, leave a clean hole, and are best if turned at moderately high speeds, ideal in an electric drill or a drill press. The bits can be sharpened with a small flat file, because their cutting edges work like a plane blade. Do not try to bore clear through; always bore back to avoid serious splitting as the bit exits from the work.

Rasps and Files

One useful rasp has a combination of coarse and fine—flat on one side, half oval on the other (Figure 1-56). Larger rasps come with a variety of teeth and have tangs for attaching handles. Have a wire brush handy for cleaning teeth. Do not attempt to use a rasp on metal—it is not a file.

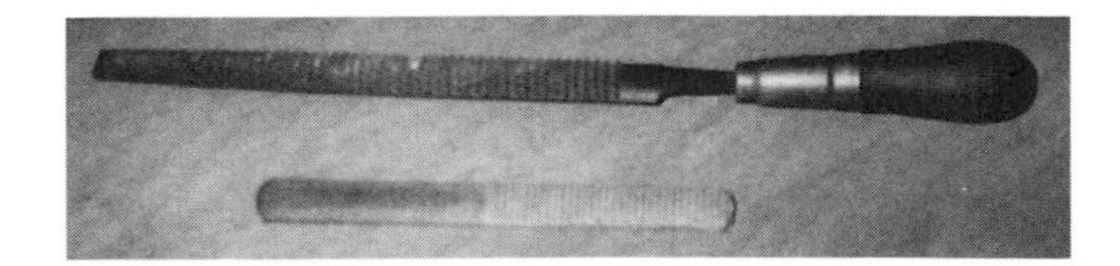

Figure 1-56. *Many types of rasps are available. Shoemaker's rasp has oval and flat, fine and coarse on one tool. Some fine boatbuilders frown on use of this tool.*

Figure 1-57. *Surform works like a plane for roughing off material. Cuts in any grain direction, even plywood. Flat and convex blades. Best on softwoods.*

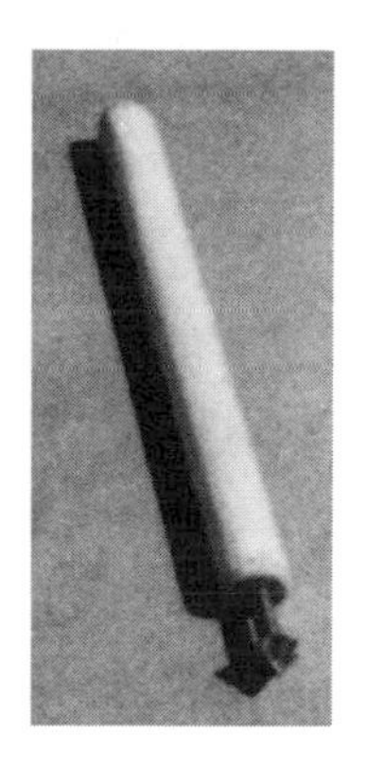

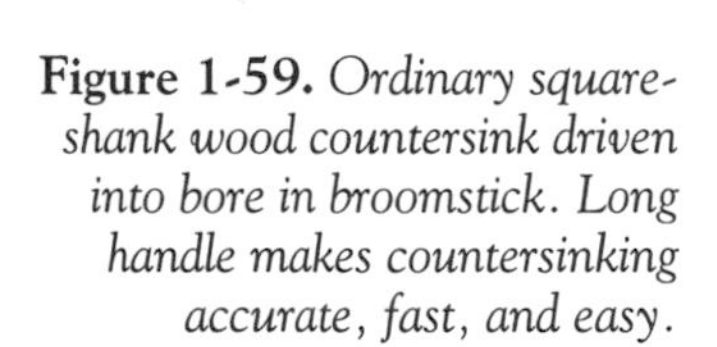

Figure 1-59. *Ordinary square-shank wood countersink driven into bore in broomstick. Long handle makes countersinking accurate, fast, and easy.*

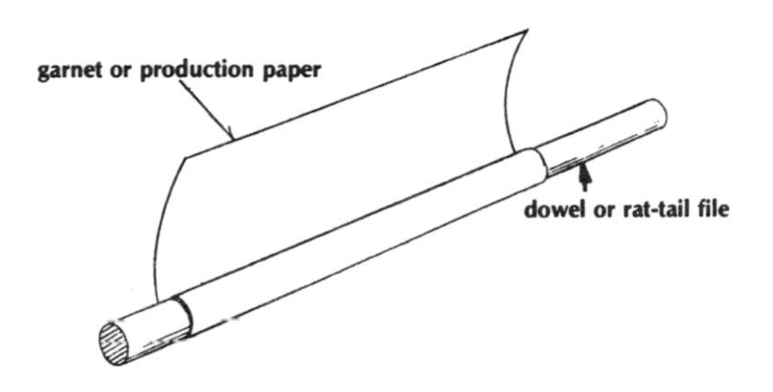

Figure 1-58. *Device for fine sanding.*

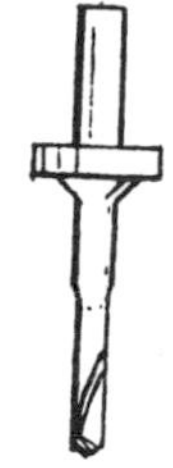

Figure 1-60. *Stanley Screw-Mate.*

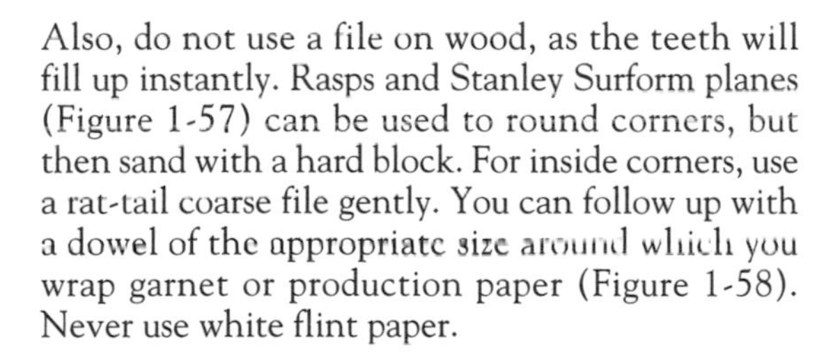

Also, do not use a file on wood, as the teeth will fill up instantly. Rasps and Stanley Surform planes (Figure 1-57) can be used to round corners, but then sand with a hard block. For inside corners, use a rat-tail coarse file gently. You can follow up with a dowel of the appropriate size around which you wrap garnet or production paper (Figure 1-58). Never use white flint paper.

Nail Sets

A kit of nail sets is required, since all finishing nails (galvanized only) must be set below the surface and the recess filled with plastic wood or matching stuff. Never drive a finishing nail right down to the surface; your hammer may make a bad indentation. Leave the head out about 1/16 inch, then sink it with a set slightly smaller than the head, as shown in Figure 1-3.

Here's a useful tip: If you mark a wood surface in any way, wet it two or three times with a drop or two of water. Yes, saliva works fine. If the dent is stubborn (such as, God forbid, from a C-clamp), place a damp cloth over it and press with a hot iron for a couple of seconds. These methods should expand the wood fibers to their original level. Let dry and then sand with a wood block.

Countersinks and Counterbores

I have found that the most useful countersink has a square shank for your brace, to be sawed off for your hand drill. You can make a hand countersinking tool, however, that will beat this for speed and convenience. Make or buy a long file handle, then drill a hole in the end slightly smaller than the square shank. Tap the handle onto the tool and it's done, or use a piece of broomstick (Figure 1-59). When you countersink, try to keep the tool vertical to the work. Two or three twists is all it takes.

In addition, I recommend Stanley Screw-Mate bits that drill a pilot hole, shank diameter, countersink, or plug counterbore in one operation (Figure 1-60). There is a type in which the pilot and shank bits are adjustable, for up to a 2½-inch screw and plug counterbore. The patent counterbore is the cleanest cutting and has a depth stop.

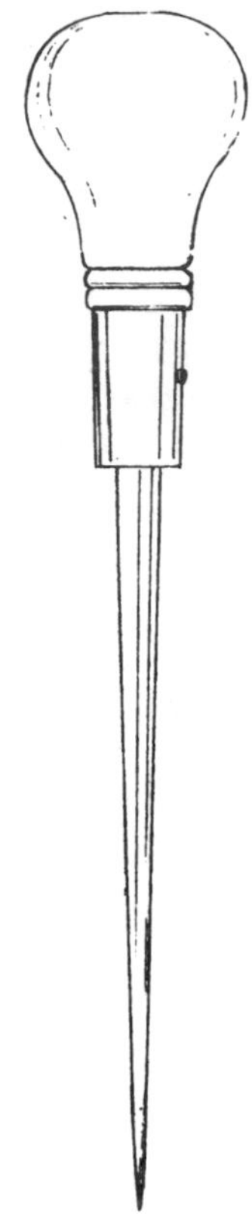

Figure 1-61. *Prick punch.*

Figure 1-62. *Brad awl.*

This tool is rather costly, usually available only at marine hardware stores, for 3/8- and 1/2-inch plugs. When using any of the others in an electric drill, watch that you do not counterbore too deeply, as they cut very rapidly.

Brad Awl and Prick Punch

For starting small screws in softwoods, I use an old, sharp prick punch (Figure 1-61). For aligning hinges, this tool is great; you can relocate the hole as little or as much as you need to, even after the screw has entered, because the wood fibers are still there, not removed, as they would have been if you had used a drill. The sharp point enables you to probe. Now, some experts recommend a brad awl, which has a flattened point (Figure 1-62). You start the brad awl with this flat across the grain. Thus it chisels the fibers as it penetrates and prevents splitting. When attaching hinges to the edge of plywood, drill lead holes after you get one screw located exactly, then go back and drill that first hole. This prevents the plies from splitting or delaminating. If the end plies are already split slightly, pry them open a bit, dab in a few drops of glue, and let it set up before you install the hinges.

2 More Basic Hand Tools

CLAMPS

I could devote an entire chapter to the use of clamps in boat and cabinet work. You can never have too many. That's true even if you are far past the type of work on a bent-frame hull that requires a half-dozen or so at a time. Just recently, for example, I laminated a small stem for a 16-foot duckboat and I ended up with 12 C-clamps so close together it was a problem to turn the screws. Resolve right now to accumulate C-clamps of all sizes. Try garage sales, flea markets, used-tool shops, the classified columns, and so on. Borrow if you can't buy.

C-Clamps

I have found the 6-inch C-clamp (that's the maximum space between the button and the pad) to be the most useful (Figure 2-2), but all sizes are handy, right on down to the 1-inch miniatures. Of course, when you get into spar construction of any size and complexity, you'll need much larger capacities. A 40-foot mast of the usual hollow box type might easily require a total of 75 clamps.

The uses of C-clamps seem quite obvious, but let's just mention a few: to hold diagonal braces while you square up structures; to secure work pieces while you plane, chisel, bore, or whatever; to hold parts in position for accurate drilling, screwing, or nailing together; to clamp for gluing; to stand boards or plywood on edge while dressing with a plane (Figure 2-3); to position your straightedge for power sawing plywood sheets. The list goes on and on, especially if you work alone. That's when a clamped-on support for a long, heavy piece is as good as another man.

There are several types of clamps to look for. Even the light 3-inch and 4-inch sizes are useful when great pressure is not required (some of the modern glues don't require pressure). The deep-C style with a deeper throat, shown at the lower left in Figure 2-1(top), is excellent, and the cost is only pennies more. Even the cheapies of the sliding quick-acting bar design are handy, because they have the reach of a heavy 6-inch forged C-clamp at about one-third the cost. True, they are cheaply made, have poor buttons, and will loosen if there is any jarring or hammering. For gluing the average assembly, however, they will do the job if you use a bit of care.

Some of the newer C-clamps have a sliding bar handle. This is good because you can set the clamp close to a corner or another part, whereas the older wing types are impossible to turn when there is this kind of interference. This is a good place to point out that the bar or wing-type thumbscrew on your clamp was designed to be turned by hand only. Do not use wrenches or pipes; if you do, you will in time spring the clamp, permanently deforming it and rendering it useless. Also, you might generate so much pressure that you damage the pieces being clamped. The circular marks left by C-clamps have a way of expanding for years after they are sanded out. It's a shame so many so-called professionals do this regularly. Avoid such damage by always using scraps of wood between the clamps and the work.

All C-clamps have one very bad drawback. They lose their buttons! A clamp without a button is highly destructive to wood surfaces. In some cases, the button can be replaced, at least temporarily. Slip it back on and fasten the clamp in a vise (Figure 2-4). Then, using a nail set or a center punch, go all the way around the edge of the cup in the button. If the button is drop-forged, the punch will gradually reduce the inside diameter of the cup enough to close it over the ball end. There

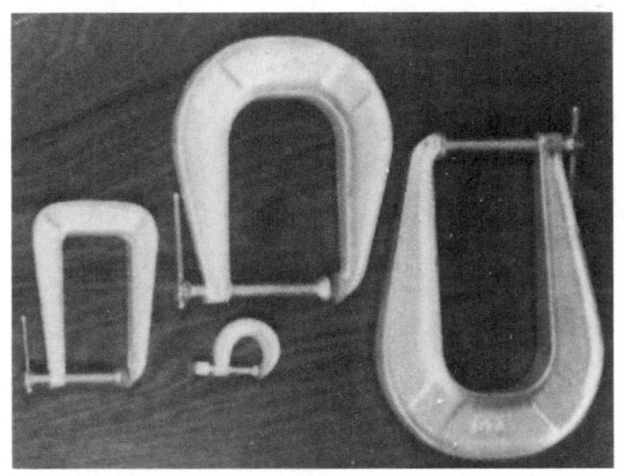

Figure 2-1. Top: *Deep throat clamps in 3-inch and 4-inch sizes are desirable. The cheap 6-inch-capacity bar clamp at the upper right is good mainly for fast action and light pressure.* **Above:** *Three deep-C clamps and a 1-inch miniature. (Courtesy Woodworker's Supply)*

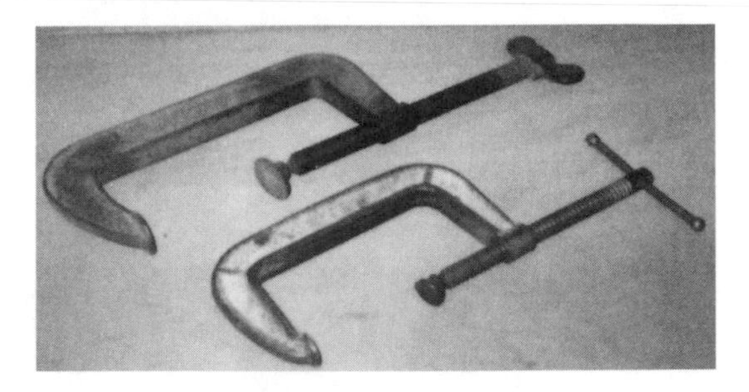

Figure 2-2. *The 6-inch C-clamp is the most versatile size. Use 8-inchers for heavy work.*

Figure 2-3. *C-clamps provide support for planks, doors, panels, and so on, on edge.*

was a time when you could buy replacement buttons at good hardware and tool supply stores. It's becoming more and more difficult, however, to find little necessities like these. If you search, take your C-clamp along to be sure you get the right size—they vary. The button should just slip on with a tap of a hammer. Keep the socket and the screw well oiled and your clamps will serve well for two or three generations.

Bar Clamps

Bar clamps are almost as indispensable as C-clamps. These quickly adjusted clamps (Figure 2-5) are used universally by cabinetmakers for drawing together cabinet joints for glue or fastenings. Just be sure you have a diagonal or two on any box shape and check

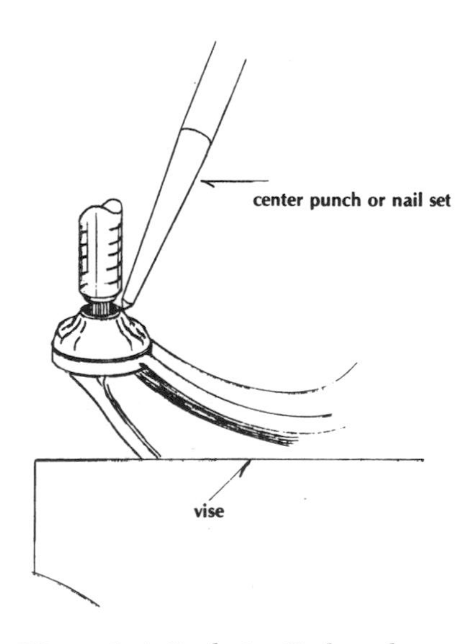

Figure 2-4. *Replacing C-clamp button.*

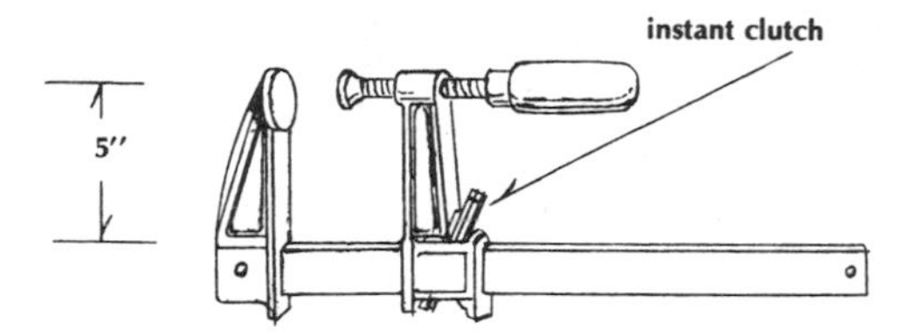

Figure 2-5. *Fast-action bar clamp.*

for square (measure corner to opposite corner), as the clamps can pull your structure far out of square. A common use of bar clamps is for edge-gluing, such as in a cabin side. First C-clamp the pieces to several cleats to hold the components flat. Then place bar clamps on each side of the C-clamp/cleat assembly to prevent buckling and damage. Remember, too, that glue will ooze out of the joints, so your cleats should have been covered with pieces of plastic such as Saran Wrap.

Bar clamps come in standard lengths such as 24 inches, 36 inches, 42 inches, and so on. A 48-inch clamp, however, is pretty awkward for a 14-inch span, and if you need 50 inches, you are licked (although there are ways of hooking two clamps together). One answer to this problem is to acquire a number of pipe clamps (Figure 2-6). A set consists of the screw end and the adjustable lower jaw end. Two sizes are available, for ½- or ¾-inch pipe. The larger size is preferable, because the jaws are a bit deeper, so it generates more pressure. The less costly one does a fine job in about 90 percent of the cases. If you collect an assortment of threaded one-end used black pipe in both sizes, you will have an endless choice. Any length of pipe can be used. I usually have pieces of black pipe from 4 feet up to 12 feet on hand at all times, as my cabinets sometimes run up to 10 feet end to end. Black pipe does not slip as galvanized pipe does, and it is sold very cheaply by the pound at scrap yards, so stock up.

The Jorgensen clamp is a very nicely made bar clamp sold by hardware stores, usually limited to 24 inches and costly by comparison with pipe clamps. Their jaws, however, are deeper, and they are instant-adjusting. The same type of quick-acting bar clamp comes in a short, heavier style, also with a deep throat. These are hard to find on the used-tool market, but they are terrific for spar work. You might be able to rent some from a local cabinetmaker over a Saturday or Sunday when his shop is closed. Cheap 12-inch bar clamps of similar appearance but sorry foreign construction are worthless.

Figure 2-6. *Gluing up a cabin trunk side is easy with inexpensive pipe clamps. (Bruce Bingham photo)*

Spring Clamps

It would be hard to get along without a few spring or pinch clamps. These are great for holding parts during drilling of fastenings as well as for small gluing jobs. They generate an enormous amount of pressure and are positioned quickly, although openings are limited.

PORTABLE VISE

For a few dollars you can acquire an extremely useful little vise that self-clamps to the end of a sawhorse, bench, or plank. It will hold a vertical piece such as a plywood panel on its edge on the floor, or a narrow part can be stood on end against the bench and clamped while you work on the upper edge. Some newer versions are aluminum, but I have a lot of faith in the old cast-iron ones.

BENCH VISES

There is always a certain amount of metalwork in a boat, so you may need a light- or medium-duty bench or machinist's vise of the type shown in Figure 2-7. Any vise must be lined with plywood to cover the jaws if it is to be used for woodwork (Figure 2-8). I would not recommend a vise with a jaw width of less than 3½ inches. If you find a used one for $25 or so, take it. Check for sloppy bear-

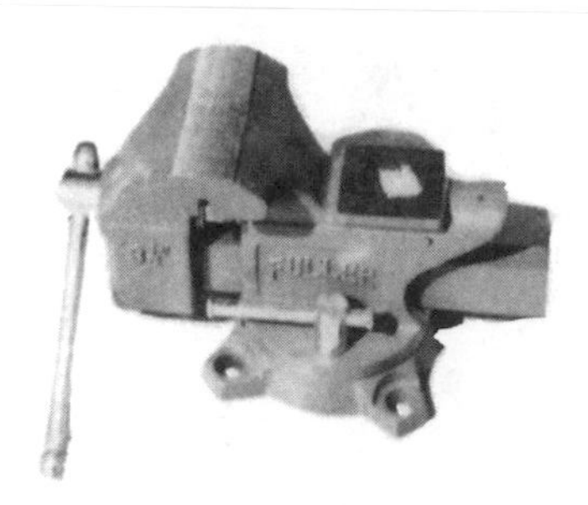

Figure 2-7. *A 4-inch machinist's vise is adequate for most small metalworking jobs, and fine, too, for woodwork with plywood inserts covering the vise's jaws.*

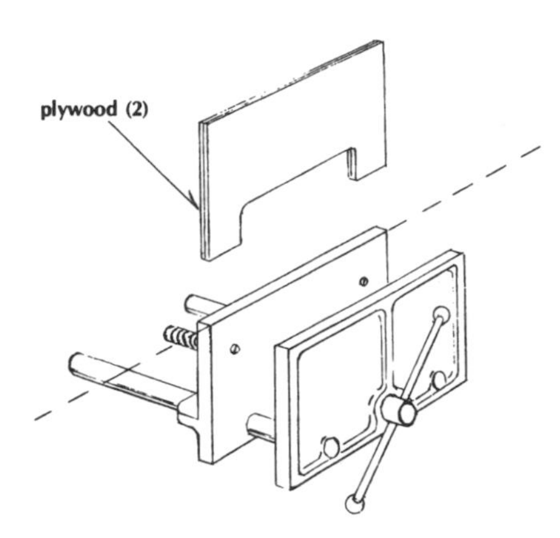

Figure 2-8. *Vise liners.*

ings, end play, beat-up anvil surface, bent screw and channel, cracks anywhere, and weakened welds. Such a vise must be bolted near a corner on a heavy bench so it can be swung for gripping vertically (Figure 2-7).

The fast-acting woodworker's bench vise is ideal (Figure 2-9). Select one with jaws 6 to 8 inches wide. Maximum width is not mandatory, for you can let your wooden liners project an inch or so at each side. The best vises have a stop in the moving jaw that enables you to clamp pieces on the flat between it and another stop on the bench. They also have a priceless feature—sliding jaws that open and close in a second with final pressure being applied with the screw. Some good ones open to 12 inches. These are quite costly, about $90 in 1992. Without the quick-acting screw, they will cost less.

In a later chapter I'll describe a split-second vise for holding planks and parts on edge. You can build one for the cost of five bolts and scrap 2-inch lumber.

CHISELS

Chisels come with three types of handles—tang, socket, and steel cap. The tang is a rectangular extension to be driven up into the handle. The handles, formerly wood, are now largely plastic. This is a light-duty chisel, intended to be used with hand pressure only, or driven by light taps of a small mallet or block of wood. But makes vary, as shown in Figure 2-10. The next heavier type is the socket chisel, so-called because the wood or plastic handle fits into a socket (Figure 2-11). These handles are easily removed for storage or replacement; if made of wood, they should be capped with leather. The best chisel is the heavy-duty type in which the steel extends right up through the plastic handle and forms a cap. It's better to use a light mallet or block with any chisel. Never use a hammer.

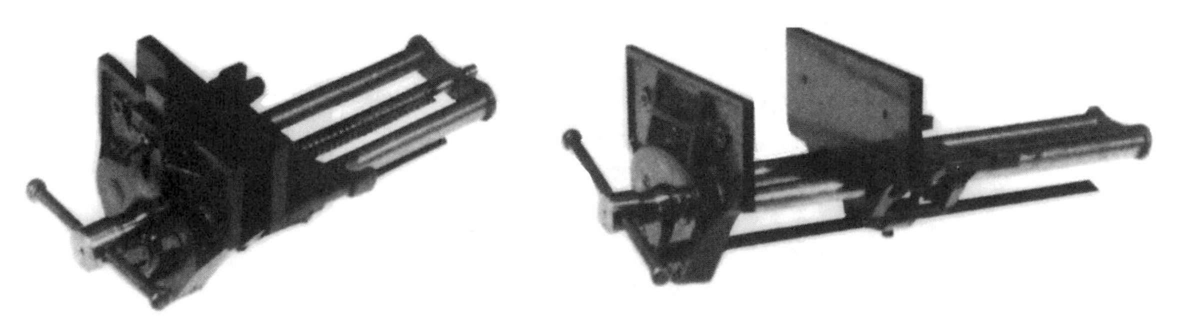

Figure 2-9. *Fine woodworker's end vise (left) has quick release for positioning of the movable jaw, and (right) is designed for face mounting on bench. (Courtesy Woodworker's Supply)*

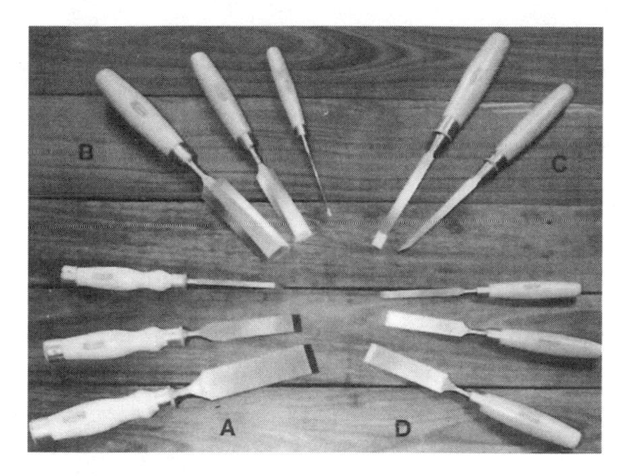

Figure 2-10. *Marples chisels: (A) Extra-thick, firmer chisels for mortising. (B) Sheffield steel cabinet-maker's chisels. (C) English mortise chisels with long blades. (D) Flat-ground, firmer chisels for general use and for hard, tough woods. (Courtesy Woodworker's Supply)*

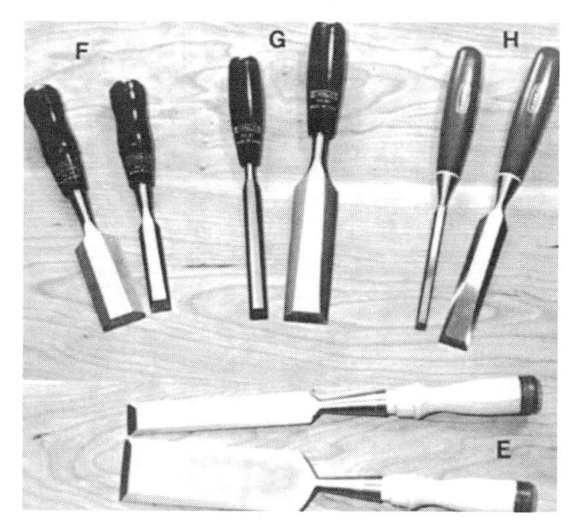

Figure 2-11. *Chisels. (E) Socket firmer chisels forged for holding edge under heavy work. These have hickory handles and leather rings. Two Stanley medium-duty (F) and heavy-duty (G) butt chisels, steel capped. (H) Two fine Marples heavy-duty socket butt chisels. (Courtesy Woodworker's Supply)*

Figure 2-12. *A worn-out slick. Slicks come in all sizes from 2 to 3 inches wide and up to 30 inches or more in length.*

Different chisel blades are used for different tasks. The tang or paring chisel, quite thin and short, allows for good control on fine work such as mortise and tenons. Butt chisels, usually socket types, are used most frequently for cutting gains, the recesses for butt hinges. They are useful for a wide variety of other work also. Long, narrow blades are best for mortising. You should have ¼-, ½-, ¾-, and 1-inch blades.

Another chisel widely used in boatbuilding and joinery is called the slick, shown in Figure 2-12. This type of chisel comes in many sizes, ranging from 2 to 4 inches wide with a blade length from 9 inches to well over a foot, plus a socket for a large wooden handle. One use is getting into corners you can't reach with a plane. You hold the flat side down on the work and slide it back and forth with some force. The blade cuts like a very low-angle plane. Sometimes the best results come from stroking diagonally across the grain, swinging from side to side and pressing down with the fingers of the left hand as you advance the blade into the work. The slick may have beveled or square sides. It is best if the upper or unbeveled side is straight for its full length, or nearly so.

Using a Chisel

The chisel should be ground just like a plane blade, and carefully whetted and honed. For rough cuts and concave shapes, work with the bevel down (Figure 2-13A). To cut a joint down to the line, rough out as much as you can with saw cuts, then chisel with bevel up (Figure 2-13B); slice in from one side to the middle across the grain, then do the same from the opposite side (Figure 2-13C). Clean the surface with a sort of paring motion until it makes a good flat surface. Do not press all the way across, as this will split out the opposite side. Always examine the direction of the grain, especially going with the grain, for the tool will try to bite too deeply if the grain is against you (Figure 2-13D). Do not try to cut to the line immediately. Work down to it in easy stages. When cutting a mortise, bore out first with a bit that is considerably smaller in size than the width of the mortise (Figure

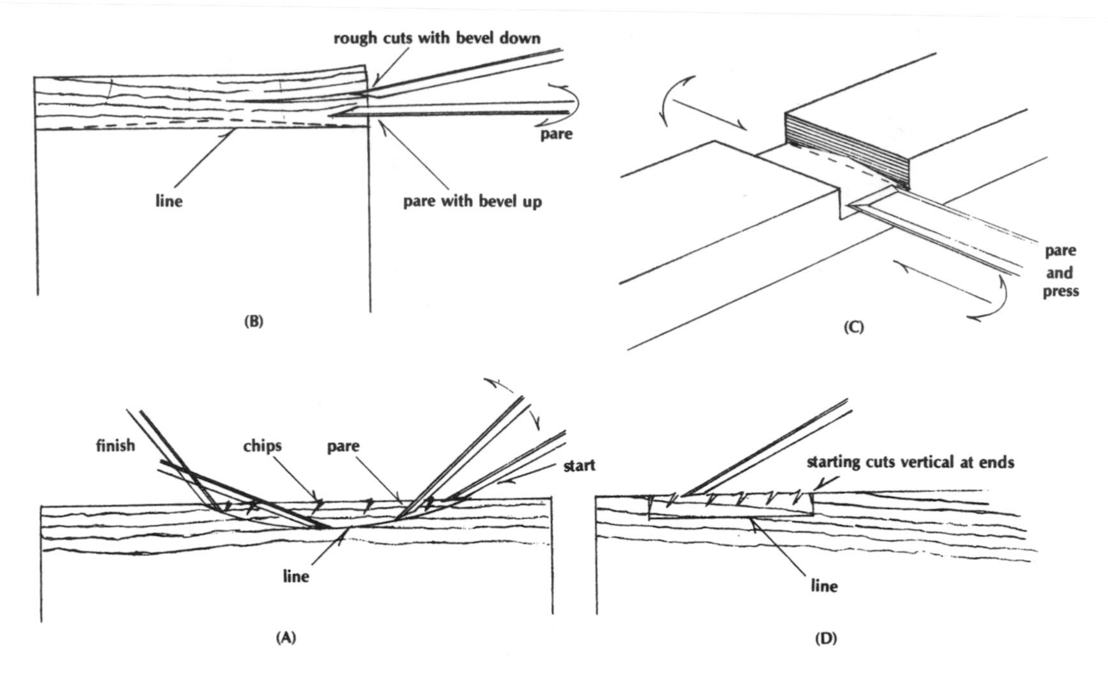

Figure 2-13. *Chiseling techniques.*

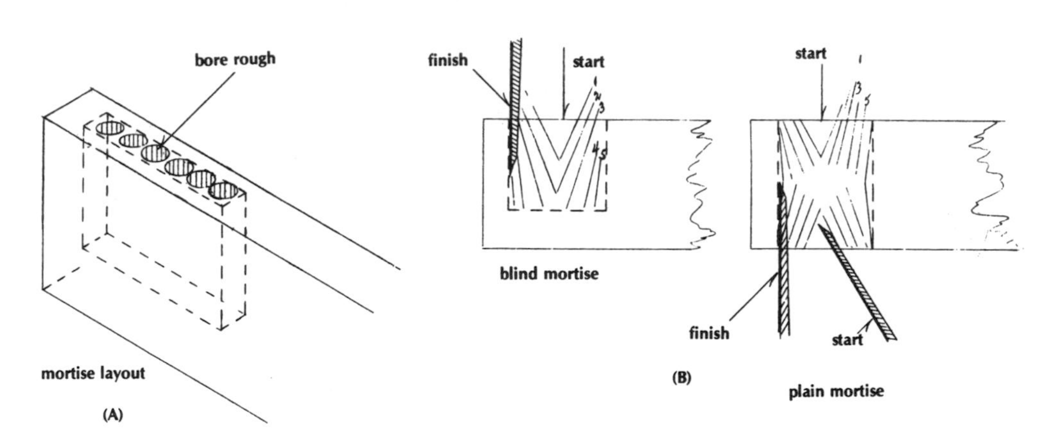

Figure 2-14. *Cutting blind and plain mortises.*

2-14A). For ½ inch in width, use a ¼-inch chisel so you don't split the piece or go past the line (a knife mark). Work from end to middle from both sides, with the bevel down (Figure 2-14B). When you near the mark on both sides, you can start nearly perpendicular strokes, bevel up, paring.

Never use your chisels for prying, opening crates, or for splitting boards or firewood. Being hard steel, they are brittle. The cutting edge will chip or nick if you so much as touch a nail. Treat your chisels with care and they will last for several generations—if they were quality tools to begin with. You can buy quite a decent set of chisels for about $25–$30. If your chisels are poor in quality, you will spend too much time sharpening—and you'll never get a good edge, anyway.

Figure 2-15. *This hacksaw has a rigid tubular frame. Blade must be set up as tightly as possible.*

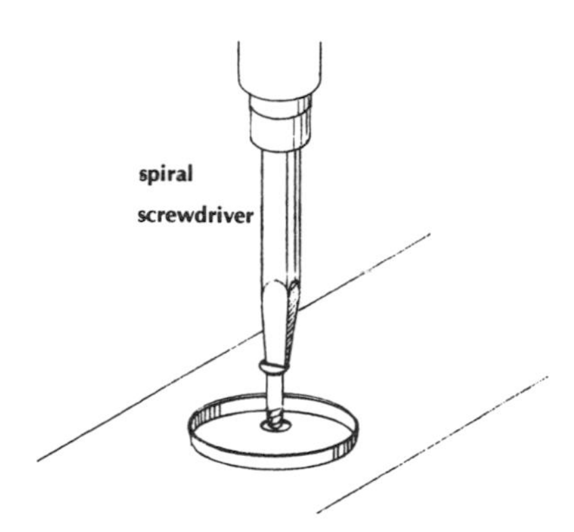

Figure 2-17. *The indispensable original Yankee spiral screwdriver. Can lid protects surface.*

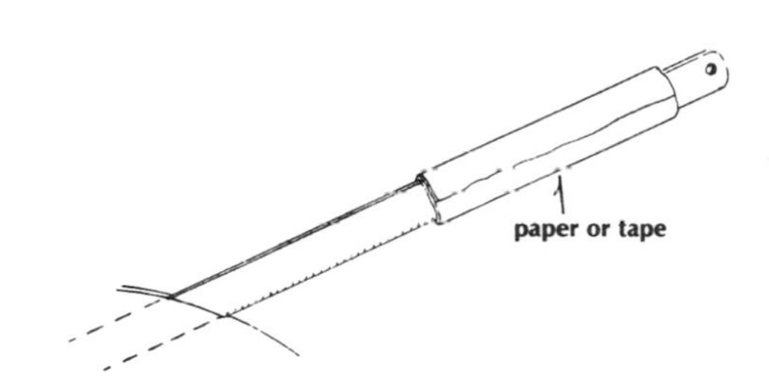

Figure 2-16. *Hacksaw blade cuts curves.*

HACKSAW

You need a hacksaw for cutting drifts, threaded rods, bolt ends, and so on. It is important that the blade be kept in extreme tension; a heavy tubular frame is best for this (Figure 2-15). I have seen hacksaw blades alone do a commendable job of cutting wood in restricted spots and on curves. Just wrap the blade with tape or paper to leave four or five inches for the stroke, as shown in Figure 2-16.

SPIRAL SCREWDRIVERS

The original and best-known spiral screwdriver, now rare, is the Stanley Yankee, now also made by Millers Falls and others. This tool provides the fastest method for driving screws short of an electric screwdriver. The ideal size is 8 to 9 inches from handle to collet, with a 4-inch wood handle (Figure 2-17). Newer models have longer plastic handles (for two hands, I suppose). The collet will take three bit sizes, both straight and Phillips head.

To operate a spiral screwdriver, release the spiral carefully, and engage the tip of the bit and the screw head after starting the screw a turn or two with the fingers. Now press, keeping the driver in line with the direction of the screw. The spiral will run back into the barrel, rotating the screw six revolutions. Let up on the pressure and the spiral will return automatically for another stroke. Or you can use a series of short strokes. You get maximum power for sinking the head by turning with both hands with the spiral latched. The spiral is reversible for pulling out screws.

The Yankee presents just one hazard. It has a propensity for jumping out of the slot, usually because the tool is not kept in line with the axis of the screw. When this happens, the driver can bore a hole into the piece. To avoid this problem, punch a hole in a paint can lid and hang this on the screw. As shown in Figure 2-17, the work surface is protected. Incidentally, every day I use a spiral screwdriver bought in the late 1930s. It has had no repairs and only a drop of oil now and then. What's more, I'm still using the original bits.

SPOKESHAVE AND DRAWKNIFE

The spokeshave is really a kind of plane. This tool (Figure 2-19) can be either pulled or pushed to make bevels or radii on either curved or straight corners. It also is good for dressing concave surfaces

Figure 2-19. *Spokeshave with flat sole used for convex surfaces, for chamfering, and so on. A convex sole is used for shaping ogees and other concave edges. (Courtesy Brookstone)*

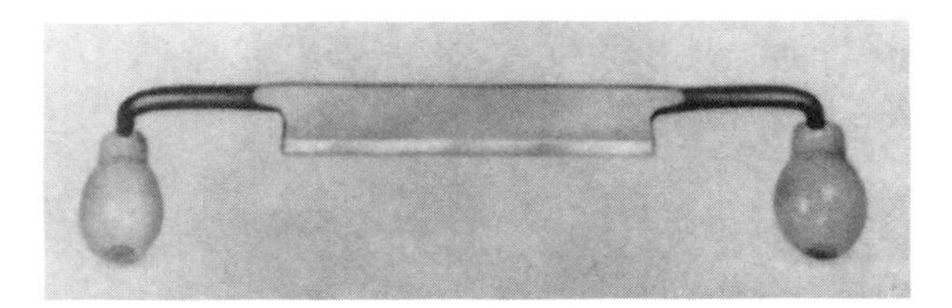

Figure 2-20. *A light drawknife. This tool is great for small spars before finishing with a plane. (Courtesy Brookstone)*

and shapes, such as the shaping on berth fronts, ogees in coamings and rails, and so on. The handles on each side of the spokeshave provide excellent control. The cutting edges must be ground and whetted just as you do with your plane blades and chisels. One I used for years had a block plane blade instead of the original. This blade was easier to grind and sharpen on the oilstone; its greater length made it a lot easier to manipulate.

The drawknife is related to the spokeshave, but it is used to remove large amounts of material, such as on solid spars. A drawknife has no sole, so there is nothing to restrict the depth of its cut (Figure 2-20). You draw toward you with the grain. Do not let the blade dig in, or you will rip off huge splinters. This blade, too, is hollow ground by drawing from end to end across the circumference of the grinding wheel. Then the blade is whetted by placing it vertically in a vise and stroking spirally with an oilstone. Some drawknives are made with adjustable handles for the most convenient grip.

MARKING GAUGE

The marking gauge is for scribing parallel to a straight or convex edge (Figure 2-21). Most of

Figure 2-21. *Hardwood and brass marking gauge (A) marks line parallel to surface and (B) has knife cutting gauge. (Courtesy Woodworker's Supply)*

these tools have a sharpened pin near the outer end and an adjustable stop to be set to the required dimension. You mark by drawing the gauge toward you. The depth of the scribe mark can be controlled by rotating the gauge slightly until you produce just a slight scratch. Always set a marking gauge by measuring from the point to the stop. Do not rely on the graduated markings.

PLUG CUTTERS AND BUNGING

If you have access to a drill press, you can save money by using a plug cutter (Figure 2-22) instead of buying plugs. The cutter shown ejects the plugs as they are cut, so place a carton on the drill press table to keep the plugs from flying all over the place. Set the stop on the press so the tip of the cutter just comes through into a piece of scrap. A lower-cost type of cutter does not bore all the way through ¾-inch stock, so the plugs remain seated, as shown in Figure 2-23. Resaw in a bandsaw or table saw and all the plugs will fall out as the saw cuts them loose. This cutter forms a radius that makes insertion into the hole easier. I usually pour a bit of glue, varnish, or paint into a can lid and set a bunch of plugs in it. This is much faster than dipping each plug. Plugs should always be inserted with the grain matching that of the work piece (Figure 2-24).

After a plug has been tapped in lightly with a small mallet (never to the bottom of the hole, as it may expand and then protrude), it must be cut off carefully after the binder has set. Use a slick or a fairly heavy chisel. Hold the tool blade bevel down with the cutting edge 1/16 to ⅛ inch above the surface, as shown in Figure 2-25. Now slide the blade along or tap it, slicing the plugs off a safe distance above the work surface, so you see which way the grain runs. Take one or two slices with the grain so it does not crack off below the surface and ruin

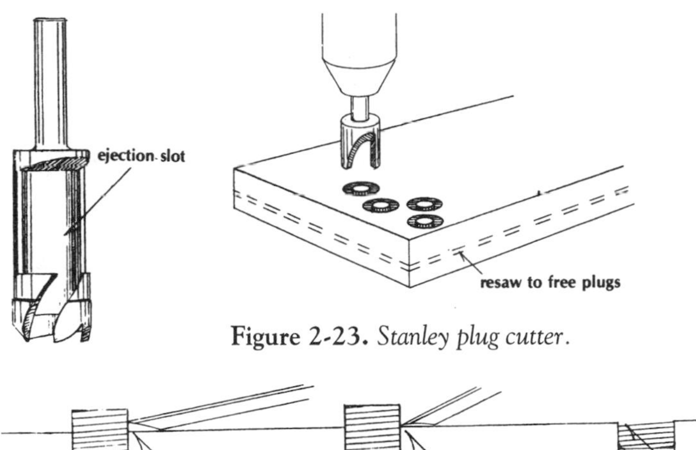

Figure 2-22. *High-speed plug cutter.*

Figure 2-23. *Stanley plug cutter.*

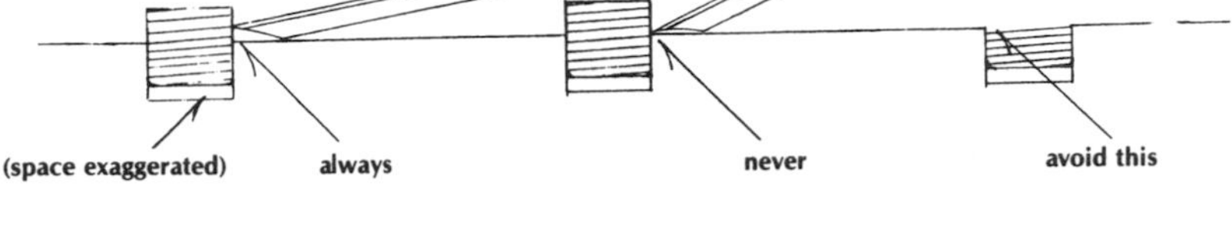

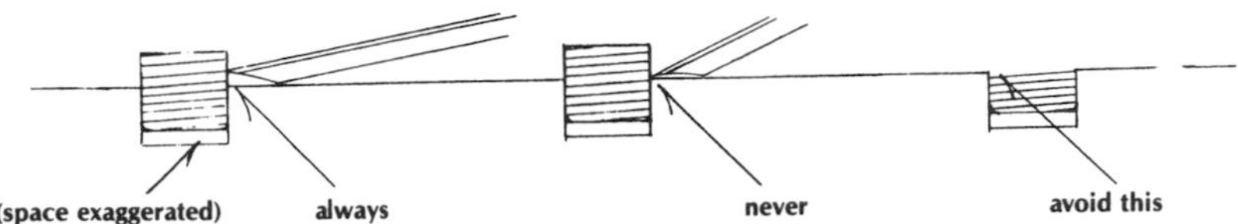

Figure 2-25. *Trimming plugs.*

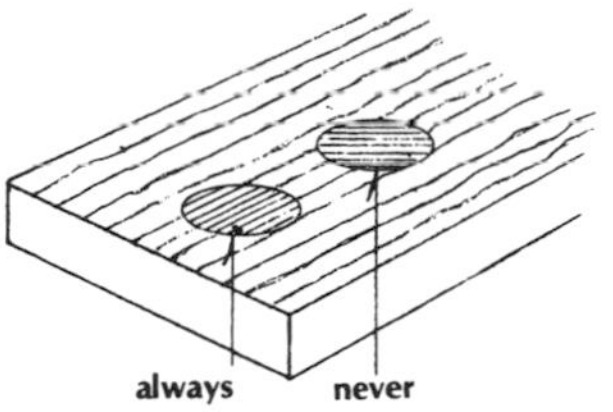

Figure 2-24. *Plugs match grain.*

always never

the appearance of the job. I used to cut off plugs in planking by walking along the hull quite rapidly. Then they were just sanded off flush with coarse paper on a block. Inserting plugs does not have to be a painstaking or time-consuming chore if you use your noodle.

SCRAPERS

Throw away any hook scrapers you may have. Cabinet scrapers are used to remove planer corrugations, for dressing irregular, wavy, or curly grain in hardwood, and especially for finishing ribbon-grain Philippine mahogany and the flat grain of fir and many other woods. Scrapers are hard steel. They come in rectangles and other shapes about 2½ by 4 inches with no cutting edge when purchased (Figures 2-26 and 2-27). They have to be worked carefully into a burred cutting edge, as below. To use a scraper, hold it in both hands, inclined at an angle of about 75 degrees to the surface, and either push it or pull it (Figure 2-28). It's easier to see the area being treated if you pull the scraper toward you. Try it both ways. Scraping produces very fine shavings and prepares the surface for fine sanding. It is a must for a superior varnish job.

A variation is the adjustable cabinet scraper (Figure 2-29). This tool has a frame or body with a flat sole and two handles. The blade is held by thumbscrews at an angle of about 75 degrees. It is usually pushed over the surface with some pressure—in long strokes and with the grain, of course. Incidentally, to remove planer corrugations, hold the scraper at an angle to the direction of the stroke so the blade bridges a number of the high spots at once. If you do not do this, the blade will simply ride up and down the hills and valleys made by the planer.

As shown in Figures 2-26 and 2-27, shaped scrapers are available for fine convex and concave work. For convex surfaces such as round spars or box spars with large radii, on the other hand, curved pieces of broken window glass do a good job at no cost. Just be sure to stick two or three thicknesses of masking tape on the edge you hold. Very light pressure will take off a nice shaving until the sharp edge wears down. Then just trash the piece of glass.

How to Sharpen Scraper Blades

Scraper blades generally are sharpened on their long sides, although there is no rule requiring this. Place the blade in a vise and file the edge by hold-

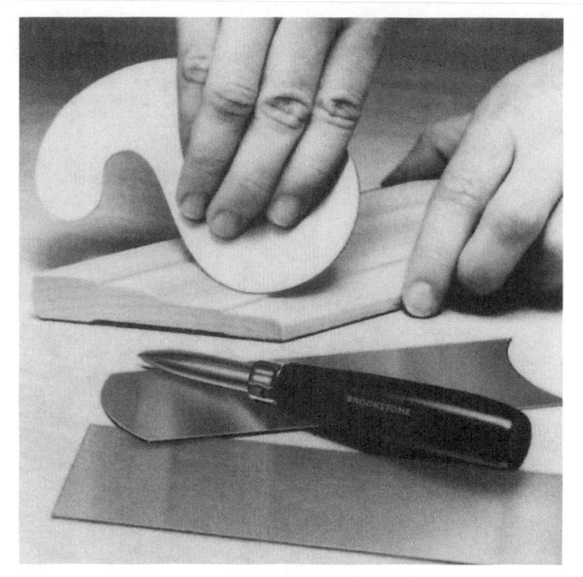

Figure 2-26. *Hand cabinet scrapers available in rectangles and concave and convex shapes. The hardened steel burnishing tool produces a fine burr on edges. (Courtesy Brookstone)*

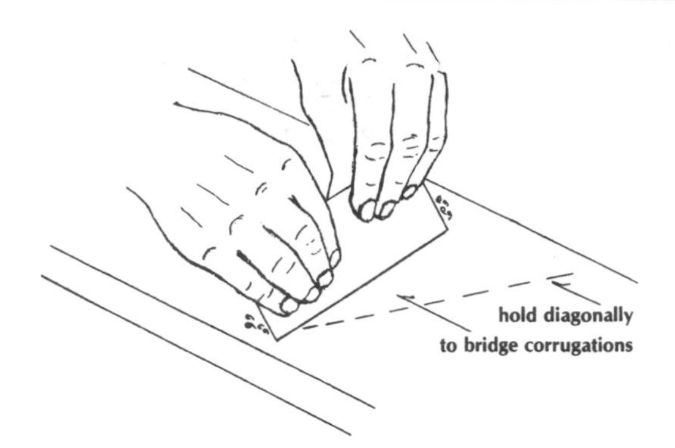

Figure 2-28. *Using a scraper.*

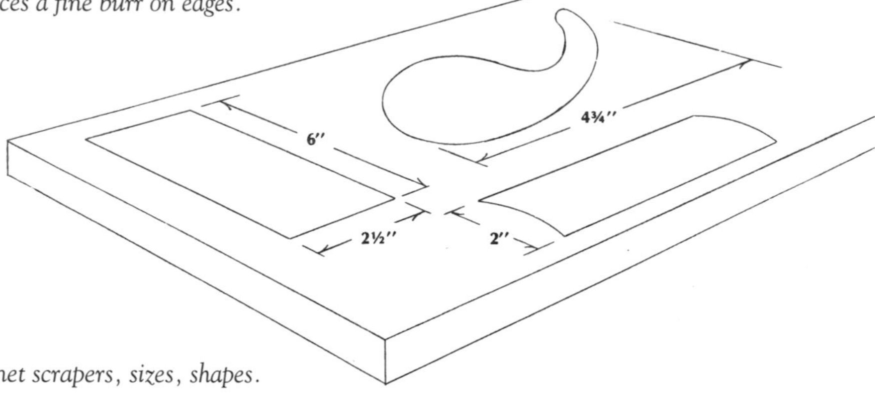

Figure 2-27. *Blade cabinet scrapers, sizes, shapes.*

ing a mill file crosswise on the blade and perpendicular to its side. Follow the steps in Figure 2-30.

Step one: Slide the file lengthwise on the blade, back and forth until the edge is bright and square. Step two: Whet the edge on an oilstone, holding the blade vertical. Remove any burrs by turning the blade flat on its side for several strokes. Step three: Now this edge must be burnished to form a hard, fine burr that is a microscopic hook. Use a burnishing tool made from a 6-inch length of drill rod or any round tool such as a drill bit. Or buy a triangular burnisher. Lay the blade flat on the edge of a bench and stroke hard over its full length with the burnisher almost flat. Do this on both sides. Step four: Clamp the blade in a vise and draw the burnisher along the edge at 90 degrees to the face, using considerable pressure. Do this 10 or 15 times. Step five: End up with a final stroke or two at about 75 or 80 degrees. Feel the edge. Is there a sharp burr? Now repeat this last operation on the other side of that edge. This gives you two cutting edges. Tilting the blade to 75 degrees will make that burr cut. When you're sure you've got it, repeat all this on the other edges. This will give you eight working edges.

The blade for the two-handed cabinet scraper shown in Figure 2-29 is filed or ground to a 45-degree bevel, then burred over in stages with a burnisher at 50 degrees, then 60 degrees, and finally 75

Left: Figure 2-29. *Adjustable cabinet scraper has blade mounted in frame similar to spokeshave. Thumb screw adjustment produces fine shavings like fuzz. This tool is needed to remove planer corrugations.*

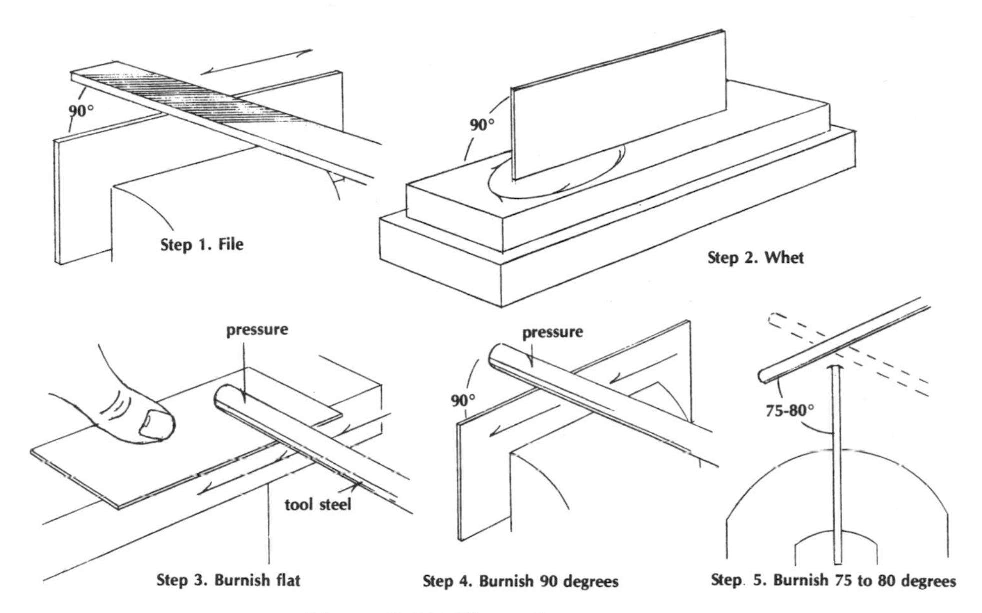

Figure 2-30. *Sharpening a scraper.*

degrees. This blade has just one cutting edge. It is clamped in the tool with the burr facing forward.

SCRIBING AND SPILING COMPASS

The scribing and spiling compass is a simple and inexpensive tool that should perhaps have been at or near the top of the list of basic tools, because you would simply be out of business without one (Figure 2-31). This tool differs from the ordinary compass in that instead of having a point on one leg, it has a rounded tip. The tip slides along a straight or curved surface while the pencil is marking the surface's shape. Always hold the scriber horizontal. Do not hold it so the mark is perpendicular (or concentric) to the surface you want fitted.

If you've been able to acquire most of the tools described so far, you are well on your way. And if you learn to handle them properly and to keep them sharp, clean, lubricated, and protected from moisture and rust, you are just about ready to turn that ordinary boat into a yacht.

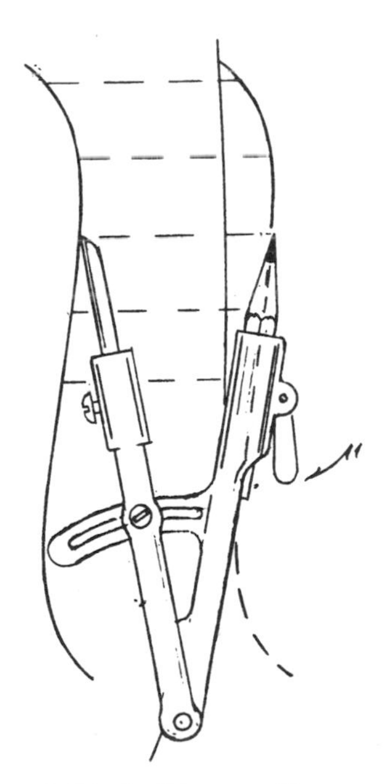

Figure 2-31. *Scribing or spiling compass.*

Figure 2-32. *Before assembly, saw the slots for specific tools.*

Figure 2-33. *Grooved strips seat backsaw and handsaw. The compartments hold planes, oilstone, drill bits, and so on.*

EVERYTHING IN ITS PLACE

While a carrying box might fit into the chapter on accessories that you can make, it seems to go well with the tools you'll be using. Of course, there are store-bought tool boxes. Any that are more than adequate, however, are very costly. And the modestly priced ones are simply wide-open trash containers. If you use your imagination and examine a few boxes used by carpenters and boatbuilders, you should be able to design and build a very satisfactory one for yourself. I have found the box described on the next few pages to be more than adequate. See Figure 2-32; alter it as you please.

A criterion for my purposes was that the box had to be light enough, fully loaded, to be placed into and lifted out of the trunk of a car. Also, I like to see at a glance that my most-used tools and those small things that are easily misplaced are right where they belong and not down among the shavings somewhere. This carrier fulfills my requirements and still has some open space for such loose tools as a level, three planes, square, pliers, wrenches, auger bits, twist drills, and a few more. Figure 2-33 shows the box loaded.

To build a tool-carrying box, you'll need about one-fourth of a sheet of ½-inch plywood, any grade you like. You will also need a piece of ¼-inch plywood about 14 by 32 inches in size for the bottom and dividers. Add a bit of scrap ¾-inch lumber, glue, a few nails, and that's it. A suggested plan is shown in Figure 2-34.

Rip out the tool retainer, a little shelf 2½ inches wide and just over 30 inches long. Using a pencil, lay out the spacing of your tool arrangement and the size of the notches. Go lightly. Tomorrow you may remember a tool that has to be there and you would have to respace everything. There may be several tools that can drop into slots in front of the tools lined up against the center panel—a try square, for example. Cut the notches deep enough to accommodate the entire tool. Chisels, for example, must have slots much deeper than just for the blade itself so they will hang from the handles. Saw multiple cuts with your backsaw to make the notches, then chisel out the remainder, or use a sabersaw or keyhole saw.

Build the basic box as shown, but let the heads of the finish nails protrude for later gluing. This makes it easier to build the little rails that retain the planes, drill bits, and so on. The gable-shaped end pieces may be nailed or screwed and glued to the box ends. (I cut a groove ¼ by ½ inch in mine to strengthen the attachment of the weight-carrying center piece.) Eventually, the bottom will be nailed and glued to the box, but now brad it temporarily in place.

The next step is to mark off on the bottom those

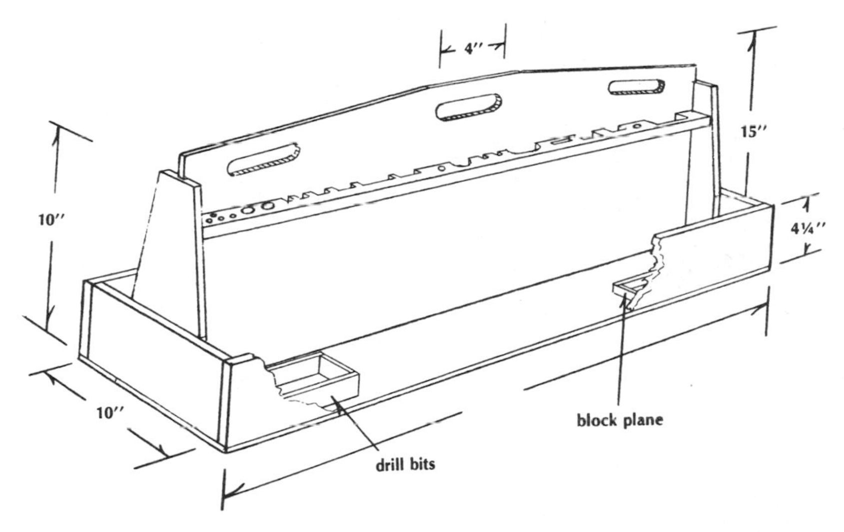

Figure 2-34. *Tool box.*

fenced areas that will hold the planes, oilstone, and drill bits. The rails can be made from any material ripped to about 5/16 by 3/4 inch. If you use 1-inch brads and glue the strips on edge, the points won't come through the bottom; they'll stay for a lifetime. The purist will miter the corners, but you're probably more interested in getting on with the boat!

To build the drill box, etc., construct the 1½-inch sides and ends, and nail and glue them together. Put these aside until the glue is set. To nail from the bottom into the sides of these little compartments, mark on the bottom of the boxes inside and drive about four brads through. Pull out these brads and insert new brads from the opposite side of the bottom piece. This way you won't miss when you glue and nail the boxes in. But you can't put these in to stay until everything is complete and the basic box is all glued and fastened.

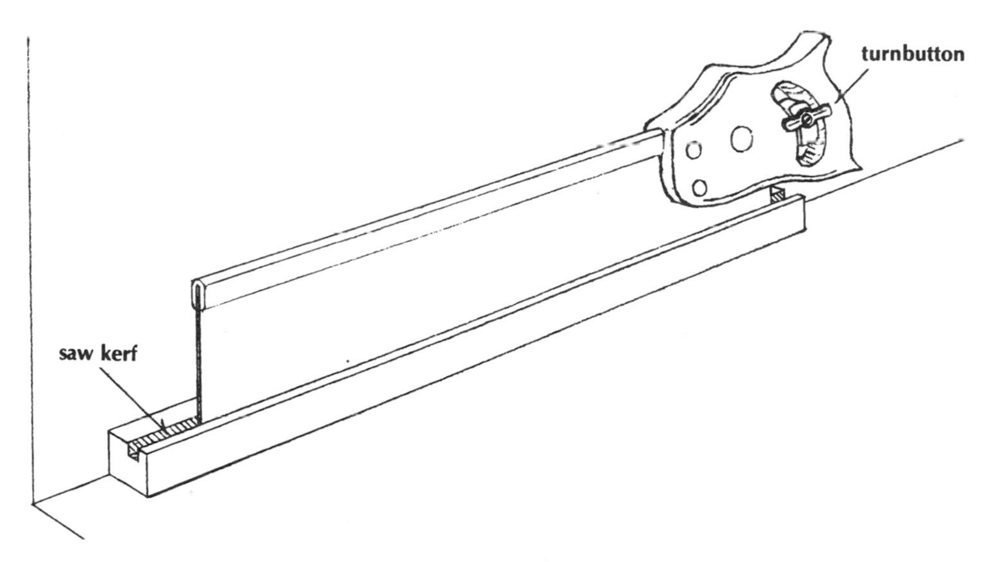

Figure 2-35. *Rack for saws.*

Figure 2-36. *To make grooves in the tool box, a make-do table saw is used. A power saw is held in a wedge-action planking vise described in Chapter Five.*

Now let's do the center piece. Draw center lines for the holes for the hand grips. Bore the holes 1½ inch in diameter and saw out the material between holes with a sabersaw or keyhole saw. Rasp or sand smooth everywhere. Cut the length; it should be about 30 inches long and fit snugly between the gable ends or mortise in. Before installing, rip a couple of strips approximately ¾-inch square and about the lengths of your backsaw and crosscut saw. Now comes a hitch: You need a groove in each of these strips in which your saws will sit (Figure 2-35). I don't suggest that you try to saw this with a handsaw, although it's possible! If you do not yet have a table saw and can't borrow one, do you have access to a portable electric handsaw?

Place the power saw upside down in a vise—gently, don't crush it. Adjust the depth of cut to about ¼ inch (Figure 2-36). Clamp a small piece of scrap to the sole plate to act as a fence. Tape or wire the switch to "on" or have someone hold it on for a minute. Just plug in and start. You now have a miniature table saw that will cut those slots in seconds. In Chapter Five, I'll show you how to build a great table saw using a portable electric handsaw as the power unit.

Notice that the groove should not be centered. This is to provide for the thickness of a saw handle and to prevent nails from entering the groove. Now that the pieces are slotted, nail the longer one in the bottom of the other half of the box and against the center piece. Position the shorter piece so it clears the crosscut saw and is at the opposite end of the box.

Drill all holes for nail sets and other small items. Make sure the notched tool shelf fits. Mark about five spots where screws can be driven from the opposite side. Nail and glue the center piece between the gable ends. Pull the temporary brads out of the main box ends, glue the corners, and fasten with 1½-inch finish nails and/or screws. Now nail and glue the bottom in place. And, finally, place the small boxes in position and brad through from the bottom to fasten.

There's just one more job to do. Place the tools one by one in their slots and mark around them with a broad felt marker or crayon. Now you will see at a glance if a tool is missing or out of position. Keeping your most-used tools organized will save you much time and give you great satisfaction.

Now let's move on to electric hand tools.

3 Electric Hand Tools

You may be one of the lucky amateur yachtbuilders who have few financial problems. If so, you are one in a hundred. Most amateurs are straining both their finances and their domestic tranquillity to bring their dreams to life. I know what they are going through. In my many years of boatbuilding and boat ownership, I never quite had the funds to equip my shop or my boats with the best available. I always had to find ways to economize (I still do). As a consequence, you'll find that many of my suggestions follow a lower-cost approach to doing things.

It may seem a contradiction, but owning a few decent power tools is the economical thing to do. Sometimes the poorest-quality electric tool is overwhelmingly superior to the finest hand tool. Let's say you are renting space on which to build your boat. How many modestly priced new or used power tools could you buy with the rent money if you postponed renting for one month? Or, if you already have a location, how many months sooner would your vessel hit the water if you had some time-saving tools? Need I go on? Yes, I had better, for I've witnessed a grown man attempting to build a $35,000 to $50,000 yacht with just $75 worth of tools. His half-completed jig now sits abandoned while the rent bills pile up and court action threatens. I might mention also that he had never heard of Howard Chapelle's *Boatbuilding*, Robert Steward's *Boatbuilding Manual*, or *The Gougeon Brothers On Boat Construction*.

I know, of course, that none of you would exhibit such foolishness. So let's see what power hand tools are suitable, and how you use them in boatbuilding, or more common cabinetmaking.

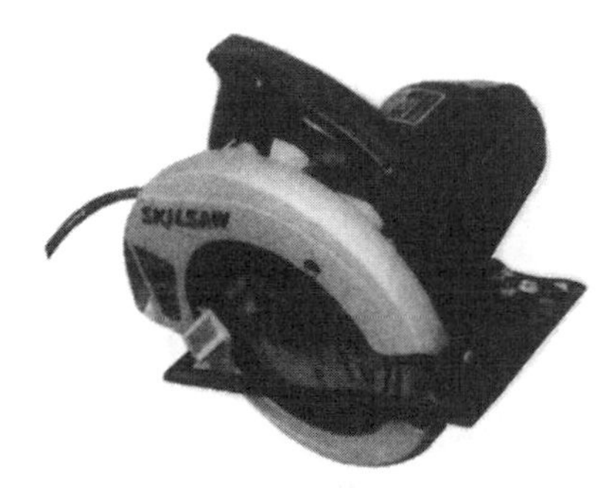

Figure 3-1. *The famous Skilsaw, the mainstay of the electric saw world. This is a 7¼-inch shop model.*

ELECTRIC HANDSAWS

I use the term electric handsaw rather than power saw to avoid any confusion with power tools or power machines such as table saws and radial-arm saws, which, incidentally, are also called circular saws. Your first acquisition must be an electric handsaw of the type called the Skilsaw. Skilsaw is the registered name for the products of the Skil Tool Company, Inc. (Figure 3-1).

Look for electric handsaws marketed by Skil, Stanley, Black & Decker, Sears, and the new companies such as Makita and Hitachi. Unless you will be using a saw hour after hour cutting two inches or deeper, as some carpenters do, you don't have to buy a heavy-duty, costly professional model. Any 7¼-inch-diameter blade capacity is adequate. Saws this size develop 1½ to 1¾ h.p., are double insulated against short-circuits, and predominantly have bearings that do not require lubrication. The gears are usually spur gears (you'll find them quite noisy), but some have helical gears. All should give you years of service for a modest investment. Avoid low-priced bargain saws made by manufacturers you haven't heard of, although even these are much

Figure 3-2. *This 7-inch saw provides excellent visibility of the blade, necessary when starting a cut or for freehand work.*

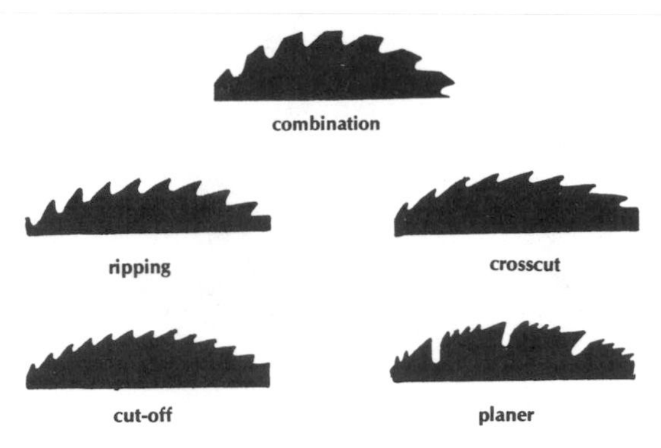

Figure 3-3. *Power saw blades. (From How to Work with Tools and Wood, 1942, 1955, 1965 by Stanley Tools, New Britain, CT)*

better than sawing by hand, especially on light work such as plywood. Moderation, of course, is the answer. So, if the case of your saw feels very hot, allow it to cool while you do something else.

Many advanced craftsmen will tell you that the type of saw I am advocating is "cheap." Yes, it is, compared to the fine heavy-duty types at the top of the price range. On the other hand, it is a low-cost tool that will do anything the costlier ones will do if you exercise reasonable care. While it is backed by guarantees, there are no warranties against stupidity. If you feel a saw running hot, stop cutting, then look for a scorch mark on the saw cut or kerf, as it is called. This tells you that your blade is dull, or that it has lost its set. Don't try to force a cut. If your project calls for extensive ripping of thick oak, or something similar, that's different. Take the job to someone with a 10- or 12-inch table saw, or rent or borrow a heavy-duty saw.

Needed in any electric handsaw (new or used) is a clear view of both sides of the saw blade, as shown in Figure 3-2. You'll be doing frequent free-hand sawing, following a line, and you must be able to see the teeth. If you are one of the lucky few who can afford two saws, consider a smaller lightweight panel saw. This tool is useful for quick, light cut-offs and ripping with a guide. It also is terrific for long curves in plywood up to ⅝ inch thick, and it will handle a 3-foot radius with ease, because of much smaller diameters available.

Saw Blades

Take a look at the drawings in Figure 3-3 showing most of the configurations of saw teeth. The combination blade, used for both ripping and crosscutting, is most useful. This blade is slightly slower than the blade designed for each purpose, but that is of little consequence. If you have a lot of ripping to do, spend a few dollars for a rip tooth blade just to make it easier on the tool and yourself. Of course, a table saw is best if you have a lot of ripping. The table saw will give you greater accuracy, too, although with practice you can get good results with the rip guide on your electric handsaw.

Combination Blades for Fine Work

Another very valuable blade is the cabinet combination. This blade is sometimes called the cabinetmaker's or planer blade. Planer is the most accurate name, as these blades cut with or across the grain and leave a surface that feels as though it had been sanded. A planer blade is excellent for plywood up to ½ inch thick, or heavier for a short cut. Because these blades are taper ground and have *no* set, you cannot adjust direction at all once you have started. No free-hand sawing is possible because of friction in the kerf. This is not a problem on a table saw because a rip fence or bevel gauge is used.

To use a planer blade on an electric handsaw, you must follow a straightedge. Clamp or tack to the work a piece of plywood or lumber that is 6 to 8 inches wide and rigid enough to guide your saw perfectly. Or use your carpenter's square clamped to the work piece. Better yet, make a square out of plywood, up to full-panel dimensions. Cut it from the corner of a sheet to guarantee 90 degrees, then brad and glue straight strips along the lower edge. Cut out the center to lighten it. Both the carpenter's square and the plywood square are great timesavers. There's one thing to watch for, however. If the planer blade saw cut suddenly wavers, stop at once. Your blade has overheated and is temporarily warped. It is not capable of doing the job. You will have to use a plywood blade or a cut-off.

The plywood blade has fine teeth and a moderate set. The cut it produces is good, but not polished like the cut of a planer blade. Most plywood blades will not handle long cuts in ¾-inch plywood, such as an 8-foot rip, but they are adequate for up to about 4 feet. The best blade for this work is a cut-off. The teeth of a cut-off are quite coarse compared with those of a plywood blade and have more set. So the surface of the cut is not quite as fine. The cut-off, however, is fast, chips very little, and is an all-around good bet, whether for plywood or hardwood. For a perfect surface, invest in a carbide-tipped blade and get your purchase cost back in the fewer sharpenings required.

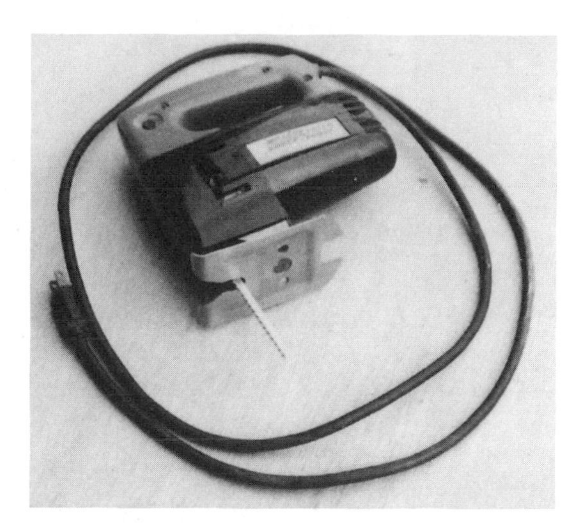

Figure 3-4. *A powerful sabersaw cuts curves as well as a bandsaw does, but more slowly. Also, material thickness is limited to ¾ inch. A sabersaw is easier on large plywood panels.*

SABERSAWS

You will need a sabersaw, scroll saw, or jigsaw, as some call it, for cutting curves and for notching and completing power saw cuts (Figure 3-4). Joinerwork curves, berth fronts, and other shapes normally handled on a bandsaw can be done with a sabersaw. Unfortunately, the thickness of material that can be cut easily is limited. In softwoods and lighter plywood, however, the sabersaw is great. This makes it the right tool for making patterns and templates from ¼- to ½-inch plywood.

The average light-duty sabersaw is reasonably inexpensive. I have seen one cut out all the ¾-inch mold sections for a 40-foot boat. It took two days with frequent rests, but it did the job. Variable-speed models are available, but I can see no great advantage in this feature. Big chain stores sell heavy-duty models at higher cost. The old reliables, however, are Stanley, Skil, Black & Decker, Millers Falls, and Bosch. Today, it's Makita and Hitachi.

Sabersaws are a bit tricky. The work must be held down, especially if it's light stock. The sole of the tool must be kept flat on the work surface, and you must bear down slightly, or there will be an unholy clatter; you could snap the blade.

With practice, you can start without first boring a hole (Figure 3-5), a plunge cut. Hold the base or sole almost perpendicular to the work surface, but keep the blade a safe distance from the line you wish to follow. Turn the tool on and *slowly* swing the blade down until its tip begins to jab into the wood. This will gradually produce a narrow groove that deepens as you lower the blade. Also move the saw forward slightly to prevent the point from striking the end of the groove. In a few seconds, the point will penetrate the work stock and you will be able to lower the sole flat on the work. Practice on scrap and you'll soon become adept, but don't try to plunge-cut hardwood.

Best to mark your line on the back side of the work, as the blades are quite coarse, and they pull upward. The splintering is unsightly, especially when cutting fir plywood. The *direction* of sawing also is important, as shown in Figure 3-6. When curves are cut across the grain or at an angle, the side of the blade going the hard way into the grain will chew it up badly, whereas the other side of the cut will be clean and sharp. You may have to reverse the sawing direction to save the piece from such

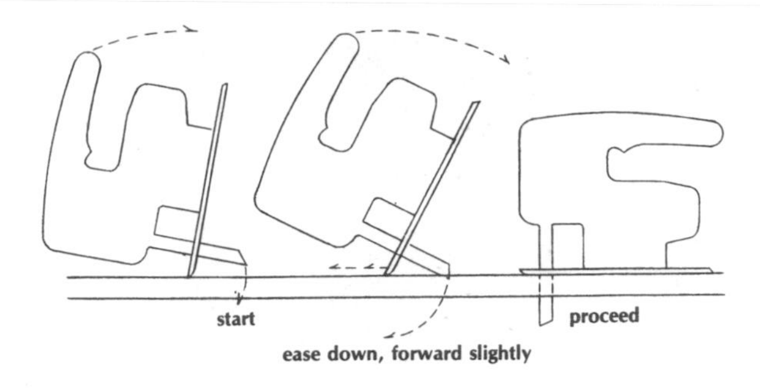

Figure 3-5. *Sabersaw plunge cut.*

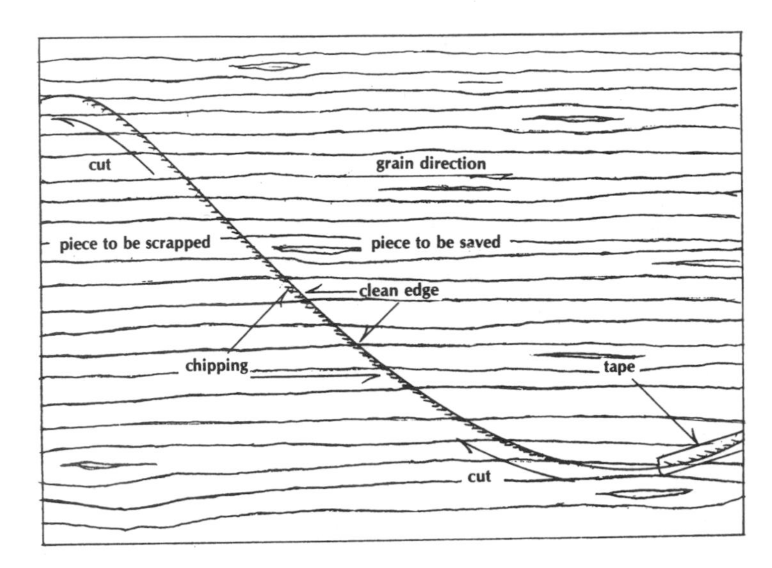

Figure 3-6. *Avoid sawing into grain.*

damage. It may be practical to put masking tape on the work piece and mark on the tape. This will prevent 90 percent of the splintering. I suggest that you not try to saw too close to the line at any time, as sabersaws rarely saw square in heavy material. Plan to plane, rasp, or sand down to the line.

ELECTRIC DRILLS

Since most yacht joinery and cabinetry is plugged over screws, it is almost mandatory that you use a light pistol-grip electric drill (Figure 3-7). This tool does a fast and neat job of boring for ⅜-inch and ½-inch bungs. You might have a hand or breast drill handy for drilling the lead holes. Chapter One covered the combination bits and counterbores available.

Needless to say, there are scores of drilling jobs in any vessel. Many require quite deep holes. You will be far ahead if you use a power tool instead of a hand tool. For all-around service, choose a good ⅜-inch drill by a well-known manufacturer. The recent battery-operated drills are practical, too, though expensive. The ¼-inch models are too limited. Stay away from high speeds. Tools of 1,000 to 1,200 r.p.m. have worked out fine for me, and variable speed is useful. If you can afford a ½-inch drill, the variable-speed types are quite efficient, especially for use with a simple power screwdriver (Figure 3-8). This size is also very good for setting up in a drill press stand (Figure 3-9). Both the ⅜-inch and the ½-inch drills are suitable for cutting plugs, although the higher speed would be better. Reversible drills have some appeal for pulling screws and backing out long bits, if price is acceptable.

If in boatbuilding you are drilling into a keel, deadwood, floor timbers, big cabin sides, or whatever with long bits, you would need a heavy-duty ½-inch drill, and it should be reversible to clear chips, stuck bits, and the like.

Look for ball bearings for thrust, helical gears,

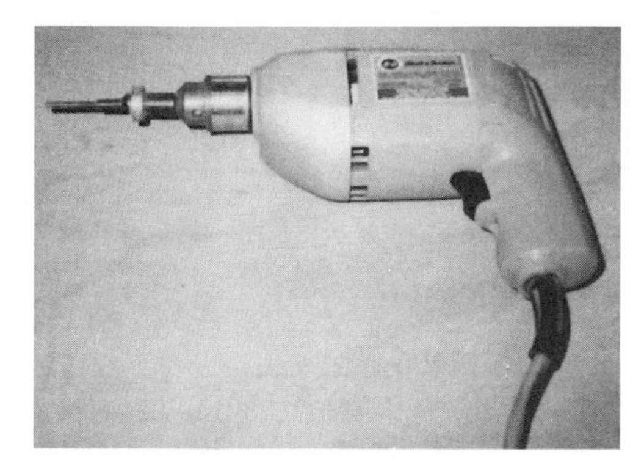

Figure 3-7. *A light ⅜-inch-capacity drill is indispensable. A speed of approximately 1,200 r.p.m. is recommended.*

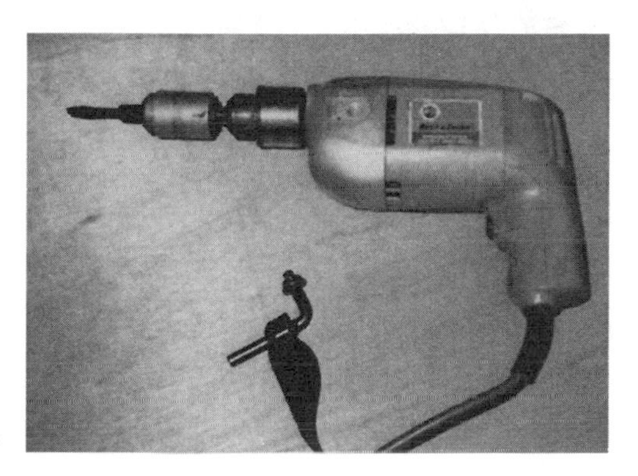

Figure 3-8. *This old screwdriver attachment in a ½-inch drill drives screws very well. Reversing variable-speed drills make satisfactory screwdrivers.*

Figure 3-9. *An inexpensive drill stand and a drill make a satisfactory press for most light work.*

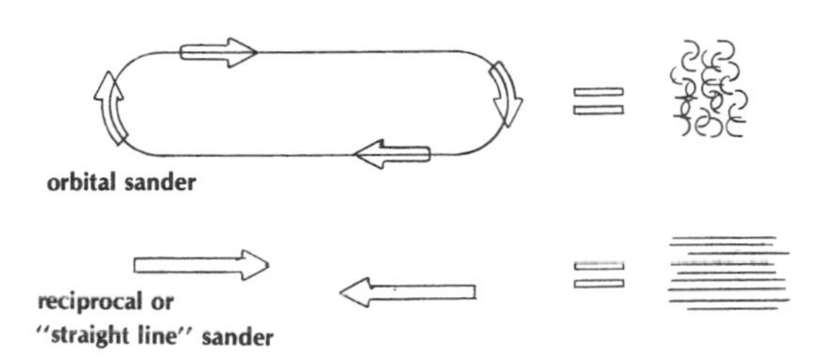

Figure 3-10. *Sander patterns.*

double insulation, and high-impact plastic and/or aluminum housings. My preference, however, is for a more or less disposable drill that can be thrown away without much feeling of loss. I bought a Black & Decker ⅜-inch drill in 1965, and I have used it to bore ½-inch holes in a cast-iron fin keel, plus hundreds of less sensational jobs. If you used one to cut wood plugs only, you would save enough to pay for it several times over.

SANDERS

Many boatbuilders advocate the use of portable electric sanders. I do not agree. I am still a believer in block sanding by hand, and especially in flexible long sanders for hulls. Let's discuss the types of electric sanders to help you make up your own mind. The reciprocating, or straight-line sander is a finish sander not intended for the removal of a lot of wood or old paint. On the other hand, you cannot do finish sanding with an orbital sander, except on work to be painted (Figure 3-10). The orbital sander may leave a pattern of tiny circular scratches almost impossible to see until you apply a stain or filler. So this surface requires much hard sanding by hand block or by a straight-line sander. Granted, orbital sanding does remove material faster than a reciprocating sander, thus it's used most for preliminary sanding, to remove old flaking or powdery paint on wooden hulls, decks, and so on. Electric sanders are tiring, too. After you have swung either kind against a hull for three or four hours in the sun, you'll know

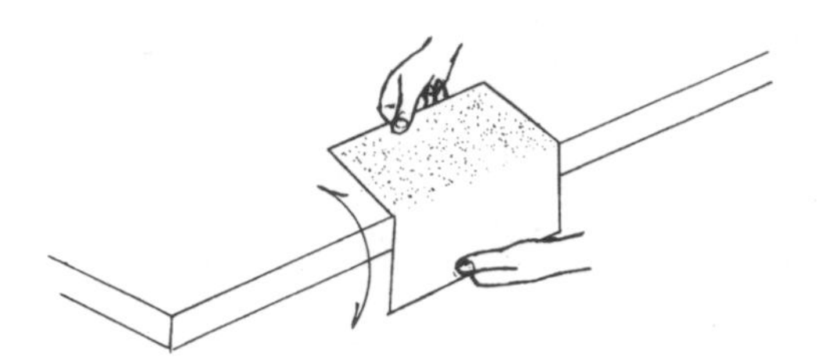

Figure 3-11. *Breaking the sandpaper sheet.*

Figure 3-12. *A belt sander must be kept in constant motion and stroked in a figure-8 or orbital pattern.*

it! Even a half-sheet sanding block is a vacation by comparison. The only place an electric sander might get the nod is on a horizontal surface.

A good electric sander ranges in weight from 6 to 10 pounds and produces from $\frac{1}{4}$ to $\frac{1}{2}$ h.p. Do not bother with anything lighter. The very light magnetic vibrator types especially are of no use around a boat.

The $\frac{1}{4}$ h.p. sanders usually take one-third of a standard sheet of garnet or aluminum oxide paper, open coat, or "production paper." Prepackaged sheets of this size are readily available, but I have found that those that contain an assortment of grits are never the ones I need. I use 60, 80, 100, and 120 grit. The finer grits will tear after a few strokes, but you should always finish up with fine paper on a block, about 220 grit or finer. Most of the $\frac{1}{2}$ h.p. sanders use a half sheet. Never use the white flint paper. You will go broke in a hurry, as it has no life and removes very little.

To cut sandpaper sheets, use the toothed side of a hacksaw blade as a straightedge. Never use a knife, as the point will be dulled by the grit if it penetrates. It is necessary to break the glue backing of the paper to prevent cracking and tearing (Figure 3-11). Pull the paper back and forth over any corner, paper side down, right up to the ends. This makes it much easier to clamp the paper in the sander. Sometimes these clamping devices have a tendency to lose their grip. Increase the holding power of the clamp by inserting a piece of wood about $\frac{1}{8}$ inch thick on the grit side along with the paper.

To use a portable electric sander, exert moderate pressure about as you would with a plane and stroke the same way, back and forth, always moving. Don't depend on the vibration alone to do the job, bear down with the grain except when removing a lot of material. Beat the dust out of the grit frequently with the palm of your hand. And when the paper begins to lose its cutting ability, *change it!* But remember, the final sanding must be with a block of wood or hard foam and fine paper, as explained above.

Belt Sanders

The second most popular type of portable sander is the belt sander (Figure 3-12). These tools start at about 10 pounds, so think twice about using one to sand vertical surfaces for any length of time. They are very satisfactory for bench work. The larger tools take off material quite rapidly. On lighter ones, a medium belt will produce a passable finish. Fine belts don't stand up very long; I find them a waste of time.

As with all tools, belt sanders require practice. Nothing is better for planking, deadwood, and so on. They have several characteristics, however, that can cause trouble. If the sander is allowed to hesitate for a split second, it will leave a groove. If you stroke as with a plane, there is a momentary stop when you reverse direction. This causes grooves. I use an oval or figure-eight motion, even at the expense of some scratches across the grain, as shown in Figure 3-13.

Another unpleasant feature is that belt widths of 3 inches, the most popular size, mean the sole is only slightly wider. In addition, the tool is rather heavy, so it is possible to rock the machine slightly as you stroke, causing disastrous grooves parallel to the stroke. The 4-inch models largely eliminate this hazard, but they are heavier still. Very probably the belt sander's most annoying characteristic is belt misalignment or tracking. The alignment adjustment, made while running, is extremely sen-

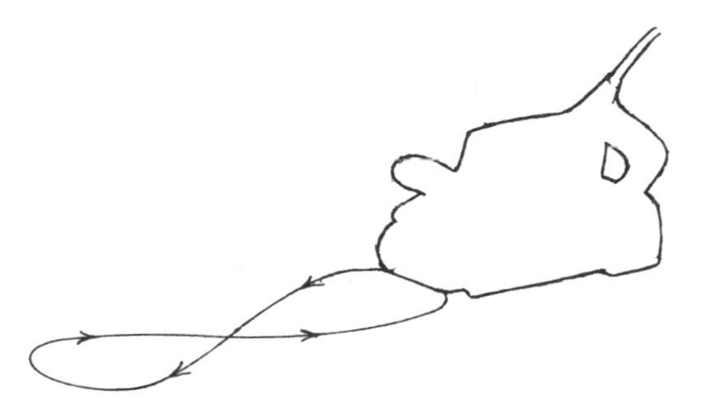

Figure 3-13. *Belt sander figure-8 motion.*

sitive. You can find yourself wasting a lot of time on maddening adjustments every two or three minutes. Also, the belts are very expensive but quite long-lived if kept in line.

If you can find someone who knows all about belt sanders, he might show you how to avoid all this unpleasantness. Although I sound completely negative, I am not, because from time to time I use a Craftsman belt sander, which is sold in large numbers. One great advantage in any belt sander is that if its back is flat it will thus rest comfortably turned on its back (Figure 3-14), it makes a good little bench sander. Tack wood blocks around it to keep it from creeping off the bench. This allows you to pass work pieces lengthwise on the moving belt, with a firm hold on the work piece. Another good method is to clamp work pieces on the bench for sanding in the usual way. Or tack scraps to the bench as retainers. Avoid running the sander over an end or edge, as this will put a radius where you do not want one.

One final word on sanding. Never sand a piece of plywood, especially fir, until it has had several coats of sealer or paint. Sanding a fir plywood surface by hand or machine only cuts out the soft fibers, leaving the harder areas standing out worse than ever. Build up the surface with several penetrating coats, if it's rough, and then fine sand with a hard block. Concave surfaces and edges may be sanded efficiently with blocks shaped to conform to the curvature. In a later chapter, I'll go into some detail on this subject, as there are no satisfactory machines to do this work.

POWER PLANES

The electric plane is an extremely useful tool in many stages of boat and cabinet construction, because heavy planing is real labor, although you don't face it every day. Consider the power plane to be an upside-down jointer, especially the longer models (Figure 3-15). A power plane is used for straightening plank edges rough from the mill (before ripping, for example), for removal of stock down close to the line, and for use as a thickness planer. Lengths of work pieces too unwieldy to handle on a jointer can be worked easily on a bench. A power plane can be used for removing old paint and varnish, but you'll dull your knives (blades) in a hurry.

If you are building a wooden boat, there are countless jobs for a power plane if you use it only for removal of stock: dressing down deadwood, center-boards, rudders, straight floor timbers, deck beams, and so on. Remember, however, that anything wider than the sole of the plane will show overlapping cuts rather clearly. If this happens, finish up with a jack or smooth plane. Power planes

Figure 3-14. *Designed to rest solidly on its back, this model is good for sanding strips and small pieces.*

Figure 3-15. A *Stanley, the last word in jointing power planes. Shorter, lower-cost planes are excellent for removing material rapidly, roughing spars, and much more.*

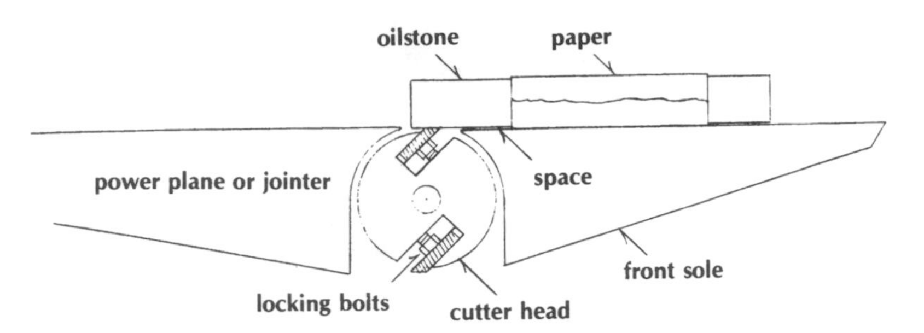

Figure 3-16. *Honing plane or jointer knives.*

are intended for edging. I have used an elderly Wen plane on solid spars to shape up the square and the octagon. For the latter, the bevel fence set at 45 degrees is pretty handy.

The greatest use for power planes is the fitting of doors. Some carpenters do dozens in a day. Some tools take off close to ⅛ inch in a pass; these are heavy-duty planes designed for that type of work. If I had a lot of spar planing, I would find a good power plane somehow. Used ones, unfortunately, don't lie around long. If you have many other pressing needs, then a lighter and lower-cost plane will do the work. It might not be as fast, but the quality of your results should be as good as the condition of the knives and bearings. The models offered today are quite short compared with older heavy-duty types, so they are not as good for jointing edges. They cost less, too, since they are belt driven. These newer tools are principally aluminum and zinc die castings. The older heavy-duty models have a lot of steel and aluminum throughout, and ball bearings rather than sleeve bearings.

Most power planes have rotary heads that turn at about 15,000 r.p.m. The knives or blades (usually two) fit into slots and are locked in with special machine bolts. The cutting edges of the knives may be touched up with an oilstone while they are in the machine, but removed for grinding. You can maintain a satisfactory cutting edge by honing them in the head. Lower the front sole to about 1/16 or 3/32 inch. Wrap your oilstone in paper except for about 1 inch at one end, so the stone does not contact the surface of the sole. Rotate the head so the stone can be laid on the bevel at exactly the original ground angle (Figure 3-16). Hold or wedge the head securely, then slide the oiled stone back and forth on the back sole lengthwise to the knife, counting the strokes until the blade feels sharp. To avoid realignment of the knives, stroke the other blade the same number of times. Now turn the depth adjustment to 1/16 inch or less and test. If the tool chatters or the surface of the work is rough, the knives are still dull. Dull knives will cause the motor to slow down and overheat. A high whine means the tool is turning at its best speed. Do not force cuts. Better ten strokes to take off ⅛ inch than three strokes to chew it off.

You must grind the knives when whetting no longer produces a good edge. Most good saw-sharpening shops have the equipment and the skill. In Chapter Five you will find two knife-holding fixtures so you can grind knives correctly, as well as instructions for installing knives in the head.

ROUTERS

Many woodworkers feel that the router is the most versatile of all power tools. It is fundamentally a simple machine, consisting of a motor and a base, as shown in Figure 3-17. Yet you can create beautiful cabinetwork with a router, dadoing, grooving,

Figure 3-17. *A router is the most versatile power tool for molding edges, rabbeting, jointing, duplicating from a pattern, coving, and so on.*

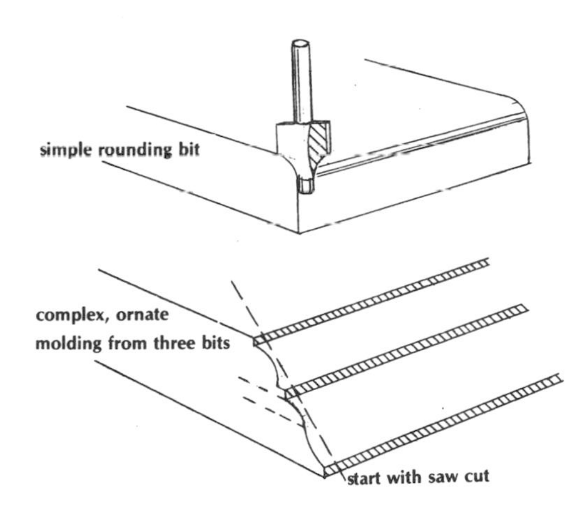

Figure 3-18. *A router creates decorative effects.*

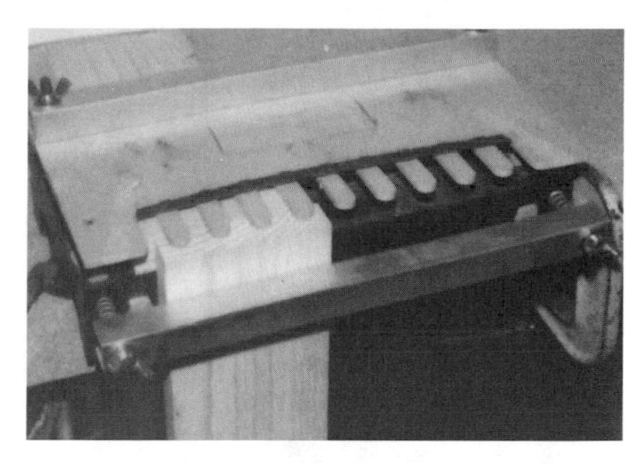

Figure 3-19. *This dovetail fixture produces accurate joints simultaneously in each part. It requires a dovetail bit and router guide bushing.*

beading, fluting, coving, dove-tailing, rabbeting, inlaying, carving, and, of course, jointing, maybe other operations. These you'll discover if you study the router in depth. In yacht joinery, with its ability to form molded edges easily, rapidly, and precisely, and work the finest of joints, it's almost a must. It's also nice to have an assortment of cutters. By combining cuts—using one cutter and then following with another—you can achieve almost any conceivable edge effect. Admittedly, unusual shapes are used in furniture construction more often than in yacht work. A judicious display of molded trim, however, adds beauty and distinction to almost any interior (Figure 3-18), and to cabinets.

Prices for routers vary considerably and are based on horsepower and engineering. The finer makes can cost more than $350 unless on sale. The Japanese-made Makita has been recommended highly to me by a number of cabinetmakers. Sears offers good tools in the one-horsepower class from under $100 (1992). You can easily add $100 more for accessories and bits. Watch the classified ads for used routers. Whatever tool you end up with, do not push it too hard. Overheating is fatal.

Accessories add to a router's usefulness. Craftsman's dovetail fixtures turn out quite acceptable joints, a quality feature in drawer construction, and the device is almost foolproof (Figure 3-19). Small steel tables are available to convert your router into a small shaper (a router really is just an inverted shaper). This is shown in Figure 3-20. The shaper table permits positioning and moving of the work piece, a desirable feature when shaping small parts.

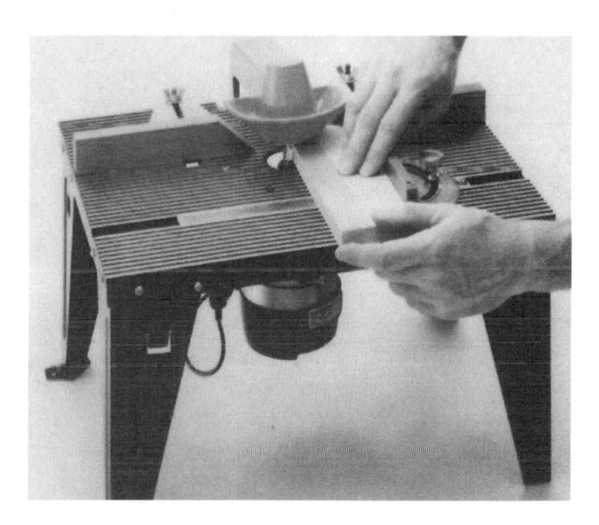

Figure 3-20. *This table is for a Craftsman router. It has a split fence, safety guard, and miter gauge. (Courtesy Sears, Roebuck)*

In Chapter Five, there's a design for a router-shaper table quite a bit larger than those you buy. I used this table with a ⅓-h.p. router to make lipped cabinet doors, cove and ogee edges, corner radii, and other items. With a more powerful machine, such as the 1½- to 1¾-h.p. Stanley, Rockwell, or Makita router, you should be able to do most of the operations normally done with a light-duty bench shaper, ½-inch spindle.

One of the most interesting accessories for routers is a guide bushing (Figure 3-21). This device permits use of a straight bit and a pattern

Figure 3-21. *Sears guide bushings are screwed to the inside of the router base plate. These devices are used to follow any pattern, dovetail fixture, and so on. They are available in steel and plastic.*

for producing repeated shapes—say, a dolphin vent in doors and locker fronts. You simply make a pattern of ¼- or ½-inch plywood, allowing a small tolerance for the wall thickness of the guide bushing. This technique also may be used to make a cutout or a carved design using a veining or V-shaped bit. I have made extremely complicated floral-design screens using this method.

A few years back I built several 36-foot rectangular masts. Each was made of two pieces glued together (Figure 3-22). I formed the two ends to the required tapers, with the joint running fore and aft. Then I clamped a short fence to the router sole and set this to leave a 1-inch wall thickness. I used a ¾-inch-diameter bit and cut down about 1¼ inches in three passes, leaving the solid areas as called for in the design. After the walls were finished, I went back and routed out the centers. I routed a groove through the solid areas for ventilation and wiring. The next step was to plane the side tapers. This was easy, for I had to plane off just the side walls and the solid at the head. (Remember that this was done on the surfaces to be glued, not the outside of the spar.) This method produced good spars, stronger than a glued-up box spar, but it did waste quite a bit of spruce. See Chapter Fifteen for details.

A router can be used for hollowing out streamlined wooden spars and for cutting a groove for the bolt rope, as shown in Figure 3-23. There is nothing better than a router for cutting out the rabbets in the staves for box masts. With the shaper table,

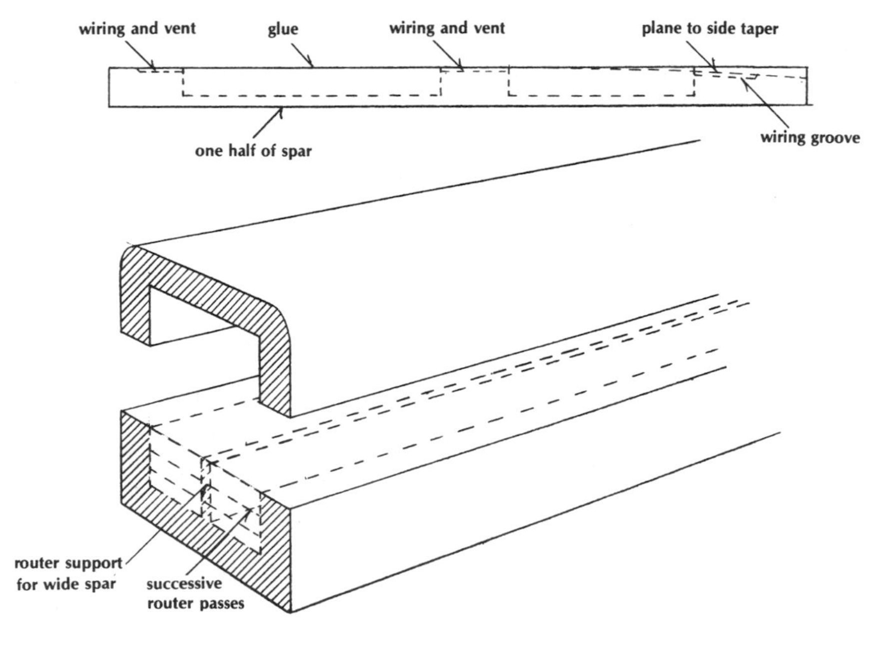

Figure 3-22. *Routed spar construction—easy, wasteful.*

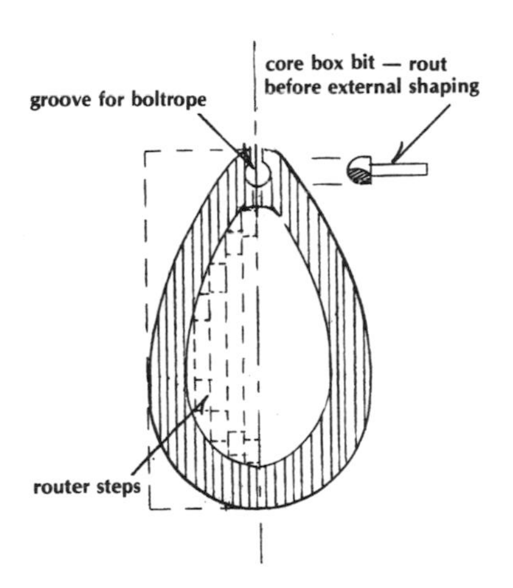

Figure 3-23. *Some steps in routing out a hollow mast.*

a router does a fine job of making tenons and lapped joints, too.

Some Router Tips

Never work on a router or make adjustments without disconnecting the cord. The router is a very dangerous machine. Once I picked one up and carelessly rested the sole for an instant against my side. At the same moment, I just happened to touch the switch. In a split-second, the bit chewed up and all but swallowed my belt, but I pulled away before it got to my skin. Stupid!

The router bit shank is gripped by a ¼-inch-capacity collet (a chuck), usually set up with an open-ended wrench while the armature is locked. Be sure that all of the shank is in the collet and cinched up tight. At 25,000 r.p.m., a loose bit can do a lot of damage. The depth of the cut is set by adjusting the base plate up or down. Be sure to lock this adjusting device firmly. I have seen the results of vibration loosening the depth control. You don't want that.

Always move the router against the cutter rotation direction, never with the rotation, for the tool will suddenly try to take charge (Figure 3-24A). Once started, the bit must be kept moving. Do not stop or slow the movement of the bit while it is in contact with the wood, or it will burn a spot instantly, and possibly ruin the bit, too. Also, do not turn the router on or off while touching the work piece. This applies especially if you are using a bit with a pilot (an extension that runs along the edge of the work; see Figure 3-24B). On the other hand, do not try to feed or push the tool too fast, for this will force the router to slow down and overheat. Keep that high whine sounding out! In most cases, you will have to make the cut in two or three passes to prevent overloading. Just set the cutter deeper each time. This is especially the case in hardwood. Always hold the router firmly against the work with both hands, but avoid pressing the pilot so hard that you burn the edge.

A router can do a decent job of dadoing (such as cutting the grooves for shelves; see Figure 3-25). I use a square made from ¼-inch plywood as a guide. The piece of 1 by 2 nailed and glued to the underside of the triangle may be dadoed by the router the first time you use it. Thereafter, this cut can be used to locate the next dado. Incidentally, you may find that the dado produced by a ¾-inch bit is too snug for some lumber or plywood. Don't try to drive the shelves in. This may cause the cabinet or bookcase side to bend (Figure 3-26). Just take a couple of shavings off the side of the shelf, preferably the lower side. The fit should be loose enough to accommodate glue. If it is too tight, all the glue will squeeze out. A depth of ¼ inch is usually sufficient for a shelf dado.

If you are running a rabbet around four sides of

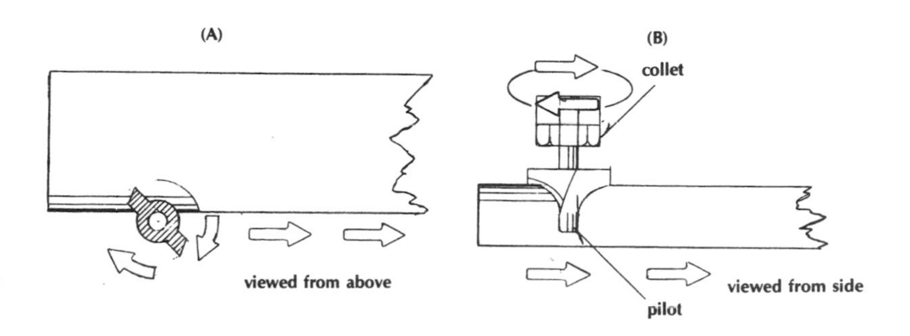

Figure 3-24. *Router feed direction.*

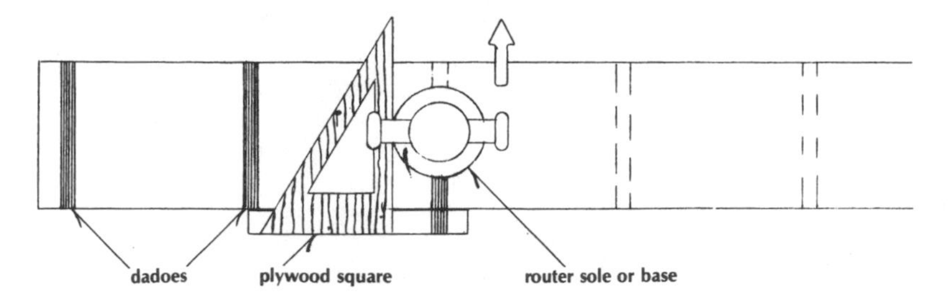

Figure 3-25. *Routing shelf dadoes.*

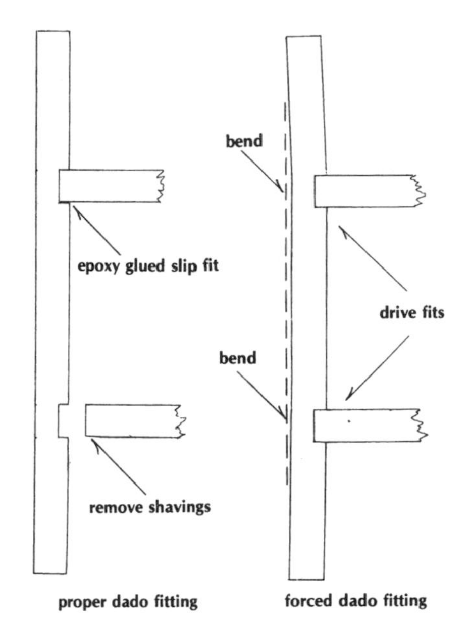

Figure 3-26. *Dado fittings.*

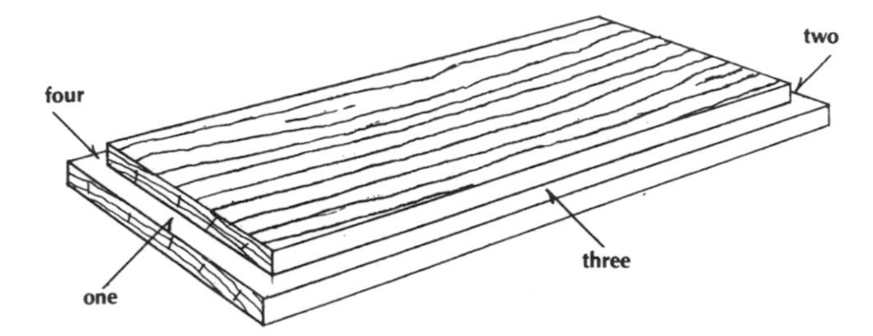

Figure 3-27. *Sequence of operations: molding, routing, sawing.*

a piece, as shown in Figure 3-27, it is best to make the cuts across the end grain first so the side passes can remove any splintering at the corners.

These are but a few of the many facets of router operation. Sears offers a book and video that cover the subject very well for the beginner. Check your library, too, for information.

4 Stationary Power Tools

In this chapter I shall discuss in some detail the types of stationary tools used in many professional boatbuilders' shops and in thousands of home shops as well. Do not be too concerned if your evaluation of a tool differs from mine. Quite aside from cost or capabilities, each tool has certain attributes that may make it valuable if not absolutely essential to your kind of work. Some of the tools discussed may be eliminated if other tools perform the same functions. Some prices will not be included because I expect them to change more rapidly than other information that follows.

GRINDERS

There's no way you can get by without a grinder for sharpening tools. I'm sure you could have someone turn a hand grinder for you, but a hand grinder can't possibly do comparable work (unless it is a wet grinder). The belt-driven wet grinder is perfect for tool grinding. Industrial-rated grinders will handle grinding of metal parts as well as tools (using the proper wheel, of course); see Figure 4-1. These tools are powered by motors of ⅓ to ¾ h.p. These are double-arbor outfits for setups with two grinding wheels, or a buffing wheel, sanding drum, disc, and so on. Well-known dependable names are Rockwell, Milwaukee, Thor, Black & Decker, and Skil.

If you're in my financial bracket, you'll find you can get along beautifully with a belt-driven double-end arbor powered by a ¼ to ⅓ h.p. used appliance motor (Figure 4-2). I recently ground a blade for a molding head to produce raised-panel doors and drawer fronts. The results were strictly professional, as you can see in Figure 4-3. I keep all my planes and chisels in top shape with such a grinder. The arbor is typical in that it takes a 6-inch grinding wheel, a buffing wheel, and a sanding drum—all at the same time—or a disc. Mine has sleeve bearings, a cheapie; the ball-bearing type costs double. The reason for the buffing wheel is a patternmaker's trick: it's great for honing the cutting edges of plane blades, chisels, and so on, after whetting on the oilstone. Load the buff with Tripoli, polish the bevel edge a bit, then just a touch on the back, and you'll have an edge you can shave with. Another tip: Mount your grinder and motor on a plank about 3 feet long so you can move it around and carry it in your car, or aboard.

In Chapter Five, I'll describe a number of homemade fixtures for holding tools so you can grind accurately.

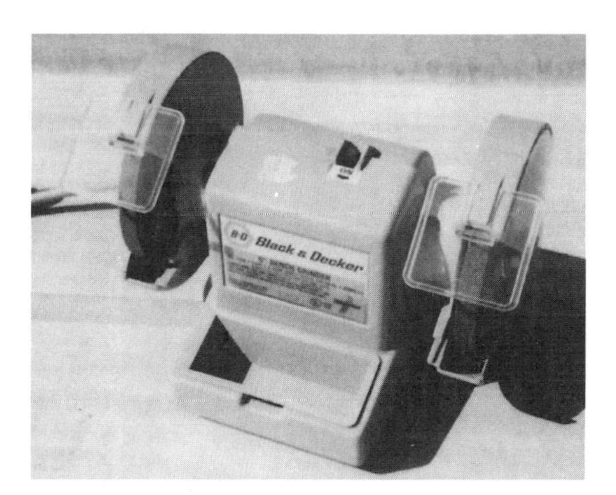

Figure 4-1. *The water reservoir in this tool grinder prevents overheating and annealing of cutting edges. Low-cost protection for your tools.*

Figure 4-2. *A narrow grinding wheel is not the best for grinding plane blades, but it can do a good job with solid tool support (see Chapter Five).*

Figure 4-3. *Panel edges produced by grinding cutter for single-knife molding head. The raised panel at the top requires a tilted table-saw arbor.*

BENCH DISC SANDERS

For many years, Delta manufactured a splendid 12-inch-diameter bench disc sander. Once in a blue moon, one turns up on the used-tool market. This sander turns at 3,450 r.p.m. and has one disadvantage: it will take off stock very rapidly, so be prepared. Rockwell made a smaller belt-driven disc sander of 8 to 8½ inches in diameter, but I have not used one.

With a miter gauge sliding in a groove parallel to the disc, you can use a sander to square up ends or fit miters. Just let the piece kiss the disc as you slide it across several times to bring it down to the mark. A disc is fine for rounding outside corners also.

Figure 4-4. *The author's low-cost 4-inch belt sander does a fine job of finish sanding on parts and strips, and removing saw marks from edges. Disc is small. The support helps handle long, flexible flat moldings and similar pieces. The motor is under the bench.*

Plans for my homemade 8-inch disc sander are included in Chapter Five.

BENCH BELT SANDERS

The combination belt-and-disc sanders available in many makes are even more useful (provided that the disc is 8 inches or better; see Figure 4-4). The belt, 4 to 6 inches wide, runs over a platen up to about 15 inches long. The work piece is pulled over this platen against the direction of the belt, so long pieces can be finished. Or the fence may be installed across the platen so very small pieces can be handled safely. A slight hand pressure is all that is required to remove planer corrugations and to do general surfacing of hard and soft woods. The piece, however, must be kept moving constantly. I usually dress edges after ripping, skipping the jointer. Because I always use coarse belts, I must hand sand with fine papers for the final finish. Fine belts do not stand up satisfactorily, in my opinion.

As with a hand belt sander, tracking or alignment of the belt is tricky. The tension is adjustable on bench models, and I have found that belts will wander if belt tension is not adequate. The projecting end of the belt is fine for sanding inside curves and shapes (Figure 4-5). If installed at the end of a bench, this is better than a drum for inside sanding.

The bench sanders that take the 6-inch-by-48-inch belts are more efficient and require at least a ½ h.p., 1,725 r.p.m. motor. The small 4-inch mod-

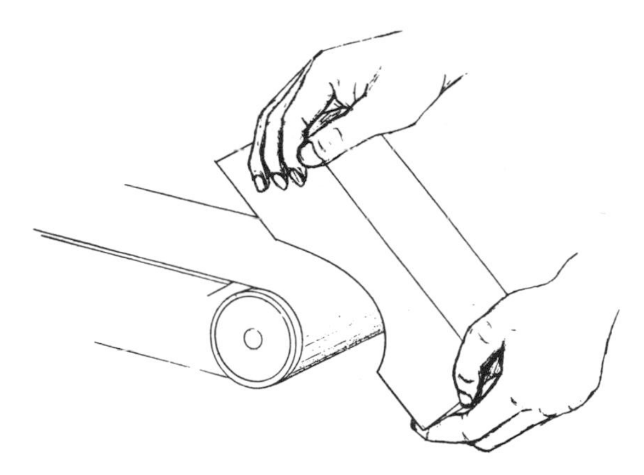

Figure 4-5. *Sanding concave surfaces on a belt sander.*

els cost in the $100 range. Both are available with or without a disc. They are sold by Sears, Western Auto, and other chains.

JOINTERS

If you have ever sweated over planing a straight edge on 2-inch oak, you'll agree that a jointer is indispensable. There isn't a lot of oak in joinerwork, but there is a lot of mahogany, pine, fir, plywood, and some teak. If you own a power plane, much of your problem has been solved. For straightening the sawed edges of pieces 2 feet or more in length, however, even the power plane comes out second best.

Here's how a jointer works. The outfeed or back table (on the end away from you) must be adjusted so that it is perfectly flush with the cutting edges of the knives or blades as the head rotates (Figure 4-6). Check this with a smooth wood block or straightedge, and bolt it. Now crank up the infeed or front table so that it, too, is even with the knives. The depth pointer should now read "0." If it does not, move it. Back the infeed table down a hair and try a piece of softwood. First, however, make sure the guard is working freely. This is a *very* dangerous tool. (A good friend has three beautifully tapered fingers because he was careless when pushing thin material through a jointer.)

If you back the table down 1/32 inch, you should now take off just about that thickness. If you feed fast, you will see many little corrugations on the planed surface. You will notice also that the high-pitched hum of the machine drops to a much lower note. These signs indicate that you are forcing the machine. This will produce a poor surface and possibly a burned-out motor, especially likely when dressing a broad work surface, as the knives are working harder. If your outfeed is set correctly, the work will pass right over the table without hesitation and the planed surface will continue straight out to both ends, with no low spots at either end. If there is a low spot at the forward end of the piece, the outfeed table is too high. If the low spot is at the back end, the outfeed is too low (Figure 4-7).

First, examine the direction of the grain in the work piece. Do not cut against the grain if this can be avoided by turning the piece end for end (Figure 4-8). If planing creates a crackling sound, you probably are cutting into the grain. This will produce a rough or pitted surface. If the grain is wavy or curly, as in ribbon-grained mahogany, taking very fine cuts may help. It would be best, however, to leave a little stock for finishing with a bench sander. On a long crook, you may have to plane from one end and then the other, finishing up with a smooth plane. Remember that no jointed edge is acceptable for finish. Always sand. Remember, too, that you can use a portable belt sander turned on its back if you don't have a bench sander (Figure 3-14.)

The jointer fence guides the work so the planed edge is true. Hence, it must be checked with a square or bevel gauge before you start to cut. Bear

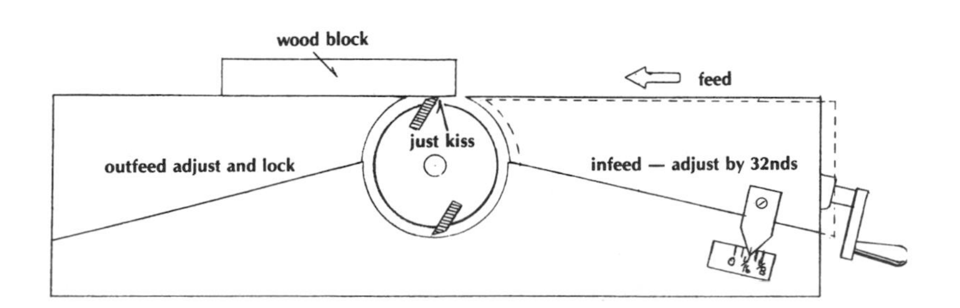

Figure 4-6. *Jointer essentials.*

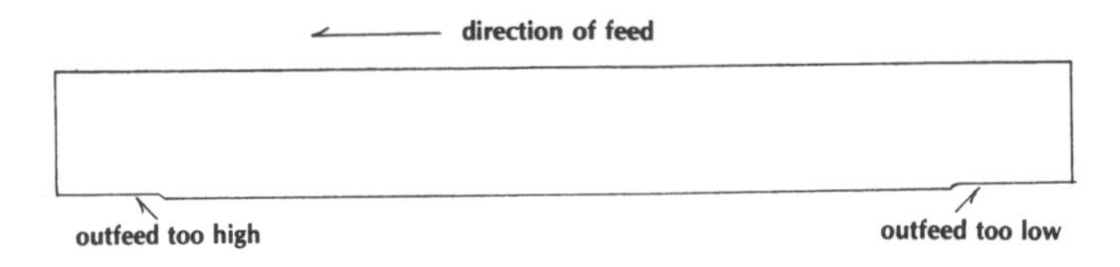

Figure 4-7. *Jointer technique.*

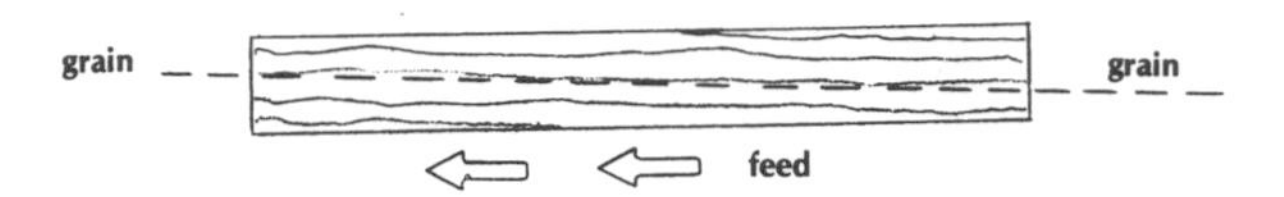

Figure 4-8. *Jointer technique.*

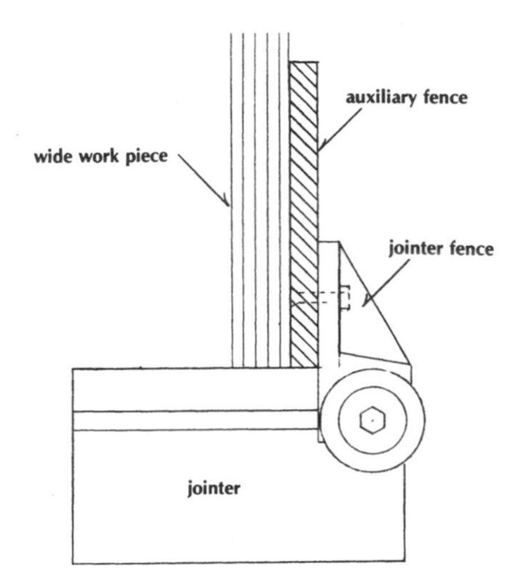

Figure 4-9. *Jointing wide stock.*

back against the fence. If you are edging a wide board or plywood (a door, for example), screw an extension 8 to 10 inches high to the fence so the work will be steady as you feed it (Figure 4-9).

If it is necessary to plane thin stock ¾ inch or less in thickness, use a pusher block as shown in Figure 4-10A. Handles for the pusher can be made from pieces of dowel or broomstick, set at about a 75-degree angle. A pusher block keeps the work from chattering (and your fingers attached to your hand). A simple pusher is shown in Figure 4-10B.

Always be sure that your fingers do not hold the edges or end of a piece. The left palm should press a wide piece flat on the table while it is moved forward by a pusher stick in the right hand. For a perfect surface, long stock must be fed alternately with both hands. Do not stop. Have the pusher handy for the last few inches. Long and very thin stuff (say, ¼ inch thick) may be planed by placing a heavier piece on top of it. Glue a little stop across one end, as shown in Figure 4-10(A).

Material wider than the jointer knives may be dressed down adequately by removing the fence and guard and making several cuts. Overlaps may show, so be prepared to finish with a jointer, jack, or smooth plane.

Long pieces are hard to handle on a short jointer, as the slightest movement due to imbalance can spoil the surface. Attach extensions to eliminate the need for a helper. If you must have a helper, instruct him to never pull or push. This applies to a table saw or bandsaw as well. I have a split finger as evidence of my failure to instruct a young helper correctly. But then, I wasn't using a pusher!

Jointers come in all sizes up to giants 36 inches wide. The best all-purpose size for the amateur yachtbuilder is 6 inches (the length of the knives) and up to 40 or 48 inches long (Figure 4-11). The much lower-cost 4-inch jointer is second best, but it should be considered carefully. I would hesitate if the overall length were under 24 inches. Lately, manufacturers have introduced built-in direct-drive motors that cost less than jointer-plus-motor, but I have no objection to belt drive. The better jointers have ball bearings. Sleeve bearings are common in the 4-inchers. The big chain stores have both 4-inch and 6-inch motor-drive and belt-drive jointers. A ⅓-h.p., 1,725-r.p.m. used appliance motor is sufficient to drive a 4-incher. The 6-incher requires ½ h.p., and 3,450 r.p.m. is preferred. Get all the catalogs and ask an experienced friend for advice, especially if you are looking at used tools. I found a somewhat rusty but relatively unused 4-inch Delta Homecraft jointer, minus fence, at a rummage sale for $5. The machine performed to perfection for six years.

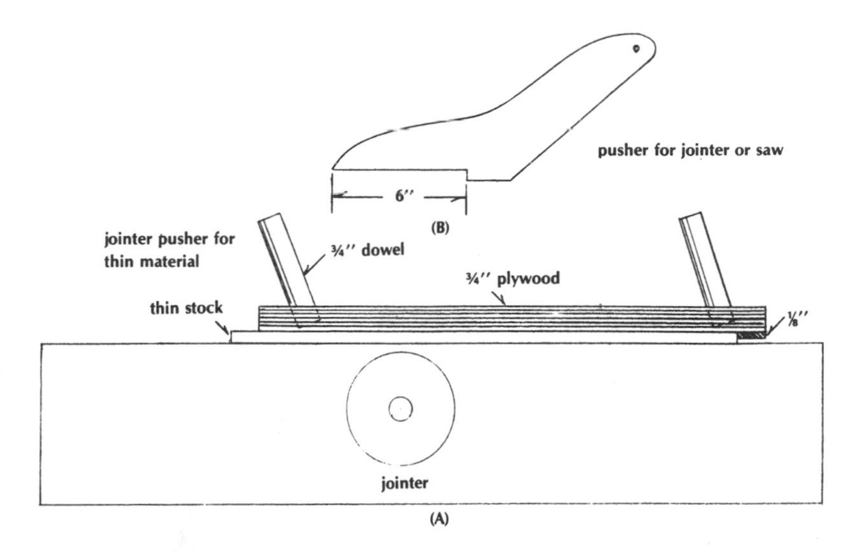

Figure 4-10. *Jointer pushers.*

Figure 4-11. *This Craftsman 6-inch jointer is capable of the finest work required in any boatshop. It needs a ½ to ¾ h.p. 3,450 r.p.m. motor. (Courtesy Sears, Roebuck)*

SHAPERS

As described in the section on routers, a shaper is basically just a large router turned upside down and mounted beneath a table. You will need a bench model for nonproduction boatbuilding or cabinets. It's good to have your shaper and jointer on stands that may be moved about or even into a boat's hull.

The shaper base is of cast iron or semisteel; most have a provision for rapid adjustment of spindle or table height (Figure 4-12).

The surface should be ground and polished. If you find a used tool, your first job is to remove chemically every vestige of rust from its surface, then sand with a very fine emery cloth. Buffing with wax is not a bad idea, since the work piece must encounter no resistance as you move it into the cutters.

The shaper fence is divided into halves so that each section can be adjusted backward or forward. This corresponds to the infeed and outfeed on a jointer (Figure 4-13). The fence or table should allow the installation of spring-steel hold-downs to prevent chatter. Small-shop shapers all have ½-inch-diameter spindles that take solid cutters from 1 inch to as much as 2 inches in height and up to 2½ inches in diameter. Many different cutters are available for each make, as are collars that enable you to mold curved pieces or to repeat shapes by riding a pattern against a collar (Figure 4-14). This is valuable for fitting scarfs and other joints, such as in covering boards and rail caps.

Very likely the commonest use of a shaper is in forming rabbet joints (Figure 4-14). A piece of work can be pushed through a medium-duty shaper at a speed of one foot per second while making a ⅜- to ½-inch rabbet. The work must be held down properly so it does not chatter, causing a wavy cut. If the hold-downs are not strong enough, you must build extensions on the table. On lighter tools, you will have to work more slowly. And if there is a lot

Left: Figure 4-12. *A shaper is indispensable for reproducing from patterns, molding edges, rounding. It takes a high-speed ½ h.p. motor. (Courtesy Sears, Roebuck)* **Below: Figure 4-13.** *The author's shaper split fence could be used on the router-shaper table or shaper you can build. Both of these are described in Chapter Five.*

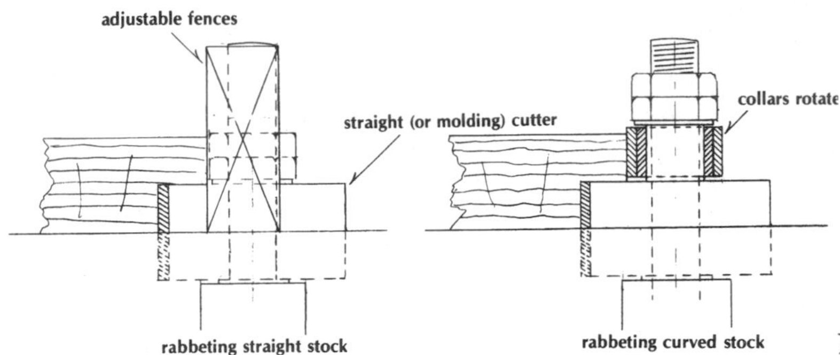

Figure 4-14. *Shaper setups.*

of work (staves for spars, for example), it would be better to make two or even three passes to prevent overheating the cutters and armatures. If many shorter shapes are required, especially in hardwoods, three-lip cutters are available. These are useful for drawer joints, tongue-and-groove, coves, beads, ogees, glue joints, flutes, raised panels, drop leaves, and so on.

It is now difficult to buy hardwood moldings, even in Philippine mahogany. You can, however, make just about anything you want with a shaper and use up small scraps and rippings to boot. You may want decorative edges for your dresser and table tops, ladder treads, berth fronts and trim, drawer fronts, doors and door jambs, bulkhead corner trim, fiddles, and so on. These are the touches that make a vessel distinctive.

Shapers are available with a built-in 1-h.p. ball-bearing motor, a table size about 15¼ by 18 inches, and a spindle speed of 18,000 r.p.m. at decent prices. The table size is minimal, but you could add permanent extensions. Sears has a shaper with a good semisteel 19-inch-by-27-inch table, but the spindle speed is only 9,000 r.p.m. This tool requires a belt drive and a ½-h.p., 3,450-r.p.m. capacitor-start motor.

Most of these tools have threaded ½-inch spindles about 2½ to 3 inches under the nut. This is long enough to set up a two-part head that holds two knives about 1½ inches wide. Using straight

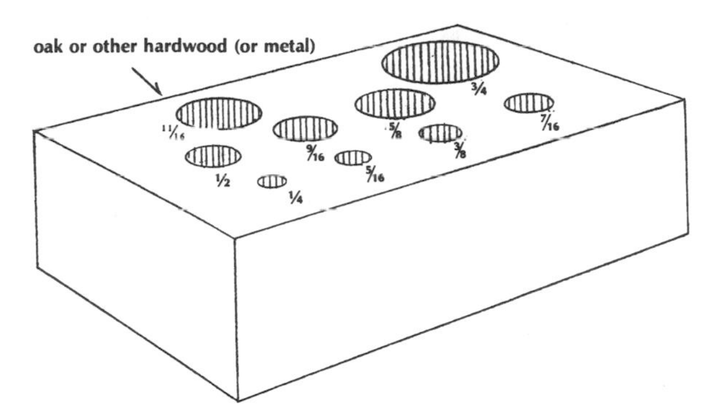

Figure 4-15. *Drill bit guide.*

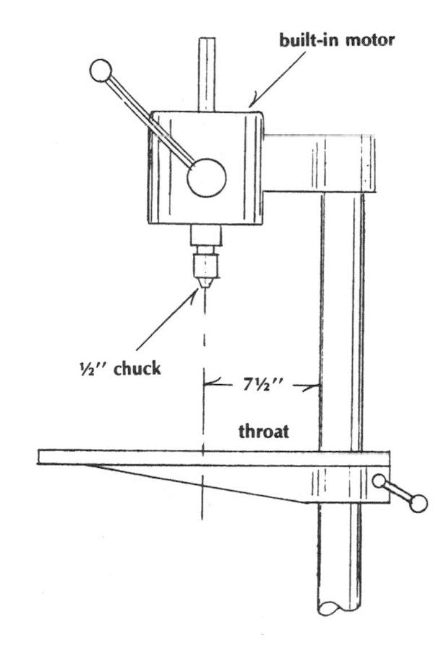

Figure 4-16. *15-inch drill press with built-in motor.*

knives, this arrangement would make a great edge planer or jointer for straight work against the fence or for square-edged shapes running a pattern against a collar. By grinding these straight knives identically to a desired shape, you can produce any reasonable molded effect. Remember, however, that these shaper heads swing a comparatively large radius, so it would be wise to set up the fence for several passes and feed slowly. Raising or lowering the spindle can control the amount of material removed by each pass with straight knives.

The versatility of the shaper cannot be described adequately in these few paragraphs; check your library for additional information. If you prefer to go the router route, see the little conversion in Chapter Five, where I also describe an efficient ½-inch spindle shaper you can build from a kit.

DRILL PRESSES

I tried for months to get along without a drill press. But there is one basic operation that is next to impossible without one: boring a hole perpendicular to your work piece. I have gone so far as to make a drill jig consisting of various sizes of holes in a block of hardwood, but even then, I had to borrow a friend's press to guarantee accuracy (Figure 4-15). Of course, there's quite a bit of metalwork in any boat. Here again, a drill press is virtually a must.

In addition to its primary operation, drilling holes, a drill press of adequate capacity can perform many other woodworking chores: routing, dadoing, planing surfaces, rabbeting, making mortises and tenons, grinding, sanding, cutting plugs, and so on. The garden-variety drill press is capable of speeds up to only 4,200 r.p.m., however, so higher-speed or variable-speed tools must be selected to perform such specialized work. Accessories make the exotic jobs practical and worthwhile.

The standard drill press is designated as, say, 12 inches or 15 inches. This means it will drill at the center of a 12- or 15-inch circle. The throat—the distance from the center of the chuck or spindle to the column—is therefore 6 or 7½ inches (Figure 4-16). The usual vertical travel of the spindle, and consequently the depth of a drilled hole, is about 3½ to 4 inches. The maximum height from the table to the chuck of a bench model is about 14 inches. Floor drill presses are equipped with a clamp-mounted table on the column. The standard chuck capacity is ½-inch drill size. The new high-speed types take up to ⅜- and ½-inch bits.

Bench models at 125 pounds are portable, whereas floor models weigh over 200 pounds. Used ones occasionally come on the market. Names to look for are Atlas, Delta, Craftsman, and Jet, a quality import.

You should look into the small presses introduced in the early 1980s that have built-in variable-speed motors, rotating heads, and solid-state electronic speed control. One end of the spindle is used for normal drilling and other operations up to 2,500 or 3,000 r.p.m., while the other end is available for shaping and routing up to 18,000 or 20,000 r.p.m. With accessories, this type can be a light duty shaper as well as a good drill press. Another

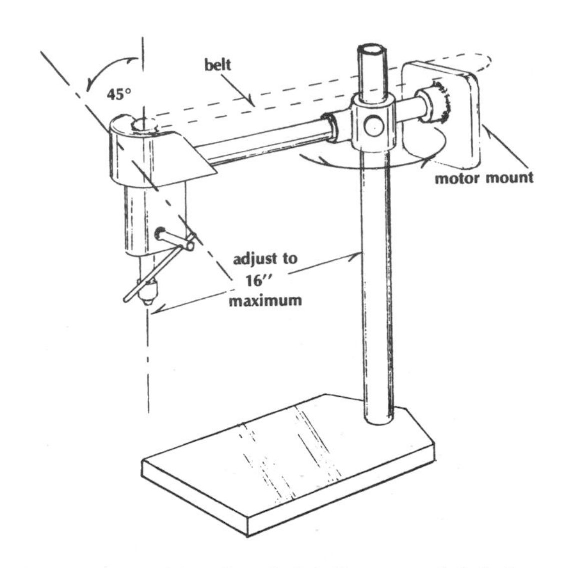

Figure 4-17. *32-inch radial drill press with belt drive.*

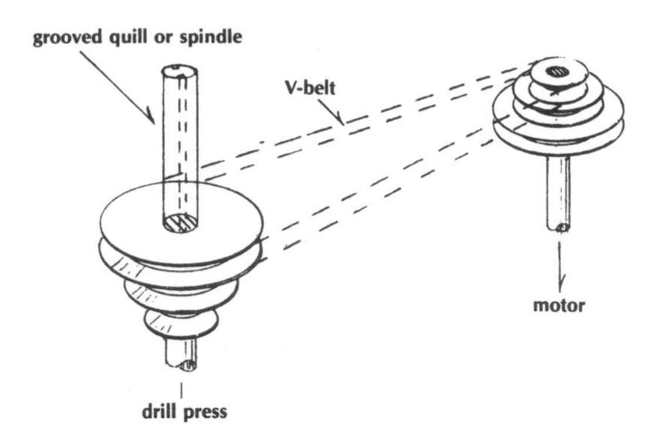

Figure 4-18. *Three-step cone pulleys.*

development in the 1980s is the light-duty radial drill press shown in Figure 4-17. Radial drills are not new, but these little bench models are. Delta manufactures one that drills at the center of a 32-inch circle and has a tilting head. This tool takes a ⅓-h.p., 1,725-r.p.m. motor for speeds to 4,700 r.p.m.; the chuck size is ½ inch. Being able to swing the head of a drill press over the edge of a bench is useful for larger jobs such as boring timbers. Be sure its motor has end-thrust bearings so it can be operated continuously with the shaft vertical.

The speeds of standard belt-driven drill presses are easily changed, being equipped with a pair of cone pulleys—three or four groove pulleys on the spindle (or quill); see Figure 4-18. To change speed, you simply shift the belt from groove to groove. This usually provides a range from about 400 to 5,000 r.p.m. Drilling holes in wood may require the lowest speed. The screw spur on wood auger bits should be filed off so there is no thread, otherwise the screw will pull the bit into the work faster than it can clear the chips. In addition, the rake angle of the cutting lips of the bit must be filed to a much lower angle to prevent digging in.

The spade bit (Figure 4-19), works very well. Spade bits cut clean, accurate holes and they are inexpensive—about $10 for a set of six, ranging in size up to 1 inch. A fine bit for wood boring is the spur bit (Figure 4-20). These come with single or double cutting edges and may be operated at 1,800 to 3,000 r.p.m. Those with more than one cutting edge may be used without clamping the stock to the table, but single-lip Forstner or expansion bits over 1 inch require secure clamping of the work. You will find that twist drills (machine bits) have a tendency to wander and to drill oversize or oval holes in wood. Of course, you should always place the work on a block of scrap wood so you don't drill holes into the table.

Accessories for the Drill Press

To use your drill press as a shaper, check to be sure that the top speed is 5,000 r.p.m. or higher. Don't expect a polished finish at under 15,000 r.p.m. Construct a heavy plywood table about 20 by 28 inches with cleats on the underside to fit the drill press base. Sears has a more-than-adequate shaper fence attachment, or you can make something similar that permits both fences to be adjusted (see Figure 4-21). Remove the Jacobs chuck (it will not hold at high-speed vibrations) and replace with a special collet chuck to take standard two- or three-lip router bits. You can also use ½-inch-bore shaper cutters if there is a tapered socket adapter for your drill press. From this point on, follow the procedures for shaping. Any $100 expenditure for accessories that enables your drill press to do the work of a shaper is money well spent.

Sanding drums, grinding wheels, buffing wheels, hole cutters, and fly cutters are but a few of the drill-press accessories available. The most important is a plug cutter (Figures 2-22, 2-23). This should be run at moderate speed, and if yours is the open-side, self-ejecting type, you'll need to rig up a carton on the table to avoid ricocheting plugs all

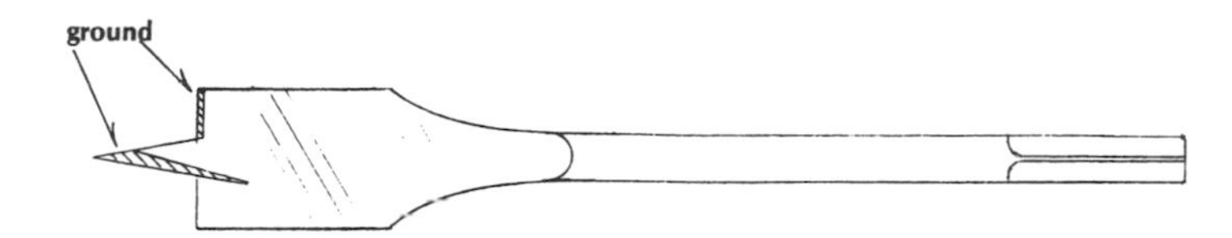

Figure 4-19. *Spade bit.*

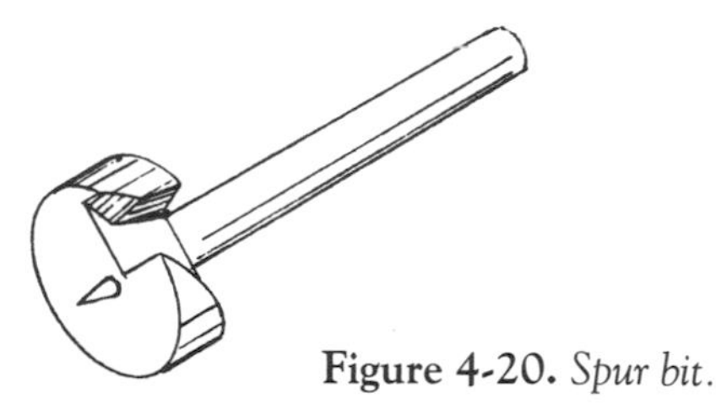

Figure 4-20. *Spur bit.*

Figure 4-21. *Typical shaper fence. Tapped holes in table take guide pins and commercially available spring hold-downs.*

over the shop. If you run this type of plug cutter into *end* grain, you get ⅜-inch by about 1½-inch dowels. Bore a number of these, then saw them loose from the stock in one stroke. Plug cutters and plug counterbore bits are available at marine supply houses or by special order from your local hardware store. The short, non-ejecting plug cutters are available everywhere at moderate cost.

Mortising may be done with a spur, dowel, or machine brad-point bit if the round end joint is acceptable—and why not? See Figure 4-23. But if not, just square up ends with a ¼-inch chisel. If you have a lot of joints of one size to make, get a drill press mortising attachment jig, bit, and square chisel (Figure 4-24). The bit turns in a square hollow tool ground sharp on the lower end. This is pressed into the work, cutting the corners square as the mortise is bored out.

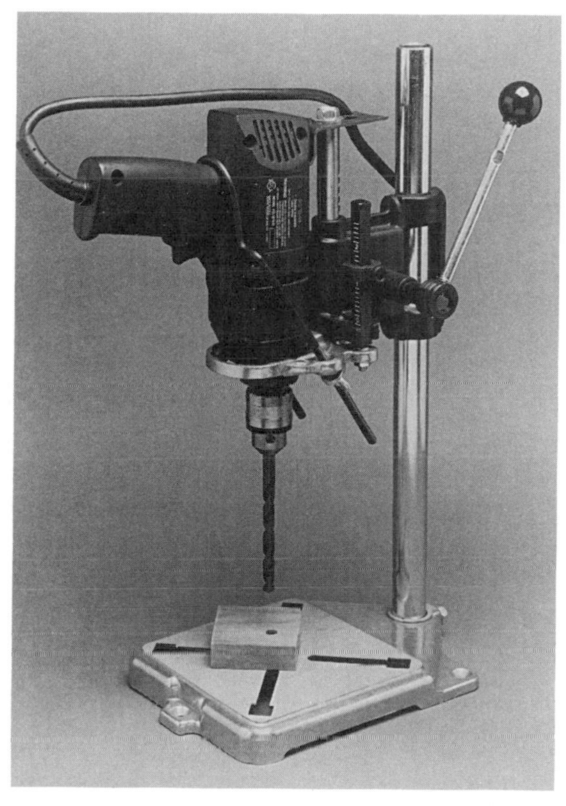

Figure 4-22. *For light wood or metal drilling, this Craftsman drill stand is excellent, especially for plug cutting.*

A Practical Light Duty Drill Press for Peanuts

Several times previously I have mentioned my dedication to simplicity and economy. Don't fail to check out the little drill stands that fit your ¼- or ⅜-inch electric drill (any make); see Figure 4-22. Allow 15 minutes for squaring and—*voila!*—you have a drill press! For cutting plugs, counterboring hardwood parts for plugs, multiple-drilling precisely spaced holes for spindle rails, fiddles, etc., and for just plain everyday drilling, I got along beautifully with a Sears Craftsman drill stand and

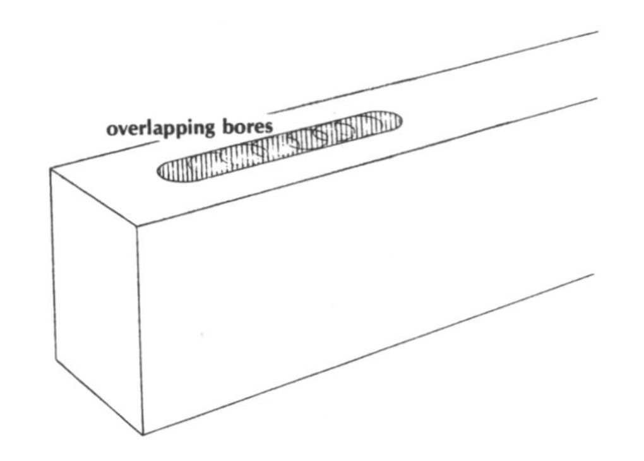

Figure 4-23. *Spur or machine bit mortise.*

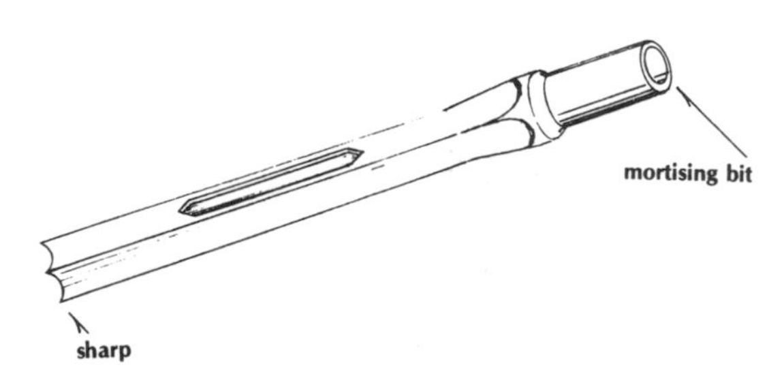

Figure 4-24. *Mortising chisel.*

a Craftsman ⅜-inch variable speed drill. The stand in 1992 was $30.

We should consider other uses of a drill press for metalworking, even though these are outside the realm of woodwork. When drilling metals, run the press at the proper speed and use a lubricant whenever it is called for. The maximum speed for a small carbon drill bit is about 1,000 r.p.m., except in aluminum, which is drilled at higher speeds. For a ½-inch drill bit, use the lowest speed on the cone pulley, 400 or 500 r.p.m. Brass and cast iron are drilled without lubricant or coolant—not even water. Steel should be kept wet with kerosene or, in a pinch, soapy water applied constantly with a brush. For aluminum, believe it or not, except for very shallow holes, soda water is recommended.

If you are using high-speed bits in metal, the speeds can be almost double those for carbon-steel bits. Keep all bits sharp. Look for chipped corners and grind them to the correct angle. If you have a lot of metalwork ahead of you, especially in stainless steel, you'd better pick up a book on the subject. Stainless presents special problems.

To sum up, a drill press is an extremely versatile woodworking and metalworking machine, even if it is only a drill stand that uses a portable drill. Next to your table saw and bandsaws, the drill press might well be your most used machine.

BANDSAWS

The first saw to be discussed in this chapter is small enough and light enough to fit in with the bench tools, but it requires a stand to bring the table to the proper height. Thus, it becomes a floor machine and a major piece of equipment. The bandsaw is a fascinating tool. It's fun to operate and full of tricks. If you build boats, you'll never regret spending your money on one.

Many bandsaws built in the late 19th century are still around and working. Of course, as they say, they don't make 'em like that any more! In fact, they do, but not for your purposes. In actuality, very little has changed. Fine bearings, accurate wheel balancing, small sizes, ball-bearing blade guides, built-in motors, and other refinements make it efficient and long lived, but the basic design is the same. Today some of the jobs traditionally assigned to the bandsaw have been taken over by the sabersaw and the portable power saw. Both cut curves. A small circular saw will cut long curves if it has a lot of set. But only one tool will cut fine curves, only one tool can duplicate shapes by stacking, only one tool can resaw 6 or 8 inches deep, and only one tool is so much fun to use.

The ideal saw for wooden boatbuilding is probably the 30- or 36-inch capacity bandsaw. (Capacity is the dimension from blade to frame.) For heavy work, such as sawing out deadwood and keel oak and other timbers, there is no other realistic answer. But there are exceptions to every rule.

If you can find an old bandsaw, 18 to 30 inches, take it. An 18-inch saw, new, may go for $1,000 and up. Look for slop around the shaft bearings (bronze or babbitt sleeve bearings are OK if relined) and for grooves worn in the rear guide wheel (this is the one the blade presses against). Bearings and guides may be replaceable, but the price should reflect this. An 18-inch saw should have a minimum of ¾ h.p., and preferably one horse, if used for resawing. For an old saw, you should pay $250 or more. Used 12-inch saws such as old Deltas, Craftsman, and Rockwell 14 (Figure 4-25) are extremely popular. Look for at least 6-inch clearance under the guide; more is desirable. Check the saw's balance by running it

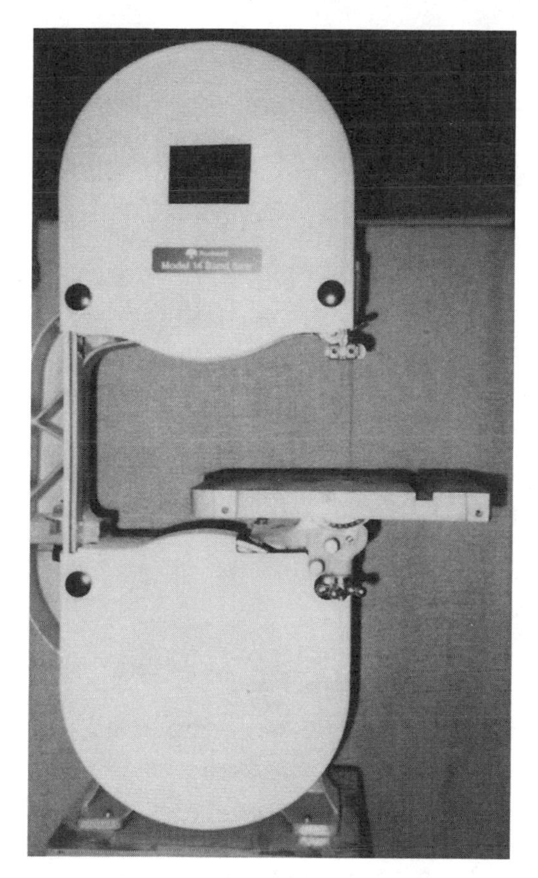

Figure 4-25. *A beautiful 14-inch bandsaw by Rockwell. The difference in utility between 12 and 14 inches is great.*

Figure 4-26. *The popular 12-inch Sears Craftsman bandsaw has a light, dust-ejection vacuum, and other deluxe features. (Courtesy Sears, Roebuck)*

before you hand over any money. Loose rubber "tires" sometimes cause jumping, making it impossible to follow a line. Replacement, however, is not too much of a task.

On a new saw, I would settle for a 12-inch throat, permitting cuts to the center of a 24-inch circle. This is for joinery as well as for general wooden boatbuilding. The Craftsman, for example, is quite typical. It has a built-in work light and a sawdust-ejection system that can be attached to a vacuum cleaner (Figure 4-26). This tool weighs only 75 pounds and requires a ½-h.p. capacitor-start 1,725-r.p.m. motor. Stands are available. The saw alone goes for about $350; recent direct-drive models cost more. Get all the catalogs before you choose. In Chapter Five, I describe 12- and 18-inch saws I have built from kits. You can do this too.

Bandsaw Pointers

The bandsaw is one of the safer power tools. If the upper guide is properly adjusted above the work, it is just about impossible to push your finger into the blade. However, never run or turn on a bandsaw, even momentarily, with the upper or lower guards (doors) open. Adjustments to align the tracking should not be made with the saw running.

First, loosen all blade guides so they do not contact the blades. Second, spin the wheel by hand three or four revolutions. (More turns could run the blade off the wheel.) You'll see how a 2-degree movement of the handwheel adjustment changes the tracking of the blade. If the blade is moving toward the back of the tire, the wheel is tilted back, so screw the handwheel slightly in the direction

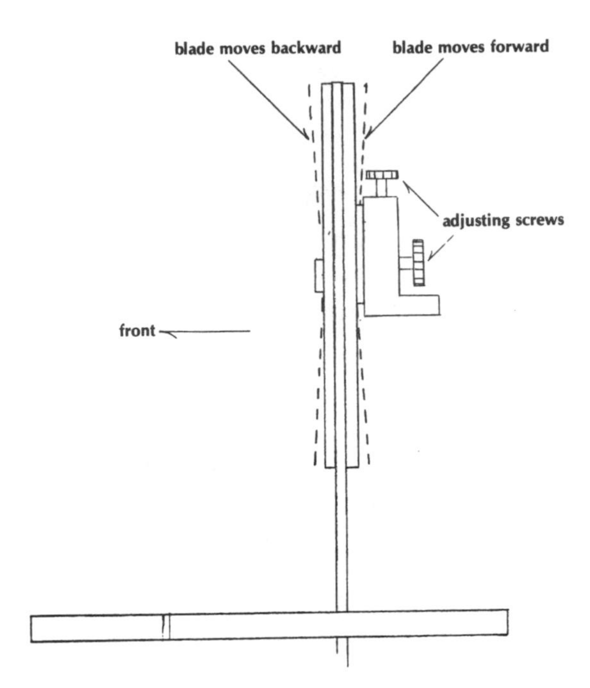

Figure 4-27. *Adjusting bandsaw blade tracking.*

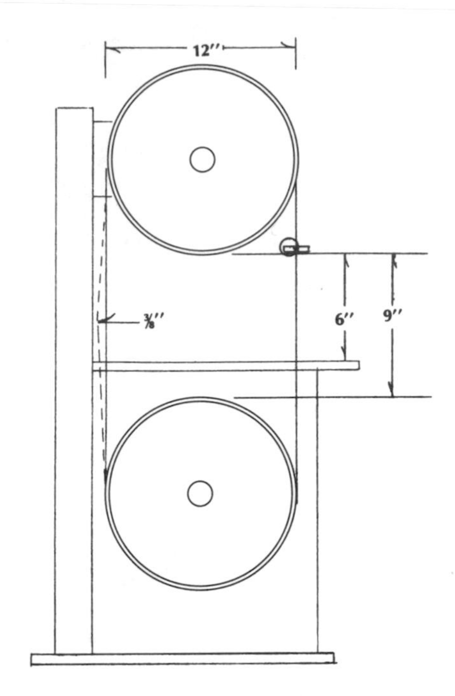

Figure 4-28. *Bandsaw tension adjustment.*

that moves the blade forward (Figure 4-27). Make these small adjustments repeatedly until there is no deviation in blade position. At this point, snap the switch on and off quickly for a trial. If the blade wanders again, turn it off and make another small adjustment. The blade does not have to ride at the center of the tire. Frequently it will ride near the forward edge. Its position relative to the side guides and back roller guides is more important.

Tension is adjusted by a handwheel mounted on a vertical shaft, usually with a strong coiled spring arrangement. Tightening the handwheel forces the wheel upward and increases tension on the blade. Press against the side of the blade on your left, usually in a recess in the frame. About ¼-inch sideways movement for each 6 inches between upper and lower wheel is about right (Figure 4-28.) Spin the wheel by hand again to be sure the blade is still tracking correctly, then turn on the switch. Observe the tracking for a moment, turn off the saw, and make the next adjustment—the guides.

The guides control the blade so there is no play from side to side or forward or backward. For accurate work, you want a minimum of movement. In some saws, the guides are small, oil-impregnated bronze blocks or rods held by set-screws (Figure 4-29A). In very old saws, these may be blocks of hardwood (which work fine). In many, the guides are often ball-bearing rollers. Regardless, they should be adjusted in or out, clear of the teeth, so they just miss touching the blade, perhaps by the thickness of a piece of paper. The rear thrust wheel or block, upper and lower, should be set so the back of the blade touches only when sawing stock (Figure 4-29B). It should not be allowed to spin the thrust wheels at all times, as this could crystallize them and wear them out. Set the side guides back so they do not touch the teeth, perhaps 1/16 inch or less from the gullets of the teeth. On some saws, the only way to adjust the side guides forward and back is by moving the entire guide assembly, which is mounted on a horizontal shaft. Some saws have no such adjustment. On these you have to move the tracking of the blade on the wheel so it runs correctly between the side guides, with no contact with the teeth. The thrust block or wheel is always adjustable fore-and-aft in all makes.

After the guides have been set, turn on the switch. If the blade slaps against the side guides, increase the blade tension slightly until the blade

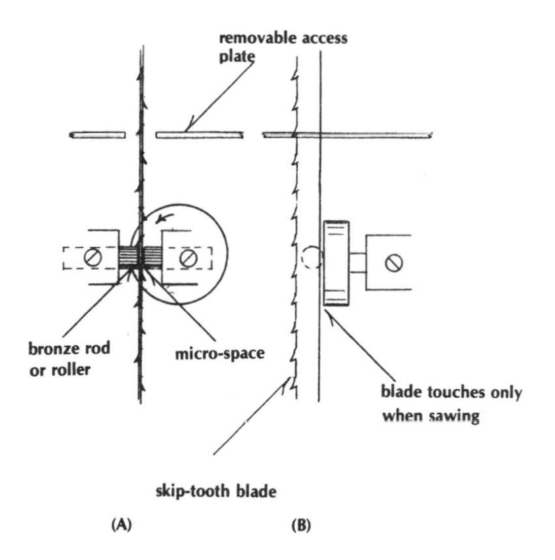

Figure 4-29. *Typical blade guides.*

quiets down. It's now ready for work. When your blade becomes dull, open both doors and slack off on the tension handwheel until the blade drops free of the lower wheel. To remove the blade, carefully pull it out from the guides and off the wheels. Take care so that you do not jam or crimp the blade, which can be resharpened. Occasionally a bandsaw blade will break. When this happens, the wheels may spin for a time. Once in a while, however, a broken blade gets jammed into the rubber tires, with sad results.

How To Coil Blades

Bandsaw blades not in use should be coiled in three loops and hung up out of the way. Coiling is one of the "mysteries" of the boatbuilder's art—one that you must learn. Hold the blade with your hands about 15 to 18 inches apart, with your thumbs out and the teeth of the blade away from you. Bend down enough to rest the bottom of the blade on the floor before you. Place your foot lightly on the blade. Now rotate both hands inward about 180 degrees. This twist will cause the upper portion of the blade to bend down toward the lower section, unless you fight it. Finally, move your hands until they cross and then let go. You will have three loops in the blade, a convenient arrangement for storage (Figure 4-30).

Most of your joinerwork can be done with ¼- or ⅜-inch blades. I have always preferred a skip-tooth blade, as it is nonclogging. I used a ¾-inch blade for occasional resawing, long curves, or straight ripping. Your ¾-inch blade is only 78 or 80 inches (for most 12-inch saws), so the gauge of the steel should be somewhat thinner than for a 36-inch wheel. If it is too heavy, the blade will harden and break. Blades are available with fine teeth, 15 per inch, for a lot of very fine work, such as on particle board and plywood. Seven teeth per inch, however, is about right for most work. A skip-tooth blade cuts faster, especially in moist or gummy wood. Because of relatively low blade cost, it is not always practical to have old ones sharpened or brazed when they crack.

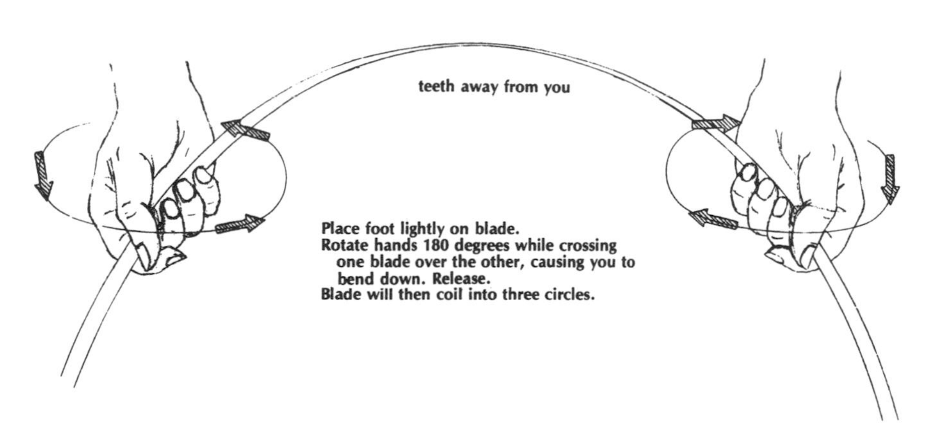

Figure 4-30. *Old bandsaw blade secret.*

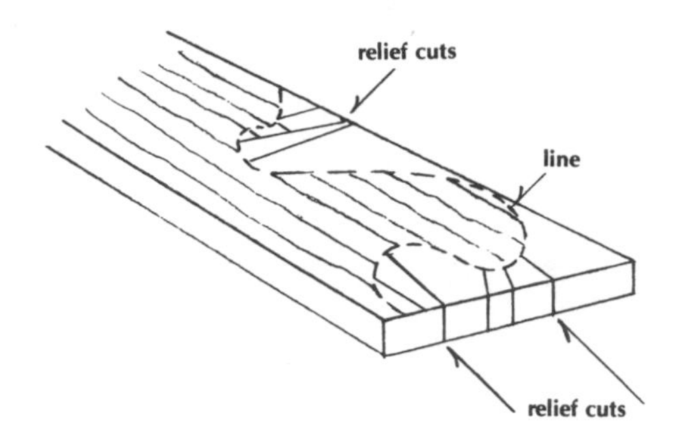

Figure 4-31. *Relieving for short curves.*

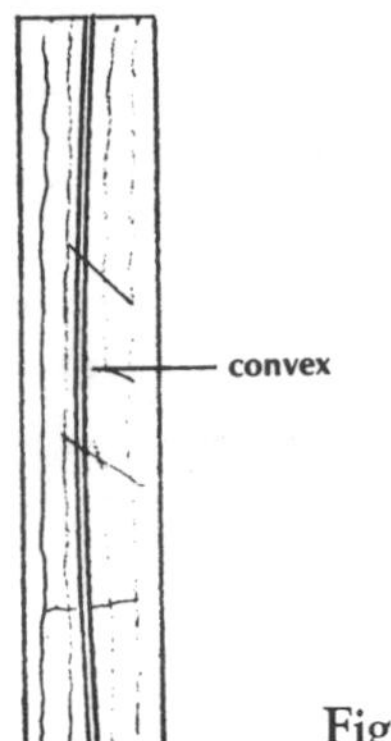

Figure 4-33. *Faulty resawing.*

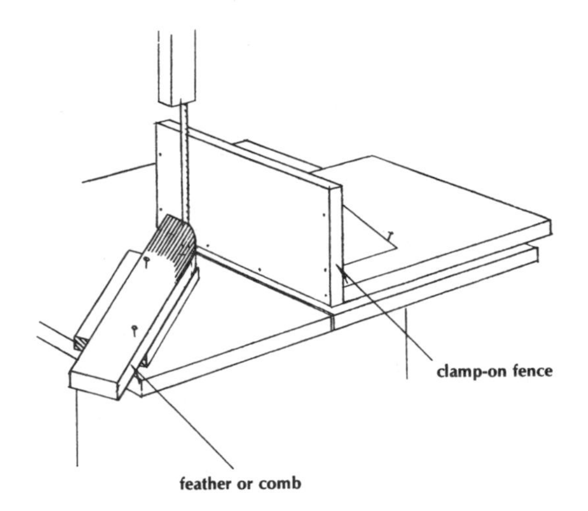

Figure 4-32. *Resaw setup for bandsaw or table saw.*

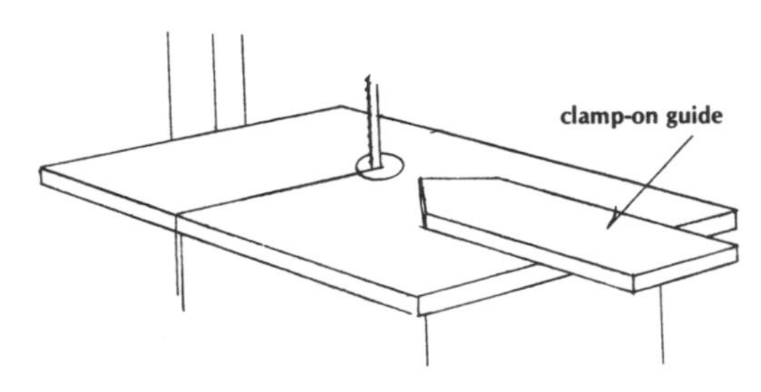

Figure 4-34. *Guide for concentric sawing.*

Using the Bandsaw

The first rule in using the bandsaw is never to try to force the blade to cut a short radius. It's better to make a number of relief cuts on the scrap side of the line, as shown in Figures 4-31 and 4-35. One way of making two pieces of identical shape—say, berth fronts, hatch beams, or something similar—is to shape the part from stock that is more than double the required thickness and then resaw. To resaw accurately on a bandsaw (or any saw), the work must be supported firmly and be exactly perpendicular to the table. Build a fence that can be clamped to the table at the necessary distance from the blade (Figure 4-32). Make a feather or comb to go on the other side of the table. This forces the work against the fence. If the resaw is deeper than 3 inches, raise the feather on a block so the piece cannot wobble. If you prefer, pass the piece between two fences. Try a piece of scrap first. If the pieces have concave and convex surfaces, as shown in Figure 4-33, your blade is not under the required tension, it is dull, it has lost its set, or perhaps all of these.

The bandsaw table is designed to tilt to angles up to 45 degrees. First check the table at its level position; adjust the screw at the left side so the cut is perfectly square. The pointer on the quadrant under the table at the right should read properly at 45-degree cuts. Check with an adjustable bevel square, then reset the pointer to agree.

To bandsaw parallel (or concentric) to curves, clamp a pointed piece at the required dimension directly opposite the blade. Let the curve follow the guide point (Figure 4-34). To make exact duplicates, or any reasonable multiple, nail pieces into a stack up to the capacity of your saw, as shown in Figure 4-35, or try sticking double-faced tape between the layers. The knee shown could well be laminated thicknesses of plywood or ash (oak does not glue well).

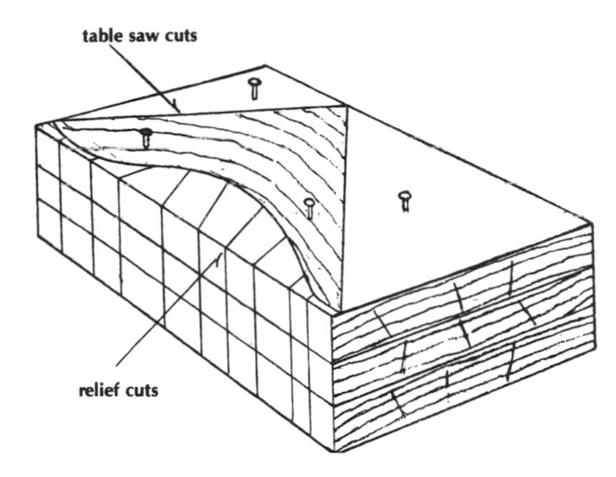

Figure 4-35. *Stack sawing.*

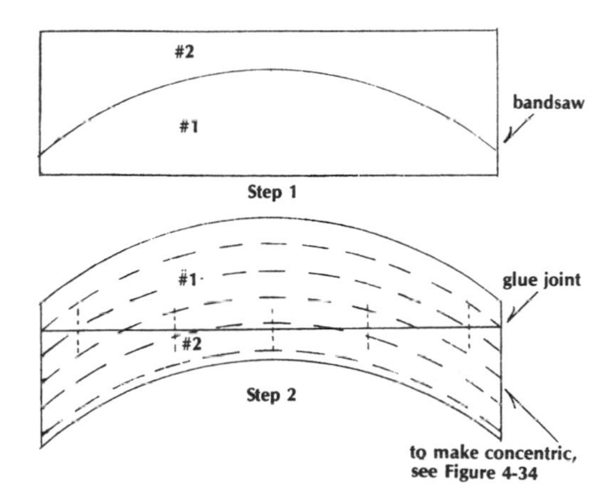

Figure 4-36. *Saving material on wide shapes.*

You can make wide curved pieces without much waste using a bandsaw. Dress edges square and straight so they can be glued, as in Figure 4-36, Step 2. Plan to use splines or dowels if the assembly is wide. Dowels are mandatory if you want to prevent cupping or warping of the final piece. Epoxy-glue piece number 1 to piece number 2. Dress the edges to the final shape. The dotted lines show the positions of the dowels.

Very old bandsaws may have large oil reservoirs stuffed with waste around each wheel shaft. Keep this sopped with ordinary machine oil. Modern saws are sealed or self-lubricated.

In Chapter Five, you'll learn how to make a simple motor mount for the $2 to $5, ½-h.p. used appliance motor. There's also a description of bandsaws you can build from a kit in less than two days. I can testify to their performance. Any halfway-decent bandsaw should last your lifetime. Moreover, the beauty of the work one can produce should inspire you to design more attractive joinery for your yacht. When your project is finished, you'll get back most of every dollar you put into the tool. And you will have had fun every inch of the way.

THE TABLE SAW

Sometimes the terms bench saw, table saw, and circular saw are used to describe the same tool. The bench saw needs a stand to bring it up to a convenient height, thus making it a table saw. The circular saw could be any portable power saw, so let's forget that name. It'll be table saw from here on (Figure 4-37).

The capacity of a table saw is the thickness of wood it will cut at 90 degrees. An 8-inch saw may have a cutting capacity of 2¼ to 2⅝ inches; most 10-inch saws cut through 2⅞ inches at 90 degrees; a 12-inch saw will handle up to 3¼ inches. A run-of-the-mill 10-inch table saw should be able to produce all the joinery for your yacht, although I can think of several things it might not handle conveniently.

The original size of the table is less important than the saw's cutting capacity. You just add extensions or wings to both sides. These may be either purchased ones or wooden ones you build (Figure 4-38). The extensions you buy are cast iron or aluminum. They usually come with various open-grid designs that provide for an extension of the rail for the rip fence. This allows you to rip to the center of a 48-inch plywood panel. Some now have stamped steel extensions.

The area in front of the blade is of considerable importance, especially if you are handling large boards or plywood. This dimension runs from 11 to 15 inches in several 10-inch makes and as much as 16½ inches in 12-inch saws. Table widths are from 17 to 24 inches. Extensions add 10 inches each. You should consider two on the right-hand side and one on the left. Saw tables incorporate two slots for the miter gauge and various devices for locking the fence. Some include micrometer handwheel adjustments for the fence, but the prudent craftsman measures from the point of a tooth to the fence every time he sets it. This is a good habit to get into. You should measure from the back of the

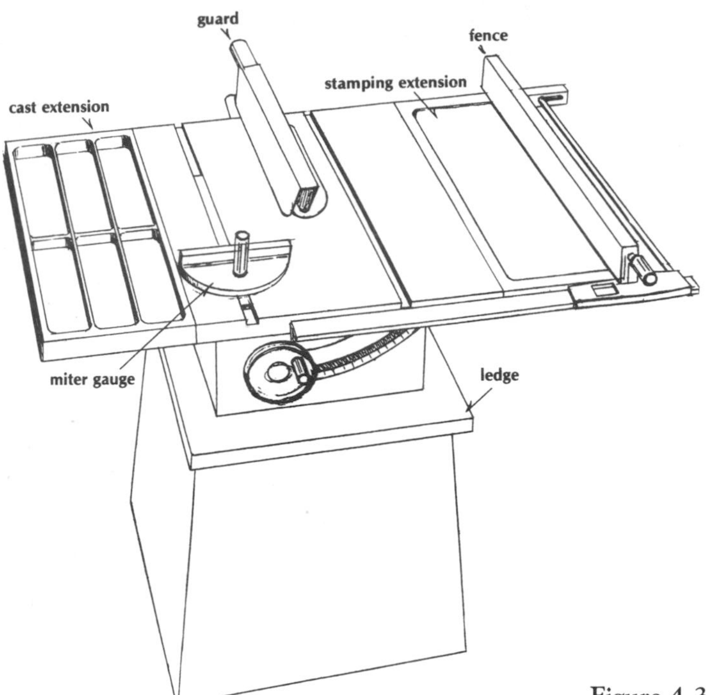

Figure 4-37. *Typical table saw with two extensions.*

blade, too, since fences do not always line up parallel to the blade.

A fairly recent development in table saws is the built-in direct-drive motor. This eliminates the slight nuisance and sound of belt drives. This feature may save you money, as you will not be buying motor, belt, and pulleys. The Sears 10-inch motorized saw with a 17-inch-by-20-inch table, two extensions, and stand sold for $420 in 1980; in 1992 a very comparable model with a 3 h.p. motor is still only $450.

There are many other good makes to check out, so read all the catalogs or find a good used saw. I have a 10-year-old cast-iron 10-inch Craftsman that does anything I ask of it. I also have a kit-built 10-inch saw that I use for precision cutoffs, dadoing, mitering, and so on. This saw is equipped with a sliding auxiliary table, which I'll discuss later.

Direct drive may appeal to you, but don't overlook an old (or even a new) belt-drive saw. Belt-drive tools for years have been excellent and almost 100 percent foolproof. Belts last indefinitely. Good used motors may be picked up occasionally at far less than new prices. You might start with a "junker" ¾-h.p. motor for your 10-inch saw for a mere $20 to $35, then move up to a better motor later on, in the $100 to $135 bracket.

Do not buy an old tilting table saw. I would take a much older tilt-arbor saw first, because this tool is safer and more accurate. Even more important are the bearings in the arbor (they should be ball bearings) and the condition of the table surface. If the surface is badly pitted, think twice. Most rust, however, can be removed. Does the arbor-raising mech-

Figure 4-38. *The author's 10-inch saw built from a kit (see Chapter Five). It is used for precision cut-offs, squaring cabinet doors, cutting moldings, dadoing, and so on. It does anything any ¾ h.p. saw can do.*

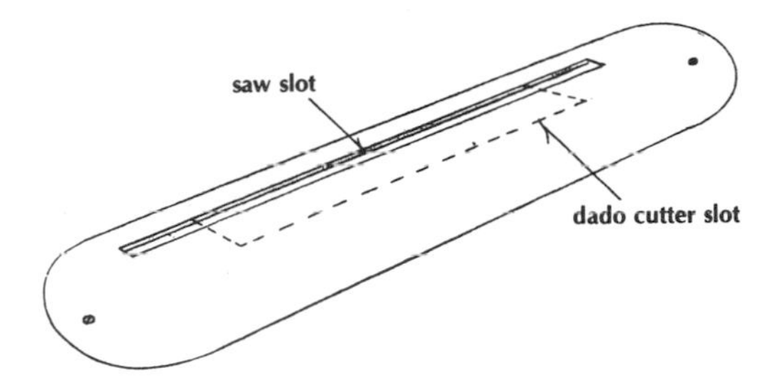

Figure 4-39. *Inserts for table saw.*

anism work freely? Is it badly rusted or can it be freed? Does the arbor tilt easily? Can you move it about in its bearings (a sign of wear)? Is the fence-locking device workable and will it slide on its rail? If the saw does not have table extensions, is it a well-known name for which extensions are available? Does the miter gauge fit its slot? Are any blades available, and perhaps a dado set? You must have two inserts for the opening over the arbor, one for the standard blade and another for the dado set (Figure 4-39).

Most saws are furnished originally with a combination guard and splitter—a vertical affair with a tapered edge that keeps the saw cut open so it does not pinch the blade (Figure 4-40). This is a common headache when sawing green, damp, gummy, or hard woods. Unfortunately, many craftsmen, especially professionals, remove the guard so they can see the blade, even though the guard can be swung back. Once a cut is started, nothing is gained by not covering the blade with the guard. In addition, it keeps sawdust out of your eyes and hair. The better guards have antikickback fingers or dogs that prevent the blade from hurling the wood like a javelin.

Figure 4-40. *Wood that is knotty, gummy, damp, or hard often pinches the saw blade. The splitter, part of the guard, prevents grabbing and kickback.*

Table Saw Alignments

A new or used table saw should be checked for several alignments. Time and rough handling can cause slight shifting of the arbor in relation to the table. The blade must be precisely parallel to the grooves in the table. Here's a simple way to check the arbor. This test checks for blade flatness as well. Measure from the corner of the miter gauge to a certain marked tooth at both front and rear positions, as shown in Figure 4-41. Use a small block as a feeler. Unfortunately, you have to turn the saw over to loosen the arbor attaching bolts. Perhaps all

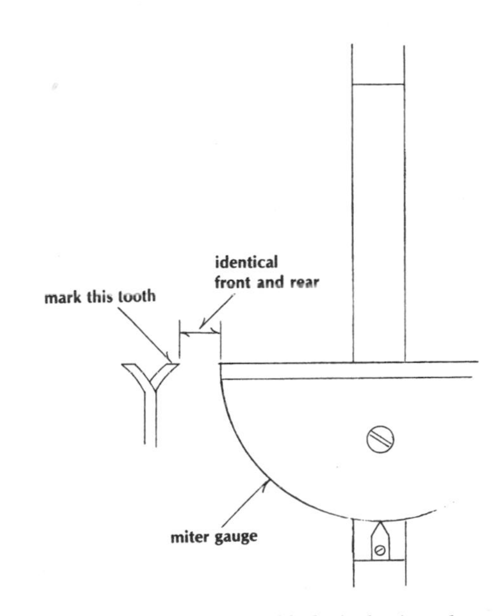

Figure 4-41. *Alignment of blade (arbor) with gauge.*

it needs is a couple of hard hammer taps against a block of wood. Do not touch the shaft; devote your attention to the arbor casting base only. Repeat the test several times.

Alignment of the ripping fence is much easier.

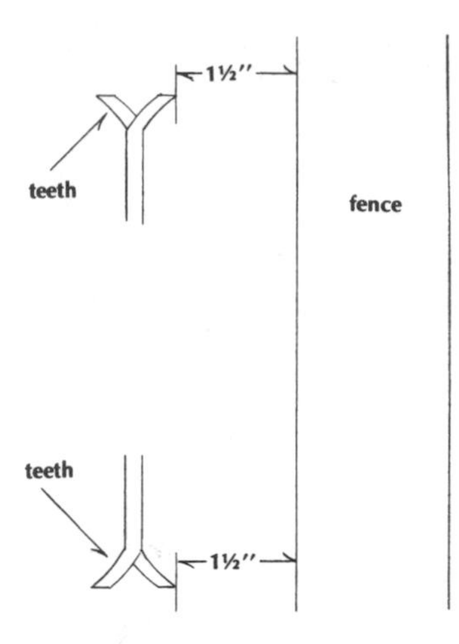

Figure 4-42. *Alignment of fence and blade.*

Figure 4-43. *Left: Planer or cabinetmaker's combination blade for polished cuts. Right: Combination rip and crosscut blade performs both tasks adequately but more slowly.*

You check it by measuring from a specific tooth to the fence, at both front and rear, as shown in Figure 4-42. Several set screws or bolts are provided for adjustments. The 90-degree position of the miter gauge should be checked regularly by sawing a piece of scrap (the wider the better) and testing the cut angle with a carpenter's square. I have a line scratched into my table against which I can set the miter gauge quite accurately in a few seconds. I rubbed a little white paint into the score for easier sighting. Actually, I never trust a miter gauge; I use a sliding auxiliary table 90 percent of the time. See Chapter Five.

Table Saw Blades

A wide selection of table saw blades is available. For doing strictly joinery (light precision work, with many glued joints), I would recommend a cabinetmaker's combination, also called a planer blade (Figure 4-43, left). This blade is characterized by deep gullets every five teeth. It is either flat ground or hollow ground with no set whatever, and it produces an edge that is almost polished, with or across the grain. It will become dull quite rapidly if used in heavy plywood and very rapidly if used in teak. Having no set, it cannot follow a line freehand and may, in fact, throw the piece at you if you try to force it. The standard combination blade (Figure 4-43, right) has deep teeth that are all uniform. It cuts fast and clean, rip or crosscut. It has a slight set so the surface is not polished. It is, however, far better than the surface produced by a standard crosscut blade. An experienced craftsman can freehand on a table saw once in a while with this blade, but it is dangerous.

For cutting plywood, there are several blade configurations. The first is hollow ground to about 1 inch in from the teeth; it has no set (Figure 4-44, bottom). Do not use this type of blade on ¾-inch plywood, as it will drag, overheat, and warp, ruining the cut. This blade is good for plywood up to about ½ inch in thickness, if you need a fine edge. The other good blade for plywood is a fine tooth, from 6 to 10 teeth per inch. The coarser one (6 teeth) is not made expressly for plywood, but it will produce a nice clean cut, not polished. The finer blade (10 teeth) has to be watched for overheating. A good all-around blade called the cut-off has 5 teeth per inch (Figure 4-44, upper right), and is excellent for plywood or any purpose. It is slightly slower than the chisel-tooth combination blade, however.

Carbide-tipped blades are great time- and money-savers. They will cut 20 or 30 times as long as an ordinary steel blade without sharpening. My 10-inch, 40-tooth blade cost about $26 at Sears in 1977 and ran for five years. A carbide-tipped blade costs about $25 to $35 to sharpen, compared with $6 to $8 for the standard blade. But there's no down time, trips to the saw shop, or whatever. And it's just plain beautiful the way a carbide-tipped blade slices through oak! An 80-tooth blade leaves a polished surface, but it may cost $70 or more. The more teeth, the more dollars!

Increasing Table Saw Utility

To increase the versatility of your table saw, several steps must be taken. First, the miter gauge that

Figure 4-44. *Bottom: Plywood blade has fine teeth, little set. Friction will cause overheating in plywood thicker than ½ inch. It produces a good surface with little chipping. Upper right: Cut-off blade will also rip adequately in any wood and produce a fair surface. This is a good all-around utility blade. Upper left: A hollow-ground planer blade produces a beautiful surface. It will bind in ¾-inch material. All of these blades are available in sizes from 6½ to 12 inches.*

comes with the saw may be only 5 to 8 inches wide. This, however, is not trustworthy for 90-degree or angle cutting. Bolt to the miter gauge a piece of good straight lumber or ¾-inch plywood about 5 inches high by 24 inches long. Position it so the blade will cut through it (Figure 4-45). Glue and nail a stiffener along the top of the near side. Now you can handle quite long material and prevent creeping by holding the work piece firmly against the fence. Always check the mark on the table to be sure the gauge is still accurate; wrestling long stuff sometimes forces the gauge out of square.

Most miter gauges have holes bored parallel to the fence. This is for a stop rod, an accessory that allows you to cut more than one piece of the same length. You simply turn the hook of the rod so the end of your material butts against it. Set screws provide for adjustment and locking. You can saw exact duplicates up to 6 or 8 feet long by making up wooden stops of about 1 by 3 inches with a wooden block on the left end. C-clamp this piece to the miter gauge, as shown in Figure 4-46. I frequently

Figure 4-45. *This long miter gauge fence facilitates crosscutting longer pieces. Note adjustable stop rods for producing precise lengths in quantity.*

cut shelves 6 feet or longer, then hang up the stop for the next use, and keep stops in various lengths. They can also be used with the sliding auxiliary table described in Chapter Five.

For resawing, make an auxiliary fence—from ¾-inch plywood or lumber—to screw or bolt to the inside of the regular fence. It should be 6 inches high and the same length as the regular fence (Figure 4-47). This will guarantee a square kerf so that cuts in wide material will meet perfectly (Figure 4-48). Always swing the piece up and over, end for end; do not simply rotate it. If the stock is too wide for the cuts to meet, rip out the remaining narrow part on your bandsaw. If you have no bandsaw, just use a handsaw.

This auxiliary fence will enable you to cut accurate tenons and halved joints. Use a cabinetmaker's combination blade, but not for resawing, as there is no set. Resawing, ripping, and all operations calling for continuous steady pressure against the fence require another simple accessory called a feather or comb (Figure 4-47). The material for the comb can be pine, hardwood, or plywood. The "teeth" can be 2 to 4 inches long by about ⅛-inch thick, made with any saw. Clamp the comb securely to the table so the teeth exert pressure against the stock. Raise it up on a block for resawing. This device is especially effective when ripping long, accurate, flexible items such as battens. It keeps the piece hard against the fence, and, incidentally, keeps your fingers away from the blade. You can concentrate on shoving the work piece through the saw—by means of a pusher, of course.

The same kind of comb or feather may be used to hold material down against the table. Clamp the feather to the auxiliary fence, as shown in Figure

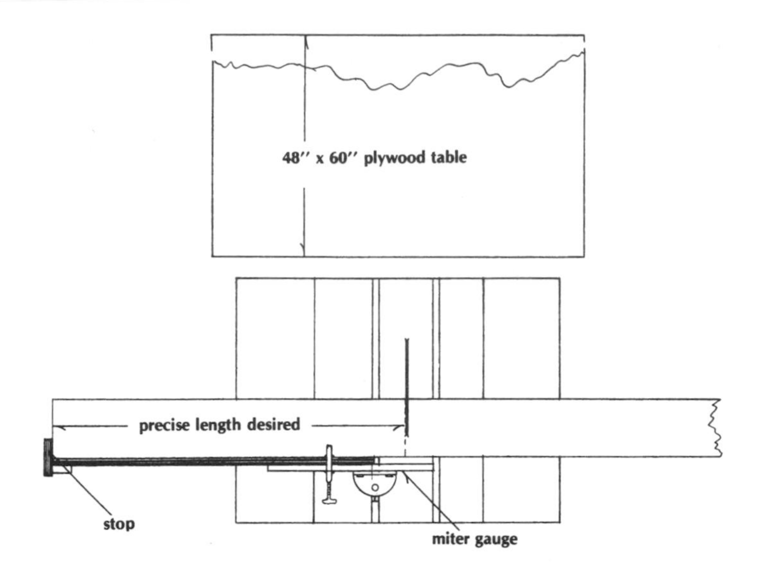

Figure 4-46. *Table saw extension stop bar.*

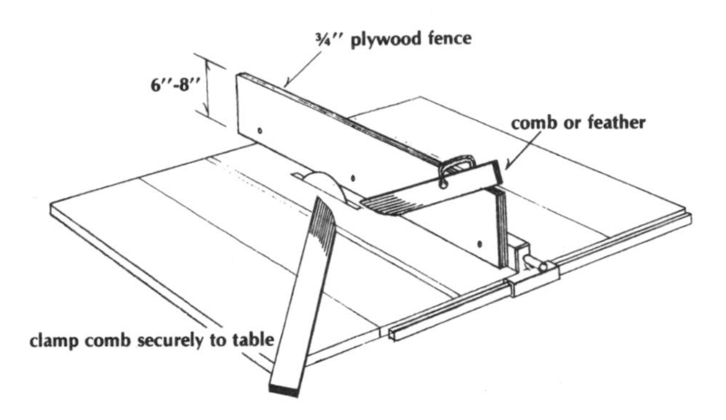

Figure 4-47. *Typical hold-down for saw or jointer.*

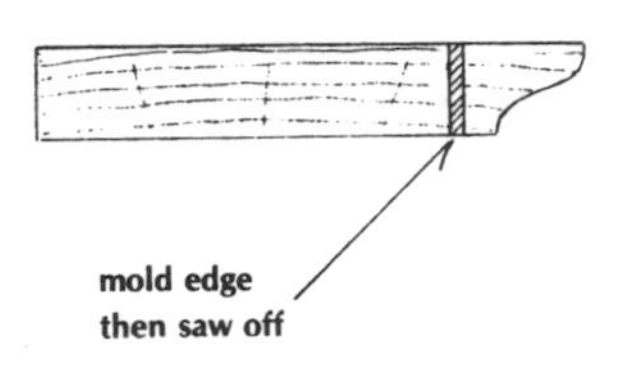

Figure 4-49. *Sawing molding.*

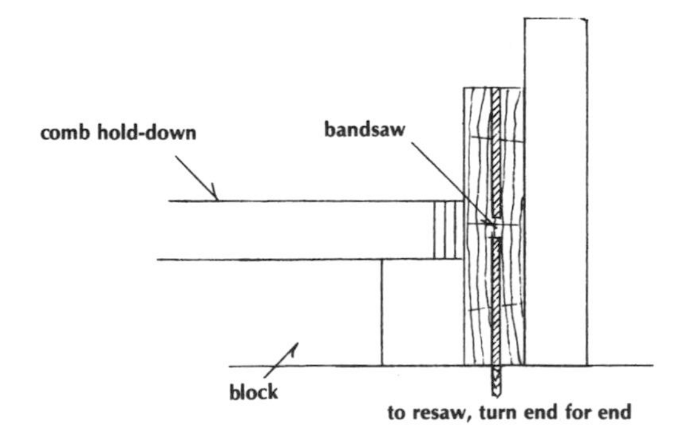

Figure 4-48. *Resawing.*

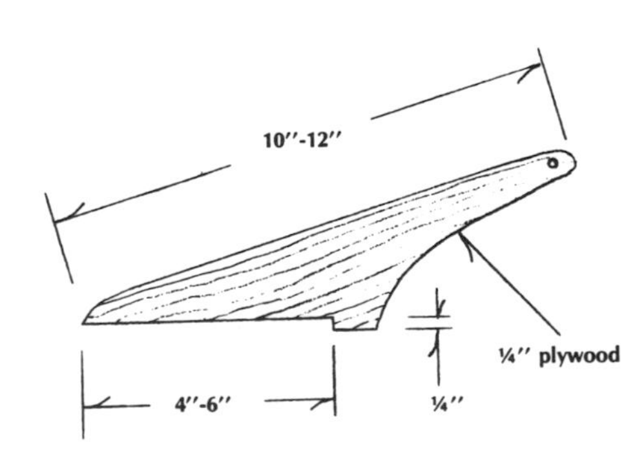

Figure 4-50. *A simple pusher.*

4-47, at an angle. With both feathers in position, neither hand comes anywhere near the blade and there's no chance that the piece will ride up on the blade. Any small stick makes a satisfactory pusher. All this applies also to your jointer.

Both of these feathers will be extremely handy if you are using a molding head or dado set on light material, when chatter cannot be tolerated. An alternative is to mold or dado the required shape on the edge of the material and then rip off the strip. This way you have the heavy stock backing up the working surface (Figure 4-49). It requires two saw set-ups.

Since I have mentioned pushers repeatedly, I suggest using ¼-inch plywood or any scrap (Figure 4-50). Bore a hole in the handle so the pusher is always hanging near your right hand, ready for use. Don't let your fingers come within 4 inches of that blade! Pushers get chewed up pretty fast, so why not gang up four or five thicknesses of plywood scrap and cut out all at once on your bandsaw?

Stock Supports

Never try to crosscut a long piece without having someone or something supporting the far end. If the piece droops, it will bind suddenly and violently, and possibly split. If you work alone a lot, you will need some kind of handy support for long stuff or plywood sheets. For now, let's say that you will need a simple lightweight table about 4 feet wide and perhaps 6 feet long. Its height should be ½ to ¼ inch lower than the saw table. If you want it to do double duty as a work table, then build it heavier. Place it about a foot behind your saw when ripping plywood and long lumber (Figure 4-46). If you have to crosscut long material or plywood sheets, locate the table on your left and toward the rear of the table saw. Be prepared for long boards to drag on the table and cause problems. The answer may be an assistant or a roller supporting stand. Such stands are available from many sources.

Dadoing and Molding

I mentioned a few pages back that your table saw has a large opening in the top, to permit replacement of blades and the installation of tools used to fashion grooves, rabbets, dadoes, and decorative edges of all kinds. (Remember: A groove across the grain, as for a shelf support, is called a dado.) Cutting a very narrow groove is called plowing. Any cabinet made with rabbets at top and bottom and snug dadoes for shelves, glued with any moisture resistant adhesive, needs no fastenings other than a few hot-dipped galvanized finish nails. Of course, such an assembly must be drawn up with pipe or bar clamps while the glue sets.

A dado set consists of two cutting blades on the outside with one to five chippers in between, from 6 to 8 inches in diameter. See Figure 4-51. Most dadoes for shelves are ¼ inch deep in ¾-inch sides. Rabbets should be ⅜ to ½ inch to provide the maximum gluing strength. The maximum width of a dado set is 13⁄16 inch, so for a halved joint 2 inches wide, you would make three passes (Figure 4-52). To produce precise widths of dadoes and grooves, it is often necessary to use paper shims between the chippers. This process usually takes 10 or 15 min-

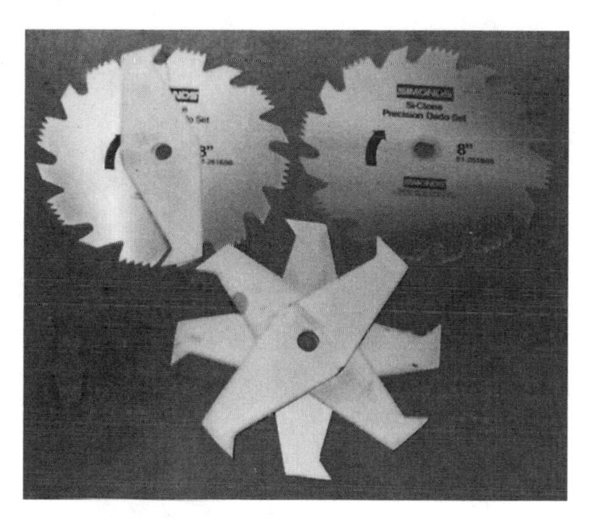

Figure 4-51. *Dado set—two outer cutter blades and as many chippers as required for groove or dado width. An 8-inch set is shown.*

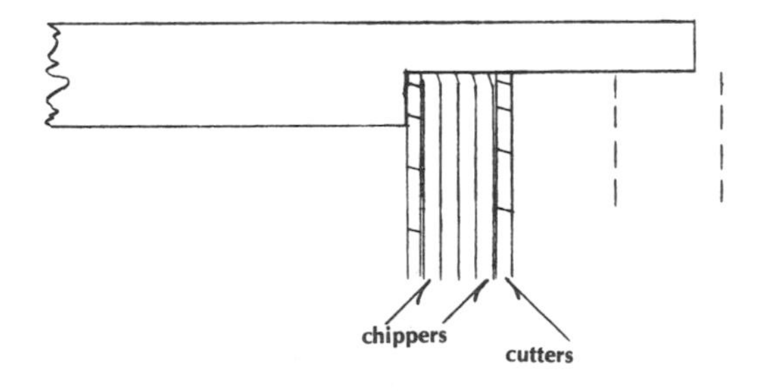

Figure 4-52. *Dadoing half lap.*

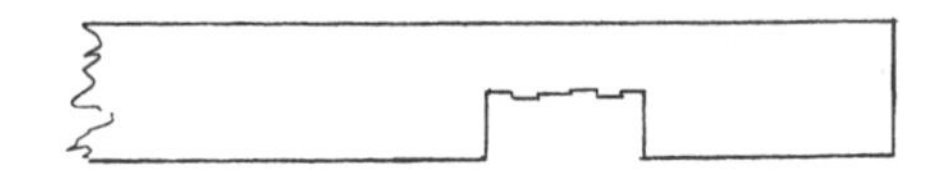

Figure 4-53. *Result of improper jointing of chippers and cutters.*

utes of experimentation on scrap wood to make the nice, easy fit required. A press fit forces the glue off the sides of the joint and weakens the structure. A number of press fits in a cabinet or bookcase side causes the side to bend.

When assembling a dado set on the arbor, space the chippers' wide-set teeth so they are opposite the deep gullets in the cutters. This prevents springing and distorting of the chippers when the set is tightened up. It makes for an accurate cutting setup, too. Dado blades and chippers must be kept sharp so that the cuts are invisible. Cutters and chippers usually are ground and jointed together in sets so that their diameters are identical and precisely concentric. Careless grinding results in unsightly and weak joints (Figure 4-53).

The hassle of setting traditional dado sets forced the invention of the adjustable or wobble dado. This consists of one heavy blade, 7 inches in diameter, usually having from 8 to 24 carbide-tipped teeth. The set includes various arrangements of large, tapered washers that cause the blade to wobble as it rotates (Figure 4-54). These washers are calibrated so the width of the cut, up to 13⁄16 inch, can be adjusted without removing the set from the arbor. This adjustable-blade dado is reliable and saves a lot of time and bother with shims. My adjustable dado has only eight teeth and does a more-than-adequate job. I see no compelling reason for buying the more costly 16-, 18-, and 32-tooth blades, except for eliminating chipping.

Can you do without this expensive tool? Yes, by making repeated saw cuts. You set the fence for each pass, with the outside cuts to extremely close tolerances the thickness of a piece of paper. This operation takes time and much patience, but it has been done for years and years. Rabbets are easier—one vertical cut with the piece running against a high fence, the second cut with the saw height set to the depth of the rabbet. Figure 4-55 shows this and the multiple-cut method of rabbeting.

There is one more important accessory for the table saw—a molding head. Definition: A head is a device that is slotted to receive one or more

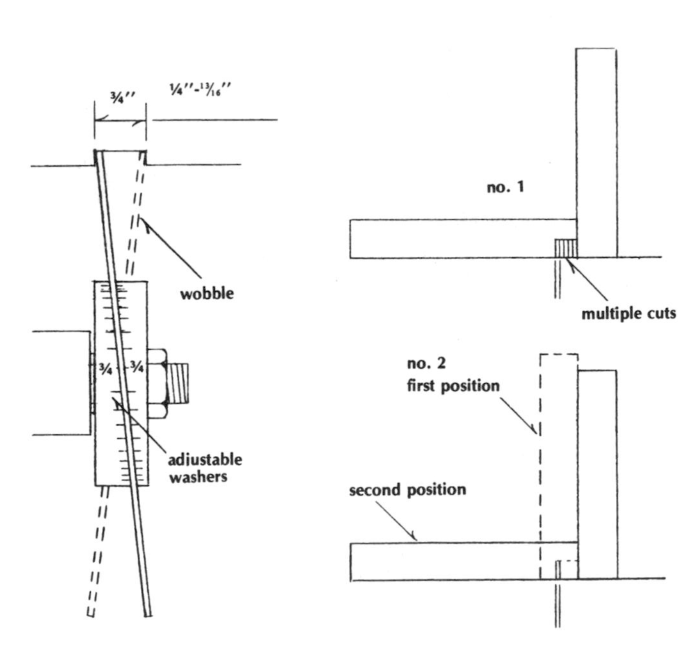

Figure 4-54. *Dado, rabbet, or groove with adjustable dado.*

Figure 4-55. *Two ways to rabbet with saw blade only.*

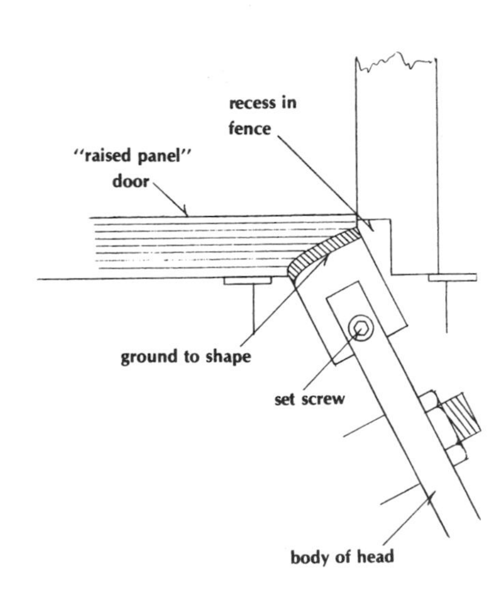

Figure 4-56. *View of molding head (blade shown reversed).*

Figure 4-57. *Triple- and single-cutter molding heads. Many forms and combinations are possible.*

blades or knives and is mounted on a shaft or arbor. Hence, the jointer head, shaper head, planer head, and the molding head for a table or radial-arm saw. A molding head is a massive steel body with set screws to lock the knives in their sockets (Figures 4-56 and 4-57). The knives are ground to various shapes that produce attractive edges, locking edges for gluing, tongue and groove, cabinet door lips, raised panels, quarter rounds, ogees, beads, V-grooves, coves, and dozens of other useful and decorative effects. The shapes are like those cut with a shaper or router. Infinite varieties may be developed by making more than one pass through different knife contours.

I chose a single-cutter molding set from Sears (Figure 4-57) because it has 18 different bits, compared with a three-cutter set making only eight shapes. Presently, there is a two-blade head with planer and five molding bits. The single-cutter set has another advantage, however. Its blades may be ground to different shapes, within reason, and considering the balance of the assembly. I recently ground a straight planer blade to make raised-panel doors (Figure 4-56); the result is beautiful. The single blade produces a fine surface if the work is passed slowly. The two- and three-cutter sets do as good a job, but a faster one. You would find it very difficult, of course, to grind three blades to identical form and weight.

To use a molding head, you will need a wooden fence attached to the regular fence. This should incorporate a recess so the cutters can't touch the standard fence. You make this wooden fence by moving it gradually into the cutters until the recess is ¼ to ½ inch deep (Figure 4-58). I must warn you that a molding head is very dangerous, perhaps more dangerous than a shaper. A molding head will refuse to take deep bites and may throw the work

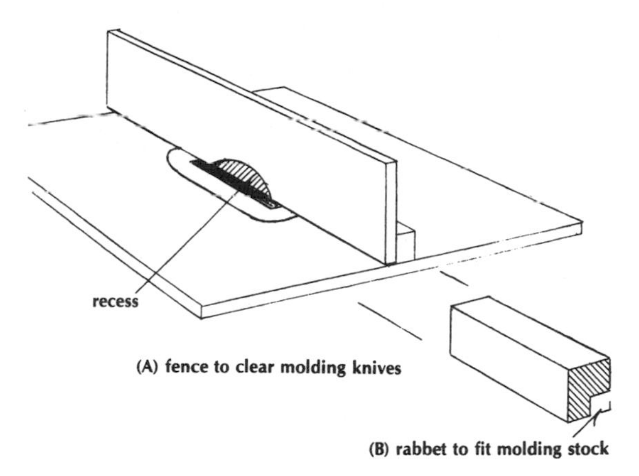

Figure 4-58. *Molding tips.*

piece at you if you try to force it. Most cuts must be done in two or three passes; you start with a shallow cut, then raise the head for each subsequent pass. Move the piece slowly; the cutters are taking off far more than a saw blade does, and they can't be rushed. Use the hold-downs described earlier, both horizontal and vertical, to keep your hands away from the head. I usually mold the edge of the stock, then saw this off if I need a molding (Figure 4-49). It is not advisable to try to mold a strip of any kind because of chatter. However, you may do this if you build a rabbeted jig to hold the strip. The aperture must fit the rectangular strip quite closely (Figure 4-58B).

Fancy molded-edge contours were a mark of quality in fine yachts built in the early part of the century. Then increasing costs of labor, and perhaps changing tastes, gradually phased them out. There is no denying, however, that the judicious use of beautifully molded mahogany and teak woodwork adds much to the beauty and value of a yacht, and to cabinetwork.

Space For Your Table Saw

A table saw requires a lot of space around it for handling material. Ideally, the saw is in the center of a large clear area, perhaps 15 feet on each side and at least 16 feet fore-and-aft. Much lumber turns up in 16-foot lengths.

You can't put a heavy 10-inch table saw down in a hull unless the hull is of impressive dimensions. You might, however, be able to lower a small saw,

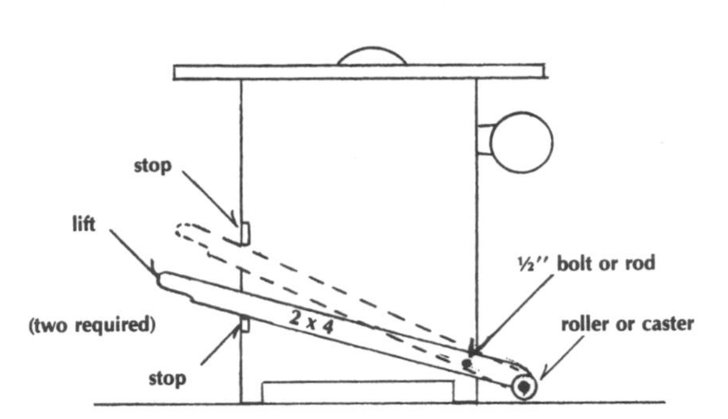

Figure 4-59. *Making the table saw movable.*

Figure 4-60. *The radial-arm saw is much more than a sawing machine. With attachments, it does almost any job except turning. (Courtesy Sears, Roebuck)*

say, an 8-inch model, into a hull. Just be sure you will be able to get it out through the companionway hatch. For easy wheeling around the shop, you might try the idea shown in Figure 4-59. When you lift the saw like a wheelbarrow, the rollers bear on the floor. With this rig, the saw won't creep as it would if it were on casters. The shape of your saw base may present some problems—but that's boat biz! A solution might be the strictly portable table saw described in Chapter Five.

I know you'll agree that a sweet-running table saw sings a beautiful song and also that it does a whale of a job!

THE RADIAL-ARM SAW

The radial-arm saw is just about the final major piece of equipment we're going to discuss (Figure 4-60). Its principal feature is that the work remains stationary while the power unit with its blade is moved to make the cut. Unfortunately, I know of no radial-arm saws at a price within reason that will crosscut more than about 15½ inches. Most will, however, rip to 26 inches wide at a depth of 2½ to 3 inches, and perform another score of operations with certain accessories. You can set the machine in the middle of a very long table, if you wish, so there are few stock support problems for either crosscutting or ripping. With regard to safety, the blade's rotation forces the work piece against the back fence, so it won't throw pieces of wood at you (Figures 4-61 and 4-62). Just keep those pinkies away from the saw kerf.

It is true that the radial is a sawing machine (just as the table saw is), but it does so many other things so well that it should have a different name. "Radial-arm woodworker" might be better. There are major and minor accessories that make this tool extremely efficient for dadoing, grooving and molding, shaping, drum and disc sanding, grinding, planing, drilling, routing, polishing, buffing, and other operations. If you have one of the machines with a variable speed control, you can benefit from the high-speed settings desirable for shaping, routing, planing, and other operations. Some of these are available because of the many positions the power unit can be placed in—with the arbor horizontal, tilted at all conceivable angles, or vertical (Figure 4-63).

I used radials in the past as sawing machines, period! I got nothing from them but sawing speed, accuracy, and safety and was unaware of their many other advantages.

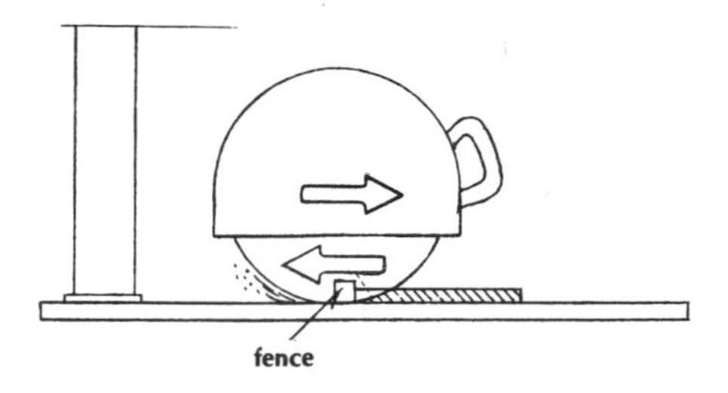

Figure 4-61. *Safe crosscutting.*

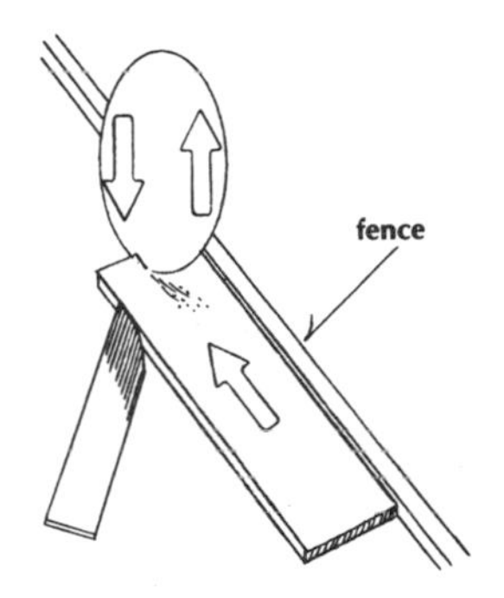

Figure 4-62. *Safe ripping.*

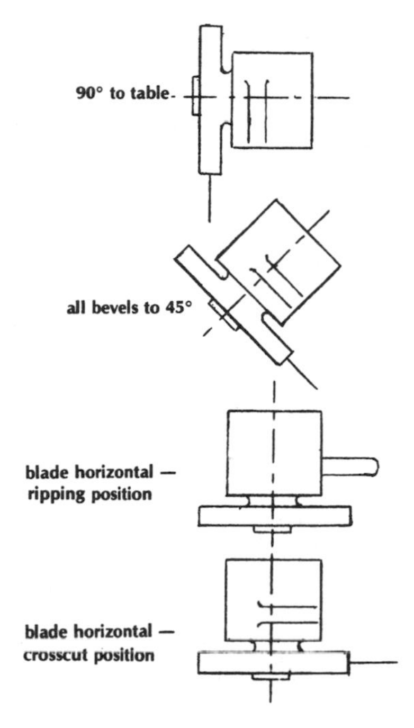

Figure 4-63. *Positions unlimited.*

Radial-Arm Saw Principles

First, when crosscutting with a radial-arm saw, you hold the work piece securely against the rear fence. The power unit and blade are behind the fence and the piece, so with the switch off let the blade touch to check its relationship with the mark. Now push the blade back, turn on the motor, and pull the unit across the work (Figure 4-61). In softwoods this operation takes seconds! If you want a fine surface, slow the feed down and use a cabinetmaker's combination blade or a carbide-tipped, 80-tooth blade. The vertical position should be such that the blade cuts a kerf 1/16 to 1/8 inch deep in the table top. Check with a square for an accurate 90-degree cut. If they and the kerf in the table are out, check the indexing device for play, the base of the column for loose bolts, and so on. If the blade is aligned properly, you can make 1,000 cuts at one second each and all will be perfect.

To rip, the stock goes *against* the saw direction. First, raise the unit, rotate it 90 degrees, and lock it. The blade is now parallel to the fence, as shown in Figure 4-62. Lower the revolving blade until it penetrates the table top 1/16 inch. The piece must be fed against the rotation of the blade (so dust will blow toward you!). Now turn on the motor, press the piece constantly against the fence, and keep it going steadily into the blade, hand over hand for best results. If you can, clamp a comb to the table to keep the work piece against the fence. By this time you know how I feel about using a pusher stick, any old stick.

This is a simple machine, but I like the advice of R.J. De Cristoforo: "Be alert but relaxed. Be master of the machine, but at the same time don't be so overconfident that you lose sight of the fact that here is a bundle of power that is completely indifferent to who is turning it on. The biggest secret to safe power-tool operation is always to be a little bit *afraid* of the machine. You'll have this awareness if you are a beginner. The danger lies in losing it as you become more proficient." So remember, no operation is complete until the blade has stopped turning and you have removed the work. Never try to remove stock while the blade is still turning, even though the power is shut off. Never feed stock when ripping, molding, grooving, or

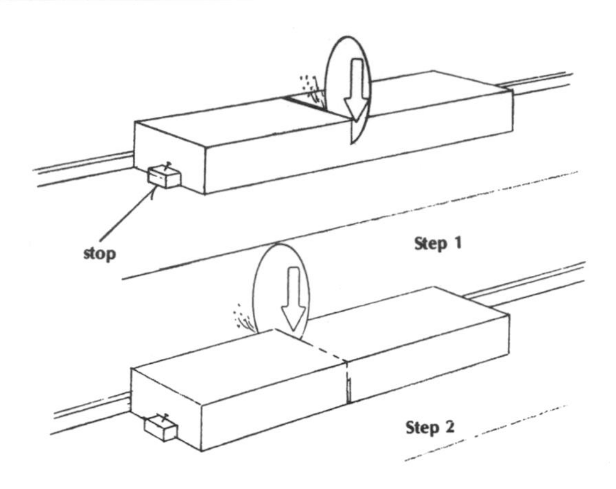

Figure 4-64. *Sawing thick stock.*

whatever, in the same direction as the blade's rotation. This could pull your hand into the blade or even wreck the machine.

If angles are to be cut, release the clamp that locks the arm to the column and swing the arm to the correct angle according to the calibration. Try the angle cut on a fairly wide piece of scrap or several pieces so they can be placed in position just as the final parts would be. If the angle is less than perfect, adjust the arm by trial and error and then reset the indicator to read correctly. Swing the blade back to 90 degrees to see if this, too, reads correctly. If it does not, you have an error, the column is loose in the base, or the arm is not locking firmly on the column. This is one of the few faults of radial-arm saws. In some models with automatic or manual stops for 45 and 90 degrees, there is sufficient play in the locking mechanism to require checking after a change of head position.

How do you cut material wider than the capacity of 15 to 20 inches? Saw across as far as it will reach, cut the switch, then return the power unit past the fence. Turn the piece over end for end, marked side down. Now shift the piece about so the blade can be brought into the fresh saw kerf. Since you are using the blade itself to position the work piece, your new kerf should match the first one perfectly. You could also locate by tacking a stop block on the table before the first cut is made. Then all you have to do is turn the piece over, but not end for end. Of course, you may mark with a square on both sides of the material if it is over 15 or 20 inches—the capacity of most radial-arm saws.

Material thicker than the saw blade depth capacity (usually about 3 inches) can be cut in two passes. Position the work piece against a stop block, make the first cut, then flop the piece and make the second cut (Figure 4-64). The second kerf will match the first one perfectly. The stop-block trick (nothing brilliant, really) is used for cutting identical lengths, which you may want to do quite frequently. Another method is to lay several pieces of the stock side by side against the fence and an end stop. Then you cut all in one pass of the blade.

Ripping extra-thick material is simple. Turn the unit to the ripping position, make one pass the full length of the material, then turn it over end for end. The original side must be against the fence. Use a comb to achieve accuracy. The kerfs will meet perfectly.

If the lumber is rough, you may have to joint an edge first. If it is badly skewed, tack a straight batten (1 by 3 or 1 by 4) to the lumber and run this against the fence. Complications arise if you must turn this piece over, of course. If the piece is wide, rotate the saw to the out-rip position. If a narrow piece is wanted, turn the unit to the in-rip position so the blade just clears your batten. You now have a straight edge to run against the fence for further ripping. If the fence is so low that the batten will not meet it, you may run the work against a temporary fence clamped to the outer edge of the table, as shown in Figure 4-65. Or you may tack an additional piece to the original fence to give it proper height.

Operating with the blade in its horizontal position, that is, with the power unit on end, offers a number of conveniences (Figure 4-63). This allows you to make a repeated-pass groove in a plank edge, or a rabbet with the blade vertical for the second cut. This is great for dadoing, rabbeting, or tenoning on a plank end. With the arbor on end, you might find it necessary to raise the stock to avoid interference from the arbor end and nut; just raise it on a length of lumber or plywood. Use the horizontal position for rabbeting and grooving as well as for dadoing across the plank end. In the latter case, the piece should be raised to prevent the blade from cutting through the fence. A horizontal blade is useful for making many of the joints described in Part Two.

Accessories for Your Radial-Arm Saw

So far I have described the radial-arm saw as a sawing machine. All of the accessories used with table saws are used on radials, so I won't go into detail.

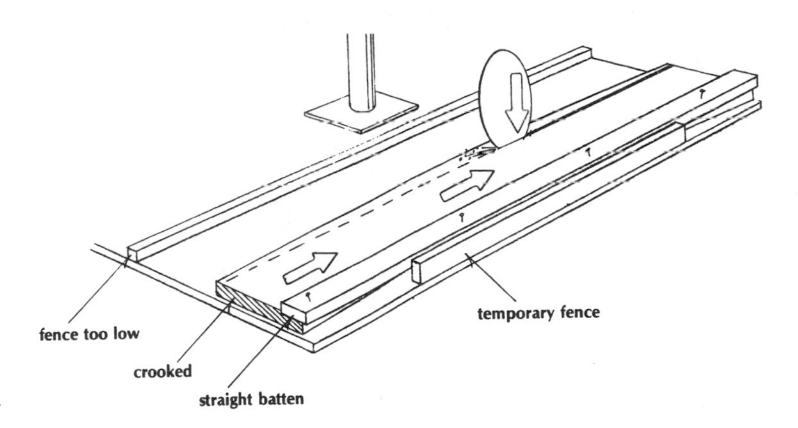

Figure 4-65. *Ripping crooked plank.*

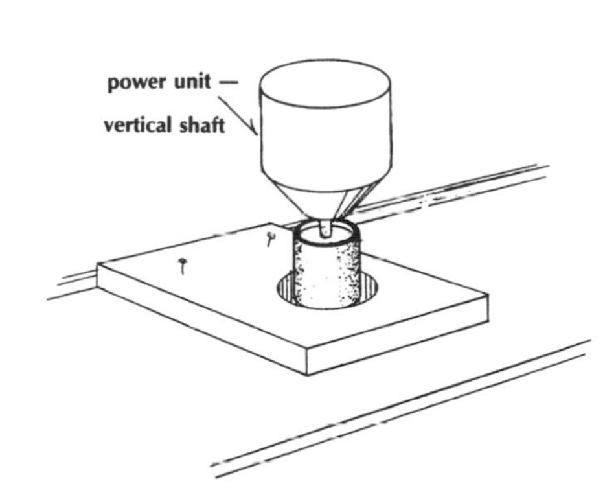

Figure 4-66. *Drum sander.*

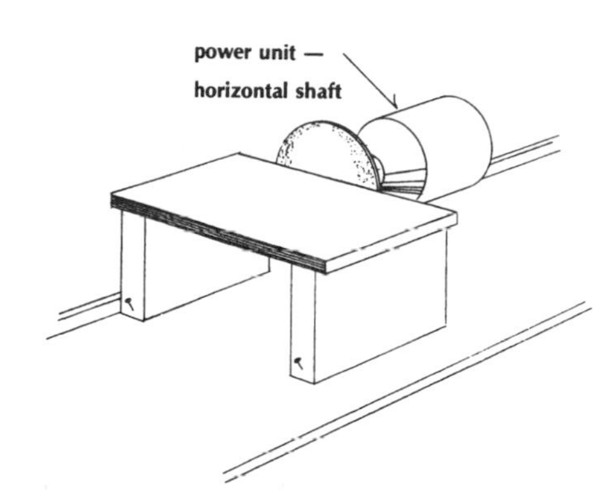

Figure 4-67. *Disc sander.*

The horizontal and angled positions possible on the radial make these attachments even more versatile. This applies to molding heads, shaper cutters, router bits, and so on. A sanding drum does a beautiful job on both inside and outside curved pieces, but apply no pressure and keep pieces moving. It works best if it turns inside a cutout in a block, as shown in Figure 4-66. This guarantees that the edges will be square. An 8- to 10-inch sanding disc is set up easily on the arbor in its normal position. Build a little table to clamp up close to the sanding surface, as shown in Figure 4-67.

There are many occasions when holes must be bored edgewise through a wide piece, perhaps a cabin trunk side. Make a horizontal boring machine out of your radial-arm saw by adapting a ½-inch chuck to the arbor. Raise the work piece on a table to approximate the height of the arbor center, clamp the piece to the table (which in turn is clamped to the saw table), and feed the drill into the work just as you feed the blade in a crosscut. I guarantee that a bit, especially a ship's auger, will bore on dead center when set up this way. Don't forget, however, to pull the bit back frequently so chips don't bind it. To do this, simply slide the power unit back and forth on the radial arm.

Need I say more about the radial arm's versatility? The only problem is the cost of the machine and the accessories. Saws for the home shop run in the vicinity of $380 to $600. I have seen numerous radial-arm saws on the used-tool market at great prices, but you should take an expert with you to check out a used machine. If you see one with a steel table and two or three accessories—dado sets, molding heads, perhaps a small assortment of blades—you are looking at an additional $75 to $100 in value, new. So anything under $200 that checks out is a buy! Among the names to look for used are Sears Craftsman, DeWalt, Rockwell, and Porter Cable.

Figure 4-68. *Ryobi $12\frac{5}{16}$-inch by 6-inch planer provides easy access to motor carbon brushes through the cap in the motor housing. Automatic feed is by front and rear rollers at 26 feet per second. (Courtesy Tools for Less, Nipomo, CA)*

PLANERS

Thickness Planers

Perhaps the first of the costlier tools I would hope to acquire for my boatshop or cabinet shop would be a 12-inch-plus planer, such as the Ryobi in Figure 4-68, or the quite comparable Craftsman. Much of the trim in yacht and boat joinery should be in the range of $\frac{3}{8}$-inch to $\frac{1}{2}$-inch thick, and for some cabinetry (without restraints), these small planers are one answer. For spar work, one with long extension tables would be indispensable. At a cost of $350 (including built-in 2-h.p., 8,000-r.p.m. motor), the Ryobi features a rapid set blade system that allows the blades to be shifted slightly so the nicks no longer leave a raised line. It weighs only 62 pounds, so it can be carried to the job site.

Open-Side Planers

In *Practical Yacht Joinery* I wrote, "If I had the money, I would make this molder-planer my next purchase, even though the need in my small cabinet shop does not compare with its many uses in a boatshop." The 1992 Sears model for home shops and other light duty is capable of taking $\frac{1}{8}$ inch off 6-inch stock, with its 2-h.p. built-in motor and power feed. It planes to 12 inches wide by reversing direction—two passes. There are 17 molding cutters available, including five picture-frame knives, in addition to the planer knife sets; all are double-bit sets. The complete planer (less molding bits) is $360.

The leader in this field for more than a half-century is Williams & Hussey Machine Corporation of Milford, New Hampshire.

Space restrictions do not permit more detailed information on all of these wonderful power tools. But believe me, there is much more to be learned. It will be up to you to pick the brains of any experts you know or to spend some time at the library.

5 Tools, Jigs, and Accessories You Can Build

Many woodworkers derive immense satisfaction from owning and using costly top-quality tools and equipment. But I never could really afford such luxuries, and you do not *have* to use only the *best* of everything to turn out super workmanship. As for longevity, most well-known tools will last your lifetime—unless you are in high-speed volume production. Only the metal parts subject to wear are likely to deteriorate over many years, and most can be replaced. The heavy bases, of course, go on and on.

So here is a suggestion for you who must stay within a budget: Get acquainted with woodworking kits manufactured and sold direct by Gilliom Manufacturing, Inc. of St. Charles, Missouri. This company has been providing the metal parts for a full line of woodworking power tools for more than forty years. All you supply are the plywood, a bit of hardwood, and a few hours of labor.

A GREAT 10-INCH TABLE SAW

Let me describe a Gilliom 10-inch table saw (Figure 5-1). The kit includes a sealed ball-bearing ⅝-inch arbor, and manual arbor-tilting system for angles up to 45 degrees. The belt-driven motor and arbor are mounted (by you) on opposite ends of a husky oak base adjusted easily by handwheel. The oak rip fence slides on two aluminum channel extrusions bolted to the front and back of the table; the oak rip-fence slides on these; there is no micrometer gearing, just a simple clamp on both ends of the fence. Fence alignment is adjustable and once set, it is reliable. Spacing the fence may be done accurately by a light tap or two. As described in "Table Saws," Figure 4-42, always measure from a tooth to the fence, to check.

I mounted extra-long fence rails to permit ripping 24 inches wide, or more, so I can cut to the center of a 48-inch wide sheet of plywood (and see in Figures 5-1 and 5-5 my 14-inch extensions for handling big stuff).

Top and extensions are covered with plastic laminate—saved from scraps of counter tops—as it wears well and minimizes friction.

Gilliom provides patterns with plans, a great time-saver, so you can trace the rather intricate plywood and oak parts. The design is for a totally enclosed, floor-model stand that holds bushels of sawdust removable through a large side door. I also increased the original top dimensions of 29 inches long by 27 inches wide by three or four inches each way. The result is a *big* saw, capable of sawing through 3½ inches (with maximum 1 h.p. motor); I found a ¾ h.p., 3,450 r.p.m. motor, based on the *necessary* long-profile frame, at a swap meet. Gilliom offers several motors for this, and other types of tools, at competitive prices. The kit, less motor, includes saw, guard assembly, step-by-step plans, and costs $103. Add about $140 for a ¾ h.p. motor or check the used market.

I used my saw every day for nine years for custom cabinet work of furniture quality, including dadoing and molding. Completing required a sheet of ¾-inch weather-resistant plywood and a bit of 2-inch oak and glue. All hardware and fastenings are in the package. I had another 10-inch saw with a good cast iron base, table, etc., but the Gilliom saw turned out large parts within 1/64 inch and precise miters. Later in this chapter, after more about kit-built tools, I'll show you how to build a sliding auxiliary table (SLAT) for this or any table saw.

Above: Figure 5-1. *Built from a Gilliom kit, this 10-inch saw has produced precision results in the author's cabinet shop for years. He considers it the bargain of a lifetime.* **Right: Figure 5-2.** *An 18-inch home-built bandsaw that performs like $700 saws. Assembly time was about 10 hours.*

TWO BUY-N-BUILD BANDSAWS

In 1968, when I bought the Gilliom table saw for my use, the company whose boat production I managed ordered an 18-inch bandsaw kit. In 1969, top-level shenanigans closed the business and I was able to acquire the unopened package for a few whistled bars. All metal parts were there, from cast aluminum balanced wheels and rubber tires, and guides with ball-bearing wheels to aligning mechanism, tilting table quadrant, and all the bolts, screws, and wing nuts needed. I bought one sheet of ¾-inch plywood and about six board feet of oak. Add two days to fabricate and assemble the plywood case with oak spine, doors, etc. (Figure 5-2).

This machine did everything that any 18-inch medium duty saw could be expected to do, and with ease, precision, and freedom from vibration. Today it is priced at $240, plus about $137 for an adequate ½-h.p., 3,450-r.p.m. motor. Since then I have built two 12-inch models and another 18-incher, later selling them all profitably.

The 12-inch bandsaw is excellent for a home shop. It will admit 6 inches under the guide and does a fine job of resawing 4-inch mahogany without so much as slowing down the ½-h.p. appliance motor. The kit calls for a sheet of ¾-inch plywood and a couple of board feet of oak. I made no "improvements" other than plastic laminate on the table. In 1992, the price was $165 including plans, patterns, all hardware, blade, etc. The ½-h.p. 1,725-r.p.m. motor is extra at $137. Mine cost $3. I refuse to make a direct comparison to today's several Craftsman bandsaws, as most have direct-drive motors and some handy features, and cost from just under $300 to about $400. Of course, you lose the fun of construction and the $100 to $200 difference is an item, but less so than in the 18-inch version where there is nothing close in cost.

To build these machines and other Gilliom woodworking tools to follow, you need a serious-type sabersaw ("jigsaw") and a power handsaw or access to a table saw. Suggestions: Mount the sabersaw upside down in a simple stand a foot high with top about 18 inches square, so you can guide parts with two hands and follow a line with precision; an even simpler dodge is to clamp the sabersaw in a vise for a quick cut, but do not crush it.

Figure 5-3A. *This shaper, from a Gilliom kit, was built in less than a day. It turns out a precision mold frame in under two minutes. The same part takes two hours to make by hand.*

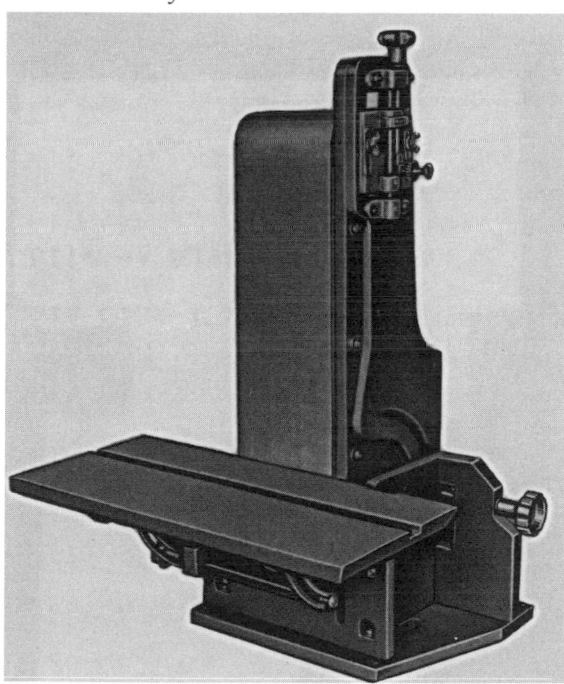

Figure 5-3B. *High-speed bench belt sander works in upright or horizontal position, with 108-square-inch sanding area.*

BUILD AN EFFICIENT FLOOR MODEL WOOD SHAPER

Gilliom's ½-inch wood shaper is easier to build than any saw. I put it together in about five hours (Figure 5-3(A)). During a week's period I shaped nearly 1,000 feet of moldings and a dozen sets of hull jig templates and a ¾-inch plywood jig frame set. My rusty ½ h.p., 3,450 r.p.m. pump motor shaped ash as fast as I could push it. The machine had a split fence for jointing and guide pins for shaping curved pieces or patterned parts against a collar. The table is a generous 18½ by 28 inches, in which the groove for the miter gauge is cleverly constructed. The kit includes templates for precise location of mandrel, all metal parts and fastenings, etc. Kit (without motor) costs less than $100 in 1992. Incidentally, it would be great for production of shaped pieces such as toys, cutting boards, and other forms.

A HEFTY 6-INCH BENCH SANDER

A good step up from the common cast iron 4-inch belt sander, this Gil-Bilt could be the most-used tool in your shop. Even with one smaller, I found that I rarely needed my jointer, going directly from the saw to sander. The belt speed of this sander is such that it can sand more than a mile's worth of wood every three minutes, in horizontal or vertical position. The flat sanding area is 108 square inches; the table is 9½ by 20 inches (or larger if you prefer), and tilts up to 45 degrees on two aluminum segments (Figure 5-3B). If you can build a box with some curlicues here and there, with care, mainly plywood or particle board, it's yours in a day or so. Metal parts kit is under $100, but the ½-h.p., 1,725-r.p.m. motor is about $135, so shop around.

SLIDING AUXILIARY TABLE—SLAT

Whatever the size of your table saw, you must provide extensions on each side, or even two of them on the right side. And then you must have a sliding table that works on top of the original table. This is essentially a much larger piece of ¼-inch plywood with a hardwood batten glued to its underside and fitted accurately into one of the miter gauge grooves (Figure 5-4). A sturdy fence on the side toward you locates everything right up

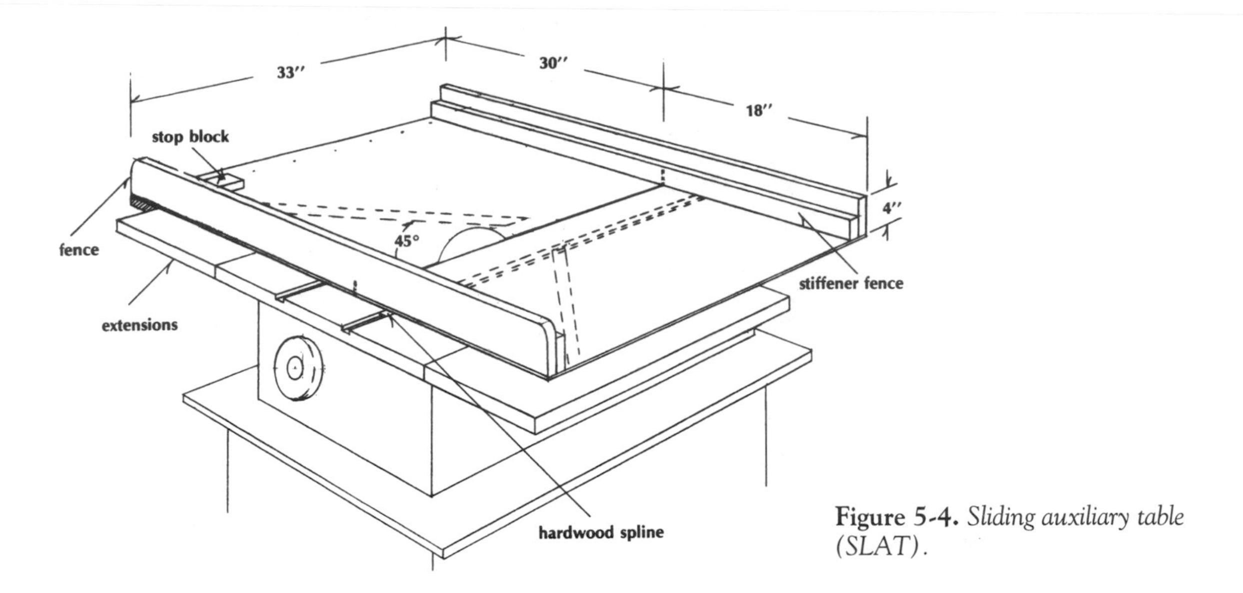

Figure 5-4. *Sliding auxiliary table (SLAT).*

to the stationary blade before the cut is begun. If the fence is square to the groove and blade, everything you cut will be right on the money!

Here's an example. Let's say you are about to fit a counter or dresser top 72 1/16 inches long between two partial bulkheads (walls). This piece of 3/4-inch mahogany plywood can be sawed quite square by clamping your 90-degree saw guide to it or by ripping on your table saw. The miter gauge is unreliable. Unfortunately, if the piece comes out 1/32 or 1/16 inch under the fore-and-aft dimension required, you can't stretch it to achieve the gentle press fit it needs. And if it comes out 1/32 inch over, it will take precious time to plane off, try, and fit. If you have a sliding auxiliary table (let's call it SLAT from here on), however, you can line the piece up to the tooth of the saw blade while it is stationary and actually split the pencil or knife line. You can even make a tentative saw cut; 1/16 inch deep is enough to show where the final kerf will be.

Here's another example: You want a number of 45-degree miters. Brad two pieces of 1 by 2 to the SLAT, but don't allow the points to come through (Figure 5-4). Experiment with two pieces of scrap until you get a perfect 90-degree joint by shifting the little fences, then add another couple of brads. Save these fences so the brads can be pressed into their respective holes for future use; mark around the fences so they can be located rapidly. If you use a cabinetmaker's combination or a multitooth carbide blade, you will get beautiful miters.

A third SLAT example is the technique for making repeated cuts to length. Tack a stop block to the fence or surface of the SLAT. Now all cut pieces will be identical. For producing pieces longer than the width of the SLAT, use a 1 by 4 with a stop block glued to one end. Clamp this bar securely to the front fence. Now it takes just seconds to make precise duplicates of any number. Here's a tip: Mark with a knife for accuracy and the cleanest possible cut when sawing across the grain.

How to Build a SLAT

The drawing in Figure 5-4 shows the general dimensions taken from my own sliding auxiliary table. The SLAT extends slightly beyond the table extensions. Make it not less than 48 inches wide. The depth should be 33 inches or more, as this enables you to saw material as wide as 24 inches (remember the blade width). The fences should be about 4 inches high, and of hardwood, fir, or 3/4-inch plywood. The drawing shows additional pieces against each fence for stiffening and to absorb the saw cuts.

Here's the assembly procedure. Make a straight hardwood spline or batten that will fit snugly but slide easily in one of the grooves in the saw table. The sliding panel itself may be 1/4- or 3/8-inch ply-

Figure 5-5. *The sliding auxiliary table (SLAT) set up especially for mitering.*

wood. A greater thickness reduces the cutting depth of the blade. Now lower the blade below the table top. C-clamp the 33-by-48-inch panel to the saw table so it is flush with the front edge. This assumes that the front edge is square to the blade (check it, to be sure). Tack five or six brads through the panel into the spline in the groove. The assembly should now slide back and forth easily with *no* side movement. If there is side movement, make a new spline. Now pull the batten off, spread white glue on its top surface, find the same holes for the brads, and drive them through and clench over. Check the sliding action. If the batten binds in the groove, check it with a straightedge and correct with a gentle tap here or there until the action is perfect. Add three or four additional brads and allow the glue to set.

The rear fence can be glued and nailed permanently in place and the stiffener fence screwed to it. Don't glue the stiffener, since it is to be expendable. Tack the front fence temporarily with brads, aligning it with the table's front edge and the plywood panel. Place the panel on the table, with the batten in its groove, located so the blade will appear forward of the rear fence. Now turn on the saw and slowly raise the blade. Push the panel back so you have a kerf or slot about a foot long. Shut the saw down. Check the slot with a carpenter's square against the front fence. If the fence does not line the square precisely along the slot, tap the fence until it does. You may now cut the kerf from where it touches the front fence to inches from the back fence.

Test with a piece of plywood scrap about 18 inches square. Hold it securely against the front fence, positioned so you are about to slice off a narrow piece. Turn on the saw and slide the SLAT back until the blade barely cuts into the front fence, then turn off. If this test does not produce a perfectly square cut, your fence is bent or out of line. Give it a few taps and try again.

Be wary of the blade cutting through the front fence if the work piece is thick. Just replace the stiffener from time to time. A dado set will cut a wide slot in the panel. This is not serious, but you can make a separate SLAT for dadoing for shelves, for example. I have an additional small SLAT for mitering moldings. I use this rather than set up two tack-on fences (Figure 5-5).

To shave an edge, say, by 1/64 or 1/32 inch, place it firmly against the saw teeth and tap it slightly against the blade. This will cause a very small deflection, just about the amount you want to shave off. Back up and start the saw. You'll see it take off a mere sliver. Short lengths may be ripped if one end is square, but you can achieve precision by using a square to line the piece up against (Figure 5-6).

The SLAT described above is a simple device. I have seen sliding tables that look like furniture, with brass corner reinforcements, handles, and so on. These are passed down through generations. When

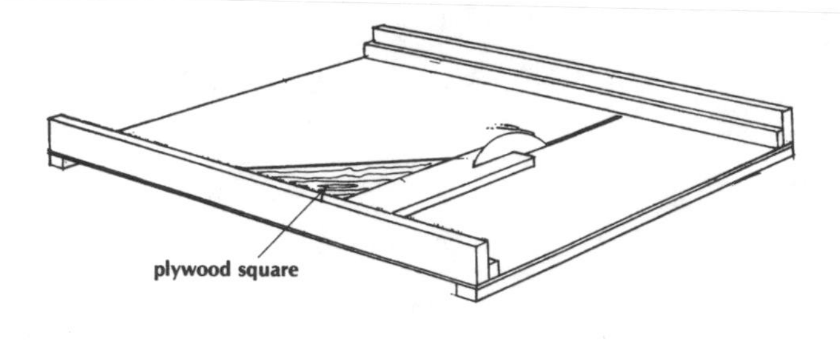

Figure 5-6. *Small ripping job on SLAT.*

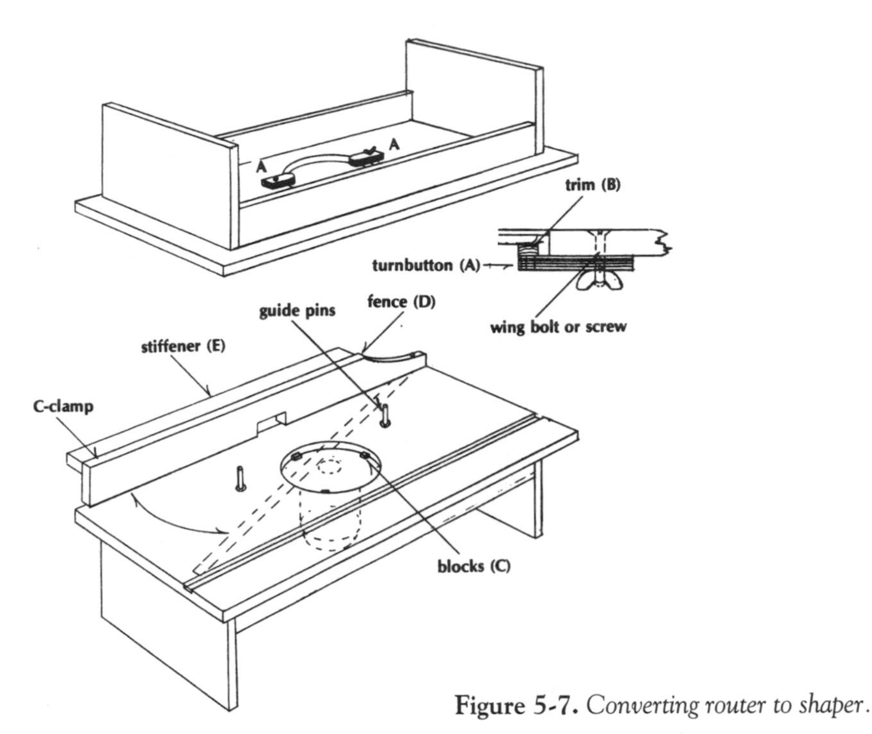

Figure 5-7. *Converting router to shaper.*

you discover how valuable this accessory is, you may decide to use mahogany or teak in your next one.

A ROUTER-SHAPER

In Chapter Three I described a router as a shaper turned upside down. Let's build a table with a cutout to receive the base plate of your router with the collet pointed up (Figure 5-7). The ¾-inch table is approximately 16 by 30 inches. The legs should be 10 to 12 inches long in order to be comfortable when in use on a bench. The rails to stiffen the top can be pine 1 by 3s. It may be nailed together if well glued. The cutout in the center should be a nice fit for the plastic base of the router.

You will probably find it convenient to remove the handles of the router before mounting it. Otherwise, it will have to be passed up through the hole from beneath. The removable turnbuttons (A) may be held by a stove bolt and wing nut or screw. Insert the router in the opening. If it is too low, glue a tiny shim (B) on the tip of each turnbutton. Trim this until the router lies flush with the table top.

Most shaper operations require a fence or other stop to guide the work piece. Figure 5-7 shows a simple one-piece fence (D) that can be adjusted by pivoting at one end. Clamped to the table, the 1 by 2 stiffener (E) keeps the fence straight. The fence is adjusted by trial and error on a straight piece of scrap. The single fence shown is good only

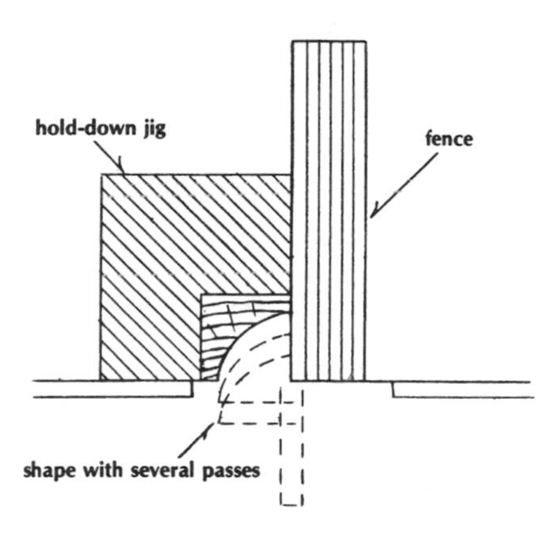

Figure 5-8. *Holding jig.*

for corner rounding, rabbets, and other shapes or straight edges where the piece is not reduced by the shaper (in contrast to jointing).

It is dangerous to run light stock in this or any shaper without good hold-downs. Moreover, you would get chatter and a poor surface. I prefer to rout the edge of a larger piece and then rip off the molding, but you may want to saw out a holding jig (Figure 5-8). Don't try to shape a molding in one pass. Raise the bit in two or three stages as shown in Figure 5-8. You can use a feather or comb hold-down, but neither of these will support the molding at the bitter end. Always try to run through as much molding on the first past as you can foresee a need for. Then run all of it through on the second pass, and so on.

Solid router bits for radii (rounding) have pilots (extensions) that are designed to ride against the edge of the piece or pattern (Figure 5-9). A pilot eliminates the need for a fence, but there is a chance you will burn the edge of the work piece.

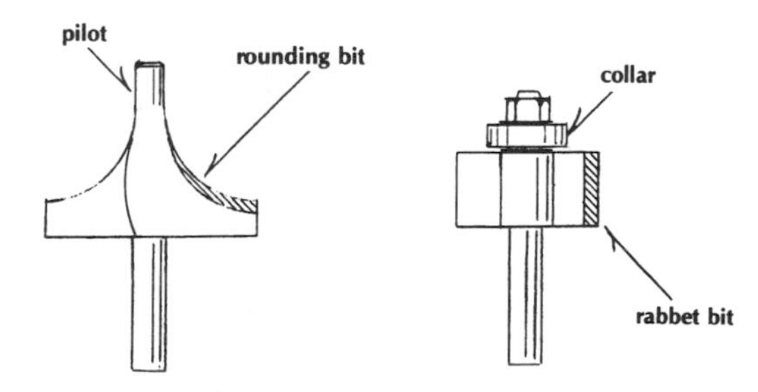

Figure 5-9. *Pilot follows straight or curved edge.*

Figure 5-10. *Collar prevents burning.*

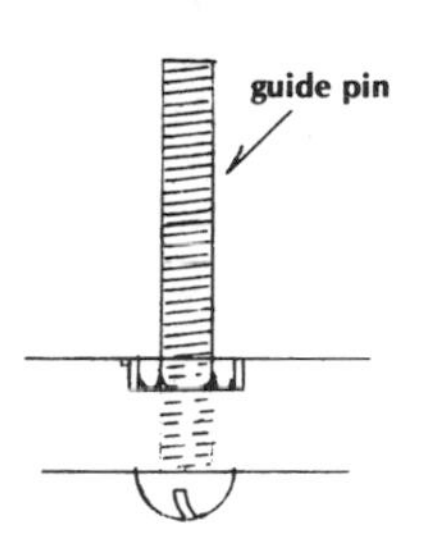

Figure 5-11. *Shaper guide pin.*

Two-piece bits mounted on a shaft, spindle, or arbor may be set up with a ball-bearing collar, thus reducing the burning problem. The collar may be used with rabbet bits and other shapes also, so try to acquire this type (Figure 5-10). Get carbide bits; they cut faster and more smoothly and do not need sharpening. Gilliom sells the castings for a good split fence. Such a fence permits using the shaper for jointing.

To get the most out of your shaper, there must be provision for guide pins (Figure 5-11). Let's say you want to shape a molded edge on a round or curved piece. A collar on the spindle is needed to carry the squared edge of the piece, but it is dangerous to simply push a work piece into the rotating bit without some means of control. Otherwise, the bit can chew into the wood and throw the piece across the shop. To prevent this, pins are inserted into the table 4 inches from the center of the spindle (Figure 5-7). The work is placed against the pin and slowly fed into the bit until it rides against the collar. Then you start moving the piece against the counterclockwise rotation and keep it going so it doesn't burn. You keep a firm grip on the work piece and move it as steadily as possible. Any imperfections in the original edge of the work piece will be reproduced in the molded edge.

The router's built-in switch may be awkward to reach. For safety, you might have to lock it "on" and put a line switch in the cord.

You may wish to add a groove for a miter gauge, a real necessity for molding the ends of narrow pieces such as drawer fronts. You can dado the groove or saw it out by making repeated cuts with your power saw, being very careful with the positioning of the guide piece so there is no slop. You can also use a router and a ¾-inch straight bit, but be sure the miter gauge fits the trial groove you cut in a piece of scrap. A third way is to cover the top of the table with two pieces of ⅜-inch tempered hardboard spaced correctly by using the miter gauge itself as a spacer. Brad and glue the cover on before any bolt holes are drilled.

AN INEXPENSIVE BENCH DISC SANDER

I designed and built the inexpensive and totally efficient disc sander shown in Figure 5-12. Anyone with basic woodworking skills can do it. The sander's cost without motor should be under $15. You can pick up a junked appliance motor for $4 or $5.

For my sander, I found a rusty motor whose base was completely eroded. It was rated at ½ h.p. and turned 3,450 r.p.m., more than enough for an 8-inch Sears sanding disc, fitted to the shaft on an arbor extension. The only disadvantage to this machine is that 3,450 r.p.m. can eat up any piece you touch to it unless you are cautious. An old appliance motor of ½ h.p. and 1,725 r.p.m. would be more than adequate for light duty. The obvious uses for a disc sander are rounding corners and convex edges accurately, sanding and squaring sawed ends and so on.

To dress ends square, place any work piece firmly against a miter gauge with the end just kissing the moving disc. Then slide the piece rapidly across the disc. If your saw cut was square, this will give the end grain a final finish. If it was out of square, a couple of strokes should do the job. Don't let the work piece shift on the miter gauge.

To round corners, mark the desired radius with a

Figure 5-12. *Constructed in hours, this inexpensive (about $10) disc sander does the work of machines costing hundreds of dollars.*

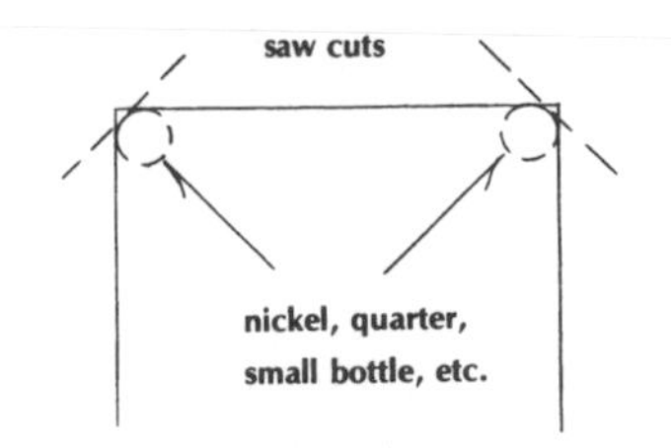

Figure 5-13. *Rounding a corner.*

dime, penny, or whatever. If a larger radius is desired, it would be best first to hack off most of the material with a bandsaw, table saw, or chisel (Figure 5-13). The disc can do that job, but you would be heating it up and filling the sanding surface unduly. Practice with this tool will enable you to take off just a smidgen and leave a beautiful surface.

How to Build a Disc Sander

Now you can construct your bench sander, just a firm support for a motor with a metal disc on its shaft against which a worktable is mounted. That's all there is to it (Figure 5-14). The tilting table is a refinement you may add if you like.

The base can be ½- to ¾-inch fir plywood. On this the side pieces are fastened with glue and nails. Leave 14 inches between the sides to receive the motor mounting base. Locate this shelf at a height that places the motor shaft just below the top of the table. Glue blocks under the shelf to support the weight of a motor.

Make your side pieces as the drawing shows. The tilt arrangement can be very handy for yacht joinery, so let's do it. Saw out the two tilt-table bearers (quadrants) to the 6-inch radius. Also mark the 5½-inch radius for the pinholes. When the two bearers are cut out, tack or clamp them together and plane off the top edges straight and square. Lay out the location of the holes every five degrees with a protractor and mark each location with a punch. Drill ¼-inch holes along the radius and another at the radius center for a pivot.

Next, make the table of ¾-inch plywood. If you want a size that will handle quite long stuff, make it 12 by 18 inches. If you want the convenience of an occasional use of a miter gauge, cut the groove with a dado set or with a router. Try it on scrap wood until you get a perfect sliding action. If the groove is wide enough to admit the gauge but too tight to allow easy movement, wrap a piece of sand-

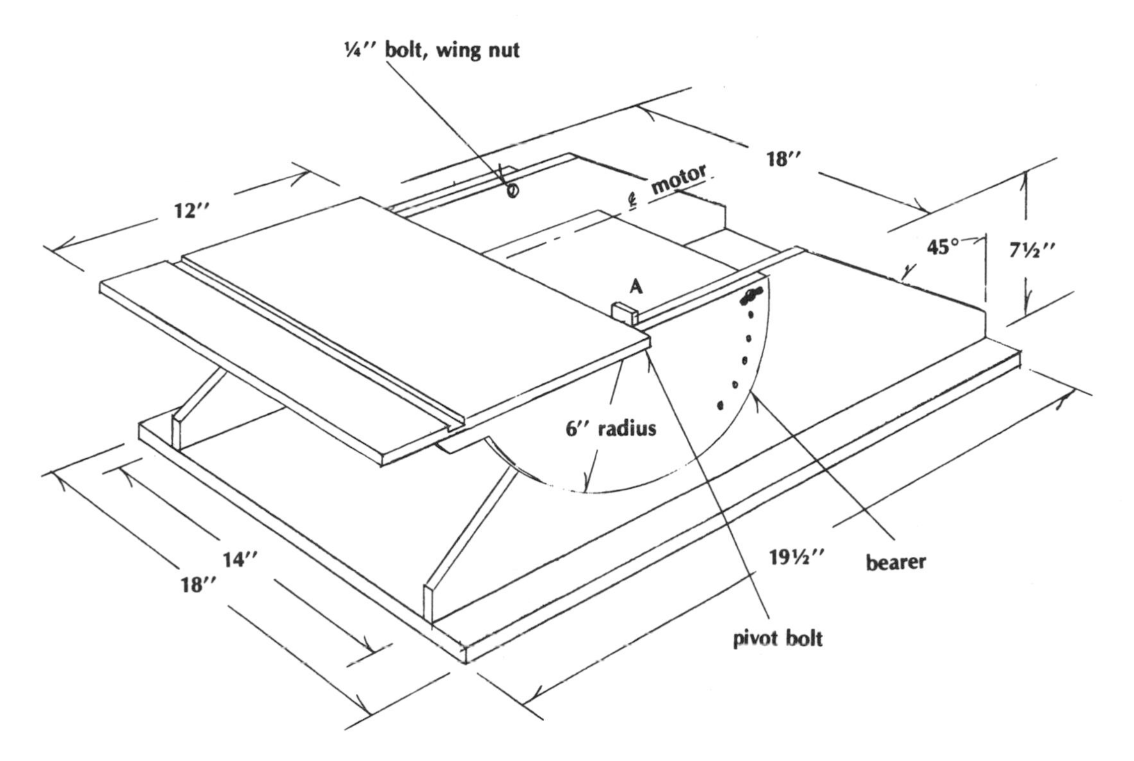

Figure 5-14. *Build a disc sander for peanuts.*

paper around a piece of plywood and sand the sides of the groove until it's right.

Clamp the table bearers or quadrants on the outside of the side pieces flush with the top surfaces. Drill a ¼-inch hole for the pivot pin or bolt on each side. Lay the table on the quadrants and square it up, then clamp it securely so it can be nailed and glued while everything is in the proper position. You'll need another ¼-inch hole at the top end of each quadrant. Put bolts with wing nuts in these and in the pivot bolt holes. Let the rig stand while the glue sets.

A SIMPLE MOTOR MOUNT

Does your motor have a base? If it is an appliance motor or a cruddy old junker like mine, you'll probably have to make a base to keep the motor from crawling around. Refer to Figure 5-15. Cut two pieces of 1 by 3 about 5 inches longer than the diameter of the motor case. Bandsaw or sabersaw to fit the motor. Mark the other piece from the first one, and fit. Then fasten both pieces to a base of ¾-inch plywood. Cut two pieces of perforated strapping long enough to reach over the motor to the ends of the base pieces. Screw these strips down securely at one end, bend the other ends over the motor; fasten the loose ends with 1½-inch roundhead screws and washers at an angle that will tighten the strips as the screws are driven. This motor base will last indefinitely.

Place this assembly on the motor support shelf with the disc mounted on the motor shaft. Bring it right up to the table. Cut a notch in the table to clear the nut holding the disc. The disc should clear the table by about $\frac{1}{16}$ inch. Check the distance from each end of the disc to the groove in the table. Put a line switch in the motor cord. That's all there is to this inexpensive sander.

A MITERING SLIDE

Moldings for doors, drawers, and other trim require clean, accurate miters. It would be well worth your time to construct a mitering slide, a smaller variation of the sliding auxiliary table (SLAT); see

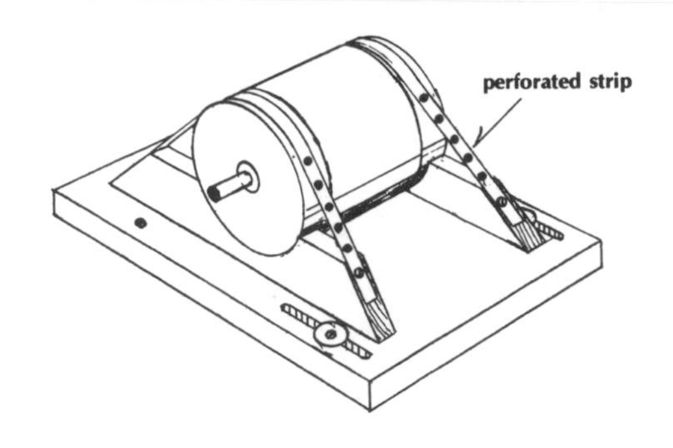

Figure 5-15. *A 50-cent motor mount.*

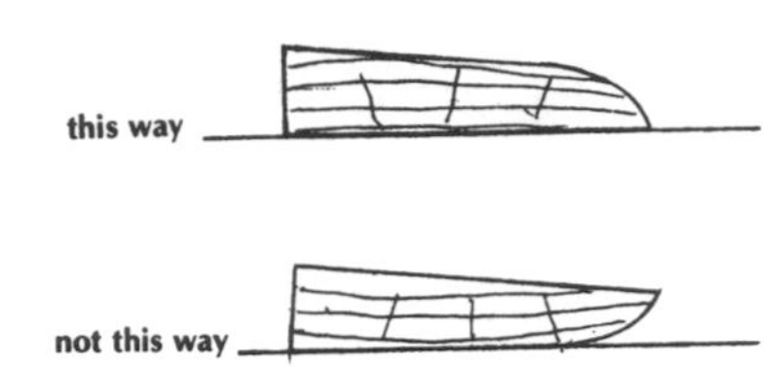

Figure 5-17. *Mitering moldings.*

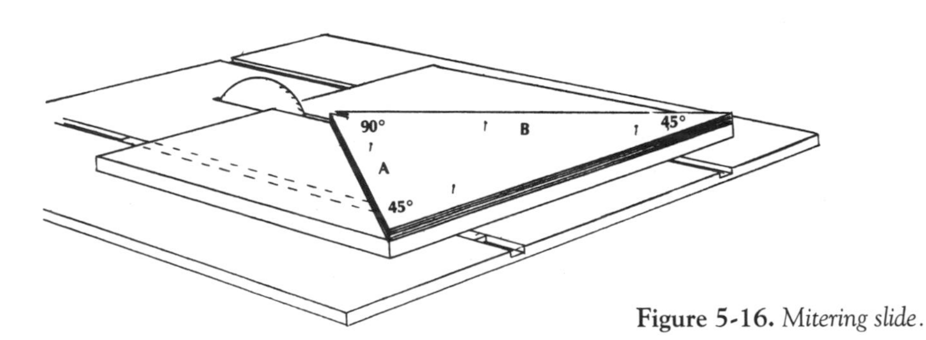

Figure 5-16. *Mitering slide.*

Figure 5-16. Saw a 90-degree corner off a plywood panel, 20 inches wide by 16 inches deep on ⅜- or ¾-inch plywood should be adequate. Precise measurements on legs A and B make a perfect 45 degrees. This must be planed carefully on the sawed side so that when it is bradded to the slider, it makes a perfect 45-degree angle to the slot. To test this, cut two 10- to 12-inch pieces of 1 by 2 scrap on the right side. Try not to let the blade cut into the triangle any more than is necessary. These two pieces should make a perfect 90 degrees when held together inside a carpenter's square. Now cut a piece on *each* side of the triangle and try these in a square. If these tests do not check out, take a shaving off the base of the triangle and check again. Keep trying until you achieve a perfect joint. Then set the brads in, but don't let them touch the table top.

The great advantage of the mitering slide is that with the blade stopped, you can position the mark right up to the tooth, leaving perhaps 1/32 inch over until you get all parts together, then come back and shave off that 1/32 inch, if needed. If you are working with odd-shaped molding like the piece shown in Figure 5-17, don't miter with the molded side down; this throws off the angle. If you are mitering deeper moldings on edge, add enough height to the A and B sides so that the pieces cannot slant or wobble. If the moldings run wider than 2 inches, you may have to use a miter box and backsaw. Or you can set the table saw blade to 45 degrees and use the miter gauge. There are always three or four ways to cut mitered joints.

SQUARE AND RIPPING GUIDES

From the complexity of the SLAT and mitering slides, we turn now to the basic square and ripping guides. The square guide is simply a triangle cut off the corner of a sheet of ¼-inch plywood with a batten glued across the base on both sides (Figure 5-18). You press this against the sheet of plywood or other material to be cut, and clamp or brad it in position. The power saw is then run along the edge of the square. I have a large one measuring 48 by 24 inches, and a small one about 14 by 10 inches for smaller stuff when I am away from my table saw.

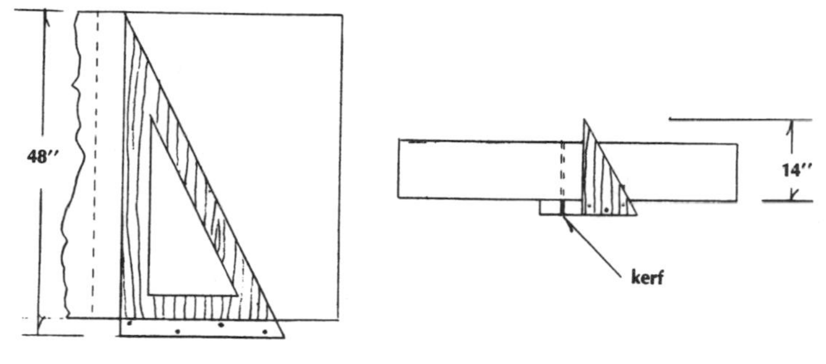

Figure 5-18. *Cut-off squares.*

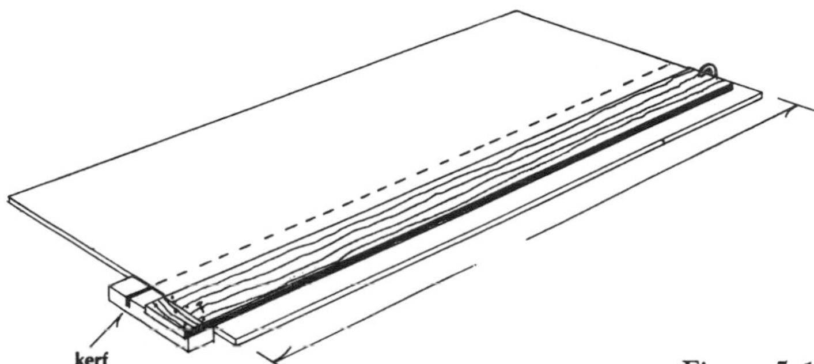

Figure 5-19. *Panel ripping guide.*

Let the end of the 2 by 2 cleat on the base be kerfed by the power saw. This provides a fast way to line up to the mark.

The ripping guide is a straightedge of ¾-inch plywood 5 to 7 inches wide and 8 feet long. Nail a piece of 2 by 4 on the underside of the near end, so it shows you where the kerf will be (Figure 5-19). To line up, measure from the saw tooth to the right-hand edge of the saw's sole plate, add that to the dimension you need, and clamp your guide in place. Repeat on the other end, using the measurement from the guide to the edge of the sheet. Double-check and adjust by light taps. A ripping guide is a great help for cutting large sheets of plywood to sizes more convenient to handle on a table saw. Sometimes I rip slightly oversize, and then rip on the table saw with a fine carbide blade. A cut-off blade will rip or crosscut ¾-inch plywood and give you a surface requiring very little dressing with a plane. A fine-toothed plywood blade will overheat and warp on anything over ½ inch, but it will produce a slightly better surface. The cabinetmaker's hollow-ground blade may heat up and bind on ¾-inch plywood, and only do short cuts on anything over ⅜ inch, but the edge is beautiful.

My apologies to those of you who find these guides elementary. They are time savers. I have seen experts follow a line on plywood freehand, then saw the piece more accurately on a table saw; this approach is a complete waste of time if the cut can be done right the first time.

BORING JIGS

Not long ago I read an article describing a method of deep hole boring through a trunk cabin side. The writer recommended clamping a batten across the work piece and having a second person visually line up the long bit with the batten, while the borer tried to keep the bit square in the other plane. Now, this might work—or it might not. Trying to hold a drill motor firmly, or, worse yet, a brace while sighting down the bit, is asking for trouble, as well as taking up another man's time. The jig shown in Figure 5-20 works, it takes minutes to construct, and it can be used repeatedly.

If the material to be bored is 1¼ inches thick, blocks B and C should be twice that, or 2½ inches wide. Find the center on one by crossing diagonals, and then punch the center. Stack the blocks and brad them together so you can drill accurately on your drill press. The blocks will then be identical. You can use a ship's auger in the jig either with a

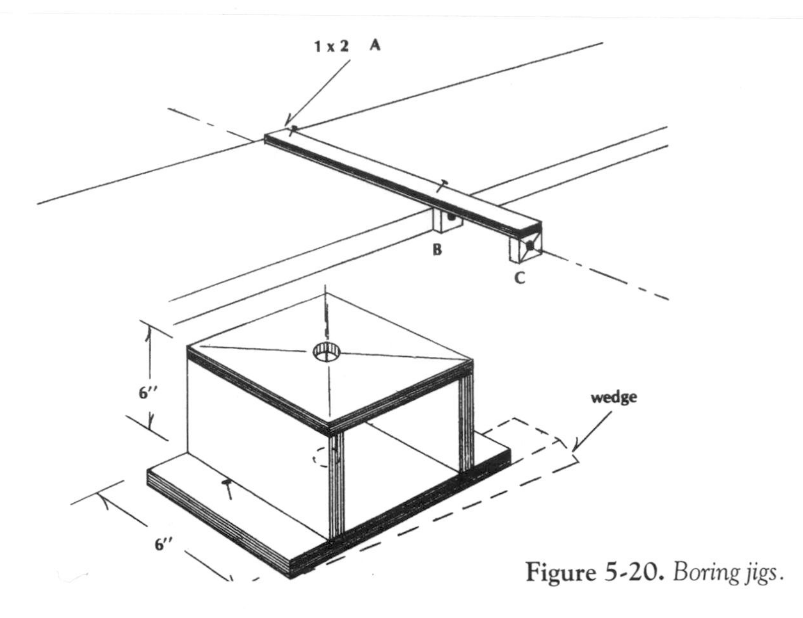

Figure 5-20. *Boring jigs.*

brace or an electric drill. Be sure this bit is backed out frequently to free it of chips. It could twist off if it jams. If your bit has a screw on it, grind this off. These screws sometimes hit hard grain or a knot and are deflected.

The lower jig in Figure 5-20 is for boring at right angles to a surface. There is a good base for clamping to the surface. It could also be used for boring at a specific angle by making two wedges to fit under it.

A JIG FOR ADJUSTABLE SHELVES

Cabinets often contain adjustable shelves. To make such shelves seat precisely, the holes must be spaced with care. Make the jig shown in Figure 5-21 of ¾-inch 1 by 3 oak or other hardwood. Use a scriber or knife to lay out the pattern of holes on 1½-inch centers, 1½ to 2 inches from the edge. Drill on a press so the holes are perpendicular. Attach the spacing leg B so the series of holes starts a uniform distance above the bottom of the cabinet. Clamp the jig in position, drill a hole at each end, and insert a dowel in each hole. This prevents any movement while the rest of the holes are being drilled. If the row of holes is to continue, again locate the jig with a dowel so accuracy is maintained. Once you have this jig, the entire layout of holes for a good-sized cabinet can be bored in 15 or 20 minutes. Cabinet hardware supply houses have metal and molded plastic brackets to fit. They

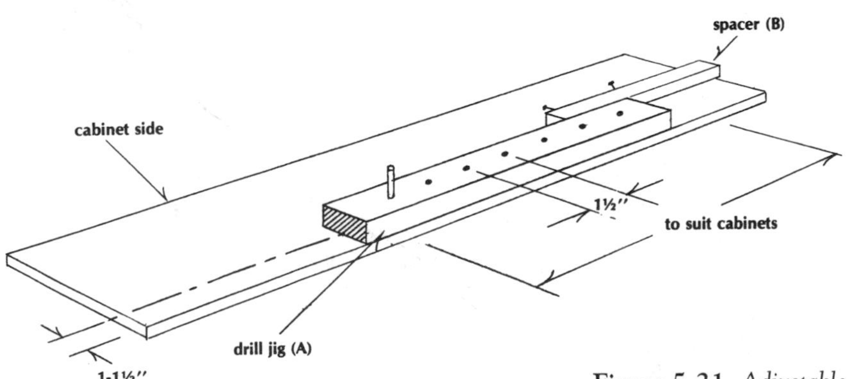

Figure 5-21. *Adjustable shelf drill rig.*

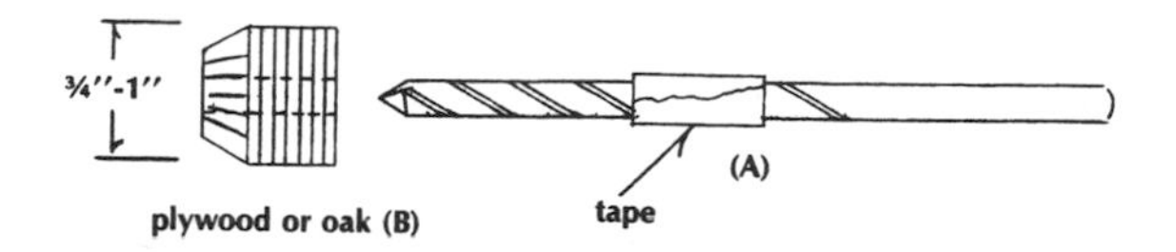

Figure 5-22. *Simple drill depth stops.*

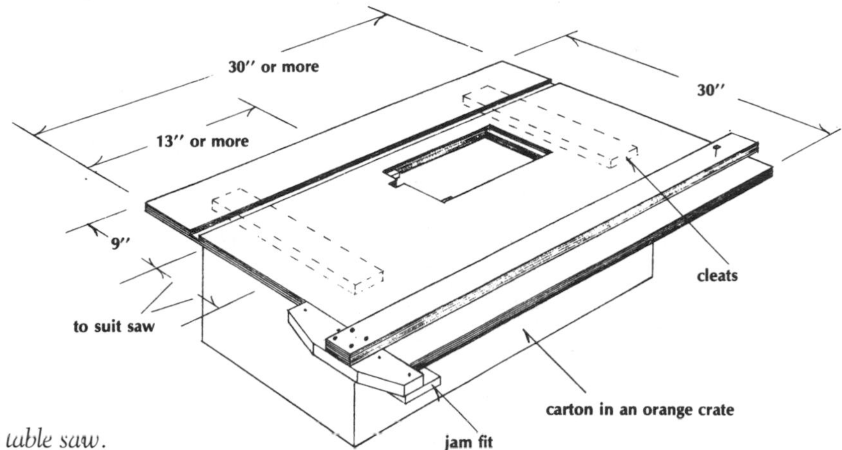

Figure 5-23. *A $5 portable table saw.*

are noncorrosive, of course, but not as romantic as birch dowels with chamfered ends.

Watch out for the possibility that your bit will penetrate the cabinet side. A quick remedy is a piece of masking tape wrapped around the bit to show when you are approaching the correct depth (Figure 5-22A). A better stop is shown in Figure 5-22B. Bore the ¼-inch hole in a piece of ¾-inch oak or plywood, then bandsaw the piece to about 1 inch in diameter. Chamfer and smooth it up on your disc sander. In use, it may creep up on the bit, so watch it! Tape above helps.

HAPPINESS IS A $5 TABLE SAW

Some years ago I needed a table saw that I could hoist up and down ladders, lift in and out of a companionway hatch, and stow in the trunk of my car. Also, I wanted to get it for nothing—or less. Perhaps I reinvented the wheel, as they say, but I had not heard of or seen the contrivance I invented (Figure 5-23). Since that time, I have built several of these tools, with improvements. If you need portability, this is it! You can use surplus materials that cost very little (about $5 in 1992) and you'll find that your saw does about 90 percent of the jobs of those costing $200 and up. (Grooving dadoes is the main exception.)

First, determine the size of the table. There's no law on this, but you will find that the area between you and the blade is important (Figure 5-23). This span provides support for your work piece and also allows a sliding gauge for cutoffs. Width, too, is important. Because you may be handling fairly long pieces, I suggest a minimum of 30 by 30 inches. If you have a larger piece of ¾-inch plywood or kitchen counter remnant lying around, just dress it to accurate squareness instead of reducing it in size.

Power saws have soles of different sizes. I do not recommend a blade size smaller than 7 inches. The cutout that will receive the saw must be accurate. It should be located approximately as shown in Figure 5-23, and its sides *must* be parallel to the sides of the table. If you can borrow a friend's table saw, the cutout will be a cinch. Try this procedure on a piece of scrap first. Place the piece against the fence, spaced correctly, then crank up the running blade slowly until you see it emerge to about ¾ inch above the work. Do not remove the piece until the blade stops turning. Now do the same with your actual table top, being sure to hold the piece down firmly. Next, move the fence to saw the opposite side of the opening and repeat the proce-

Figure 5-24. *Portable table saw top. Note notches and supporting blocks glued and bradded in.*

dure above with no sloppiness. The cuts will be precisely parallel. You can connect the cuts at the ends of the opening with a sabersaw. The sole of the saw should fit snugly, and the blade should come out parallel to the left side of the table.

If you are unable to borrow a table saw, you can make these cuts by clamping a straightedge to the table and lowering your power-saw blade with care just within the lines. This, you may recall, is a "plunge cut." You'll get the best results if your guide piece is quite thick, perhaps a length of 2 by 4.

When your saw fits the opening, you'll need to hold the surface of the sole flush with the table top. If the saw has projections, such as a wing nut for the rip gauge, for example, notch the opening as needed (Figure 5-24). Rip out little pieces of pine to about ¼ by ½ by 1 inch. Glue and brad one in each corner of the opening so that it holds the saw sole flush. This facilitates juggling the saw into the opening, from the underside.

This is the time I admit that the portable table saw has one disadvantage. You can't operate the built-in switch while you saw. Thus, each time you set it up, you must tape, wire, or tie the switch to "on." It's disconcerting to have the saw start howling as soon as you plug it in, and it's dangerous! For safety's sake, install a line switch in an extension cord where it can be reached quickly, or install a reachable switch.

You'll find that the saw's movable guard is a nuisance, especially if you are ripping strips narrow enough to catch inside it. An uncovered blade is dangerous, however, so you must learn to respect—even fear—this machine just as you respect your jointer, bandsaw, or sharp chisel. Fear generates caution. To hold the guard down and keep little cutoff pieces from falling into it and jamming the blade, tack a piece of ¼-inch plywood to the underside of the table so there's just clearance between this piece of scrap and the blade. This holds the guard down and catches any cutoff pieces so they can be removed after the saw is shut off.

Now that your saw is working, how about a stand? A carton inside an orange crate worked well for me. The carton catches the sawdust. Place the crate on a couple of sawhorses to support it at the right height. Cut an aperture for the projecting motor housing. Nail two cleats on the underside of the table to fit inside the crate and keep the thing from creeping around, and lightly nail the top to the crate to prevent tipping.

Figure 5-23 shows a simple rip fence. Fasten the T crosspiece with finish nails until you get it truly in line with the blade and have tried it a few times. Then glue and screw or nail it. Don't overlook the piece glued on the underside of the T. This piece gives you a jam fit that can be augmented with a C-clamp. A clamp at the far end is good insurance, but if you made it out of ¾-inch plywood 3 inches wide, I don't think there will be visible deflection. You can use a manufactured miter gauge if you want to tackle the job of cutting out the groove. This procedure was explained earlier in this chapter for a router-shaper.

This little saw's usefulness is greatly expanded if you equip it with a SLAT (Figure 5-25). However, you don't need a groove in the table if the sides are exactly parallel. Just glue battens to the underside of the SLAT so they slide nicely along both outer edges of the table. There should be no binding, no slop. Glue the fence on after the action is perfected.

A JOINTER HOLD-DOWN

In order to joint edges truly square, the entire length of the work piece must be held securely against the jointer fence. A feather or comb does this job very well and has the advantage of freeing the hands to feed the work (Figure 5-26). The guard must be removed. In most jointers it is impossible to bolt

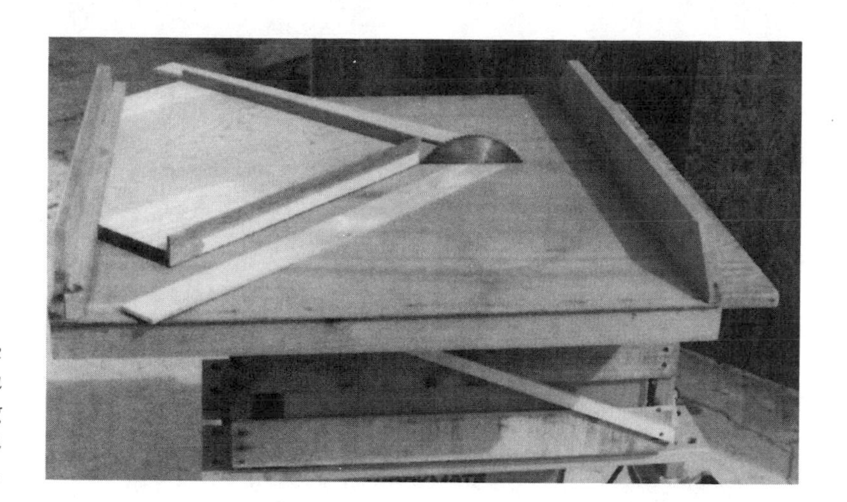

Figure 5-25. *An inexpensive portable table saw based on a hand power saw. The sliding auxiliary table (SLAT) has an additional mitering jig in place.*

through the hole for the guard pivot pin. Bore a 5⁄16- or 3⁄8-inch hole through the table at a point where a wing nut can be set up on either end of a hex bolt and as near the edge as feasible. You may have to tap a threaded hole. The feather may be done on the bandsaw or table saw. The splines should be about 1⁄8 inch wide. Place a couple of large steel washers and several cuts from coarse sandpaper between the piece and the table (this helps prevent movement), and clamp if possible. You will be surprised, however, at how little pressure is needed to hold the work piece properly against the fence and how easy it is to feed the work through.

When jointing the edges of wide boards, avoid wobbling. Most jointer fences are not high enough to prevent movement, so bolt on a 3⁄4-inch plywood extension to increase the height of the fence to 10 or 12 inches. (This was shown in Figure 4-9.) The combined pressures of the feather and your hand, holding the work piece against the upper part of the fence extension, will guarantee accuracy. To joint smaller stuff, bolt an additional hold-down to the tall fence. Now the job is simply to feed the piece through with a pusher. This idea is adaptable to your table saw as well.

To joint small rectangular pieces with safety, make a jig similar to the one shown in Figure 5-8. Use the same jig if the dimensions are correct, or add a shim to the top and/or side of the rabbet if that's faster than making a new jig. There is almost no

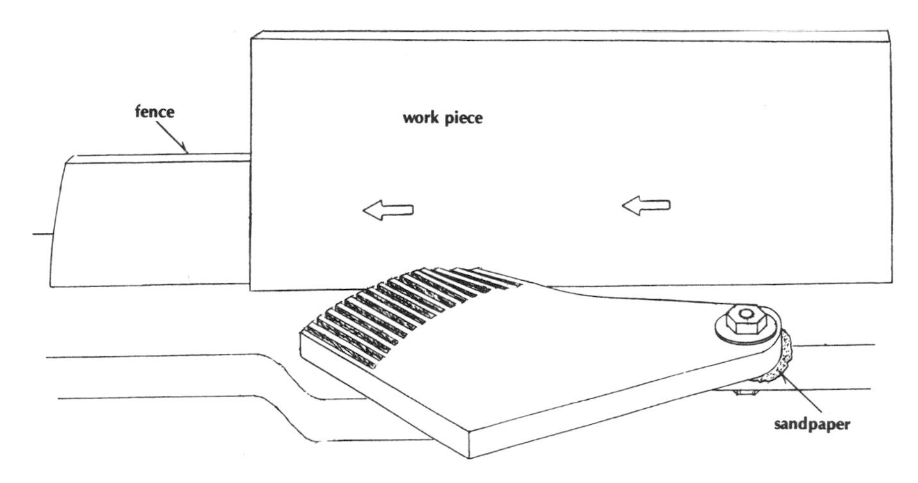

Figure 5-26. *Jointer hold-down replacing guard.*

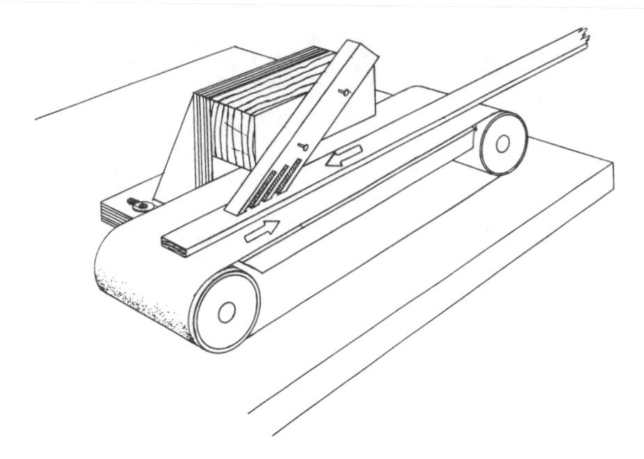

Figure 5-27. *Belt sander hold-down.*

limit to the small sizes that can be jointed, routed, or sawed by this method, except that the tools can tear the pieces apart if you try to cut too fine.

A BELT SANDER HOLD-DOWN

Surfacing on a bench sander, even a light 4-inch type, can produce very satisfying results if uniform pressure is maintained for the entire length of the work piece. This is not always easy, for one hand pulls and the other holds the piece down on the belt. It takes a lot of practice to avoid a fits-and-starts pass, especially if the pieces are more than 2½ to 3 feet in length. To solve this problem, make the device shown in Figure 5-27.

The 2-inch block places the feather near the center of the belt and over the center of the work piece. The whole affair is screwed to the bench. Leave enough clearance under the block so the jig can be left in place when sanding other work. Of course, the feather may be C-clamped to get the right pressure.

A BOX JOINT JIG

A box finger joint is strong, easily made, and has many uses in yacht joinery (Figure 5-28).

Next to a dovetail, it is the strongest for drawer construction if glued with Aerolite, epoxy, or plastic resin (Weldwood) glue. It would be superior to the half lap for hatch coamings and frames. If you were to run a dowel through a box joint, it would be second to none.

To cut a box joint, it is best to use a dado set adjusted to the desired slot width and depth. You could, however, set up two or three identical saw blades with plywood or doorskin shims in between like big washers. On the other hand, I have seen narrow box joints cut with spacing made to match a wide saw kerf, perhaps 3⁄16 or ¼ inch for small boxes. You might try to have extra set put in an old blade if you have many such joints to run.

Refer now to Figure 5-28 (left). Pick a piece of scrap (A) to screw or clamp to the wooden face of your miter gauge. Let's try a dado set adjusted to the material thickness of ¾ inch, and ¾ inch high. Cut the notch (C) in the backing (A). Glue into this notch a block (B) ¾ by ¾ by 3 inches, letting it project 2¼ inches. This is to space repeated cuts and duplicate pieces for an accurate fit. Now shift

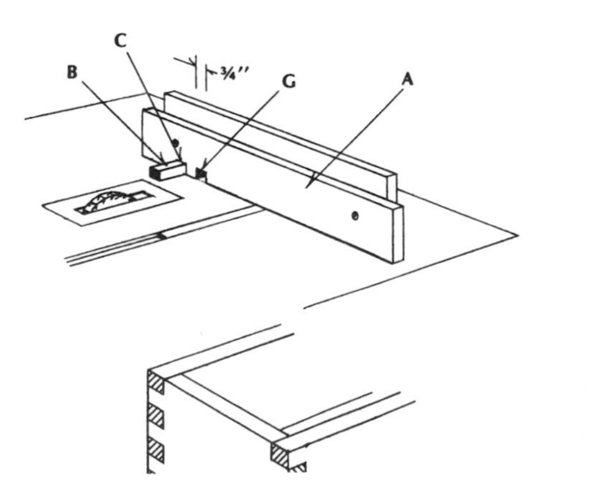

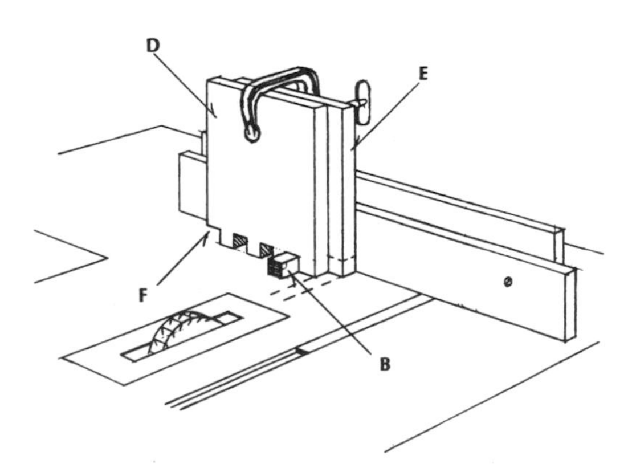

Figure 5-28. *Box or finger joint jig.*

the backing piece to your right (Figure 5-28, right), as you face the saw table, twice the width of the cut. Use two thicknesses of the material as a gauge. Screw or clamp to the gauge face and make the trial cut (G). Try this measurement with two nice pieces of scrap (D and E). Hold D over the slot (G) and take out corner F. Now place piece E against the block and corner F over the block so the two pieces are in staggered arrangement. Clamp them together. Now this pair can be notched simultaneously by making a pass over the dado cutter, shifting that notch over the block, cutting another notch, and so on, for the width of the piece. The drawing shows the final notch about to be cut (dotted lines). Try these pieces for fit. The widths of the notches can be increased or decreased by adjusting the dado set. Drive three or four screws through A to ensure precise replacement next time. The height of the notch is the height of the dado set, of course. When you are satisfied with the fit, cut four scraps to identical lengths, number them to avoid error, notch them out, spread glue between the fingers, tap all four together, check with a square, then tack a diagonal across from corner to corner. In a few hours this should prove to be tremendously strong and, after sanding, very handsome.

Making this box or finger joint jig may sound like a very complicated process, but it is not. It does require accuracy to create something you'll be proud to display. If you don't have the patience, take heart, because a glue such as Aerolite will fill cracks without losing strength. The right way, however, is one of the little differences between a boat and a yacht.

BENCH HOOKS AND STOPS

The bench hook (Figure 5-29) is a simple device for holding small work pieces being planed. A 3-foot length is handy. So is a short one for using *across* your bench. The stop (A) need not be more than 1/4 inch thick. Piece B, however, should be a 1 by 2. Bench hooks can be clamped or bradded to the table to prevent creeping on the back stroke.

Figure 5-30 shows two variations, the taper stop and the wedge stop. These are used for planing pieces on edge. They should be made substantially of 3/4-inch or more plywood or hardwood and be nailed and glued to a rather longish bench hook or screwed down to a bench.

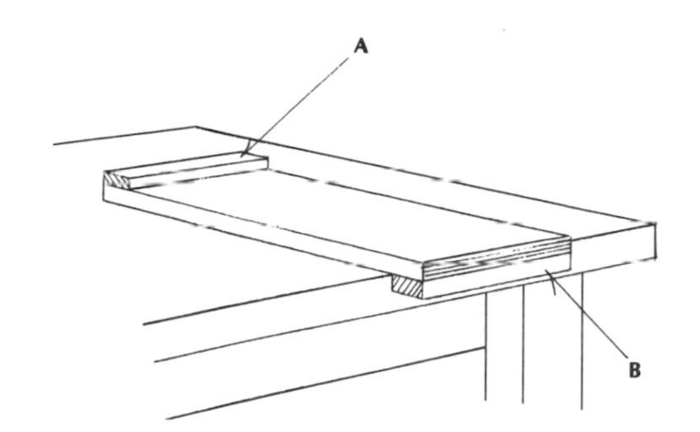

Figure 5-29. *Bench hook.*

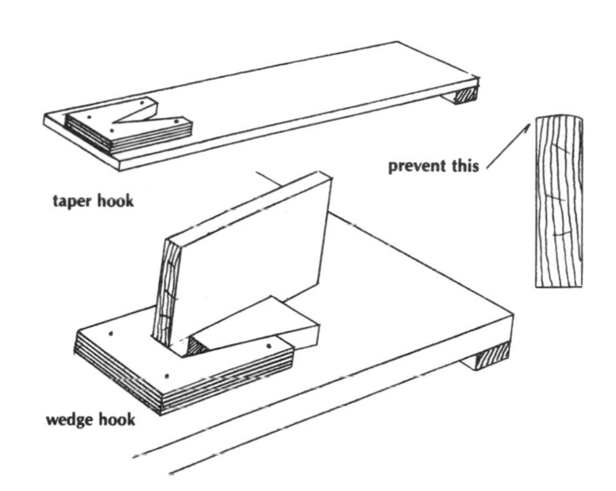

Figure 5-30. *Wedge and taper hooks.*

SHOOTING BOARDS

I wish I knew the origin of "shooting boards," (Figure 5-31). It's a simple little bench jig that enables you to plane the edges of a piece square with almost no thought. A shooting board can be tacked to a bench or clamped against a stop to prevent creeping. The work is held flat on the shelf (A) and you stroke the plane back and forth on its side on the board (B). Make A of 3/4-inch material. It is sufficiently above B so that it will work well with a smooth, block, or jack plane. The surface of B should be waxed. The stop (C) should be removable so long stuff can be handled. I keep a 40-inch-long shooting board handy so it can be used across a couple of sawhorses.

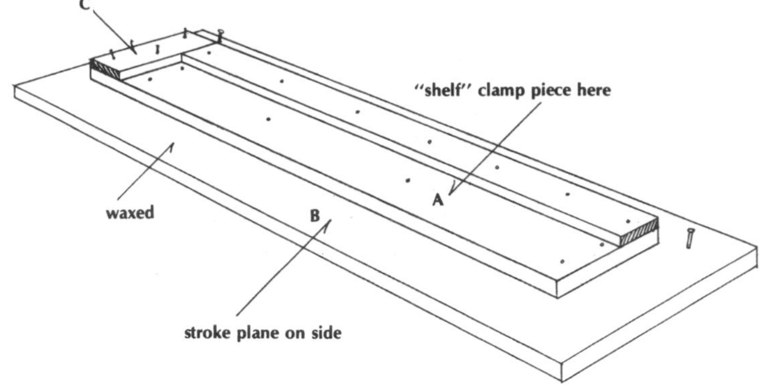

Figure 5-31. *Shooting board.*

CUTTER GRINDING JIGS

I promised to describe several ways to hold power plane and jointer blades or knives for grinding. A grinder must have an adjustable tool rest that allows you to bring the tool bevel to the grinding wheel at the proper angle. But this works only with tools long enough to be grasped. The tool holder shown in Figure 5-32 (left) is an extension to the narrow jointer blade. It can be moved up to the wheel properly and slid back and forth across the face, guided by a stop. Pick out a block (A) about 1 inch thick and perhaps 4 inches wide. Saw out a slot (B) about one-half the depth of your blade so the blade has a depth stop. Then cut a finer kerf (C) 2½ inches deep with a bandsaw or backsaw. Bevel under the blade holder to clear the wheel. Drill for a ¼-inch flathead stove bolt, with washer and wing nut. The hardwood stop (D) is bradded square to A so that the blade touches the wheel at the approximate original angle. Fine adjustments can be made with the original adjustable tool rest. Start the wheel and run the blade about four strokes end to end. If the original bevel was flat ground, it may take a few more strokes for the hollow grind to reach the cutting edge. Take off no more than absolutely necessary. Whet this new edge on the oilstone as described in Chapter One.

The tool grinding stand shown in Figure 5-32 (right) is quite simple to make, but some of the angles must be worked out by trial and error. The vertical block (A) is loose and free to slide against the batten (B). Its height is determined by the height of your grinder above the bench. Saw a slot to receive the jointer knife (C). Drill down into the flat grain for a No. 10 wood screw or stove bolt (D),

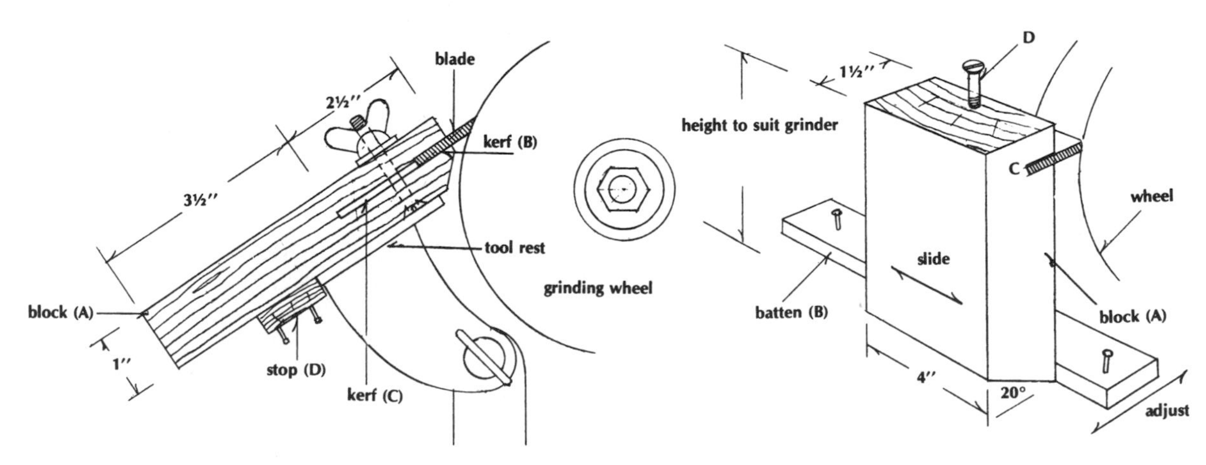

Figure 5-32. *Jointer blade grinding fixtures.*

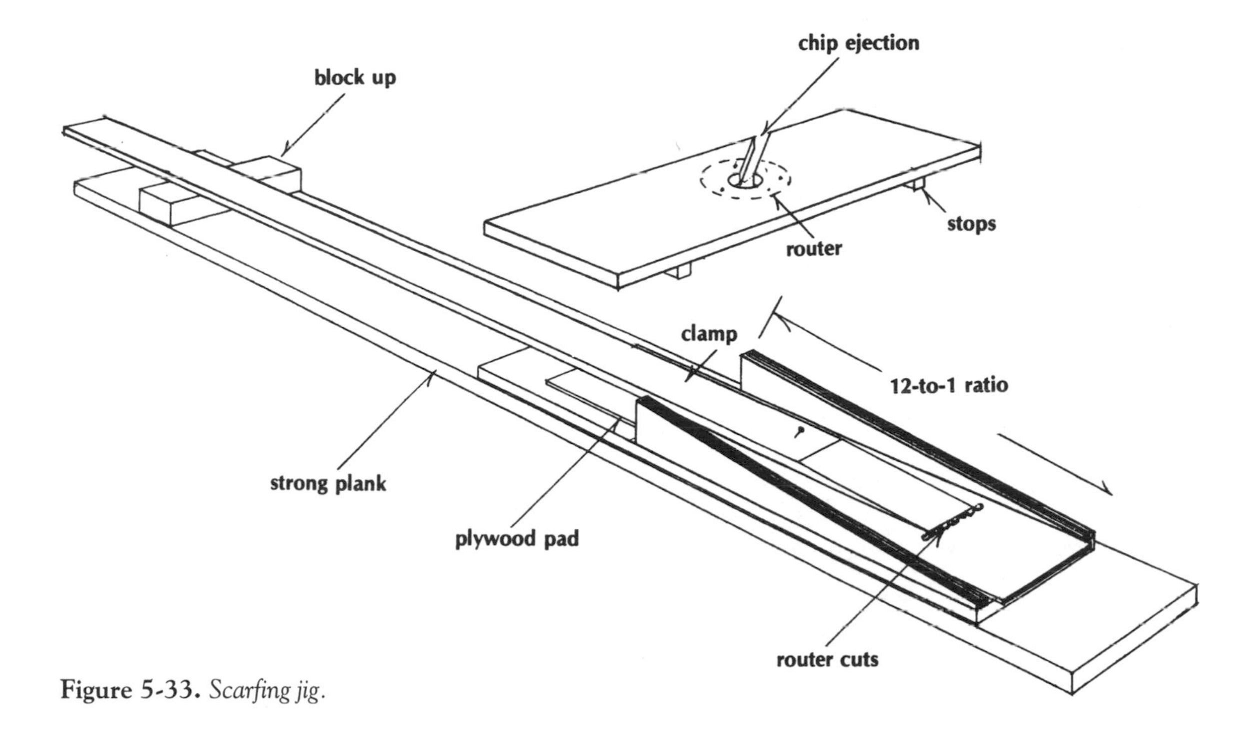

Figure 5-33. *Scarfing jig.*

located so its point bears against the blade, locking it in the slot. The lower end of the block should be cut off at a slight angle so it can be rocked against the wheel and away. To make the knife contact the wheel at the proper bevel, move the batten (B) until you are satisfied. The tool bevel is run across the wheel lightly by sliding block A back and forth. Rub a little soap or wax on the block and the batten.

A SCARFING JIG

When you get into constructing hollow box spars, you'll become an expert at making long, thin scarfs. A properly made scarf on a ratio of 12 to 1 is stronger than the original material, so it is practical to join shorter lengths of costly material. This applies only to spruce, fir spar stock, or mahogany, not oak. Glues like Aerolite set up clear, so discoloration should be minimal. However, I would not advise scarfing a piece where the joint will jump out at you.

Sawing and planing to a paper-thin feather edge is a job no man in his right mind wants. This makes a scarfing jig (Figures 5-33 and 5-34) a necessity. The one shown is dimensioned for a router mounted on a plywood base with a length about three times the width of the jig. On the underside are stops to keep the router bit from running into the sides. This job can be done with a jack or jointer plane also, but the jig would have to be much wider. Otherwise, you would simply plane it away inadvertently. Rent a router and a straight bit up to ¾ inch and avoid a headache.

Make the jig of hardwood or good plywood,

Figure 5-34. *This 12-to-1 scarfing jig is made of three pieces of ¾-inch plywood glued up.*

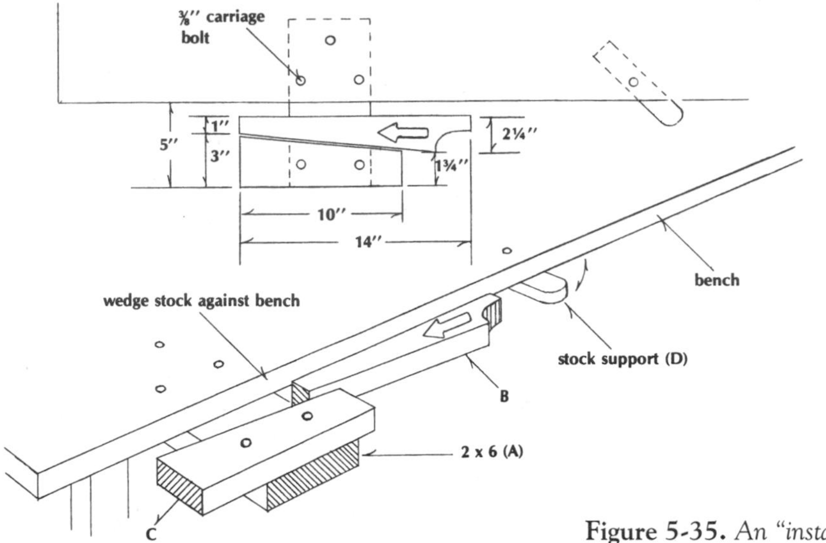

Figure 5-35. *An "instant" planking vise.*

screwed and glued to last. See that the inclined faces are dead straight. The chip ejection chute is a questionable refinement. Remove the plastic base of the router and mount the plywood base with three screws centered over the hole. Clamp the jig to a strong plank or bench. Place several blocks to raise the stock to the level of the jig and its pad. Note an expendable piece of plywood added to absorb the router cuts as you form the feather edge. Place the work piece so the router will take off a good ¼-inch bite in the first pass. C-clamp it securely equidistant from the sides, with no vertical spring. Now just run the router back and forth or with the grain, no matter. Lower the bit until you get close to the feather edge. Run a cut along each edge to eliminate splitting at the corners, and then run a fine and final cut. Rout off the rest to a nice flat surface, with no grooves or ridges. Do not plane or sand this surface, as the "tooth" produces the best glue joint.

AN "INSTANT" VISE

This vise, usually called a planking vise, is an old-timer. It is great for dressing the edges of anything. I ran a boatshop once that had no other vise, and we got along just fine. Your bench is probably constructed of 1½-inch material. To the underside of the top, bolt and glue a piece of 2 by 6 fir (A) about 14 inches long, as shown in Figure 5-35. Let this project from the edge about 5 inches, square to the bench. Now make a wedge or tapered piece of fir or oak (B) 14 inches long, 2¼ inches at its greatest width, and 1 inch wide at the small end. Smooth this up. Then a piece of 2 by 6 (C) is bolted and glued to piece A projecting from the bench. The taper of C must match the taper of the wedge (B). So clamp the wedge plus a ¾-inch piece of scrap to the projection, then clamp the 2 by 6 block while you bore holes for three ⅜-inch carriage bolts. Glue and bolt piece A to stay. Much of your work material will be ¾ inch or less, so the angle of the block now matches the wedge. To use, place your work piece on the projection and against the bench edge, slip the wedge in the gap, and give it a light tap with a hammer or block. The work piece will be held absolutely rigid. For stock much lighter than ¾ inch, insert a piece of scrap to make up the difference. To handle long work pieces, provide one or more supports (D), 4 to 8 feet from the vise. To remove work from the vise, just wiggle the wedge or give it a light tap.

RADIAL-ARM SAW STOPS

For crosscutting multiples to exact length, you can just tack a bit of scrap to the table or fence. But here's an idea from De Cristoforo's book *Fun with a Saw* that looks neat and efficient (Figure 5-36). This is a grooved block that straddles the fence.

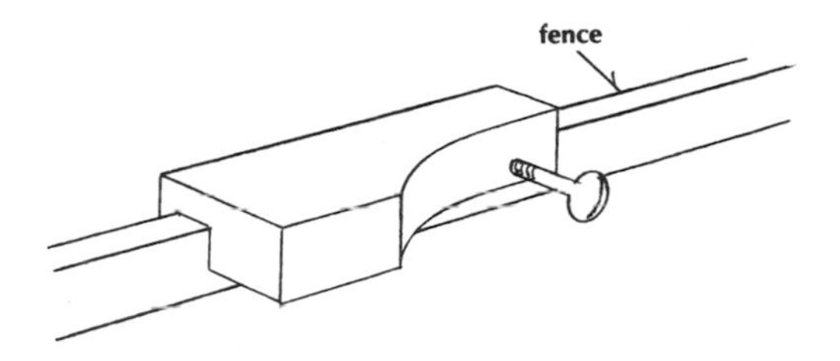

Figure 5-36. *Radial-arm saw stop.*

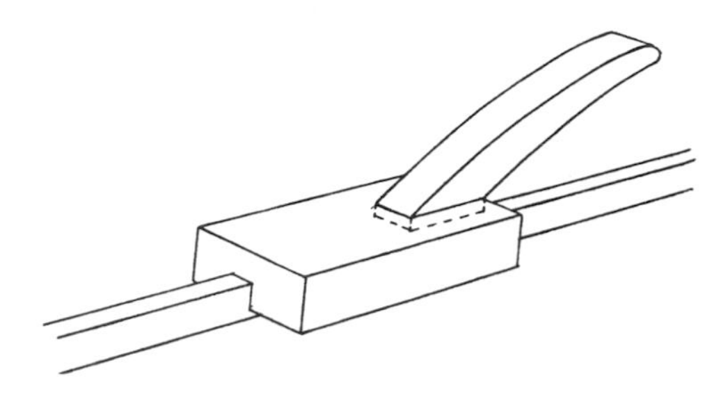

Figure 5-37. *Sophisticated stock pusher for radial-arm saw ripping.*

The wing bolt or set screw is threaded through the wood so it hits the fence. De Cristoforo also shows a refined pusher stick for ripping (Figure 5-37).

HINGE JIG

There probably will be a score or more cabinet and passage doors in a 35-foot yacht, and as many cabinets in a small house. This makes the installation of hinges a major task, and a real problem unless you have skill and patience. Plywood doors in particular are a nuisance because it is more difficult to chisel out the gain (the recess into which a butt fits) and achieve the exact depth (Figure 5-38). If the gains are too deep, the doors will refuse to close. If they are too shallow, there will be an ugly gap. Gains routed in a jig, on the other hand, are uniform in size and done quickly and easily. Some hinges have rounded corners that fit a routed gain. If your hinges are square-cornered, you'll have to chisel out the corners of the gains.

The jig shown in Figure 5-39, made of scrap plywood and a piece of 1 by 2, is easy to put together, but watch the arithmetic. The guide piece (A) can be ½ inch by 8 inches by 14 inches or more, cut carefully with a sabersaw or bandsaw. The router is supported on B, which is bradded and glued to A. Piece C, cut from 1 by 2 stock, enables you to clamp the jig securely to the door. Pay attention. The arithmetic that follows can be confusing.

Dimension D is the sum of the width of the hinge leaf (1 inch) plus the measurement from the cutting edge of the router bit to the outside of the router base plate (2¾ inches), a total of 3¾ inches. Dimension E is the length of the hinge (2½ inches) plus 2¾ inches twice, a total of 8 inches. Warning: If you use a larger or smaller bit, the gain will be changed accordingly.

The same jig can be used for several sizes by making the cutout suitable for the largest hinge, perhaps 3 inches. For smaller hinges, just tack in filler pieces or shims to reduce the sides and back. Save the shims, marked for size, for future use. Of course, you can buy butt guides that require manufactured guide bushings. These are quite reasonable, but they are not adaptable to various hinge dimensions, and they won't do work any better than this gadget that you can make in just a few minutes.

Incidentally, if you are going first class, you'll rout gains in the jambs as well as in the doors. Set the router bit only as deep as the thickness of the

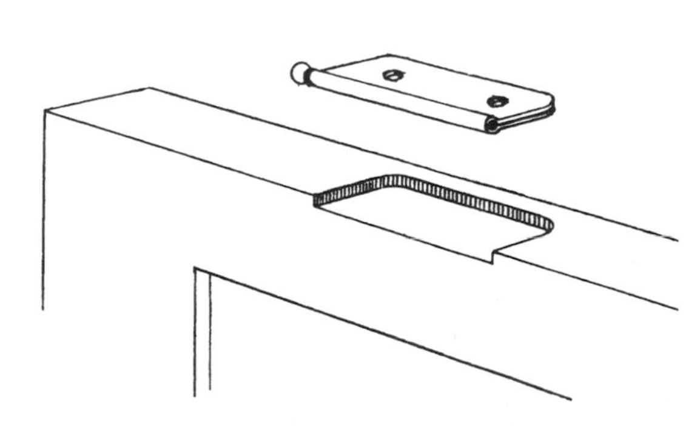

Figure 5-38. *Typical hinge gain.*

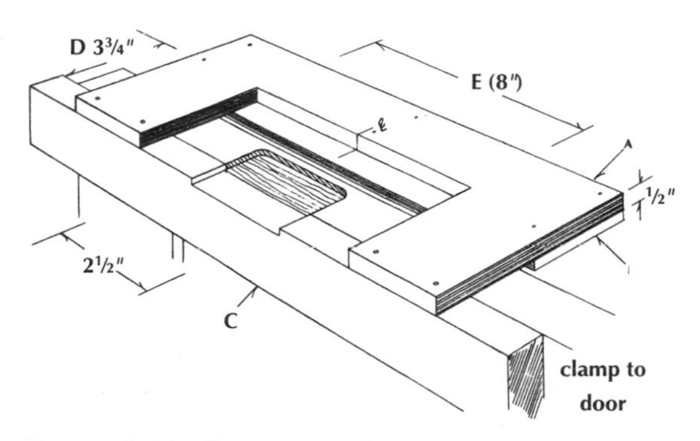

Figure 5-39. *Gain routing fixture.*

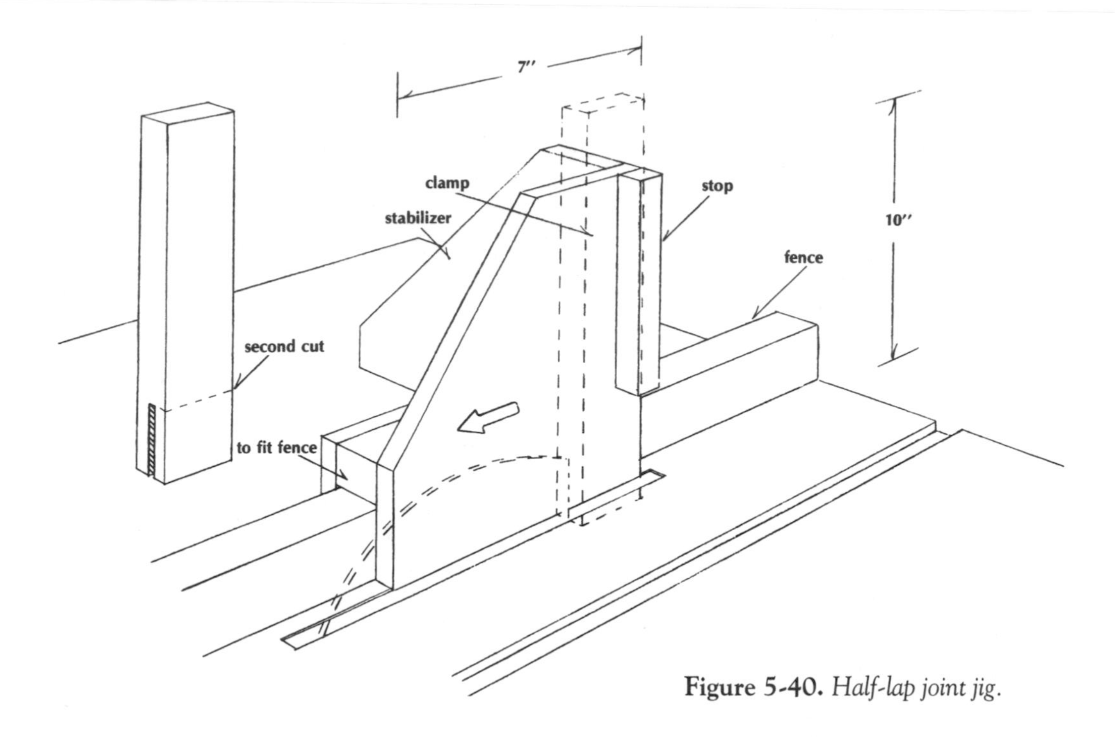

Figure 5-40. *Half-lap joint jig.*

metal. (Always use brass hinges in a boat, not plated steel. Check with a magnet, to be sure.) If all the gain is to be in the door, set the bit to slightly less than the thickness of both leaves of the hinge when they are in the closed position (Figure 5-38). Once your jig is clamped to the door, the job takes about 10 seconds, running the bit back and forth or lengthwise.

HALF-LAP JOINT JIG

There are many uses for the strong and versatile half-lap joint. Half laps are easy to make, but they must be precise. All you need is a way to hold the parts firmly and parallel to the table saw blade. Double-check the blade setting. The jig (Figure 5-40) works very well. You should be able to put one together in about an hour. It's designed to slide along the rip fence without visible play; the stabilizer maintains squareness to the table while it slides. The parts to be cut could be C-clamped against the stop. With a little practice, however, you should be able to hold them in position while sawing.

The second cut, too, must be accurate. This is simple if you have a sliding auxiliary table with a stop tacked to its fence to ensure precision. The second cut may also be done by tacking a stop block to the face of your miter gauge. You'll get good, clean cuts if you mark these shoulder cuts with a deep knife mark.

BEAM MOLDS AND PATTERNS

The pattern for decks and cabin trunks is a "beam mold." It is used to lay out the camber on the lumber for sawing beams to shape. It is used also for clamping carlings in the deck framing so these members match the designed camber in the deck. Thus, the beam mold must have considerable strength. You may find two or three handy so that the full-length carling can be forced into the sweep caused by the sheer of the vessel while the short deck beams are being fitted. First, however, the proper camber or crown must be laid out on the pattern material. This can be done either mechanically or geometrically.

The first system is quite simple (Figure 5-41). Use seasoned pine or fir of double thickness so the possibility of distortion is eliminated. Select two straight pieces of material ripped to about 1 by 6; plywood would be fine for beam under 8 feet. Drive

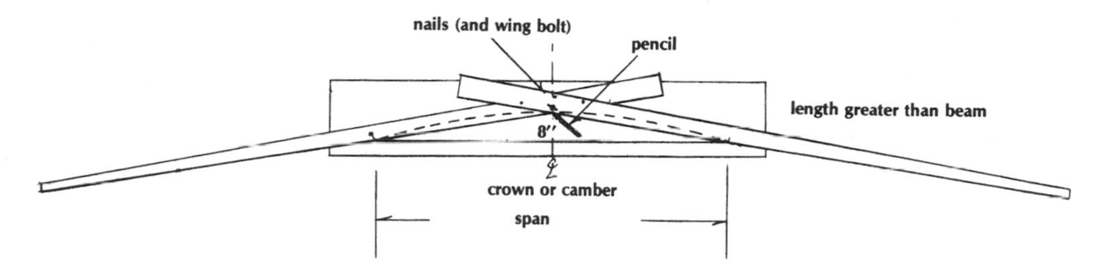

Figure 5-41. *Jig to scribe any camber.*

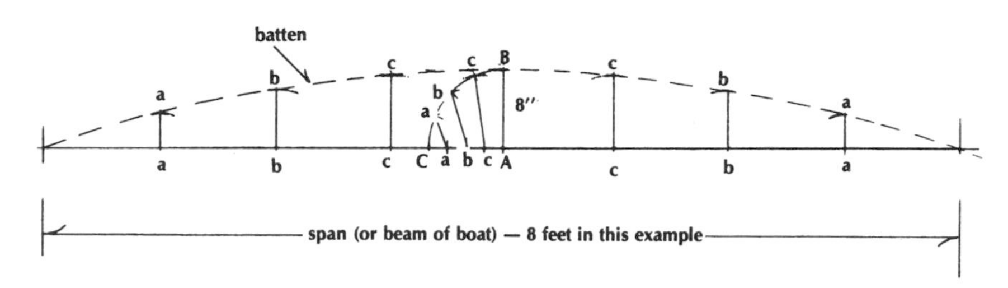

Figure 5-42. *Geometric layout of any camber.*

nails near the ends of the pattern material and another at the midpoint, this one being the height of the crown specified by the designer (so many inches per foot of beam). Lay the two straight lengths so they cross at the centerline and bear against the three nails. Nail securely. Remove the center nail. Hold a pencil at the intersection and swing the two battens from one side to the other, keeping them tight against the nails. It may help to backsaw a tiny nick at the intersection so the pencil can't creep. This produces a perfect arc. Saw close to the line and dress very carefully so the beams will be precise. Make several duplicates from the first one finished.

The second layout method, shown in Figure 5-42, is a bit tricky; your lofting, that is, drafting, must be accurate. On the centerline swing an arc equal to the height of the camber A–B to point C; divide this line by four. Divide the arc (B–C) into four. Connect the points (a–a, b–b, and c–c). Now divide the baseline into eight, and at each point erect the heights of a–a, b–b, and c–c. Drive small brads in at these points, and bend a good batten around the nails, using weights to hold it. (Never drive nails through a batten.) Mark inside the batten and you have it.

Because many deck beams could be laminated, the beam mold would be used for locating the blocks or forming the mold to which the laminations are to be clamped. Remember that this is the pattern for the upper surface of the beam. Make another pattern for marking lower and upper surfaces, the total depth of the beam. This must be rigid, so double it. A 2½ -inch-wide pattern is quite flexible otherwise.

The arithmetic method of finding the radius to produce a certain camber between two known points must have been worked out by Archimedes (or was it Noah?); see Figure 5-43. Square the span, add to this the height squared multiplied by four. Divide this total by eight times the height. The result is the required radius. If you had to build a set of forms for a cabin trunk top, no two alike, you could use this method or the beam scribers in Figure 5-41. But sometimes the dimensions of the

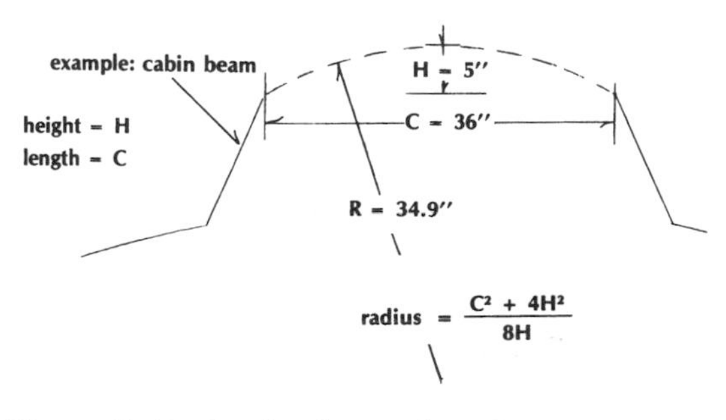

Figure 5-43. *Camber by simple arithmetic.*

shop interfere. In this case, calculate by the arithmetic method.

WOODEN PLANES

There are a few uses for wooden planes in yacht joinery. Spar work and backing out of hatch rails are examples. John Gardner, technical editor of *National Fisherman*, wrote a series of articles on the origins, uses, and construction of wooden planes (November and December 1972, and January 1973). Reprints might still be obtained from the magazine.

Gardner describes one secret of old boatbuilders that I feel bound to pass on to you. Working woods with cross grains is extremely difficult, for even the sharpest blade may tear the surface. In Gardner's first article, he says:

He honed the cap on the oilstone until it was sharp and fitted the surface of the plane iron perfectly. Then the edge was turned down with a burnisher as if it were a scraper. The cap was then fastened to the iron with its burnished edge set very close to the sharpened edge of the iron, within 1/32 inch or thereabouts. A plane set in this manner is harder to push, but it will smooth the most difficult and treacherous cross-grain wood without catching, digging in, or tearing.

I can report to you that this is almost as effective with a smooth plane blade ground almost perfectly square, but with an arc of less than 1/32 inch. It's great for smoothing joints such as the meeting of stiles and rails, since even planing across the grain takes off a shaving you can read newsprint through. I have used this technique on flat-grained fir and oak to remove planer corrugations.

Figure 5-44 shows just three of the many configurations of wooden planes. One of my old craftsmen had a trunkful of planes he had made or otherwise acquired in his 40 years. And he performed wondrous magic with them. But perhaps we should be content with a little less than magic.

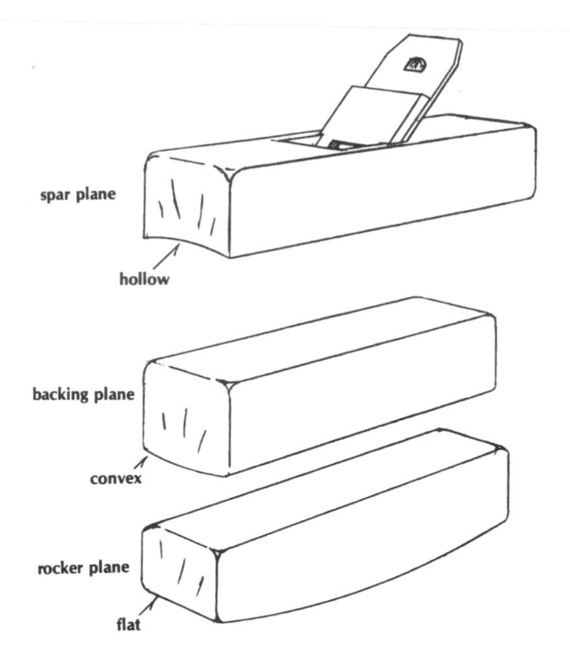

Figure 5-44. *Planes.*

How about cleaning up a rabbet after you have dadoed or sawed it out? The tool in Figure 5-45 is based on an article by Jim Emmett in the November 1962 issue of *Rudder*. You can make this rabbet plane if you have just a little skill and a lot of patience. Lay out the angles of the two sockets with a sharp knife. Saw with a backsaw to a depth slightly less than the width of the chisel. Clean out the side of the socket with a 1/4-inch chisel. The hardwood wedge must be a perfect fit for tapping in to hold the chisel. Minute depth adjustments are made by tapping on the appropriate end with a mallet or hammer. Live oak would be the ideal wood, but you may have to settle for beech, birch, maple, or ash.

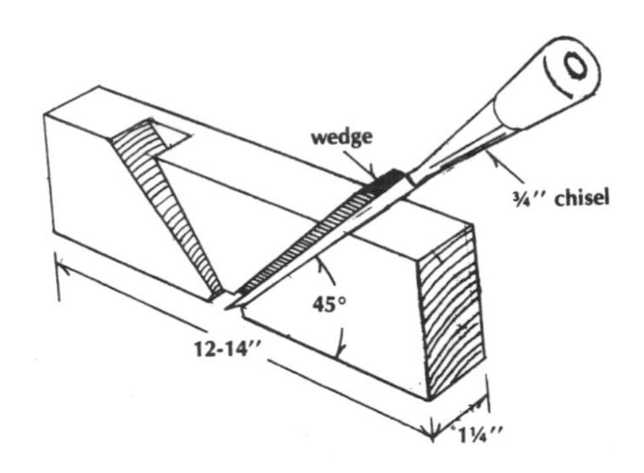

Figure 5-45. *Rabbet plane.*

I could go on for a few pages more about accessories, jigs, gadgets, and gimmicks. Similar information, however, will emerge as we go through the actual construction of doors, drawers, tables, hatches, decks, rails, spars, and all the rest. So let's drop anchor here for now.

6 Tips, Techniques, Facts, Opinions

This chapter is a slop chest of tricks, standards, common practices, and advice on several subjects important to the builder of wooden boats, as well as carpenters and cabinetmakers. To some of you, what follows may seem elementary. I must assume, however, that other readers are less familiar with the subject. A good part of the chapter consists of facts and opinions from experts far better qualified than I to instruct less-knowledgeable wood-lovers. So please bear with me.

POWER TOOL SPEEDS

To achieve maximum efficiency, every shop power machine must make the correct number of revolutions per minute. Adapting used motors (as is often suggested) is easy if you know the r.p.m. required, and the pulley sizes needed to achieve the desired r.p.m. Here are recommended speeds:

Machine	Size (")	Motor Speed	R.P.M.
Table saw	10	3,450	3,100
Table saw	8	3,450	3,400
Bandsaw	12	1,725	700
Bandsaw	14	1,725	600
Jointer	4–6	3,450	7,000
Jointer	"	1,725 (less efficient)	5,000
Drill press	all	1,725 (with 2-, 3-, 4-, 5-inch cone)	600–5,000
Bench disc sander	8	3,450 (direct)	3,450
Bench disc sander	10–12	1,725 (direct)	1,725
Shaper	½	3,450	8,000–10,000

SPEEDS AND PULLEY SIZES

It is simple arithmetic to work out the pulley sizes required to give reasonably accurate tool speeds. Just apply these formulas.

$$\frac{\text{tool r.p.m.} \times \text{tool pulley dia.}}{\text{motor r.p.m.}} = \text{motor pulley dia.}$$

$$\frac{\text{motor r.p.m.} \times \text{motor pulley dia.}}{\text{tool pulley dia.}} = \text{tool r.p.m.}$$

$$\frac{\text{motor r.p.m.} \times \text{motor pulley dia.}}{\text{tool r.p.m. wanted}} = \text{tool pulley dia.}$$

THE TWISTED BELT REVERSE

Frequently you will find that a motor turns opposite to the direction needed. On a 4-inch jointer, I use

Figure 6-1. *The twisted belt reverses the arbor direction on the jointer.*

a ⅓-h.p., 1,725-r.p.m. washing-machine motor mounted on a shelf directly under the tool pulley. The motor is set at an angle of about 30 to 40 degrees so the belt can be twisted. This reverses the direction of the jointer. Friction is minimal; the jointer has been running for four years with no sign of wear (Figure 6-1). Some motors can be reversed by changing the wiring, but cheap motors are not worth working on. Besides, you still have the original direction if you want it for another kind of tool.

SANDING TRICKS

You may find a long sanding board preferable to a belt sander wherever a fair surface is required. Try a flexible board, as shown in Figure 6-2. The board should be of ⅜- or ¼-inch fir plywood just long enough to take two halves of a sheet of sandpaper. Make a number of shallow saw cuts about one-half the thickness of the board if flexibility for quick bends is necessary. Secure the two halves of the sandpaper sheet with staples into the end blocks. Then fold the paper over the back and staple. Or use Glop, the aptly named cement used for sanding discs. It permits immediate use and rapid replacement of sheets.

If you can pick up a remnant of a sandpaper belt,

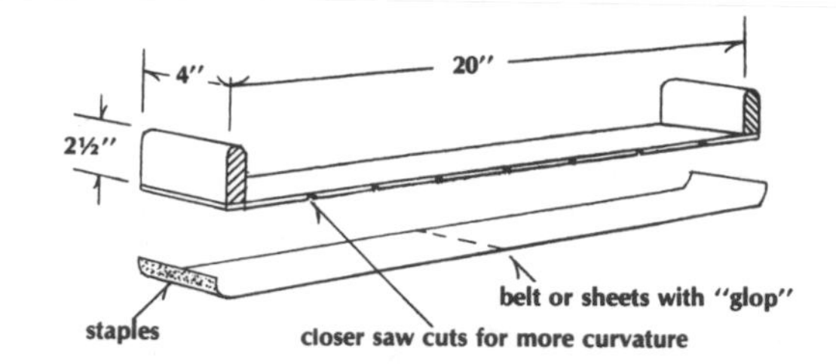

Figure 6-2. *Flexible board for sanding.*

say, 80- or 100-grit open coat garnet or aluminum oxide production paper, you'll need staples only at the ends. This paper lasts longer, too. I use about 36 inches of such a sanding belt stapled to a board clamped on a nearby bench. It's great for instant sanding of small pieces such as cabinet stiles and rails without taking them to the bench belt sander.

Sanding concave curved edges using a drum mounted on your grinder shaft is all right, but a drum in the chuck of a drill press is better. Place the work piece on a block fitted to the drum so you can control the cutting action and always maintain a square edge—not easy on a shaft-mounted drum. To get into concave spots when the piece is too large to handle, make a cylindrical block out of a length of carpet-roll tubing. Such tubing is about ¼ inch thick and holds staples quite well. (Figure 6-3). This, too, is a good place to try Glop. The tubing idea works well for sanding small spars also. Simply shape the sandpaper to the inside of a split tube.

Tooling foam makes great sanding blocks and shapes. Ask fiberglass buffs where to buy it. You can plane, rasp, saw, or sand this foam to almost any shape. Use it to sand round or rectangular spars, moldings such as ogees and coves, and so on (Figure 6-4). Do not sand a flat surface without a block of some kind. When you get into fine sanding (220- to 400-grit), use a rubber block whenever possible. Never sand plywood until it has had several coats of primer, sealer, or thinned-out varnish.

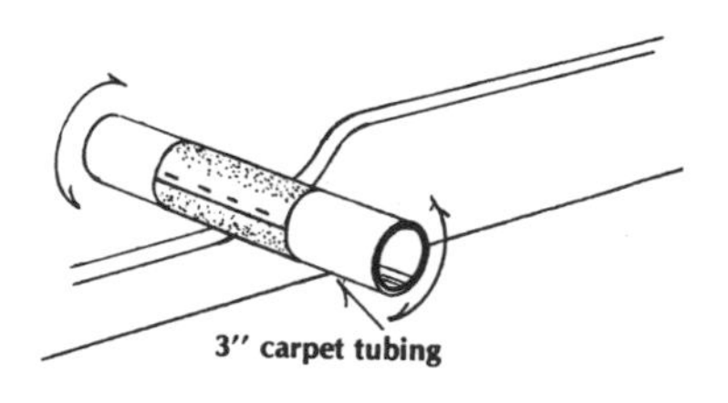

Figure 6-3. *Tubing for sanding.*

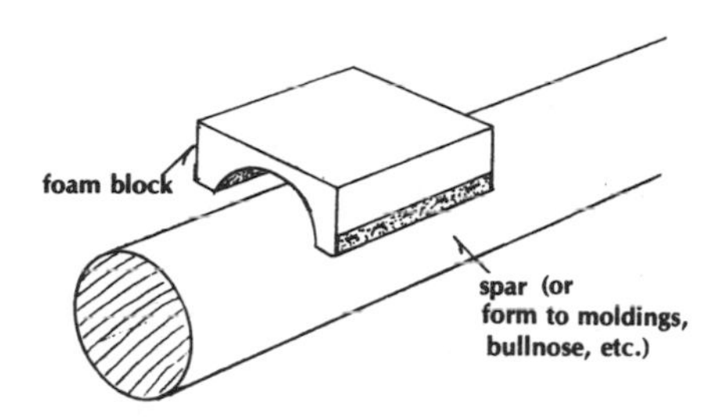

Figure 6-4. *A sanding device.*

SAW KERF BENDING

There are occasions when it is necessary to make a sharp bend in thick material. If strength is required, the piece may be slashed with saw kerfs across the inside of the area to be bent (Figure 6-5). The idea is to space these cuts so they close up completely when the bend is made. A thickened epoxy/filler glue fills up the wedge-shaped spaces. Here's a formula used for centuries in furniture building.

First determine the approximate radius of the inside of the finished piece. Let's say 3 inches in ¾-inch material. Find a piece of scrap 1 inch thick. Set your table saw to about a ⅞-inch depth of kerf. Make a cut in the piece of scrap, then clamp it to a bench with the open kerf up. Raise the free end until the kerf closes fully, and block the piece in this position. Now place a mark 3 inches from the kerf on the raised portion. The distance between the bench and the raised piece at that point is the exact spacing required for a 3-inch radius. Experiment with scrap to determine how many cuts will be needed.

If your piece is hardwood and visible, sanding with a hard block will fair out any irregularities on the outside surface. This method can be used for forming lightweight blocking for laminated cockpit coaming corners, for table framing, and for bending dummy frames to take ceilings in fiberglass hulls, among other uses.

FASTENINGS

This will not be an exhaustive discussion of the many types of fastenings used in a wooden vessel. Your best sources are the excellent books mentioned earlier. Reliable guidance is available in Howard Chapelle's *Boatbuilding*, Robert Steward's *Boatbuilding Manual*, L. Francis Herreshoff's *Common Sense of Yacht Design*, and others.

For exposed work that is to be finished bright, it is best to use bungs (plugs) over screws. Galvanized hot-dipped flathead screws are more than adequate, for their life and holding power are excellent. They drive harder, however, because of the rough zinc surface. Also, the slots are not always clean. Stainless steel screws are used in large numbers by production shops because they drive easily, the slots or Phillips sockets are clean, and they are available with pan-type heads. As shown in Figure 6-6, these have less tendency to split the wood than cone-shaped flathead screws. Many situations require glue even with screws. This is true for the cleats supporting berths and settees, for example, and when the fastening is in shear, it should always be glued.

Some cabinet cleats can be glued and nail-fastened. The glue does the actual work; the nails may be hot-dipped finish, Anchorfast, or headed nails if the spot is not seen. (If the bulkhead is thin, fasten

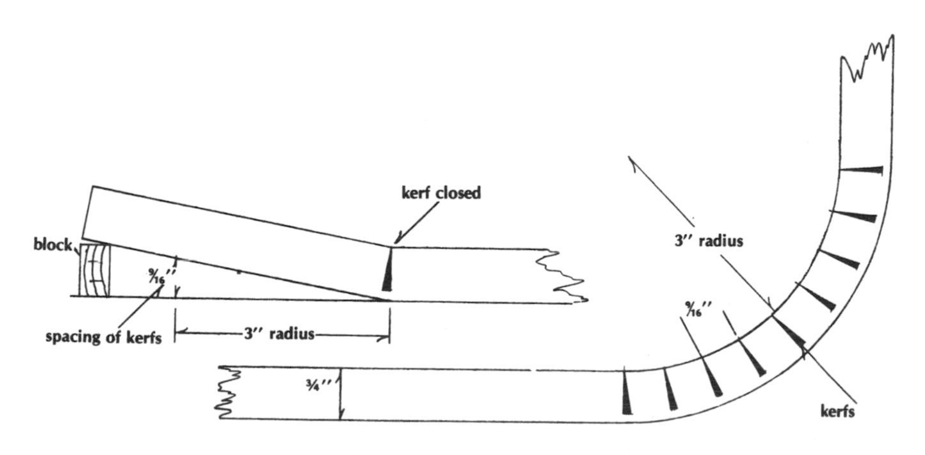

Figure 6-5. *Bending to given radius by kerfing.*

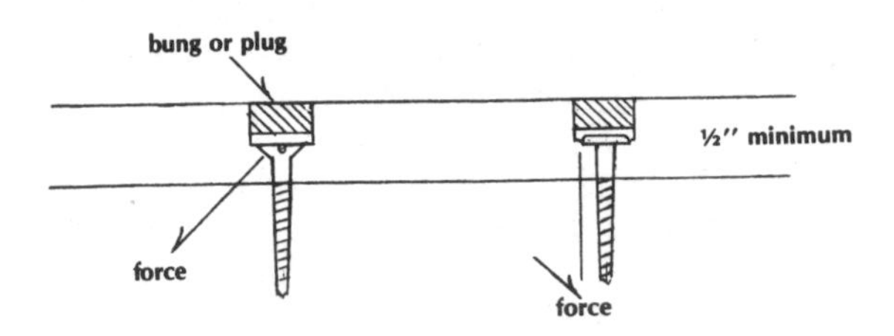

Figure 6-6. *Flathead screw versus panhead screw.*

from it into the cleat.) Serrated nails of bronze or Monel are great almost anywhere, but their cost is high. They drive fast in softwoods and hold forever! These nails require drilled lead holes in hardwoods and sometimes in softwoods. Where the joint is visible, serrated nails must be bunged just as screws are. It's best to set them with a large nail-set into the same ⅜-inch counterbore you use for screws.

The table in Figure 6-7 shows the screw sizes to be used when bunging is required. Robert Steward recommends using white laundry soap on screw threads; this certainly makes large screws drive more easily. However, paraffin and tallow have none of the chemicals found in soap. They, too, make excellent lubricants.

Light trim and moldings of mahogany and teak may be fastened with hot-dipped galvanized finish nails set in and filled after staining and one or two coats of varnish. The thinnest material you can bung is generally about ½ inch. Even this leaves very little to secure the bungs. You might have better luck with pan-head stainless steel screws, for this head does not have the conical shape of the flathead screw and thus has less tendency to split the trim when driven in hard. If you are careful, you might successfully bung pan-head screws in ⅜-inch material. Where the trim or molding is being applied to a glass liner or other part, your best choice, in my opinion, is the pan-head sheet metal screw, which cuts a pretty fair thread. For full information on bunging, see Chapter Two.

DECK FASTENINGS

The most common deck built today is fiberglass over double plywood. This can be fastened with galvanized or bronze screws sunk just below flush. I would use serrated bronze nails, set in slightly. Hot-dipped headed nails are less desirable for this purpose, but they will hold if toenailed. Galvanized finish nails can be used only if the plywood is glued to the beams. In fact, it is preferable to glue all plywood decking to fir or spruce framing, regardless of the kind of fastening used. The sunken heads of the screws or nails should be glazed with a filling compound compatible with polyester and/or epoxy resins.

The best-looking deck is probably white pine, fir, or teak sprung to the curvature of the deck edge or covering board. Traditionally, this type of decking would have to be 2-inch strips a minimum of 1½ inches thick, each fastened with two bronze screws into the beams, with one or two hot-dipped nails holding each pair of strips. The seams would be caulked and filled with a black compound, as shown in Figure 6-8. There is much detail on decks in Chapter Nine.

Material Thickness	*Screw*	*Size*[1]	*Screw Dia.*	*Shank Drill*	*Lead Drill*[2]	*Plug Dia.*
⅜"	¾"	#7	.150"	9/64"	#44	none
½"	1"	#8	.163"	5/32"	#40	none
⅝"	1¼"	#9	.176"	11/64"	#37	⅜"
¾"	1½"	#10	.189"	3/16"	#33	½"
⅞"	1¾"	#12	.216"	13/64"	#30	½"
1"	2"	#14	.242"	15/64"	#25	½"
1⅛"	2¼"	#16	.268"	17/64"	#18	⅝"
1¼"	2½"	#18	.294"	9/32"	#13	⅝"
1½"	3"	#20	.320"	5/16"	#4	¾"

1. May be reduced one gauge for decking
2. For hardware

Figure 6-7. *Sizes for screws in planking and joinery.*

Figure 6-8. *Deck seams caulked and filled with compound. Note handmade locust or ash halyard cleats. (Bruce Bingham photo)*

GLUES

John Gardner wrote about glues in a 1976 *National Fisherman* article called "Excellent Foreign Glue Liked by U.S. Builders." The comments about Aerolite 306 from several authorities who tested it were overwhelmingly favorable, even enthusiastic. This glue has been used for over 30 years in England and Canada in both aircraft and marine applications. It has been distributed in the United States only since 1974. Classified as a urea-formaldehyde resin adhesive, it is a product of Ciba Co., Ltd.

Aerolite 306 differs from all other formaldehyde systems. The Aerolite liquefied resin (one part dry powder to four parts water) is applied to one side of the joint, and the activator liquid is applied to the other side. Curing does not occur until the two are brought together. No heat or pressure is required, just firm contact. Glue line failures that occur with other urea-formaldehyde glues are prevented.

In an article called "Gluing Can Be Easy," published in *Sport Aviation* in 1965, Clint Lillibridge said that the advantages of Aerolite are:

1. It fills gaps up to 1/16 inch wide and still produces a strong bond.

2. Firm contact of joint faces and pressure of a pound or two are all that are required.

3. Assembly time may be extended as long as coated surfaces remain moist, and they may be remoistened. Curing time is uniform.

4. Excessive moisture in the wood slows curing but does not weaken strength.

5. Aerolite will cure at temperatures as low as 50 degrees F.

6. Joints are entirely waterproof and will withstand exposure for many years.

A couple of advantages Lillibridge did not include are: (1) The glue is crystal clear when cured, and (2) there is very little waste, as pot life of the resin mix is several days.

Comparisons of Aerolite with other glues showed it to be superior to all. Weldwood Plastic Resin, a product of DAPP, was a close second. The cost is considerably lower than epoxies. Distributors are Aircraft Spruce and Specialty Co., 210 W. Truslow Ave., Fullerton, CA 92632.

I used Aerolite 306 to glue the laminated stems to the keel batten on a very lightweight duckboat, intending to add bolts later. Completion has been postponed. This assembly has been kicked around my shop for several years, really mishandled, and it appears that no bolts are necessary. One day, however, I shall add fastenings to calm the doubters. I have used Aerolite in a dozen critical spots with complete satisfaction, more recently in an entire yacht interior and hatches.

I have been using Weldwood (also a urea-formaldehyde) for many years. Before that I used its predecessor, Borden's Casein, mainly for spars. All my plywood boats were glued everywhere, including plywood planking butts, which were also bolted. Charles G. MacGregor, the designer and pioneer in plywood construction, said one could remove the fastenings in such a boat and it would stay together. I never had the guts to try it. A number of sailboats up to 27 feet long built by me in the late 1930s and early 1940s, however, were still sailing some forty years later. Of course, plastic resin requires a lot of pressure, lots of clamps, and fast action. The great advantage of Aerolite is that nothing happens for quite a while after the parts contact.

A fine article, "Adhesives and the Boatbuilder," by Gerald Schindler, published in *WoodenBoat* in 1975, says

. . . all joints are variations of four basic types, as defined by the manner in which they are stressed. These are shear, tensile, cleavage, and peel. Since the nature of the stress in a joint has a great influence on the characteristics of the adhesive required to function most effectively in that configuration, before we talk much about adhesives we ought to consider the stresses that must be withstood.

The four stresses are shown in Figure 6-9. Schindler continues:

1. SHEAR. Force is exerted across the adhesive bond. The bonded surfaces are being forced to slide over each other. All of the adhesive contributes to bond strength.

2. TENSILE. Force is exerted at right angles to the adhesive equally over the entire joint. All of the adhesive contributes to bond strength.

3. CLEAVAGE. Force is concentrated at one edge of the joint and exerts a prying force on the bond. The other edge of the joint is theoretically under zero stress. Only a portion of the adhesive resists load.

4. PEEL. One surface must be flexible. Stress is concentrated along a thin line at the end of the bond where separation occurs. A very small portion of the adhesive carries all of the peeling force.

Most adhesives perform better when the primary stress is shear or tensile. However, in most applications a combination of stresses is involved . . . other factors influence the selection of an adhesive. Environmental exposure is an important consideration—moisture, temperature extremes, impact, vibration, chemical contact, dimensional changes, and biological attack all impose added requirements on the adhesive.

Next, application requirements or process limitations must be examined. Is the adhesive easy to use? Are special tools required? A weight scale? How much working life is needed? What is the range of application and curing temperatures? How critical is the condition and fitting of the surfaces? Is clamping pressure important to the ultimate strength? Can dissimilar materials be bonded?

Finally, the nature of the cured product is of concern. Is color important or will staining detract from the finished product? Will the adhesive be compatible with materials applied later? What is the anticipated life under the expected service conditions?

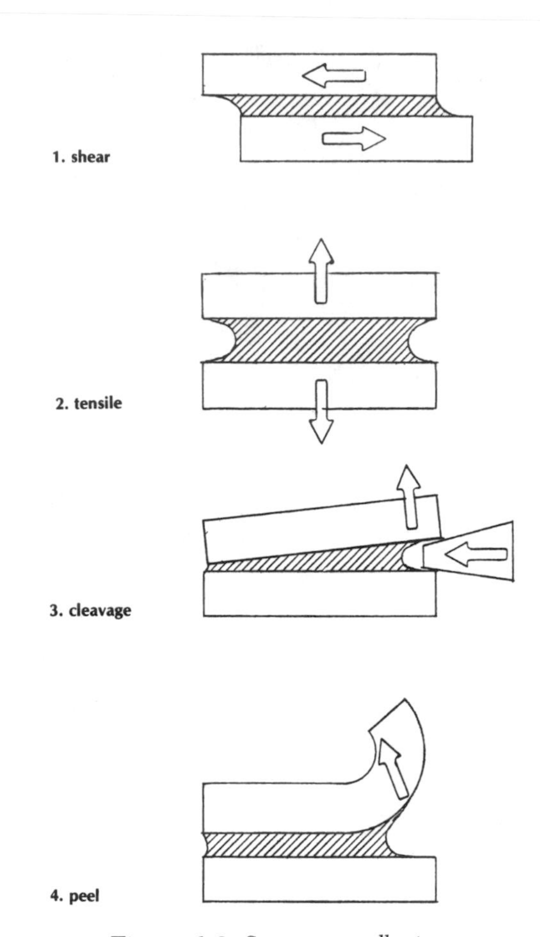

Figure 6-9. *Stresses on adhesives.*

Schindler's article reiterates most of what has been stated above about urea-formaldehyde adhesives, but he includes information about an interesting epoxy, Chem-Tech T-88.

Although T-88 is tailored to the needs of boatbuilders, and particularly amateur boatbuilders, many of its properties are shared by other epoxy adhesives, although no other epoxy duplicates all of its characteristics.

T-88 is supplied in kits consisting of two bottles of rather viscous liquids designated A (resin) and B (hardener). These are mixed in equal amounts by volume. . . . At 70 degrees F, the user has about 45 minutes to apply the adhesive if he has mixed a total of 8 ounces or less. . . .

The epoxy is easily applied with a flat stick, and because resin and hardener are mixed in the same cup, it need only be applied to one of the mating surfaces. . . . T-88 differs from most adhesives in that it may be applied to damp surfaces, which is helpful in making laminated curved structures where soaking or steaming is required.

Fit and clamping pressure do not present any problems either, since epoxies are truly gap-filling and require only sufficient pressure to hold the surfaces together. . . . Where gaps are really bad, a thickening powder furnished with the adhesive can be used to keep it from running out of the joint. . . .

T-88 will remain workable in a joint for two hours or more, allowing for more leisurely clamping and fastening of large panels without concern that the joints may not fully close. . . . T-88 is light amber in color, won't discolor wood, and becomes invisible when varnished.

T-88 can be applied in temperatures below 70 degrees F, down to 50 degrees F, but the curing time is extended from six to eight hours to as much as a full day. This temperature range, however, adds many working days during the winter, an important consideration. Chem-Tech, 4669 Lander Rd., Chagrin Falls, OH 44022.

WOODS

The following comments may be a bit late if you have already built your wooden hull. On the other hand, if you bought it, or you are rebuilding an older vessel, or you are about to finish a fiberglass yacht with decks and structures of wood, then what follows might be useful at this point.

All yacht joinery woods should be air dried, not kiln dried. Lumber right out of the local yard (Oregon fir, Douglas fir, pine, white or red oak) is apt to be high in sap and oils. Such wood is bound to shrink, warp, and check as it dries. Warping and shrinking place great stresses on fastenings, including glue; serious splits may occur. Moreover, wide-open joints are next to useless. Of course, appearance suffers. If your lumberman knows anything about boat construction, he should be willing to find the quality of air-dried material you need. If not, settle for green lumber and stack it yourself for proper air drying. Even the kiln-dried stuff can be made reasonably serviceable by stacking, so purchase everything you can far in advance of your needs. Paint the ends of boards with any old paint to prevent drying and checking of the end grain.

Here's how to stack lumber. Find a flat spot where you can cover the material with a plastic cover, sheets of plywood, or a simple framed-up shed roof covered with plastic, roll roofing, or whatever. Space out four or five strong timbers. These should be 4 by 4, 2 by 6 on edge, or heavier. Level these timbers with blocks (Figure 6-10). Now put down a layer of lumber with 1-inch spaces between the planks. Next, lay down four or five sticks ½ to ¾ inch thick (laths, tomato stakes, or scrap) across the lumber at right angles to it. Repeat this layering for your entire stock. Finally, arrange the cover loosely so it sheds rain, but leave the ends and sides open. The lumber should be shielded from direct hot sunlight, too.

Drying must be slow and uniform, so every two or three weeks unstack the pile and turn each piece over. Watch for crooks. Wedge badly bowed boards between straight ones or stand them on edge so the weight of the pile forces them straight. Don't be too concerned about the graying, especially if the lumber is still in the rough.

Avoid lumber called construction grade. This is used to frame houses and buildings and has no place in a vessel. It will rot in several years in a moist atmosphere. Clear vertical-grain Oregon fir is excellent lumber. It is heavier than spruce, but very strong. Thus, lighter sections may be used. Clear Douglas fir is good spar lumber. Fir is excellent for framing, and it is used for keels and deadwood also. Fir should be treated for rot resistance, particularly if the vessel is going into warmer climes.

I assume that all modern wood boatbuilders will treat all materials against fungus, perhaps by epoxy saturation techniques, or by direct application of Cuprinol, or by other preventive measures (more on this later). However, the wood should be quarter-sawn. This is called vertical grain or edge grain also. Look at the ends of quarter-sawn boards. The grain

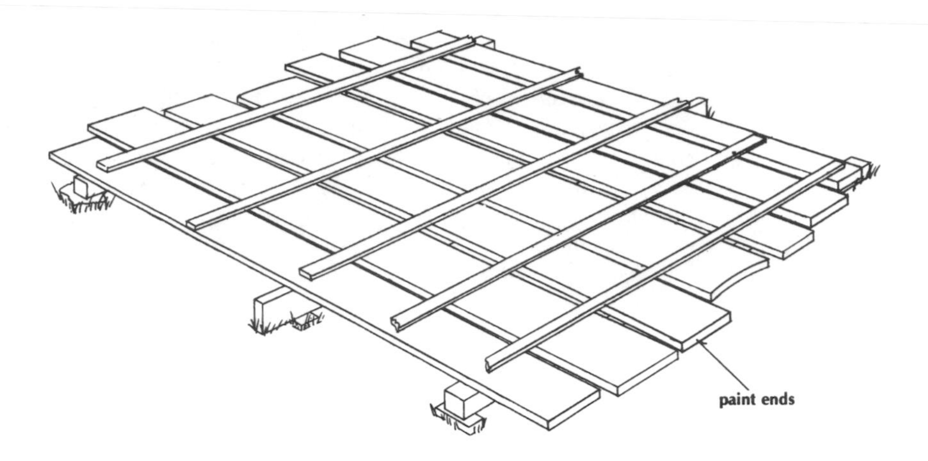

Figure 6-10. *To dry lumber, begin by "sticking."*

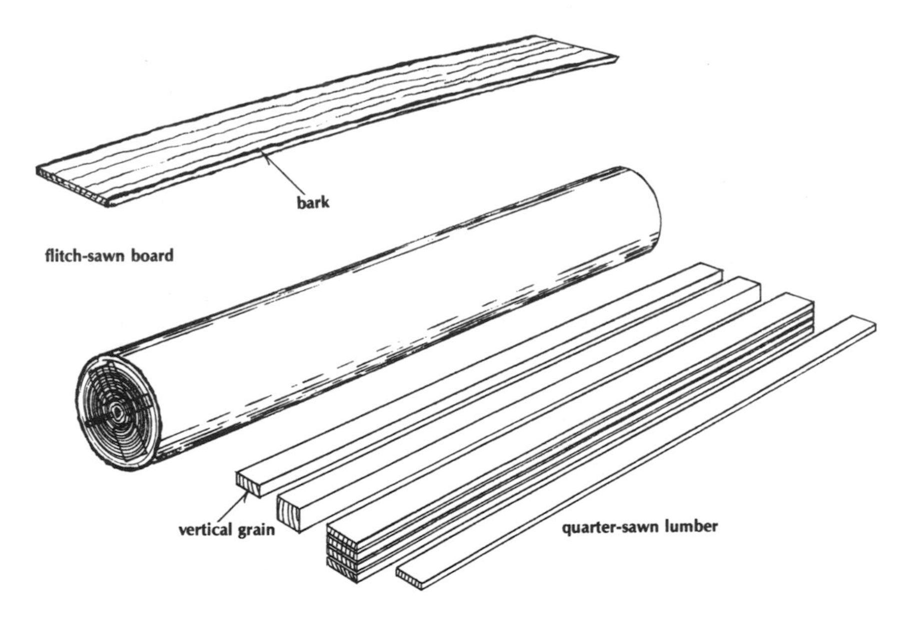

Figure 6-11. *Lumber cuts.*

runs more or less vertical to the wide surface. The boards were sawed from the log in quarters, as shown in Figure 6-11. This grain pattern increases a board's stiffness, compared with a flat grain. It shrinks and warps less and is also easier to plane.

Oak or fir may be used for structures below the waterline, such as sole beams and supports, but these woods are too heavy for deck beams and framing. Spruce is best here. Laminated beams are excellent and conserve lumber. Moreover, they look rich with the entire beam finished bright. A lower grade of spruce other than Sitka or aircraft

Figure 6-12. *A bare teak deck. Note beautiful Dorade vent. (Bruce Bingham photo)*

Figure 6-13. *A white pine deck with contrasting compound. Note traditional pinrail. (Bruce Bingham photo)*

would be perfectly satisfactory. Long lengths of any grade are now rare. That's why you have to learn how to make strong scarfs.

Western red cedar is sometimes used for relatively light planking. It is more than adequate for laminated or cold-molded planking saturated (or sealed) with epoxy. Philippine mahogany is almost perfect for any use, but it is much heavier than any cedar. While planking is not the concern of this book, you should remember that laminated systems can be used for streamlined deck structures, bent coamings, and so on. For such purposes, if economy is vital, you can use cheaper grades of wood such as the so-called knotty pine and even construction-grade fir. Just cut out the knotty areas, and rip the wood in between into strips and/or resaw for laminating.

Some of the so-called hardwoods may actually be soft. Philippine mahogany, teak, and Honduras and African mahoganies are all softwoods. They are delightful to work, although the last two are rare indeed, and teak is deadly to cutting tools. The most desirable Philippine is lauan, which should be selected for color (a deep red), straight grain, and perhaps some ribbon grain for beauty. This grade is called first, meaning "clear of knots on one side and free of rot, shakes, and checks and other faults on the other side," according to Howard Chapelle. For steam bending, use oak firsts only. This will have to come from young green trees, with no drying at all. Ask your lumberman for bending stock. Or try to find a country mill that is cutting oak.

DECKING

The best deck for footing, wear, and appearance is teak left bare (Figure 6-12). Teak weathers to a beautiful silver gray color, or it can be oiled. It is costly, but because it requires no finish, even on interior trim, the final cost is moderate. Sometimes exterior trim, grabrails, hatches, moldings, and even cabin trunks of teak are varnished. Many swear there is nothing handsomer. You'll see varnished teak on some of the world's finest yachts—especially those with crews. For decks, however, bare teak is superb. You can't varnish any deck, as it would be too slippery. This means Philippine mahogany is out for decking; it can't be left bare.

Clear white pine also is a beautiful decking wood, especially with its seams filled with a black compound (Figure 6-13). White pine must be sanded to keep it clean, and it does absorb suntan oil and grease. Vertical-grain Oregon fir makes a great deck, and it may be treated with sealers such as Rez, Firzite, or special oils made for this purpose. You can buy kiln-dried Oregon fir at most yards as stair tread, dressed from 1 inch to 1⅛ inches. All decks are assumed to be laid over plywood, so 1½- to 2-inch pine is not necessary. All in all, this reduces total weight, saves precious lumber, and makes a watertight deck—something most traditionally laid decks cannot claim. Details on decking follow in Chapter Nine.

PART 2

7 Nineteen Useful Joints and How to Make Them

Some of the following information may not be new to those of you with extensive woodworking and yacht joinery experience. On the other hand, the information presented should be useful to readers with less skill and limited tool inventories. Included are descriptions of the construction of scores of components from sole beams to spars. I'll show you several ways to make parts, taking into account tool availability and differing degrees of skill. In general, I prefer a simple method if it produces a strong part, even though appearance may suffer a bit in rare instances. In other cases, a complicated procedure is more likely to create the better result. I'll attempt to show why these steps are preferable, if not mandatory.

In my opinion, a simple, low-cost approach often can create an equally functional and eye-pleasing effect. Thus, I hope to help those who want to save time and dollars and still turn out a respectable yacht.

A THOUSAND WOODEN JOINTS

If you have never spent time in a hard-working wooden vessel, your life is incomplete. The sound of a ship as her hull drives to windward against a sea is music. You hear the soft sighing and bubbling of foam caressing her planking, a muffled thud far up in her bows, and the whoosh of spray and green water shooting over her bulwarks. Then there's the hurried gurgling of water cascading across her decks and pouring from her scuppers. In addition, there are the voices of the ship as she butts her path through the seas. You hear the creaking, cracking, and groaning of thousands of joints protesting the forces battering the hull. No orchestra can match this music. Moreover, the tighter the ship's joints, the sweeter the symphony.

It is sometimes said that it isn't the fastenings that hold a ship together, it is the joints. If this is true, then each joint must be appropriate to its function, and it must be fitted with care. Of course, these comments pertain primarily to traditional wooden vessels. There is, however, much joining of wood in modern laminated, fiberglass, metal, or ferrocement vessels. The sounds of such a vessel when underway may not be the same, but the satisfaction to be found in quality workmanship surely is.

The fitting of a bulkhead into a hull is a joint. A drawer contains many joints, and a deck consists of hundreds of joints. This chapter will tell you how to fit many common joints with a variety of tools and techniques. Three or four ways are often described. Method A may take five times longer and be painfully difficult because it uses only hand tools. Method B may show how the use of hand power tools improves matters. Method C will show how speed and accuracy are attained with sophisticated tools and accessories.

Do give this chapter some attention. Later, when I describe yacht joinery problems, I shall not repeat recommendations given here. It would help, too, to review the suggestions given earlier. Not all of these will be repeated.

Because naval architects and designers do not always spell out the details of joints, you may have to decide which joint to use. Often there are several choices. The joints shown in this chapter are basic. Many variations are commonly used. To supplement what follows, I recommend *Woodwork Joints*, by Charles H. Hayward (published by Drake), and *How to Work with Tools and Wood*, by

Robert Campbell and N.H. Mager (published by Pocket Books).

Keep in mind that you must use tools that are sharp. Your backsaw must be fine-toothed and properly sharpened and set. Your table saw must be equipped with a cabinetmaker's combination or a fine carbide-tipped blade. Your jointer knives should be honed just before you tackle wild grain hardwoods. Your plane blade or chisel should be razor sharp. Sharp tools save hours and produce fine workmanship with less effort.

Now let's fit those basic joints.

BUTT JOINTS

Butt means to join end to end, or end to side. End-to-end butt joints are used in yacht construction only when backed by a long block of similar thickness, as in planking. Should you elect to butt-block a sheer clamp or bilge stringer (they really should be scarfed), you can achieve a perfect fit of the ends. Clamp up and then run a handsaw between the ends two or three times, closing up with a crack of a hammer on the outer ends after each try.

The 90-degree side or T butt shown in Figure 7-1 has some uses. This joint is extremely weak, however, as it is entirely dependent on fastenings. Glue is of little value on end grain. Toenailing, in addition to nail or screw fastenings from the back side, adds some strength. This joint could be supported by a shelf or riser.

DADOED BUTT

The dadoed butt is a joint that provides greatly increased gluing area and quite rigid construction if carefully made (Figure 7-2). Fasten the same as the T butt joint. Excessive tension and/or working can make this joint fail, as it is greatly dependent on glue.

If you are using method A, mark the dado with a knife and make several cuts with a backsaw to a depth equal to a third of the thickness of the material. The material between the cuts is then removed with a sharp chisel narrower than the dado groove. Swing the chisel from side to side in a paring motion, coming in from both ends of the dado. If the dado is more than ¾ inch wide, make three or more cuts so the chips are small and you have control. Cut off the end of A in a miter box.

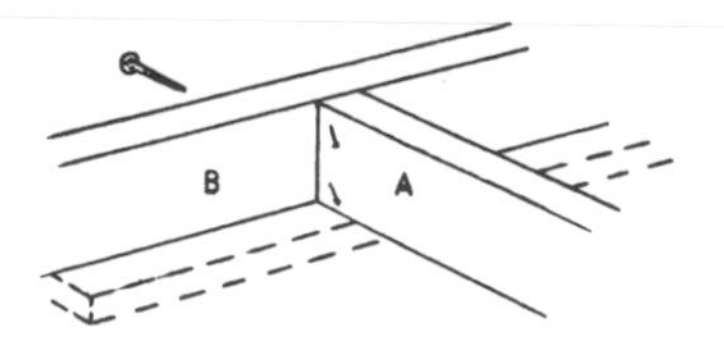

Figure 7-1. *T butt.*

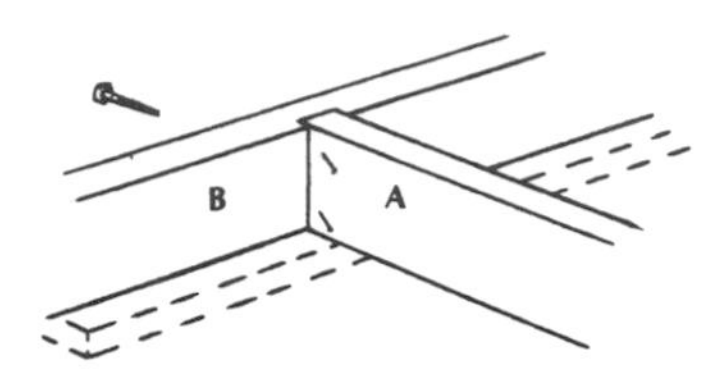

Figure 7-2. *Dadoed butt.*

Method B is identical except that the cuts are made with a hand power saw set just less than the wanted depth. Use a square guide clamped to the piece to control the cut. Or use a router. Clean up the dado surface with a chisel.

Method C is done on your table saw with the blade set to the depth of the dado. The knife marks should be on the edge of the piece so the line can be sighted along the blade while it is stationary. If you trust your miter gauge, make repeated cuts to clean out the dado groove. A SLAT would be more accurate, for when one is used, there is less tendency for the piece to creep. Even better is a dado set or adjustable dado cutter used with either a miter gauge or a SLAT (Figures 5-4, 5-5, 5-6).

Glue the dadoed butt joint by coating the sides and bottom of the groove with crack-filling Aerolite 306. Paint the catalyst on the entering piece. Do not make a squeaky fit, and clamp only enough to hold the pieces in position while the glue is curing. Do not squeeze the glue entirely out of the joint. It's OK to add the fastening through the back. Epoxy glue would be strong. If Weldwood Plastic Resin is preferred, coat both surfaces, clamp, and screw fasten. Glue and C-clamp to the underlying shelf, if one is to be used.

DOVETAILED DADO

The dovetailed dado is an extremely strong joint that has many uses—for example, joining the short side-deck beams to the carlings—but it is one of

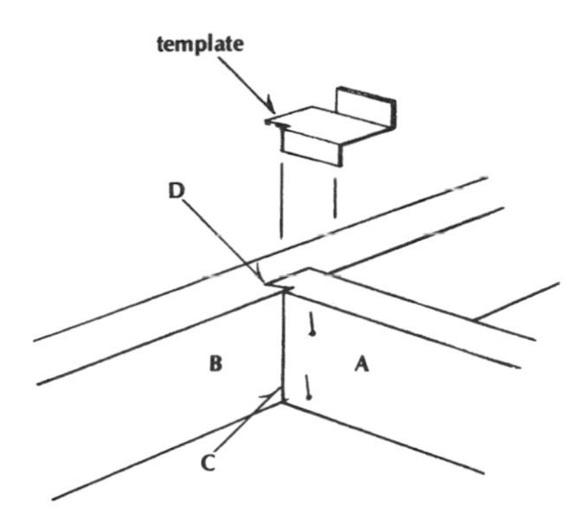

Figure 7-3. *Dovetailed dado.*

the more difficult to construct with accuracy (Figure 7-3). If you plan a number of these joints, I recommend that you first make a sheet metal pattern of aluminum. Use this template to mark off with a knife the angle and other dimensions of pin A, the part fitting into the recess—the dovetail. Save the template. The depth of the socket should be one-third the thickness of the piece and the depth of the shoulder (C) in A should be one-third or less. Fashion pin A first.

Method A squares the end of part A, then clamps the template to it so you can mark it accurately, top and bottom, with a knife. Use a miter box to backsaw the shoulder (C) just shy of the line. Secure A in a vise vertically so you can saw the angle that locks to D. A dovetail saw (a small, fine backsaw) is almost a must. Try to leave a little stock for trimming back to the line carefully. The best way is to clamp the beam flat on a bench and chisel back. Check repeatedly with the pattern and a try square until you shave down to the knife marks.

Now for part B. C-clamp the template over the desired location and mark. Square across B and mark on the bottom from the template. Backsaw just inside the marks with repeated cuts to ease removal of the material. Your chisel should be a socket firmer (with beveled edges), for this shape enables you to get into the corner (D).

Try dropping part A into the dovetail, marking with a pencil where more wood has to be removed. Pare in carefully, cutting from both ends of the dovetail at both top and bottom. In theory, you come right down to the knife marks, but life is not like that. It will take a great deal of fit and try to attain a snug fit. A squeaky fit here would be best to resist tension. If you see some daylight, however, don't despair. Aerolite glue is coming to the rescue. It fills cracks wider than 1⁄32 inch with full strength (but don't let your workmanship slide). Coat the dovetail sides, back, and shoulder with the resin, and wet the pin with the catalyst. Press the pin in and fasten with a couple of galvanized toenails.

If you are fitting from sheer clamp to carling, don't glue any joints until all are completed. You may have to disassemble the whole thing a dozen times as you fit one after another. It's far better to fit all into the clamp, then clamp the carling to a couple of strong beam molds while you mark the locations and lengths. Then you can do the actual cutting of dovetails on a bench. By the time you finish four or five, you'll be a pro.

Method B is a bit different. Use a square guide on part B and a fine-toothed blade in your power handsaw. Most of the material can be removed with repeated passes, but not much faster than with a backsaw. The last cut is the shoulder angle (D). Once you get the angle right, saw all of it. Then it's back to the chisel for cleaning up. I would not try to use a power saw for shaping the pin on part A.

Method C uses your table saw to cut the rough socket, but of course, only if it is not installed. Thus, this lets out the sheer clamp. It's all right for the carling, though, and also for the pins. I would much prefer to use a SLAT for this job, but this is a handmade joint. The rough cut could be made with a dado set, but the setting-up time is not justified unless you have a number of cuts ready to go.

There is a method B, however, that uses a router with a dovetail bit. In this method, you follow a square guide clamped to part B. Be sure that the bit angle and the cut angle of the pin coincide. Make the dovetail pass, then dado out the remaining material to the mark. This can be done precisely on a SLAT (table saw), method C. None of this, however, is practical unless you have a sufficient number of joints to fashion in one setup.

MORTISE AND TENON

A mortise and tenon is one of the strongest joints. It is used more often in furniture than in yacht joinery, but you will find applications. This joint is a real test of your skills. There are many variations. See Figure 7-4, parts A and B, a plain mortise and tenon. Some of these include wedges inserted into saw kerfs in the tenon. Some are pinned by a

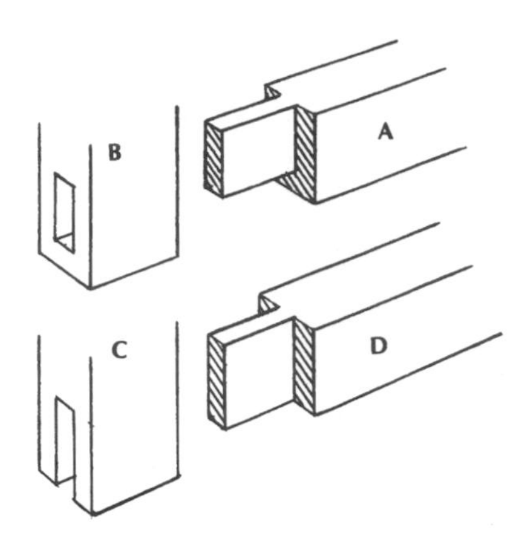

Figure 7-4. *Mortise and tenon.*

dowel. Others have an extended tenon through which a wedge is fitted. This is sometimes seen in yacht table construction.

An easier joint is the open mortise and tenon, parts C and D in Figure 7-4. Not being totally enclosed, this joint lacks some of the strength of the plain mortise and tenon, but it can be made on a table saw. The open mortise and tenon is excellent for supports for cabin soles, for transom berths, for hatch screens, and so on.

The plain mortise and tenon can be made by method A. First, you backsaw the tenon to knife marks. The mortise (Figures 2-13, 2-14, 4-23) may then be cut out roughly by boring four or five holes with an auger or spur bit smaller than the width. Open up the rectangular socket with a thin, narrow chisel, working in from opposite sides of the opening. A wider chisel is best for cleaning up to the marks. Again, this is try and fit. Insert the tenon and mark and trim the shoulders to a square fit. Then finish off the exposed end of the tenon.

Routing, a method B, is commonly used to cut mortises, provided the machine's depth adjustment is sufficient for the bit to reach the bottom of the mortise. Extra depth can be gained by moving the bit out of the collet somewhat, but be sure it is locked securely. Clamp a piece on the side so the router has a wider base to ride on. Rocking will ruin the work. A rather deep guide must be attached accurately so the bit starts right on the line. It takes practice and skill to start and stop the slot within the end marks. A fast-turning drill press with a router bit simplifies this job, if you can keep the bit secure in the chuck. Vibration and high speed have a way of loosening bits. Clamp a fence to the drill press table and slide the work piece slowly the length required. Now that you have rounded ends in the mortise, round the tenon to fit.

The open mortise and tenon also may be made with a backsaw and chisel, but its design cries for a method C approach using a table saw. Add an extension to your fence, a piece 6 to 8 inches high, square to the table. The work piece (D) for the tenon must be truly vertical when the cuts are made. Make both cuts in all your pieces right up to the line, then set the fence back that dimension and lower the blade so it just cuts out the piece to form a nice shoulder. The thicknesses of all tenons will be identical because they were made by rotating the work piece.

Now make the open-end mortise the same way. Practice on scrap, adjusting the fence as needed, until you get a snug fit on the tenon. Make all these cuts in sequence, resetting the fence once or twice so the multiple cuts remove all the material.

This entire joint can be made on a bandsaw using a temporary fence, but it is less likely to be accurate with smooth surfaces. Whether you use a table saw or a bandsaw, push the work through with a square block sliding along the fence to hold the piece vertical. And set up a feather or comb to hold it against the fence so it is vertical in both planes.

Both of the mortise-and-tenon joints should be glued, especially if screws would be visible. Coat the mortise with Aerolite and the tenon with catalyst. A dowel through will treble strength. Clamp lightly.

DOWELED BUTT

A doweled butt is a form of mortise and tenon that eliminates the difficult hand work at a sacrifice of some strength (Figure 7-5). It is a fair joint for hatch screens, paneled doors, and so on. You'll need a drill press (for method C), because the holes must be precisely parallel and on center. The layout of the centers must be done with a knife, not a pencil. It is essential that the dowel holes in part B be drilled first, for drilling in end grain frequently causes wandering, unless you use a Forstner or machine spur bit. Never use a twist drill.

There are several ways in addition to a scribed layout to transfer center locations from B to A. The best is the use of dowel centers. These are button-like devices available in sets to fit several sizes of

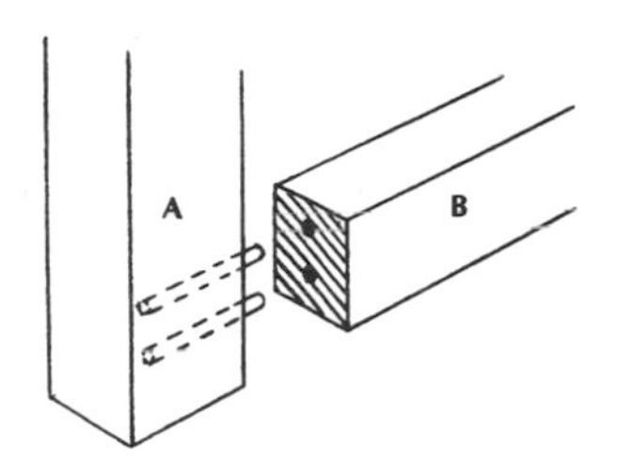

Figure 7-5. *Doweled butt.*

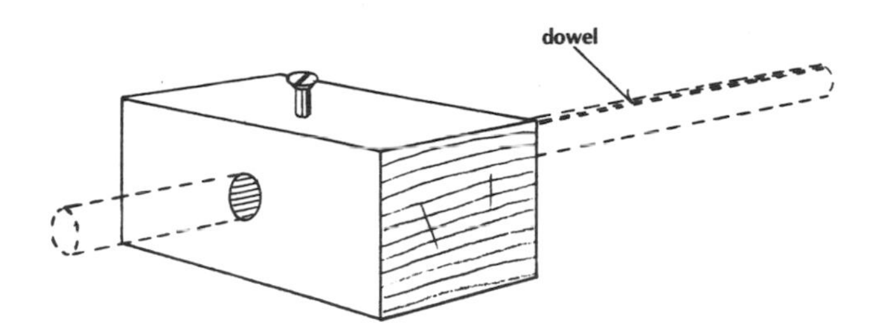

Figure 7-6. *Dowel groove jig.*

dowel holes. The dowel centers are inserted in one set of holes, then the parts are pressed together so the points make an impression in the opposite part. Try clamping both pieces flat and square on the saw table or a bench, then tap sharply with a mallet. Drill the holes about ¼ inch deeper than the length of the dowels so glue and air have somewhere to go.

There's another method for marking dowel centers. Grind the heads off brads, and then grind the ends to a sharp point. Next, drive another pair of brads or small nails into the dowel locations, leaving the heads out for easy removal. Remove these brads. Now insert the sharpened brads in these holes, leaving enough protruding to be grasped with pliers. Press the two parts into careful alignment so the sharpened brad tips press in just enough to make a center mark for your bit. (Don't press the pins below the surface of the wood. You'll never get them out.)

If you don't have a drill press, you can make a boring guide on the order of the one in Chapter Five. Or use a metal dowel jig, which is available wherever tools are sold.

A third way is strictly method A. I have used it many times because it is fast and simple. Clamp the parts together exactly in position on a flat surface, then bore through part A into part B. This eliminates all the problems of surface matching, hole centering, boring alignment, and so on. In my opinion, the exposed ends of the dowels contrast nicely with mahogany, teak, or walnut, creating a pleasing effect.

You can make your own dowels, if you don't mind the effort required. Dowels must have a couple of grooves running full length so air and excess glue can escape. Otherwise, part B of the joint might split. You can build the simple grooving jig shown in Figure 7-6. Make the hole an easy fit for the dowel. Turn the screw until its point enters the bore in the block $1/16$ inch or better. Now just drive the dowel through two or three times so the screw scores grooves along its full length. Cut the dowel to length and chamfer one end on your belt sander or even on a sheet of sandpaper on a bench.

But let's face it—ready-made dowels are perfect with their spiral grooves. You won't be using them by the hundreds, so save time and invest in what you need.

MITER JOINTS

Miter joints (Figure 7-7) are extremely weak in themselves. However, there are dozens of uses for them in cabinets and yacht joinery where their neat appearance is desirable. Door frames, trim, and drawer fronts are examples. Miter joints are quite easy to make with a miter box and, preferably, a miter or dovetail saw. This is method A. If trim goes around four sides of a door or its opening, the fits cannot be too long or too short. Miters must be clean and sharp, with no visible opening.

I cut the first piece about $1/32$ inch long; tape, tack, or clamp it in position; then try the corner with a short piece I call a fitting piece (Figure 7-8). Your first length can be reduced with a sharp block

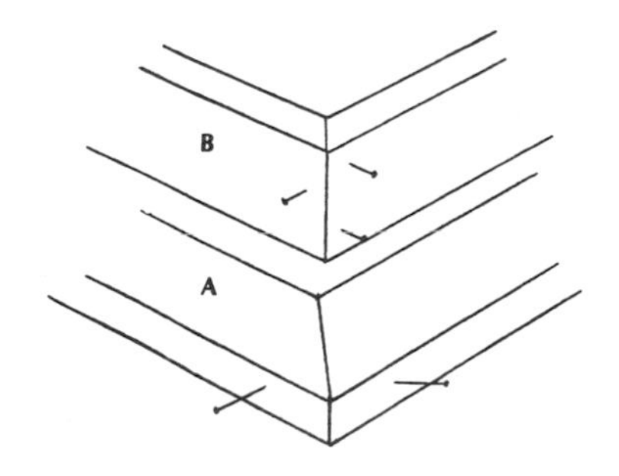

Figure 7-7. *Mitered joint.*

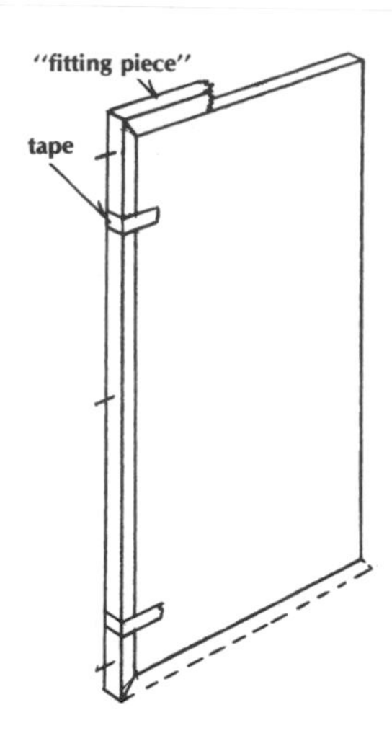

Figure 7-8. *Mitering moldings.*

plane until the fitting piece meets the corner perfectly on both ends. You may need a right and a left fitting piece. Tape, brad, or clamp the first length in position. Start your second piece by placing the corner miters as they belong while you mark with a knife where the next miter is to be cut. Again cut slightly long in the miter box, unless you have a lot of confidence. Tape the second piece against the first at several points along its length, if needed, and again try the fitting piece. Trim off if needed. Once more, tape, brad, or clamp this piece to the door or drawer, or whatever.

Fit the third and fourth pieces in the same manner. When all the miters look perfect, glue and fasten everything in place permanently. I use pipe clamps if I can't use brads or finish nails, sometimes three or four across one face of the door, and as many across the other. It's possible to use nothing but tape, but be sure the joints are fully closed if you go this route. Tape and clamps work well together, and so do brads set in. The miter joint is a good place for Aerolite glue.

Method C uses your table saw miter gauge set at 45 degrees, or the miter gauge at 90 degrees and the blade set to 45 degrees. Using a SLAT on the table saw, however, is so superior in speed and accuracy, and also in its ability to take off 1/64 inch, that it is the only way to go; see Chapter Five. You can tack 45-degree guides right on the sliding panel itself and get perfect joints (Figures 5-4, 5-5, 5-6, and 5-16).

There is a fine method B that uses your power handsaw. Remember the little portable table saw described in Chapter Five? For this mitering job, that saw will produce the same accuracy as a table saw, at the same speed. But with a simple SLAT, of course. So build one now (Figure 5-25).

Let me interject a couple of hints. Always use a cabinetmaker's combination blade to get a polished miter surface. Otherwise, the miter will absorb a lot of filler and stain that will show. Second, position the piece about to be sawed on the right- or left-hand guide so the edge you want to keep does not splinter. The blade must not go against the grain in the piece you keep. See Chapter Six on saber-sawing a panel. The same rule applies.

SPLINED MITER

A splined miter is made and fitted in the same way as the plain miter, but the spline (Figure 7-9) adds greatly to the joint's strength. While this is not a small task, it should be done where there is stress. For example, hatch screens wracked by careless stowing. The splined miter is one way to reinforce such structures. Also see the splined joint, which is described later in this chapter (Figures 7-18, 7-19).

Method A is the same as for the plain miter. Then, after the miters are done, lay out a groove on the joint surfaces (1/4 inch wide by 3/8 inch deep in 3/4-inch stock). Use a dovetail or fine backsaw to saw to the full depth, then chisel out the material between the kerfs, as shown in Figure 7-10. There is a danger of splitting off a chip as the chisel approaches the thin end, so back up this edge by clamping it to a piece of scrap. Make the splines by sawing a 3/4-inch strip off the end of a sheet of 1/4-inch plywood. A plywood spline is less likely to split than a strip of wood. Fit the splines, cut the V on the inner ends, and let the outer ends protrude. Now assemble the entire rig with glue and pipe clamps. Check with a square as you go, and secure by tacking on a light diagonal. Weight the assembly on a flat bench or the floor to hold it flat.

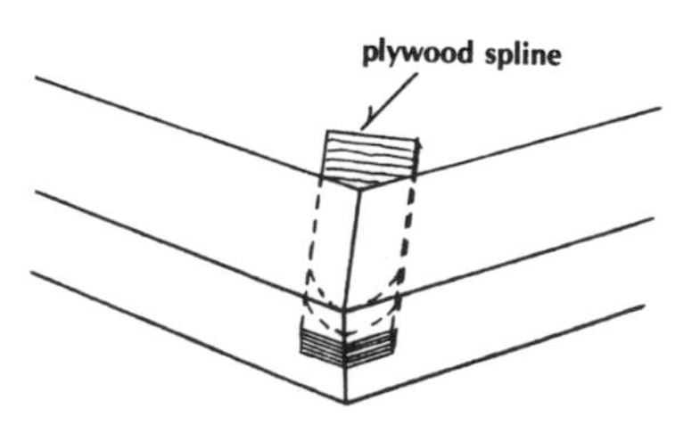

Figure 7-9. *Splined miter.*

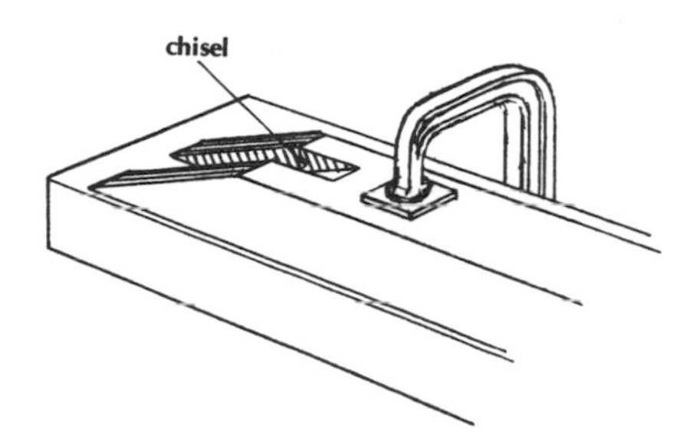

Figure 7-10. *Miter spline groove.*

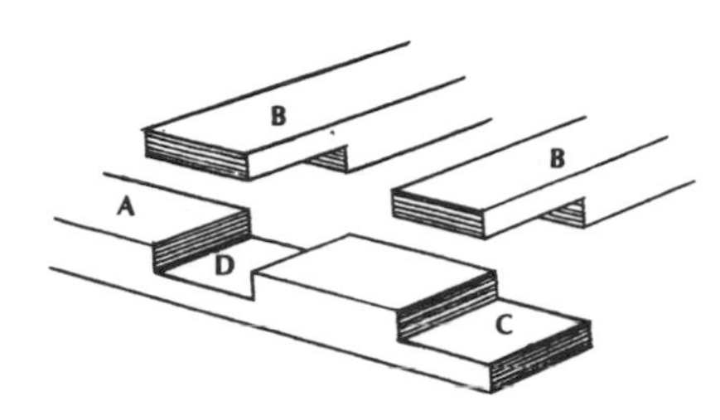

Figure 7-11. *Middle and end half lap.*

The only method B that I can suggest is to mold the grooves with a ¼-inch router bit, or on a portable table saw above. In method C the grooves are sawed or dadoed on a table saw.

LAP JOINTS OR HALF LAPS

End and middle laps are used frequently in yacht joinery (Figure 7-11). The end lap is a substitute for the mortise and tenon, doweled butt, and mitered joints. The end lap is not as strong as the first two, and perhaps not as attractive as the miter. End and mid-lap joints would be excellent for supports for sole beams, settees, berths, and so on. The end lap will support a weight. If used in a frame, however, such as in hatch screens or in a paneled door, it would be entirely dependent on the glue for strength, if the material thickness won't take mechanical fasteners. This calls for epoxy thickened with limestone flour or silica fillers.

Method A for these joints is a backsaw job. Mark with a knife and make repeated kerfs in the portion to be removed. Use paring strokes of the chisel, working in from both sides. Use care so that the mating pieces match up to make a smooth surface. You can use five ¾-inch brass brads in a pattern if the stock is ¾ inch or more in thickness, or you can set the brads at a slight angle. Plugging over screws in a ⅜-inch thickness is too close for most craftsmen.

Method B is to cut repeated kerfs with your power handsaw after the square cuts, followed by cleanup with a chisel.

Method C is to cut the B parts and cutout C using the lap joint jig described in Chapter Five (Figure 5-40). Cutout D can be done on your table saw, using a SLAT, or with a miter gauge and repeated cuts, or with a dado set. Allow for fitting with a chisel if appearance is vital. Much of the cutting for this type of joint can be done on a bandsaw, with practice.

SINGLE DOVETAIL

A single dovetail is essentially the same as the dovetail butt joint, but the long dimensions are horizontal and there are two shoulders (Figure 7-12). The joint is easy to make. In method A, the pin can be cut with a backsaw or dovetail saw; in method B, with a sabersaw; in method C, with a bandsaw. Pare the edges with a sharp chisel. Clamp the pin square on the part (B) to be dovetailed, and mark with a knife. Cut out and trim to fit part A. Secure with epoxy, Acrolite, or Plastic Resin glue. This joint is good under tensile stress or wracking.

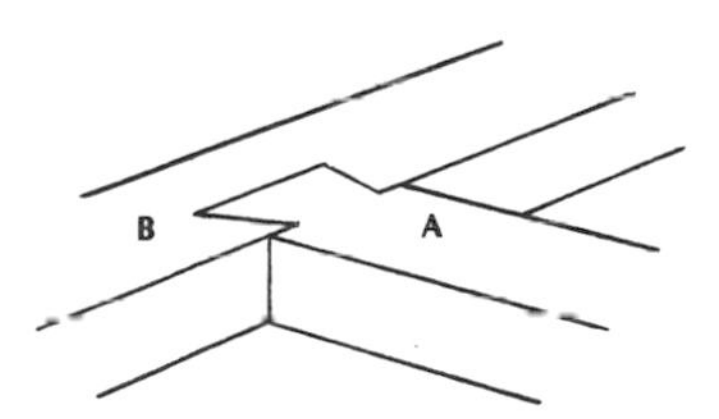

Figure 7-12. *Single dovetail.*

TONGUE AND GROOVE

The tongue-and-groove joint (Figure 7-13) is used more and more these days because of the scarcity of wide lumber. It is used also to join pieces with alternating grain. This helps prevent warping. A wide structure such as a cabin side, if built of one width, would be subject to a lot of shrinkage and possibly even splitting. A tongue-and-groove joint here, however, would remain watertight even

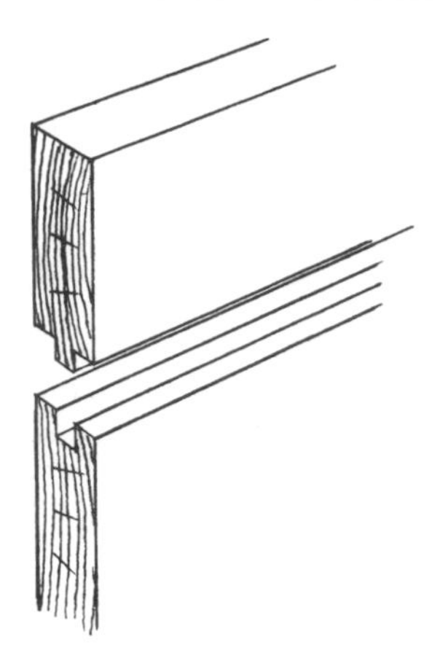

Figure 7-13. *Tongue and groove.*

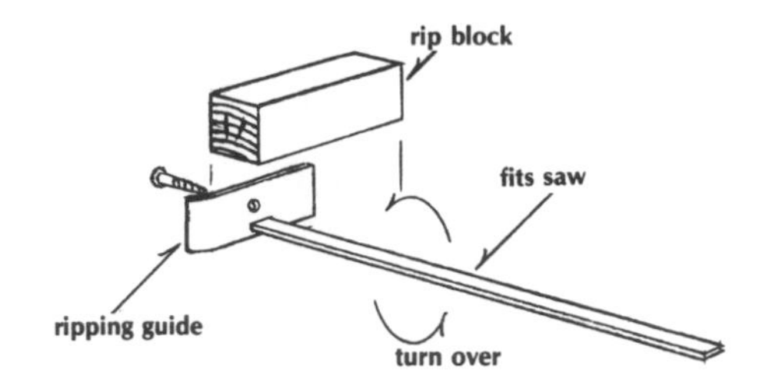

Figure 7-14. *Jig for tongue and groove.*

though the joint separated. Of course, this applies also to the splined joint (Figure 7-18). Tongue and groove is standard procedure for table-top construction if solid hardwood lumber is used. Such long joints can be doweled, as discussed earlier.

Any method A for building a tongue-and-groove joint is likely to be reminiscent of the 17th century. Even with a special combination rabbet plane, it would be an extremely laborious task. Modern rabbet (or filister) planes have matched tongue-and-groove sets, including blades and fences that adjust for the material thickness. These must follow a guide such as a clamped-on straightedge. If the cost of such specialized tools does not suit you, try to make do with doweled joints.

Method B is less laborious but still primitive. First, you must pad the ripping guide of the power handsaw if it can't be adjusted close to the blade. This is done by screwing a small block 1 inch thick by about 2½ to 3 inches long to the T on the guide (Figure 7-14). Try the tongue cuts on scrap. Cut the side of the tongue first. Your saw will need a broad base to ride on, so clamp a 2 by 4 on one side, as shown in Figure 7-15. Use two if you don't feel confident. Make the second cut to leave a tongue ¼ inch thick by about 7⁄16 inch high. Cut steadily, for parallel sides are needed. Now lay the piece on its side and rip out the second dimension to form a ¼-inch by 7⁄16-inch rabbet. Repeat on the opposite side.

To saw out the groove, repeat the 2 by 4 step, but make the depth a bit more than that of the tongue to allow space for glue and air. Experiment on scrap. The tongue should slip without pressure.

Another method B uses a ¼-inch router bit and a guide attached to the sole plate. You can make both the tongue and the groove this way. In addition, you may use router tongue-and-groove bits or glue joint bits, if desired. To do this, you will need a router-shaper table, as described in Chapter Five.

Method C, the table-saw approach, offers several ways to go. First, you can get by with any reasonably fine-toothed blade and a tall fence attachment. Flip the piece end for end so the tongue will be centered. Do the groove the same way. Be sure to make these cuts on scrap first, so they fit. Second, you can set up a dado cutter for a ¼-inch groove, and take out both the shoulders of the tongue and the groove itself. The third way is the best. It uses a molding head with matching tongue-and-groove or glue joint blade. In all of these methods, be sure to use a comb hold-down to keep the piece against the fence.

The ultimate method C is to form with a shaper, using matched cutters. These can be found at most tool shops. These cutters are not cheap, so be sure

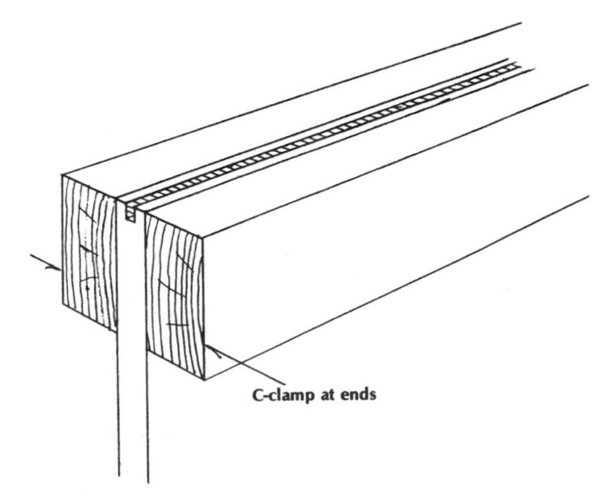

Figure 7-15. *One or two 2 x 4s make base for saw.*

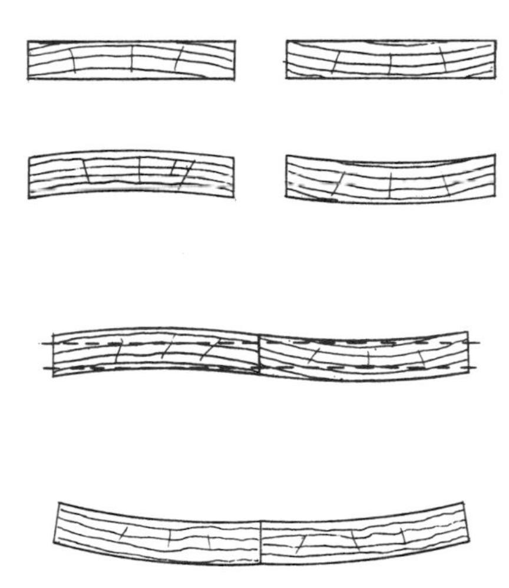

Figure 7-16. *Alternating grains.*

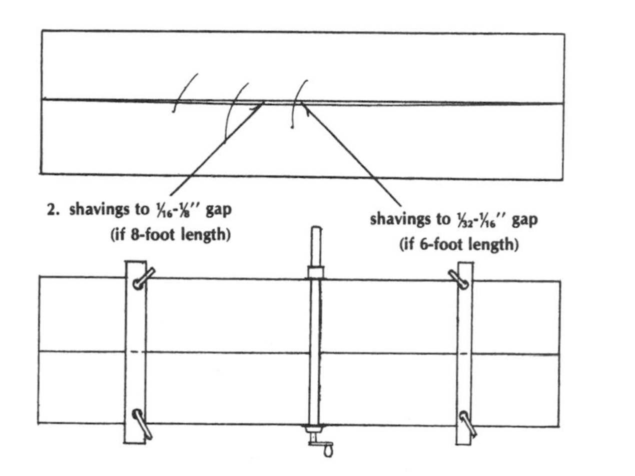

Figure 7-17. *Long pieces joined lengthwise.*

you have plenty of work ahead of you before buying. If you build a shaper table for your router, you'll get the same results. This kind of molding requires only a straight fence, or you can shape against a collar. If your tool inventory includes a head and set of molding cutters for your table saw, a third good method is available to you. With these tools too, be sure to use adequate hold-downs so the grooves don't wander.

Long pieces joined lengthwise have two characteristics you should be aware of. They warp and tend to open up at the end of the joint as the pieces shrink. The answer to the first problem is alternating the grain, as shown in Figure 7-16. This is mandatory if the pieces already show warping. The second problem is solved by planing the edges so they meet evenly full length, as shown in Figure 7-17. You then take off a couple of shavings so the ends close up first. Thus, the pieces are able to shrink slightly without showing an open joint. The drawing shows an example of clamping. In practice, you would need three or four pipe or bar clamps, rather than just one. The clamps should be alternated. The cross-wise pieces shown keep the assembly from buckling. Thus, there should be one on each side, like the bread of a sandwich.

If you are using plywood, all of this is unnecessary. Two good plywood remnants, however, can be joined by tongue and groove or splines. The gap procedure described above can be skipped, for plywood is dimensionally stable.

SPLINED JOINT

There are other methods for forming good tongue-and-groove joints, but the complexity and cost of tools may lead you to a simpler joint. The splined joint (Figure 7-18) has all the strength of a true tongue and groove, yet it can be made in a few minutes with a rabbet plane, a power saw, a table saw with a regular blade, or a dado set. You don't need a router or a shaper. Another advantage of the spline is that the groove and spline need not be visible at the finished ends of the joints. You simply stop the planing, sawing, or dadoing before the tool reaches the end of the work piece. This produces a blind groove, as shown in Figure 7-19. A 1/4-inch spline is fitted in and glued with Weldwood, Aerolite, or epoxy glue, then the opposite groove and surfaces are coated in the same way. For indoor cabinetry, a good white glue is fine. Pull the assembly together with bar or pipe clamps.

Similarly, the use of a spline can make a miter joint quite strong. In a wide joint—say, for a panel door—the groove can be blind.

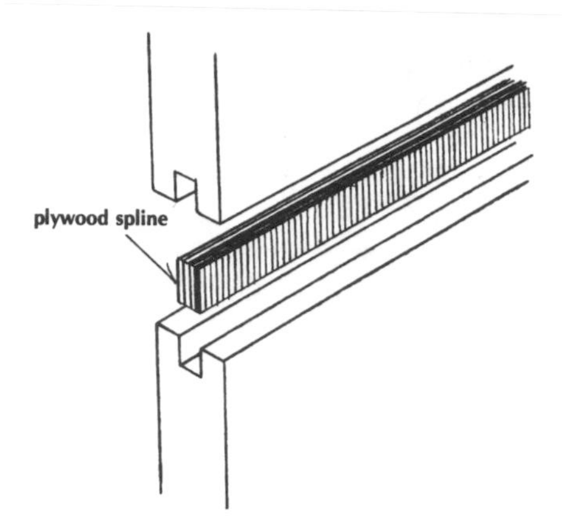

Figure 7-18. *Splined joint.*

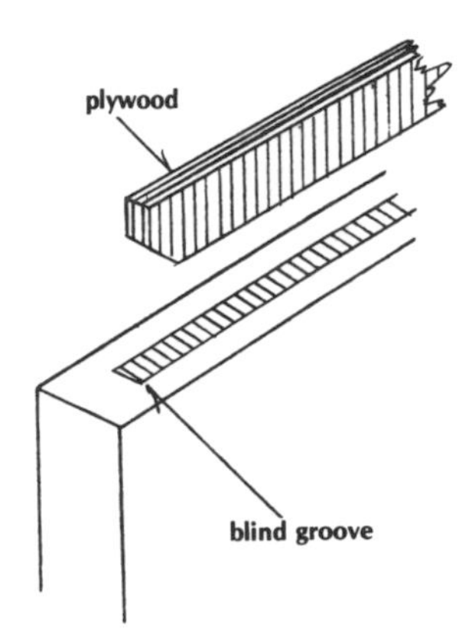

Figure 7-19. *Blind groove.*

RABBETED CORNER

Bulkhead corners and others, such as on dressers and lockers in a yacht, must have strength. A body can be thrown about a bit on occasion in a vessel. If this projectile meets a stationary object, both might be wrecked. The basic corner (Figure 7-20) is rabbeted and glued and screwed or nailed. (Later I will describe a more involved type of construction in corner posts.) The procedures for rabbeting are similar to those for tongue-and-groove joints.

The rabbet can be made by a method B using a power handsaw with a rip guide. To cut surface A parallel to the sides of the plywood, clamp on a straight 2 by 4 so the saw has a reasonable platform to run on without wobbling. Surface B is no problem, but accurate setting counts. Another method B is making a couple of passes with your router rabbet bit. To ensure a nice clean edge on B, score a knife cut to eliminate splintering. Fasten the joint with Aerolite and galvanized finish nails. If the corner is not covered by a hardwood molding, round it with a router after installation, but be sure the nails are set in sufficiently. Alternatively, you can hand plane the radius before installation.

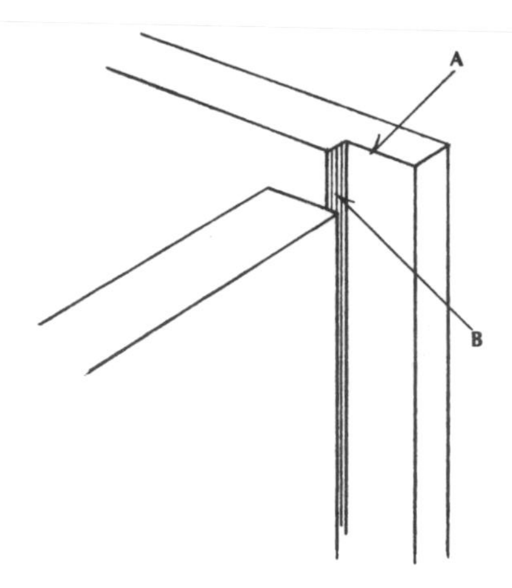

Figure 7-20. *Rabbeted corner.*

GLUE BLOCK CORNER

This joint (Figure 7-21) does about the same job and forms a structure equally as strong as the rabbeted corner. The block may be of mahogany or teak. It increases the effective gluing surface 100 percent or more, depending on the size of the piece. There is no method A, unless you are masochistic enough to want to rip out the block with a handsaw. With your table saw set to 45 degrees, it's a cinch. The appearance and strength of the glue block can be improved by cutting off the corners of a rectangular piece, as in Figure 7-21B. Glue as above, and finish nails or screws driven from the outside of the bulkhead. Or fasten from the inside with bunged-over screws.

The outer corner looks best if covered with a corner molding of mahogany or teak. A spartan yacht, however, can have the sharp corner routed to about ½-inch radius. A rabbeted joint backed up by a glue block is considerably stronger. A third option is to miter the plywood panels, but this is not an easy task in lengths greater than six feet. Such a miter can be done successfully with a power handsaw and a clamp-on guide, or on the table saw

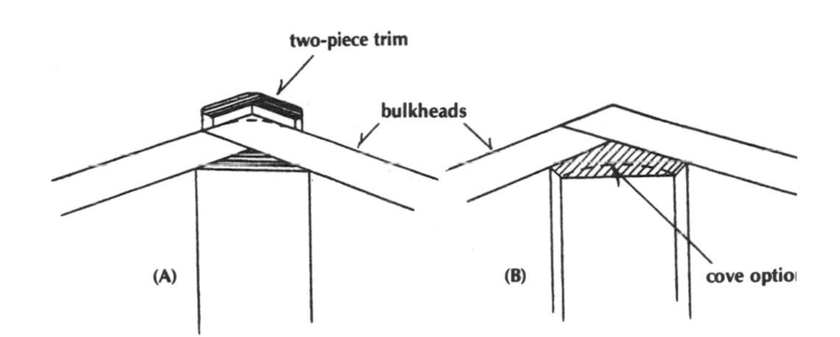

Figure 7-21. *Glue block corners.*

or radial-arm saw if the panels are held down flat while cutting the miter. Any slight upward flexing will cause the mitered edge to wander. I have used weights and an assistant with satisfactory results.

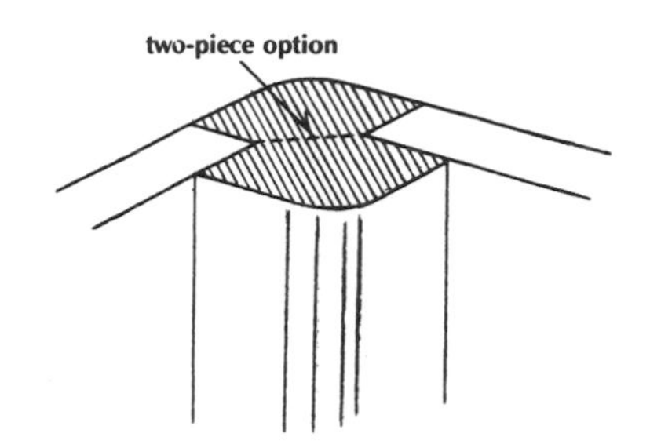

Figure 7-22. *Rabbet block corner.*

RABBET BLOCK CORNER

This joint is deluxe construction, and very strong. Figure 7-22 shows a two-piece option with the rounded outer corner fitted after the inner piece is glued and screwed. Actually, while this saves a bit of material, the one-piece block is easy because the two rabbets can be sawed or dadoed out in minutes on a table saw. These could be fashioned with a power handsaw, a method B, and, of course, laboriously with a rabbet plane in a method A. Be sure to allow ample room for the bulkheads and plenty of stock for screws into the rabbets, or a split could result. A nice touch is to make the rabbet ⅛ inch deeper than the material thickness, so the corner block stands out from the bulkhead. Finished bright, this makes a very handsome corner.

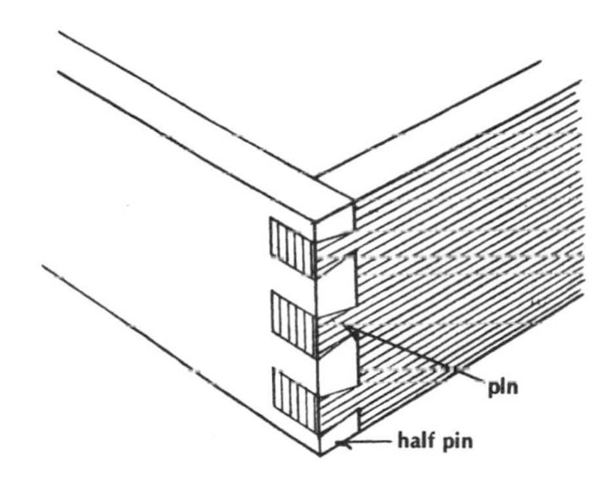

Figure 7-23. *Dovetailed corner.*

DOVETAILED CORNER

This joint (Figure 7-23) is extremely strong because of the wedge effect of the pins locked into sockets (the dovetails or mortises). Slamming drawers full of heavy stuff quickly destroys most other joints, but not the dovetailed corner. This is a very difficult joint to make, however.

The first step is to divide off the height of the drawer (or box) into a number of dovetails. Let's assume this front will be covered by an attractive overlay. If the front and sides are ½-inch material, mark the dovetails with a knife, ¼ inch wide at the thin end and 7/16 inch wide at the outer end on the side pieces. The material left over at the top and bottom is called a half pin. In fine furniture, the pins are sometimes very narrow. For our purposes, however, the base of the pins should not be less than ¼ inch wide.

For method A, you'll need a fine dovetail saw, ¼- and ½-inch chisels, a light mallet, and, of course, a bevel gauge. Saw the angled sides in the dovetails and random kerfs in between, but stay within the knife cut! Chisel out to the lines, paring off the sides until you "split the line." Be sure the socket ends are square and clean. Now clamp the side against the front and mark around all the pins. Put an X in each. Repeat the sawing and chiseling as above. Try for fit and pare off any small

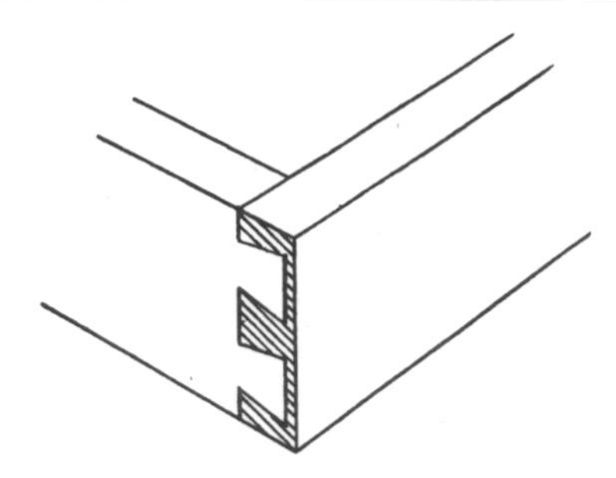

Figure 7-24. *"Secret" or "blind" dovetail joint.*

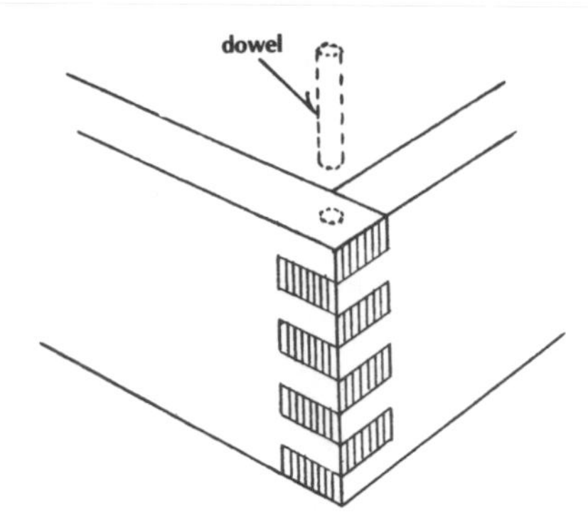

Figure 7-25. *Box joint.*

corrections. If you choose to make many dovetails by hand, I suggest that you make an aluminum sheet-metal template.

Method C, using a bandsaw for the picky sawing, would save much time. And so would the table saw set to length and angle of the pins. Method B, using a dovetail fixture and router, has been satisfactory. Several companies manufacture jigs and templates for a special router dovetail bit and guide bushing that follow the form with both pieces clamped together in the device. It takes about 20 seconds to dovetail both (Figure 3-19).

Dovetail fixtures are available to accommodate various thicknesses of material. I've seen one costing about $50 to $75 (1992) that can handle stock from 7/16 inch to over 1 inch thick and up to 12 inches wide. A lower-cost model works for wood up to 8 inches wide. The necessary bushings for the router are included, but the special dovetail bit is a little extra. Practice for a few hours on cheap scrap. Once you have it down pat, you can dovetail both parts literally in seconds.

An interesting feature is that the dovetail fixture allows you to fashion secret or blind dovetails (Figure 7-24). Such a dovetailed drawer front does not have to be covered by a false front. Thus, it can be fitted flush with the cabinet face. In Chapter Twelve I describe several other ways to build drawers. You don't really have to learn to dovetail with your eyes closed.

BOX JOINT

This strong, simple joint (Figure 7-25) was described in Chapter Five, together with detailed instructions for building a jig to produce it. In Chapter Thirteen you'll find a photograph of a beautiful example of box joints in a deck storage box.

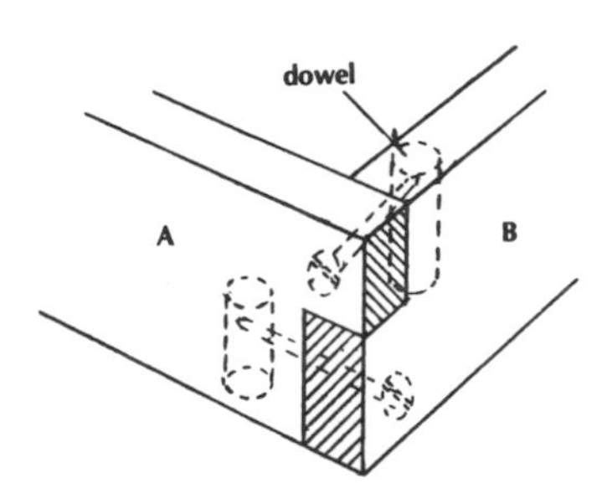

Figure 7-26. *Half-lap corner.*

For centuries craftsmen have made this joint with hand tools, much as detailed above, Method A. Dovetails are only slightly more difficult. Method B is using a router table set-up. The section in Chapter Five mentioned above covers very thoroughly method C, using a table saw and dado set, so there's no point in repeating it.

HALF-LAP CORNER

The corners of hatch coamings and hatch frames quite frequently are made with half-lap joints (Figure 7-26). Keeping such joints from working as a result of expansion and contraction of the parts is difficult. A screw into end grain does not hold well, and a single screw down into good grain is inadequate.

The half-lap joint is used by furniture and cabinet builders. I see no reason why it should not be used in yacht joinery. Moreover, it is not as difficult

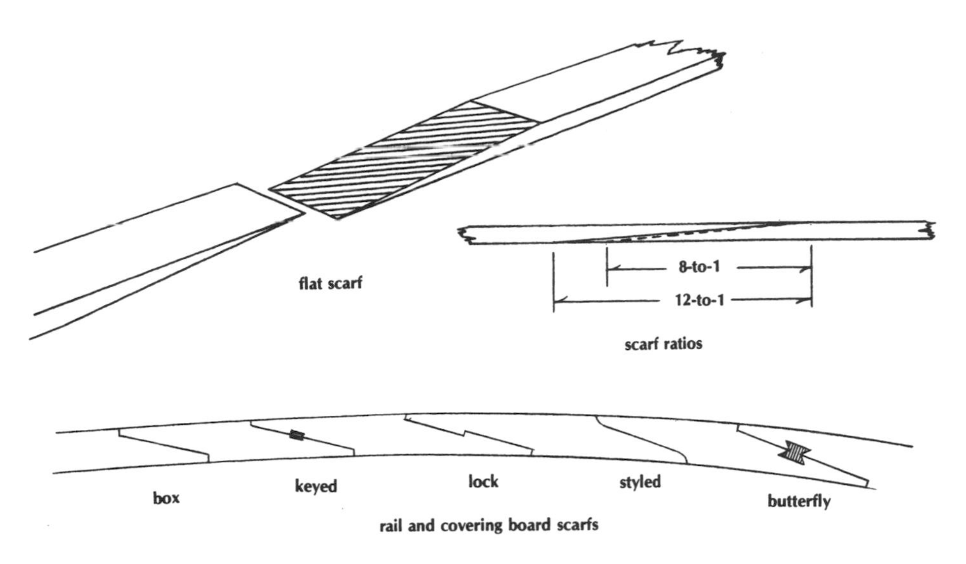

Figure 7-27. *Rail, covering board, and flat scarfs.*

to make as the sketch makes it appear. As shown, the whole frame or coaming would be assembled, completely fastened, and then installed. The dowel in part A, however, could be bored for and inserted from the top so the coaming could be assembled right on the deck. (I have not shown an additional screw that is driven from the top of A into the lower lap of B.) Screws passing into or through the hardwood dowels will draw up tight and stay that way. Use epoxy, plastic resin, or Aerolite. All plugged over, of course.

SCARFS

This section will be brief, for a router scarfing jig is described in Chapter Five (Figures 5-33, 5-34). The principal purpose of this type of scarf is to join short lengths of good material to make a longer piece, such as a stave for a hollow box spar. The shortage of Sitka spruce for spars makes a knowledge of this process mandatory. There is a method A, but what man in his right mind would saw out a 10-to-1 or 12-to-1 scarf and plane it to the paper-thin feather edge needed? Would you? However, there is a fine method B, so why ask for more?

A method B using a router allows the construction of scarfs that join short lengths around a curve. Two examples are covering boards and rail caps (Figure 7-27). These joints must be very strong, because the topsides of a vessel sometimes receive stresses, even blows, that strain the frames and clamps inside the hull. A covering board must be strong enough to absorb such stresses, yet flexible enough to yield without splitting and still return to its original close fit. Many such scarfs are hooked or locked together. See Chapter Fourteen.

8 Tying It All Together

Unless they are somehow held to their designed shapes beforehand, foam core and fiberglass hulls and many plank-on-frame ones go as limp as a handful of spaghetti when their male mold sections are removed. Glass hulls from a female mold have several bulkheads glassed in to hold the hull's lines. Both conventional and laminated wooden hulls present little problem. In the former, cross spalls and diagonals can be fastened athwartships from frame to frame and bulkheads built right in as the mold sections are removed. Laminated hulls often have built-in bulkheads as part of the mold or jig.

RECEIVING CRADLES

One-off glass and foam-sandwich hulls require special treatment. In both types, if you wish to preserve the major parts of the jig and will not be installing bulkheads, you have to provide a receiving cradle (Figure 8-1). There are two ways to go. Tear out some of the expandable mold ribbands and glass in diagonals or cross spalls at two or three favorable locations. Then turn the hull and place it in a receiving cradle fitted and padded to the hull. Or fit the cradle, as shown in Figure 8-1, and turn. Another way is to build a receiving cradle on the inverted hull, without the carpet padding, and then drive six or eight #12 screws and washers from the inside into the plywood or lumber of the cradle. Use large washers. Actually, there isn't much stress, because the hull is light, flexible, and easily pressed to the form of the cradle.

If you plan to save the mold, lift it out with a crane or chain falls. If you insist on pulling the inverted hull off the mold without cross spalls and diagonals, don't depend on the screws to lift it. Use at least four hydraulic jacks to break it loose, and don't let one jack get ahead of the others. Once loose, the hull also may be lifted by glassed-in iron rings or chain glassed on the keel. It's not good practice to lift the hull or break it loose by tugging on the receiving cradle. That puts a lot of stress on the screws. If you decide against a crane and your beams won't support a chain fall, you can roll the hull and cradle with the help of a few friends.

AN IN-HULL SCAFFOLD

I've tried laying loose timbers, planks, and plywood in deep wine-glass hulls, but it's not worth the time. Instead, hang a scaffold from the sheer as shown in Figure 8-2. Make the vertical 2 by 4's long enough to reach into the keel cavity, and bore plenty of bolt holes for vertical adjustments. The 2 by 4's or 2 by 6's on which your planks rest should be short so they can be placed low in the hull. Nail longer pieces on top of these when you have to reach out to the greater width of the hull. When installing a bulkhead, place this rig a foot or so away. This will let you sit on the plank end and/or work down in the bilge without hanging by your tail.

WHICH COMES FIRST, SOLE OR BULKHEAD?

Some designers prefer to erect bulkheads on top of a continuous cabin sole. This makes everything easier and faster, and permits alterations in the future. Other designers, however, value the strength of bulkheads extending right down to the keel timber or bonded into the fiberglass hull all the way to the covering over the ballast. If your hull is built on the cold-molded or laminated sys-

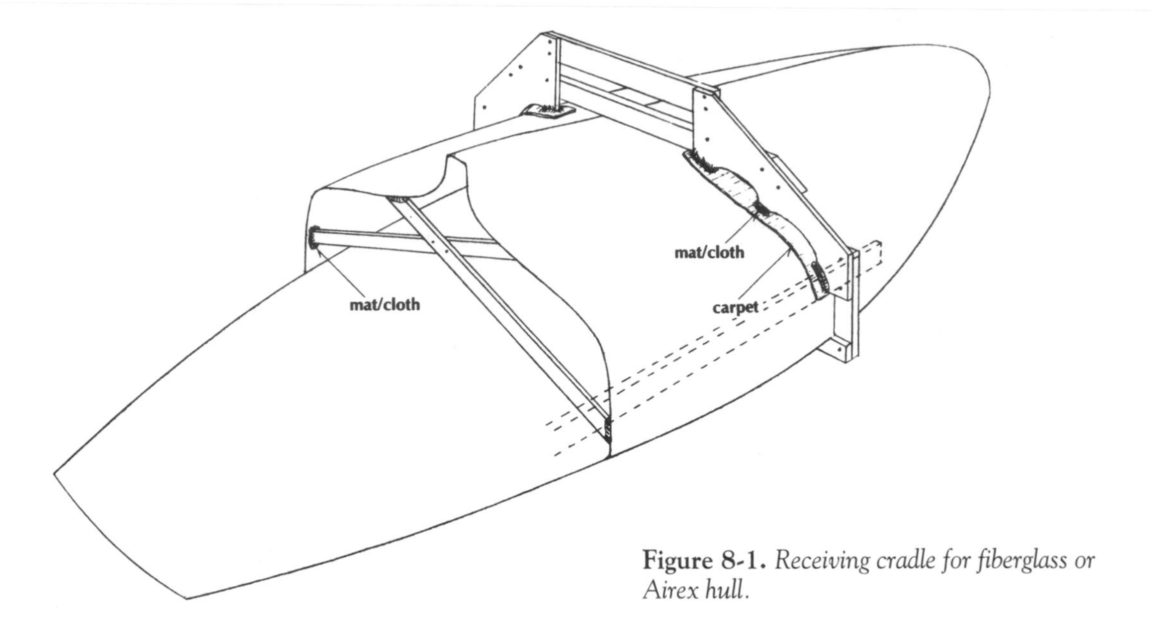

Figure 8-1. *Receiving cradle for fiberglass or Airex hull.*

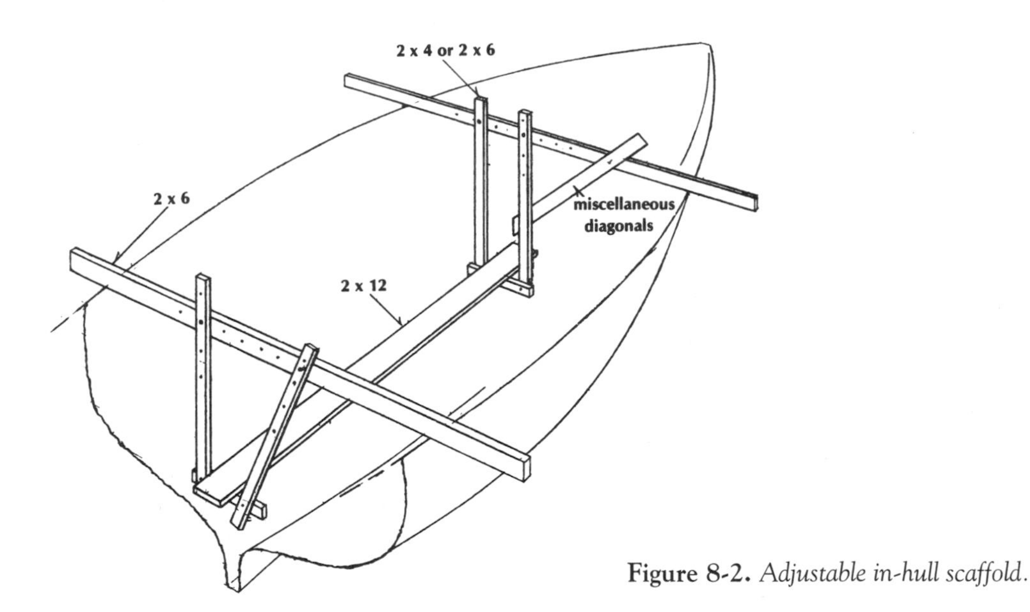

Figure 8-2. *Adjustable in-hull scaffold.*

tem, the bulkheads may be part of the mold, with the laminated planking built over.

Your first step is to provide a positive location at the sheer batten or clamp. Nail together a straight 2 by 4 and a 1 by 6 to form an L-shaped beam. Nail or clamp this across the hull on the bulkhead marks, so many inches from station such-and-such. Use a plumb bob or a level and straightedge to find the position of the bulkhead at the turn of the bilge and on the keel timber. Make four or five marks. If your hull is wood, tack three or four blocks against these marks on each side of the hull (Figure 8-3). If the hull is glass, bond these blocks lightly. They'll be chopped off later.

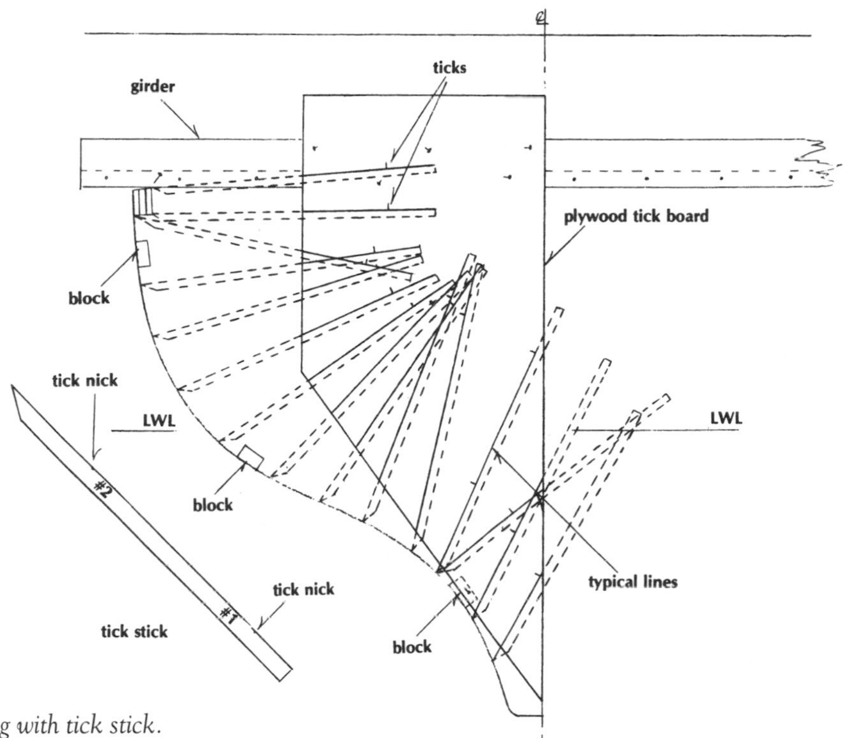

Figure 8-3. *Bulkhead fitting with tick stick.*

The Tick Stick Method

Don't fool around with dividers or scribers to shape the bulkheads. Find a straight piece of plywood that reaches from the keel to the top of the cabin trunk. The width of this piece should be sufficient to come within 12 to 18 inches of the hull when the inboard edge is placed vertically on the bulkhead location. If necessary, hack off a corner or two so it will enter the bilge area, as shown in Figure 8-3. Tack or clamp this tick board to the locating girder and to a block low in the hull or on the keel.

Now for the tick stick. This can be a straight 1 by 1 long enough to reach from about the center of the tick board to the hull. Cut one end to a sharp angle. Hacksaw two nicks to hold a pencil point. You make the ticks by holding the stick firmly against the board with the sharp tip touching the hull, shelf, clamp, bilge stringer, and so on, down the surface to the keel. Mark along the side of the stick and tick your pencil point in the nick and at the inner end of the stick. The more tick marks the better. If your stick is too long for some areas, just use the second nick. Mark on the board its position relative to the centerline of the hull. Now it is a simple job to transfer these ticks to your plywood bulkhead material or pattern. Be sure to leave ample room above the sheer for later shaping of the cabin trunk.

Full-Width Bulkhead

Most of the preceding instructions apply to partial bulkheads. For a full-width bulkhead, set up a tick board to port first, then to starboard, positioning it relative to the centerline. Or you can use a wide board or two cleated together to pick up all the ticks simultaneously. Before ruining good material, make a pattern of cheap plywood, transfer the marks, connect them with a flexible batten, and cut out with a sabersaw. If your hull is accurate, there will be small differences only. If the differences are greater than 1 inch from the lofted shape, you no longer have the lines designed for the hull.

The bulkhead can be built of two thicknesses, perhaps fir plywood on one side and teak plywood on the other. To save this precious material, be very careful when fitting the template or pattern around

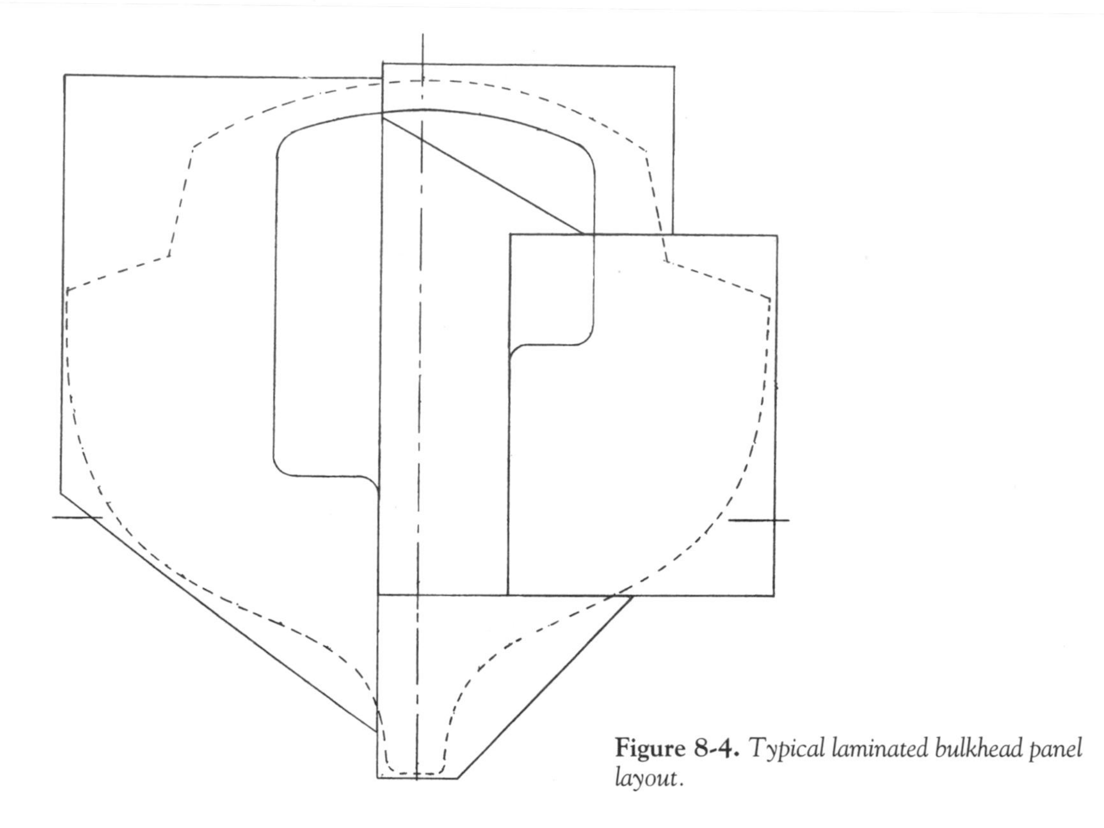

Figure 8-4. *Typical laminated bulkhead panel layout.*

the intricate clamp and bilge stringer areas. Drop or swing it into place against the locating blocks. You'll see immediately where it is too fat or too thin. Saw, plane, or grind away until the template fits accurately. If this is one-half of a full bulkhead, leave the template in place and fit another to it and to the other side of the hull. Nail substantial 1 by 4 or 1 by 6 cleats across the two sides. If necessary, pull nails to remove the template from the hull, then replace the nails, lay the template on the bulkhead material, mark, and saw out.

It may be necessary once again to insert the bulkhead against the blocks in two pieces, joining them after positioning. There's always a smarty who suggests that we could have just made the starboard template and flopped it over to make the port side, thus saving a lot of work. The two sides of a hull, however, are almost never truly identical. If you want to sleep soundly, don't even think about this, because the differences will probably shock you. The point is that while the sides may vary slightly, each one is fair and the "slow" side always seems to keep up with the "fast" side.

These large pieces of template material, of course, can be used for fitting the smaller bulkheads. This applies to the tick boards, too, so there is little waste.

Laminated Bulkheads

Many designers prefer to use a double thickness of plywood for bulkheads. Obviously, it takes no time to make a second thickness from the first, allowing for the tapering of the hull. However, lay out the second lamination so the panels overlap the joints in the first thickness by not less than 4 inches. Shift the pieces around so you don't have to saw out large door openings (Figure 8-4). Do not permit joints to form a cross. Brad the two laminations together, making sure that all joints are snug. When you are satisfied with the fits, lay down a sheet of polyethylene plastic so your glue won't bond the whole thing to the floor.

At this point, you have a decision to make. Can the entire bulkhead be located in the hull in one piece, or must it be joined after it is in position? The assembly may be very heavy, and there will be interference from the clamp and bilge stringer in your wooden hull. If the two templates cleated together can be horsed into position, you can assume that the lamination will go in. If some of the locating blocks are in the way, knock them out. If there's no way to fit the entire bulkhead in, just plan to join the pieces at a convenient spot. We'll discuss fastening the pieces later on. Let's make it in two parts.

Coat the joining edges with Weldwood Plastic Resin or epoxy adhesive. Presaturate the panel with penetrating-type resin if using epoxy. Hold the parts together by nailing blocks around the perimeter of the bulkhead. Since this is a two-piece bulkhead, do not coat those edges and overlaps that make the final assembly joint after the parts are in the hull. You can tape plastic strips to those spots to be safe. Now pour cups of the adhesive here and there and spread it evenly, preferably with the type of serrated tool used to glue down floor coverings. Work fast, especially if you're in sunlight.

Next, coat the edges of the second lamination piece (also presaturated) and lay these on the first panels. Fasten with a dozen brads so nothing can shift, and wipe away all excess glue. The entire assembly must now be weighted down with as many concrete blocks as you can crowd on top. The weight will force the glue out through the joints, so stand by to wipe it up. It may be necessary to fasten the piece with four-penny nails and blocks here and there. Be sure to put bits of plastic under the blocks. The two halves of the bulkhead should lift up easily in about 24 hours.

The matter of excess glue is extremely important. If you are laminating costly material such as marine teak or mahogany plywood, you'll have to be careful. The glue must be spread carefully, with no excess. Do not glue the joining edges, unless you are prepared to clean up small dabs immediately with a damp sponge or rag. Glue in the grain can cause permanent disfigurement—a spot that will not take stain. If in doubt, try applying masking tape along the surfaces next to joining edges. Aerolite glue cleans up with water before it cures and it can be removed completely if thoroughly wet. This is costly for large areas, however. Experiment with scraps, if you can, and glue up one or two of the small bulkheads first, just for practice.

Joining Bulkheads

You have tried the two parts for successful fit at the hull and at the inner joint. This may or may not be on the centerline. Have on hand a long pipe clamp, about six C-clamps, two 2 by 4's, and the necessary glue (Weldwood or Aerolite). Apply the glue to the part clamped in position to the girder. Place the second part in position and quickly and lightly C-clamp the joint, using long blocks over pieces of plastic to flatten and protect the plywood. Now place the 2 by 4's across from side to side, and C-clamp at the opening so the assembly is rigid. Position the pipe clamp athwartships so it pulls the joint closed. Move the C-clamps and blocks so you can see that the joint is flush and smooth on both sides. This is where a careful laminating job shows up.

If you want to play safe, you can drive flathead screws or short nails from the side you don't care too much about. Space the fasteners vertically about 4 inches apart in two rows. The points must not come through on the good side. Actually, these may help to pull the laminations together and hold the parts securely until the glue is cured. In themselves, however, they do not add much strength because they have little to hold to. Years ago, when building plywood boats, I placed many small bolts through plywood butt blocks and planking. These were removed in some cases after the glue cured. Often we used brass or bronze stove bolts or flathead machine screws that stayed in the hull. I never had such a butt fail or be detectable from the exterior after finishing. Your joint, too, should remain invisible from one side.

BULKHEAD-TO-HULL FASTENING

If the designer of your wooden hull specified fastening the bulkhead to the side of a frame, you can do it by simply driving screws. I assume the frame is straight, so this will place no strain on planking fastenings. If the frame is not straight, the bulkhead cannot be fastened securely to its side. The gaps may be shimmed where the screws are driven, but don't try to close gaps with screw pressure alone, you'll bend the bulkhead. A stronger method is to fill the gaps with a mush of epoxy resin thickened with pulverized limestone—a soil additive, called flour, you can find at nurseries very cheaply. (Equally strong is Cab-O-Sil, a well-known silica, but it is much more expensive.) Press the goop into gaps with a putty knife and form a fillet with a bottle or spoon. For neatness and incredible strength, immediately cover the fillet with fiberglass tape

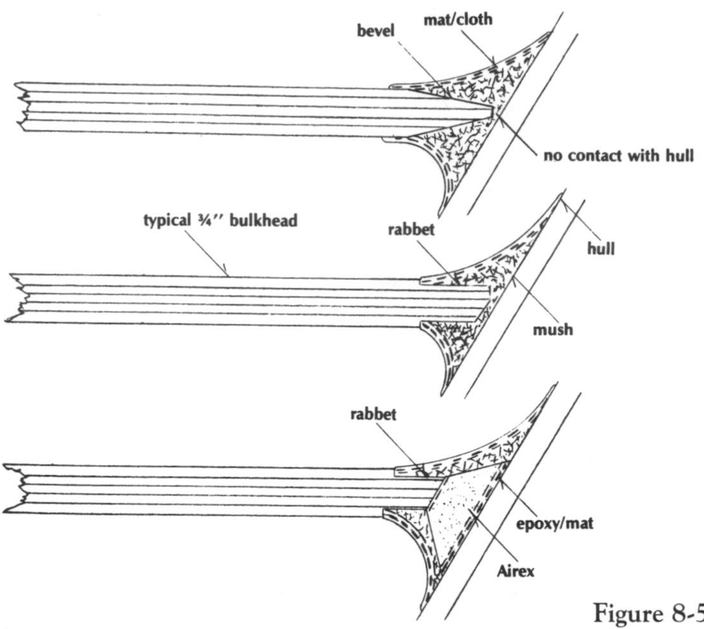

Figure 8-5. *Bulkhead bonding ideas.*

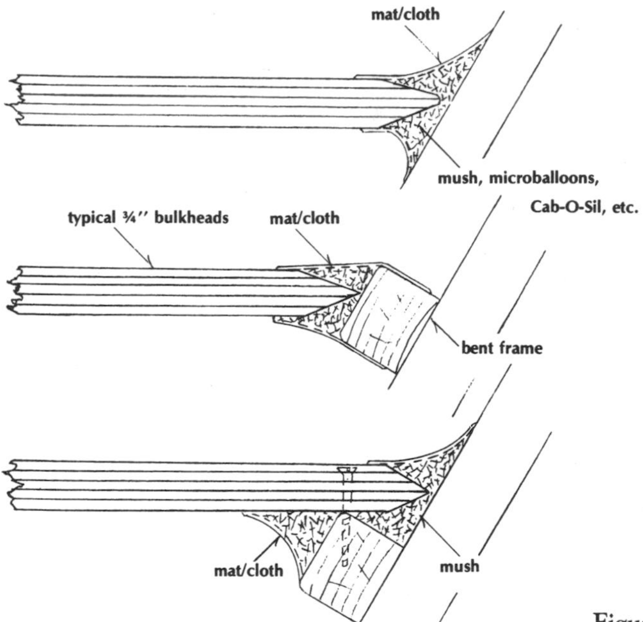

Figure 8-6. *Bulkhead bonding in wooden hulls.*

wet-out with epoxy. The only way to separate wood surfaces bonding with epoxy is to tear the wood apart. If you are using oak frames, wash them first with acetone to remove surface oil. Remember that polyester resin is not reliable and that epoxy is not compatible with fiberglass mat—use fiberglass cloth only.

The beveled edge is generally quite strong, but the routed rabbeted corner permits a flush finish and better appearance. To make these bevels in a hurry, rent a high-speed disc sander, put on one of those extra-coarse discs that look like gravel and just eat away the plywood. This leaves a rough surface that is best for bonding.

You can combine any of these ideas, of course. Don't overlook the tooling foam pad against a fiberglass hull. I strongly advise that you use only *epoxy* adhesive between the pad and the hull. There should be similar adhesive between the bulkhead and the pad also. A rabbet is an unnecessary refinement here, if you don't mind the look of glass tape all around the bulkhead.

Anyway, all this mess should be covered by a nice mahogany or teak frame that follows the contour of the hull. Need I suggest that you study one of the excellent references on fiberglassing procedures? Note also that epoxy is always preferable.

Don't forget that such methods, or even simpler modifications, should be used to bond smaller partial bulkheads, and also such horizontal members as berth and counter tops, long bookshelves, and so on. These become structural and contribute substantially to the strength of a hull, especially hulls of light fiberglass, wood laminations, or foam-sandwich designs. Horizontal bonding is difficult and messy in a plank-on-frame hull. Nor is it as essential, for the inherent strength of planking generally is generous. Always build up a broad pad along the juncture of any part bonded against a glass or foam-sandwich hull. If you don't, a disastrous "print through" will show up on the hull exterior.

BULKHEADS ON THE CABIN SOLE

Clearly, there are fewer problems when bulkheads are simply mounted on the cabin sole. Fitting is identical, using the tick stick and the tick board. The bulkhead can be fastened to a substantial cleat or into a rabbeted sill. The latter method is good practice, and the sill is made easily with a table saw. Be sure the rabbet is deep enough for adequate fastening with screws and glue (Figure 8-7). If the sill is finished bright, like a baseboard, the bulkhead can be fastened from the exposed side and bunged over.

Much might be written on bulkhead options, especially if we discussed the nostalgic paneled bulkheads of an earlier era. Do look into plywood faced with hard surface material. Think about panel effects made with delicate moldings. Don't forget that mirrors can be mounted on bulkheads to

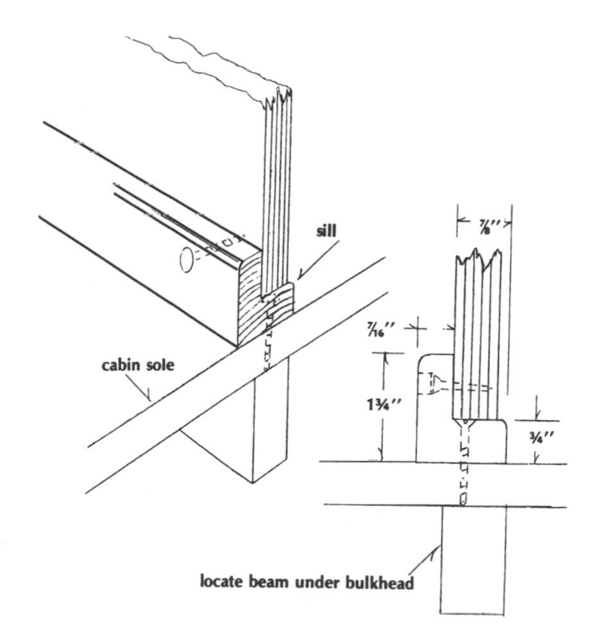

Figure 8-7. *Bulkhead in rabbeted sill.*

brighten an interior. They don't have to be large, either. I had one about a foot wide on a forward bulkhead that just happened to be located so that I could admire myself as I lounged at the tiller. Uplifting and useful, that mirror.

CABIN SOLE BEAMS

Let's go back a bit to where we talked about deep bulkheads reaching down into the bilge. Here's an easy way to line up your cabin sole. Having found the sole height on the bulkheads, glue and screw a strong cleat or beam at this level clear across the hull (Figure 8-8).

Incidentally, have you learned the trick of leveling with a transparent tube or hose and water? Find a tube quite a lot longer than the space. Have a pal hold the tube against the cleat (and tape it securely), while you fill the tube with water until it reaches the height of the cleat. Now move this over to the next bulkhead and mark at the water level. That's the level of your next cleat. This trick works anywhere in the boat—on opposite sides of bulkheads, way up forward, it's all the same. I'm assuming, of course, that your vessel is dead level, not floating.

With the cleats securely in place, you'll next

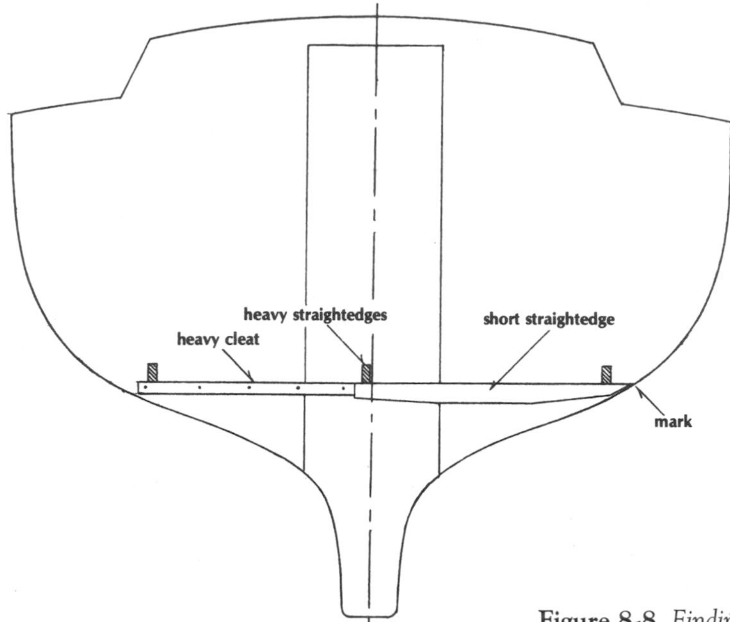

Figure 8-8. *Finding perimeter of cabin sole.*

need three good lumber or double-plywood straightedges reaching fore and aft from bulkhead to bulkhead, with one on the approximate centerline. Clamp these or nail blocks to keep them vertical and secure (Figure 8-8). To find the locations of the beams at the hull, you use a smaller pointed straightedge along the underside of the first three. Mark all along the hull where the straightedge touches. Try a sharpened piece of chalk, a crayon, a grease pencil, or a fine felt pen—except on wood, of course. As an alternative, use the same transparent hose all around the perimeter of the sole.

Much of the following applies to yachts with beams of 9 or 10 feet or more, so use your judgment on the details that fit your circumstances. For example, *Allegra* has a deep sole 24 inches wide; only a couple of bilge access openings are needed. The sole needs a solid structure in both the wood and fiberglass versions. This is what I want to discuss, as the information may be helpful.

SOLE BEAM SUPPORTS

Openings in a deck or sole are framed at the fore-and-aft ends by heavy beams that span the hull. These heavy beams receive and support the timbers running fore and aft along the openings. They are called carlings, headers, and other names. Some builders use simple half-lap joints. Because it is best to have the supporting leg rest on a frame or floor timber, the beam itself may be shifted according to the way it is joined to the hull. Thus, the leg may be let into the beam or not, as shown in Figure 8-9, A and B. The joint in Figure 8-9A weakens the beam somewhat, and its lower end would tend to creep down the frame if the angle were acute. An alternative, then, would be to cant the leg as shown in the dotted lines. The leg in Figure 8-9B serves the same function but is not lapped. This leg is secured by nailed and glued cleats or even doubled full length. The lap could be in the leg only, as in Figure 8-9C. All of these joints can be made with a backsaw, power saw, or table saw, as previously described.

If the heavy beam is located where it cannot land on a frame or floor timber, do not place the leg directly on the planking or the glass of the hull. Bond down a large pad shaped to the hull, and toenail the leg to it or to a block on the pad, as shown in Figure 8-9C. Or span the block across two frames. This is done because weights or heavy stresses on the sole could be transmitted and concentrated in a small area of the hull. Whenever possible, stand a leg on the keel. It's all right to shift the location of a tank slightly to permit this. You never know when

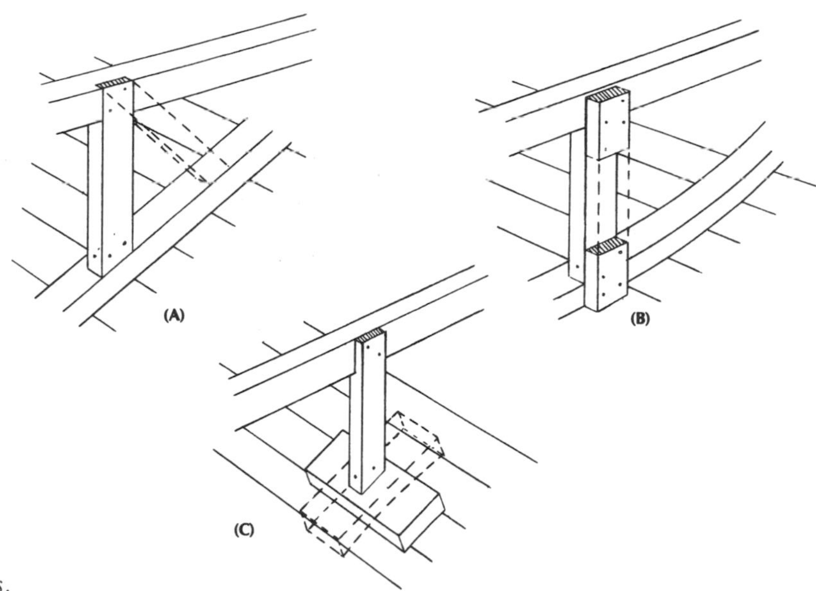

Figure 8-9. *Sole beam supports.*

a 200-pounder is going to jump off the companionway ladder. Solid support for the sole could avoid a lot of destruction. Glue every joint, preferably with epoxy. Make them strong, and these structures will even help to stiffen the hull.

It's assumed that you have installed all plumbing, tanks, batteries, and so on, before the framing goes in, at least loosely. Keep in mind that openings must be ample to service these areas and also to maintain the hull.

BEAM-TO-HULL ATTACHMENT

The heavy beams must be tied to the hull. In a fiberglass hull, this problem is quite easily solved by bedding the beam in a mush of epoxy/filler and then covering it with a nice fillet, as above. It would be good practice to lay the beam end on a ¾-inch plywood pad formed and bonded to the hull or a built-up fiberglass area. This will prevent print-through from the beam.

You have several choices for installing sole beams in your wooden hull. Ignore my suggestions, however, if your designer has provided specifications. My subject is not design, even though there are many options in actual construction. Many designers leave much to the discretion of the more experienced professional builder. And that's the way it should be. Otherwise, the fees for custom yacht designs would be beyond reach of *all* of us. Detail is costly!

If you are free to choose, fit the beam right on the frame, as shown in Figure 8-10. This is fine if it locates the heavy beam where the opening is desired (Figure 8-10A). You can attach a cleat to the side of the beam (dotted lines) with screws and glue. I would bond all this to the frame with epoxy and cloth over a couple of screws, or the gusset if you want greater strength. Believe me, a series of beams all tied in this way can give you a nice, warm feeling in a really severe grounding. Half-lapping the beam over the frame would be a refinement of some value, but do not drive fastenings into bent frames unless they are at least 1½ inches in depth, as in Figure 8-10B. Any beam end termination becomes immensely strong if there is a substantial triangulated leg under the beam.

Another way to land beams on a wooden hull is to build in a pad spanning two or more frames—or, better yet, a continuous stringer, as shown in Figure 8-10C. This has little to do with the bilge stringer in a wooden hull. It's actually a shelf to lay the beams on, but there is no denying that it adds strength. Because of the great curvature of this shelf, it may be advisable to build it up strip-fashion, edge-nailing and gluing each strip to its neighbor. Bonding all the short beams as well as the heavy

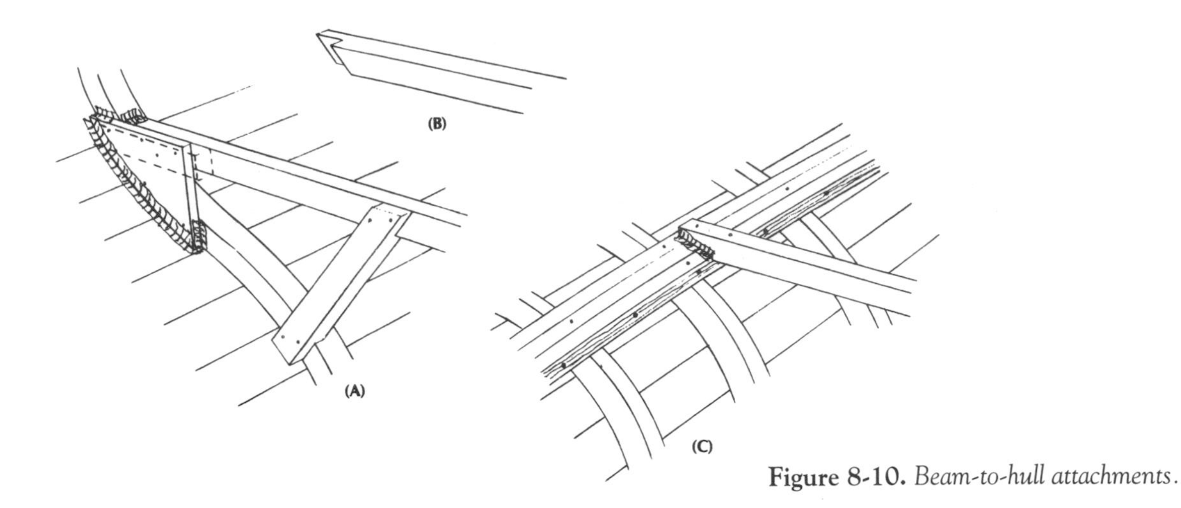

Figure 8-10. *Beam-to-hull attachments.*

beams to the hull in this manner would make your cabin sole a major structural unit. This permits locating the beams anywhere—a definite plus.

You have noticed, I'm sure, that I like to reinforce many joints with fiberglass fillets of epoxy mush covered with cloth. I haven't shown this in detail in my sketches, but where I indicate it, I mean that the mush is between the parts as well as in a fillet around the exterior. Repeated wetting and drying is one of the promoters of dry rot. I try to avoid if not prevent this by making it difficult for water to find a place to collect. I don't want it, for example, in a pocket between the gusset and the frame in Figure 8-10A. Polyester is not impermeable to water, so let's take full advantage of the good characteristics of epoxy resins wherever we can.

CARLINGS

The beams running fore and aft, called carlings, must be attached securely to the main or heavy beams. These carlings then receive the short side beams extending out to the hull, as just described. Most builders and designers specify an ordinary half lap (Figure 8-11A). This is sometimes used in the main deck from clamp or sheer batten to the cabin trunk carlings. This joint has several obvious weaknesses. First, the side beam has a tendency to split from the corner where it rests in the carling notch. Second, one-half of the carling has been removed, creating a weak spot, unless the area is supported by a leg. However, this is not 100 percent true if the side beam (E) is a snug fit into the half-lap socket, since the socket would not be able to close up under stress if filled completely. I prefer Figure 8-11B, a haunched dado. This is a much more difficult joint to make, but it retains three-quarters of the original strength of the short beam and weakens the carling very little. It would not resist tensile stress as well as the joint in Figure 8-11A, but these situations rarely experience that kind of force. All of these joints are highly resistant to compression.

The joint in Figure 8-11C weakens the carling very little and the short beam not at all. It must rely on the compression of the wedge form, and the fastenings and glue to resist excessive downward stresses. Note, however, that a glued decking over the joint would eliminate weakness almost entirely. This is much easier to construct than the joint in Figure 8-11B. I think it would be very adequate. The joint in Figure 8-11D, a single dovetailed dado, resists tension well. It is a bit more difficult to fit closely and it also has little resistance to downward force. Like the joint in Figure 8-11C, however, it would be amply reinforced by a glued deck on top. This was described in Chapter Seven. Be sure to use Aerolite or T88 if your fit is much less than perfect—or even if it is.

CABIN SOLES

One owner I know built his sole of beautiful vertical-grain fir with strips of dark mahogany glued in between, and then varnished with about six coats.

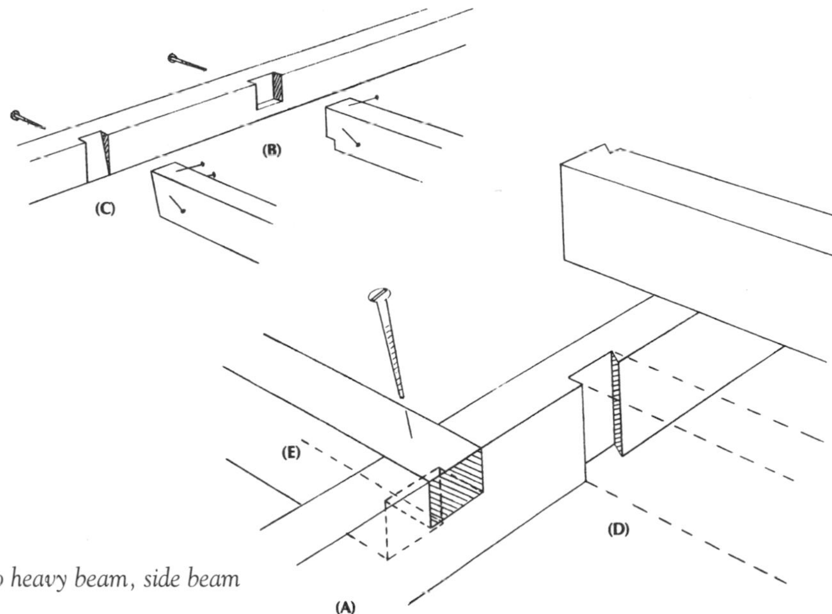

Figure 8-11. *Joints: carlin to heavy beam, side beam to carlin.*

This is great to look at, but that skipper will need three hands and a tail when the boat is heeling and pitching and a few drops have come aboard. Bare wood would be good, but that's difficult to keep clean even if it's teak. There is a teak plywood with holly strips that make it look like the costliest sole available. Some production yachtbuilders use it. You might be able to pick up the few pieces you'll need at a reasonable price.

To go this route, first fit your underlay of ½- or ¾-inch plywood all the way out to the hull. Bevel the top corner back about 1 inch as you did with the bulkheads so it can be glassed to the hull (Figure 8-5). This means the underlay must be notched around frames in your wooden hull, but do it rather casually to allow for filling the gaps with an epoxy mush. Be sure to provide about ¾ inch around the access openings so the hatches rest securely on the frame (Figure 8-12). Make the hatches with ⅛-inch clearance to avoid swelling and jambing. Juggle the plywood so all second-layer joints also fall on beams, and do not allow joints to meet to form a cross. There should be T's only. Keep in mind also that the teak overlay joints should form pleasing patterns equidistant from the centerline, and so on, so underlay joints must be staggered accordingly.

You can nail and glue these sections to the beams and frames with bronze, galvanized finish, or box nails. Pack the resin goop all around the perimeter of the sole, forming a nice fillet with a spoon or bottle. Cover this with fiberglass cloth tape, as above. You have just added tremendous strength to your hull if you used epoxy resin.

When this has cured, sand off high spots and clean up with a vacuum cleaner, if possible. Be sure the surface is fair and smooth. Assemble the teak overlay. It is not necessary that these pieces be fitted close to the hull. Brad each piece so it cannot shift. Lift one piece at a time and spread epoxy or Weldwood evenly over the area (but rather thinly near the edges) with a brush or toothed squeegee. Replace the brads. When two pieces have been laid in the glue, cover the area with plastic, walk all over it, and then lay concrete blocks on the plastic. Be sure all joint edges meet flush, and immediately sponge off any glue that oozes out. No other fastenings are needed, but you can use more brads if you wish. Repeat this process with the remaining sections. Give the sole about a day to cure. This can be replaced only by demolishing it.

If you are willing to settle for a plain, painted, plywood sole, use ¾-inch marine-grade plywood, screwed down. The exterior grade has voids in its core that will pick up moisture in time and thus start to decay. In addition, such voids create weak

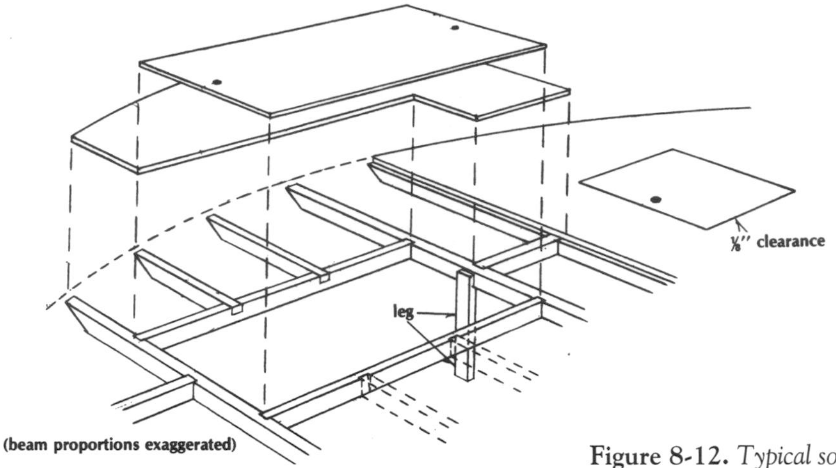

Figure 8-12. *Typical sole aperture framing.*

spots. Do not for a minute consider covering with linoleum, carpeting, or so-called indoor-outdoor stuff. Linoleum is exceedingly treacherous when damp, and moisture always creeps under it. This also applies to any kind of carpeting. All coverings prevent evaporation and drying for weeks at a time. Nylon carpeting may be laid for use in port only.

Cabin soles vary with the type of vessel. On one of our old boats, we had ½-inch plywood pieces laid loose on top of the floor timbers. We liked this because we were able to clean the bilges often with ease. Smaller areas in your boat should be planned this way—that is, simple, inexpensive, and handy. Regardless of the type of sole, miter a small decorative molding to cover the deck corner joints at bulkheads and berth fronts. Finish it bright for a nice touch.

Strip Cabin Soles

While teak plywood with artificial holly strips may satisfy many people, there are those who favor a genuine teak sole with raised holly strips (Figure 8-13). First of all, the narrow decking generally specified needs support from an underlay as described. Even with that, I would not try to deck with less than ⅝-inch teak because of fastening problems.

Figure 8-13 shows three systems. The first, on the left, uses strips about 1¼ to 1½ inches wide. These are toenailed into the beams through the edge (drill pilot holes first) while clamped or weighted to the beam. Then 2-inch or 2½-inch finish nails are placed horizontally, two or three between beams, from each strip into its neighbor. The ½-inch holly strip with its upper surface slightly rounded and raised goes in between. Epoxy glue is required, along with 1- to 1¼-inch finish nails through the holly. Both the teak and the holly must be bedded in viscous epoxy glue spread on the underlay. The second system, at the center, shows the teak rabbeted to take a smaller holly strip about ⅜ by ½ inch wide. Pilot holes are drilled for nailing the teak to the underlay and to its neighbor. It is necessary to drill completely through both pieces of the holly because the nail locates right on a joint. Use the type of long, flexible drill bit common in aircraft assembly. As an alternative, you can lay all the rabbeted teak strips, clean out the resulting grooves with care, then press in the holly strips. If you omit glue under the holly strips, they can be replaced much later when they wear. The third approach is the easiest. Using a dado set on your table saw—or a router—groove out ½-inch slots into which the holly strips are then pressed. You can use pieces of teak from 6 to 8 inches wide.

This laid sole will not need finishing, but if it is left bare, it will show grease spots. Perhaps a teak cleaner will take care of that. If the teak, or even mahogany, were varnished, the holly would provide good footing and reduce wear on the base surface. All this looks like a lot of work, but isn't it worth it to make your boat a yacht?

THE BREASTHOOK

It is assumed that your design shows the breasthook details. If it does not, study the fine boatbuilding

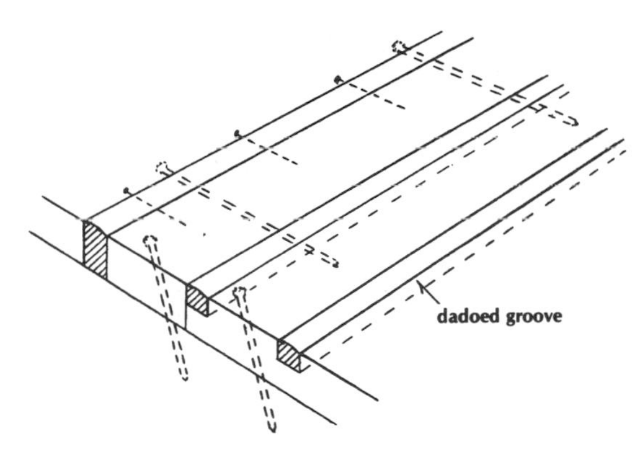

Figure 8-13. *Three ways to deck with holly strips.*

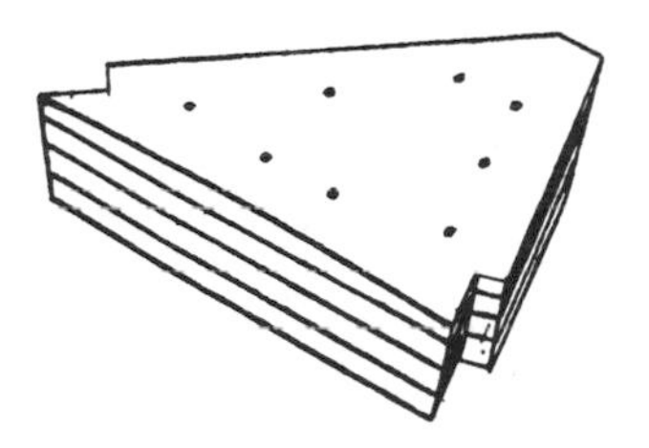

Figure 8-14. *Built-up breasthook.*

references mentioned earlier in this book. Traditionally, the breasthook is oak, but you can build up a perfectly satisfactory one using three or four thicknesses of ¾-inch fir marine plywood—glued with epoxy and galvanized nails (Figure 8-14). The angle of the topside flare can be bandsawed if there is a slight curvature. Most of the slight camber should be dressed before it is installed. This applies, too, to the notches to receive a sheer batten, if any. If there is a clamp over bent frames, spring it up slightly and fit it snug to the aft edge of the breasthook. In a fiberglass hull, bed the breasthook with thick epoxy adhesive. Wooden planking also must be screwed and glued. Fill all open joints with epoxy or Aerolite, then grind off the top surface smoothly to receive decking.

SHEER BATTENS

I assume that sheer battens were installed previously in your fiberglass hull. Of course, no builder would allow a hull to leave his shop without sheer battens and several bulkheads or numerous cross spalls in place. Nevertheless, here's one way to install a sheer batten. First saw out the bulkhead apertures to size if not cut previously. The sizes of the sheer batten members will vary, of course, with the size of the hull. There must be ample height—about twice the molding of the deck beams would be safe—but follow the designer's scantlings if they are in the plans. The siding (thickness) should be enough for the canted or half-lap joint and a bolt or heavy screw, or scarfed 1 to 8.

The batten can be built up of two or three thicknesses of clear fir in long lengths. This timber could be replaced by three or four thicknesses of ½-inch plywood for comparable strength, but more work (Figure 8-15). The fir can be scarfed full length on the floor, then sprung to the sheer the hard way, one at a time. Try a shorter length first to be sure it can be urged the hard way into this curve. If not, you'll have to use wider material and cant the scarfs as shown in Figure 8-15 to save material. Each component (especially if plywood) may have to be sawed to the sweep of the sheer before joining. In this case, it would be easier to glue the scarfs with the batten parts clamped temporarily to the hull. Fit to notches in the breasthook and to the transom frame quarter knee. When scarfed from bow to stern, lift it up, and apply epoxy glue to the surfaces of hull and batten and to the scarf areas. If you are short of clamps, install one piece at a time. Bore for flathead stainless steel through-bolts or use temporary short screws. This is insurance—you can rely on the thickened epoxy to hold. The second and third laminations should be fitted the same way but fastened with ringed nails and epoxy glue (or through-bolted after pulling the temporary screws).

A power saw should be able to make any of these easy curves, especially with a small-diameter blade. The finished shapes may be more exotic than I have sketched, depending on the flare of the hull.

I have shown the sheer batten upper edges in a step arrangement so the rise for the deck camber can be accommodated. This can be planed down or ground off with a high-speed disc sander using 20-grit paper. Be careful not to cut below the line of the sheer.

DECK BEAMS

Framing the deck is a large subject that is covered extremely well by the fine books by Chapelle,

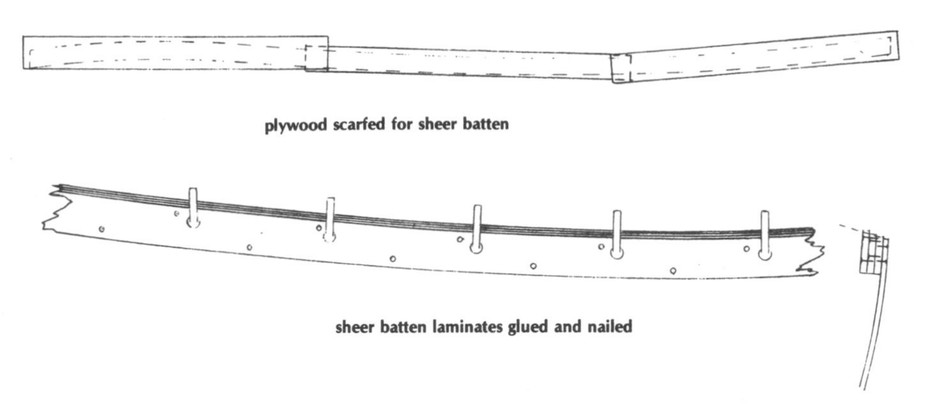

Figure 8-15. *Sheer battens.*

Steward, Herreshoff, and others. However, it is a joinery subject, too, and for that reason is included here. I have several ideas on simplification and the use of glues to save materials.

First of all, avoid beams sawed out of a wide plank, as this consumes horrendous quantities of costly wood, and the beam is very weak because of the run-out of the grain. If the beams are normal camber, that is, ½ inch per foot of beam, or somewhat more, they may be sawed from a broad epoxy-glued plank as shown in Figure 8-16. Second, for all purposes a laminated beam is stronger, costs less, and can be extremely handsome. Laminated beams can be finished bright (Figure 8-17). Don't use oak. Oak is far too heavy for this location so far above the waterline. This tenderizing effect is multiplied if oak is used in the cabin trunk beams. In addition, oak does not take glues well.

Spruce is by far the ideal beam material, but it is costly. Ash is a good substitute for oak. It is strong, lighter than oak, takes glues, and finishes beautifully. It is not known for its resistance to rot, but today it can be sealed with a penetrating epoxy. Oregon or Douglas fir makes a very good beam. It is midway along the weight range, very strong, glues well, and finishes nicely to a deep golden yellow. It does split, however, if not properly bored for fastenings. All in all, next to spruce, Oregon or Douglas fir would be my choice. Spruce or fir with alternating laminations of mahogany or ash makes a beautiful beam, but my only detractor called it a barber pole. The traditional beam has a mahogany or teak cap along its lower edge.

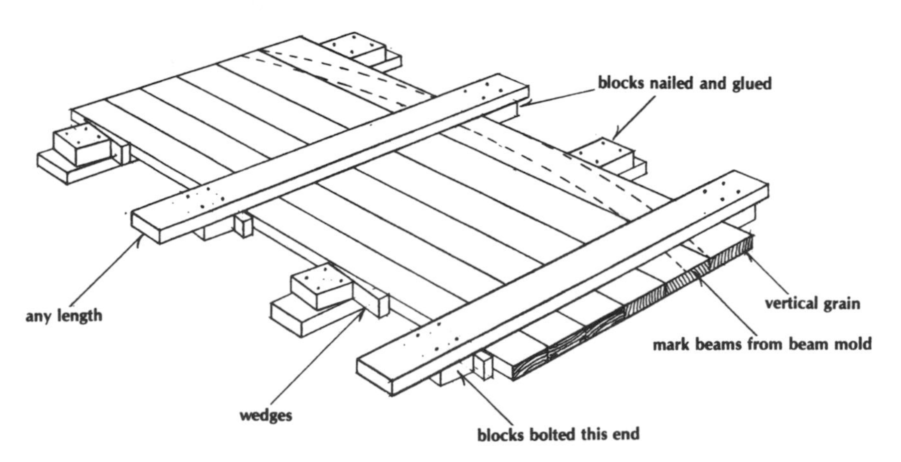

Figure 8-16. *Make-do clamps for sawed beams.*

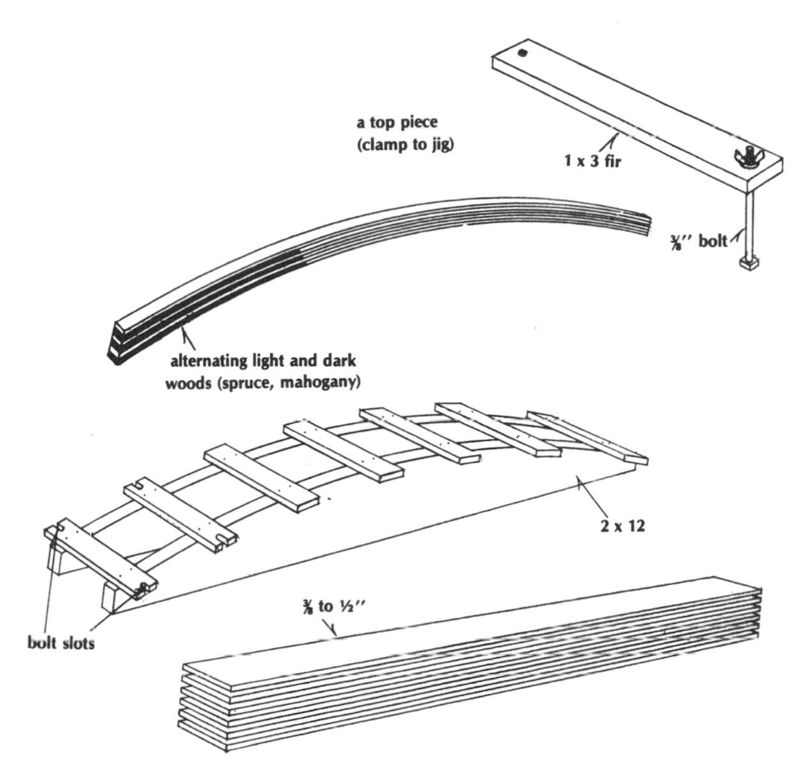

Figure 8-17. *Laminated beam jig.*

Sawed Beams

Let's assume you have some good clear fir on hand with vertical grain. If in the form of 2 by 4's or 2 by 6's and your beams are to finish to 1 by 2½ (a good ratio), the boards may have to be run through a planer to bring them down to about ⅛ inch over the final size. Or you can reduce them after sawing, but this is a lot of work with a plane or jointer. The job is to edge-glue these boards.

Figure 8-16 shows an easy and inexpensive way to clamp up a very wide structure, using wedges. The crosspieces may be of any convenient length. The blocks on the wedge end should be bolted for adjusting to another use. The fixed blocks should be nailed and glued. I think 2 by 4 material would be adequate, but use 2 by 6 for a very heavy beam, and 1 by 6 for light beams. You may need more clamps than I have shown to spread the pressure equally. I prefer epoxy glue here, but Aerolite would be good, sinces it's water-clear. Place plastic sheets between the fir and the clamps. Try to line up the edges as you wedge up so you won't have too much to plane off later. If the pieces are edge or vertical grain, but show some cupping (warping), alternate them as explained in Chapter Seven. If they are flat grain, they must be alternated. Let the glue ooze out for a few minutes, then scrape and sponge off the excess. Let the assembly cure for about 24 hours. Chisel or plane off the hardened glue.

When the surfaces are clean, lay out the beam with a beam mold (see Chapter Five). If you have a template or pattern of the upper camber only, don't forget that the bottom curve is different because it has a shorter radius. You either make a beam mold or you set up your bandsaw for sawing concentric to the upper curve. Again, see Chapter Five. This requires considerable care. To plane the under concave surface, stroke your plane at an angle of about 45 degrees, then finish with a belt sander or coarse (60- to 80-grit) sandpaper on a curved block of foam.

Production is done by a shaper. If you have one, run your beam mold against collars with a straight cutter on the spindle. If the beam is sided too thick for your cutter, raise the spindle after all the first passes. Take care when the cutter begins to shape against the grain. You may have to slow the feed to avoid chipping the surface. It's best to bring the beams down to their specified siding on a jointer or let a mill put them through its planer. Of course, you can use a roughing plane, a jack, and a 24-inch jointer plane, with a final run over a bench belt sander. The appearance of any beam is enhanced by routing a small radius, ogee, or chamfer on the lower corners. Or you can cap with hardwood in traditional yacht style.

Laminated Beams

I recommend strongly that all beams be laminated, so let's build the form now (Figure 8-17). Start with a couple of 2 by 12's cut to the necessary crown. This means the camber of the beam mold, less the molded depth of the beam, less ¾ inch—the thickness of the crosspieces or staves. By laminating wide pieces together, you wind up with a bent plank that you then rip into three or four beams. The more laminations, the stronger the beam. For a small beam, the laminations should be ¼ or 5⁄16 inch thick; for a large beam, ⅜ to ½ inch thick. This means that your spruce, fir, mahogany, or whatever may have to be run through a planer, or possibly resawed first to save on material and planer time.

If you want four beams 1 inch wide (sided), use laminate 5 inches or more wide. This should rip out to at least 1⅛ inches. If you need four 1½-inch beams, use 7-inch material. The wider the laminates the better, for the form staves tend to bend considerably with narrow stock. You may have to place C-clamps so there are no air pockets in the laminate. The staves should be about 8 inches longer than the material, leaving the ends for clamps or bolts. Screw the staves to the 2 by 12 so they overhang 2 inches. The slots permit rapid setting up with bolts and wing nuts, but you'll need a wrench, too, for power.

Try a dry run. Stack your material over a center mark. Clamp the staves down one by one, tacking lightly near center. Light laminates can be sprung down easily by hand. Remove them and get the glue ready. I prefer Aerolite here because the glue line is crystal clear and nothing happens until the two parts come together—and then it happens slowly. Epoxy adhesive is the most powerful, especially if you presaturate. Working time with either is ample. As you tighten the clamps, be sure to align the sides of the stack of laminates and lightly finish nail the bottom one to the form to keep the pieces centered.

Bolt up the center stave first, tacking to avoid creeping, then work toward the ends, clamping loosely. Watch the side alignment closely. The laminates will want to slide about. Place C-clamps near the center of each stave and pull down all the bolts and clamps. Wipe and wash off all the glue for an hour or so, and then cure for about 24 hours before releasing the beam from the form. If you remembered to cover the forms with plastic, the lower side of the bent plank should be fairly clean. Plane one side so it is fair and rip slightly oversize on your table saw or bandsaw. The final dressing to the required dimension is best done on a jointer, with a finish sanding; but see Planers, Chapter 4. Once again, a nice touch is a routed radius or ogee or a hardwood cap installed after the beams are set (Figure 8-18).

Special Laminated Beams

Sometimes a special beam is needed in the deck just abaft the breasthook, or the cabin trunk may require each beam to be cambered to a special radius. To make a series of special beams, first cover an old bench, plank, or heavy plywood panel with plastic sheet. Nail blocks along the beam pattern (or a pen-

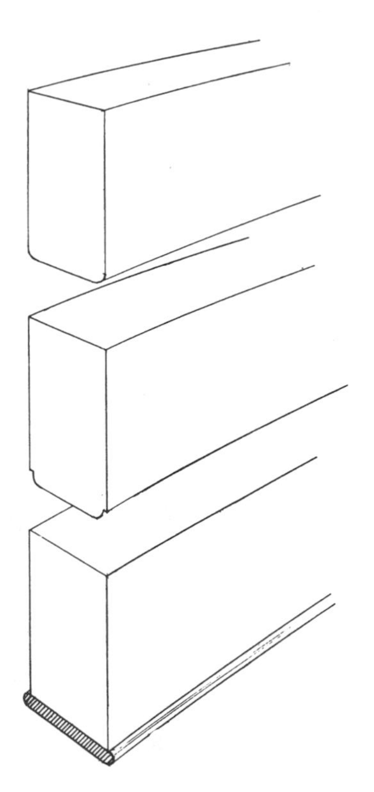

Figure 8-18. *Beam treatments, yacht style.*

cil line under the plastic) about 6 to 8 inches apart. Make them of ¾-inch plywood or full 2-inch lumber and high enough so you can turn the clamp screws (Figure 8-19). Go through a dry run. After the glue has cured, save the blocks for the next special beam, and then the next, and so on. Of course, there's the tiller ("S" curved?) and very light beams in hatches, too. Nothing equals laminating for strength. Alternating laminates of light and dark wood are very attractive. You'll never regret taking the time to do a proper job of laminating.

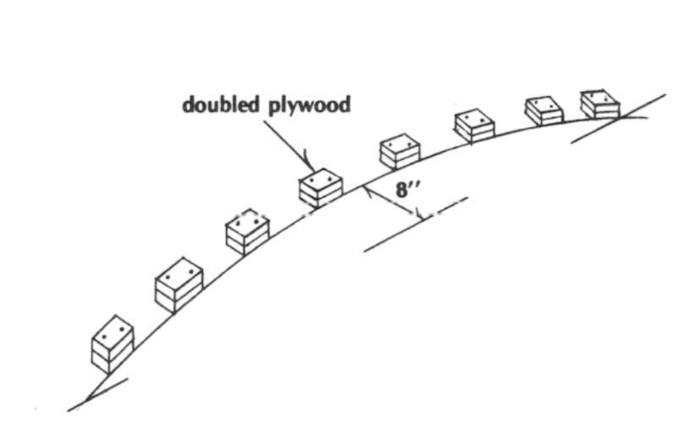

Figure 8-19. *Form for laminating special beams.*

Short Beams

You'll need beams extending from the carlings—to which the cabin trunk is eventually fastened—out to the sheer plank. These will be mortised into the sheer batten if there is one, and bolted to the clamp or shelf.

The easiest practical way is to saw these out of short pieces left over from other work. You can also laminate a couple of long beams and saw them up into the lengths required. There is less chance of the laminated beams being split by bolts near their ends.

CARLING AND BEAM ALIGNMENT

I have already discussed the sweep of the beams in the foredeck. If there were no opening for the trunk cabin, all beams would have to line up similarly, forming a very fair line down the center. This would be much flatter than the sheer, depending on the amount of camber. The short side beams and all other beams must be cambered to meet the invisible centerline. Likewise, the carlings must be sprung down to match the cambers.

To start with, the main beams at each end of the opening in the deck have been installed. Let's say they are 2½ inches molded (depth). The beams in the foredeck have been lined up and installed as above. I would now stretch a 2 by 4 or 2 by 6 strongback from bow to stern under the main beams, either doubled or with long butt blocks to keep it fair at the joints (Figure 8-20). The forward end must be clamped under the breasthook with blocks to position it 2½ inches under the decking. The aft end must be clamped under the transom frame, cleat, or beam the same 2½ inches under the deck level. Clamp the strongback to the main beams of the trunk opening also. If the strongback does not form a fair sweep, prop it up or pull it down until it does.

I know that some boatbuilders lay this strongback over the beams the easy way, but now you have a useful shelf to support other beams while they are being fitted. You also have a sure way of measuring the required cambers anywhere by stretching a taut chalkline across the hull and measuring up to the strongback and shim. All you have to remember is that some beams are molded to different heights so shims will be necessary. Just tack them to the 2 by 6.

Where bulkheads are in, but not the main beams, clamp 2 by 4's across at the designed height for each beam (beneath the beam mold). Place a removable shim between the strongback and the 2 by 4. Remove this when fitting a main beam into the sheer batten, or on the clamp or shelf.

I mentioned earlier that you will need several strong beam molds. The height of the carlings is fixed where they pass through bulkheads, but elsewhere they must be forced down into a fair curve, using the beam molds (Figure 8-20). Clamp these to the sheer battens or clamps and to the strongback, to pull the carlings down to their proper sweep. The carling notches through the bulkheads should have been cut out from the lofting before installation. If not, mark the locations and dimensions from a beam mold and saw out with a keyhole saw or sabersaw. The face of the carling usually is vertical, the upper surface matching the camber of the beam mold.

The carlings must be jointed into the main beams fore and aft and carefully fitted, allowing for the sweep. Once they are installed, tie them down to the hull and to any partial bulkheads with a couple of 1 by 3's. The tiedowns may be glassed in temporarily and chopped off later, if your hull is glass. The beam molds may then be removed, if you wish,

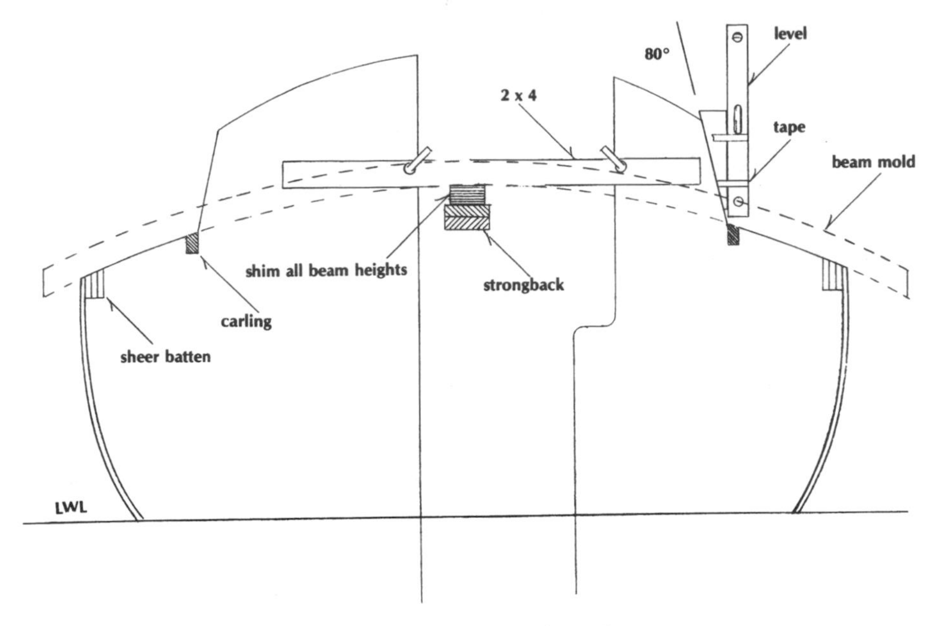

Figure 8-20. *Springing the carlins.*

but they help to stiffen the carlings while they are being notched, as shown in Figure 8-21.

Some designers prefer half laps, others dadoes and dovetailed dadoes. If the canted half-dado is strong enough for one end of a beam, however, it's strong enough for the other. Note the simple notching of the beam into the sheer batten.

SETTING THE BEAMS

I assume you know that beams must fit nicely to the inside of your hull, whether it's wood or glass. This prevents the hull from springing in (when rolling against a dock?) and cracking near the beam end around the fastenings. In addition, if there is a sheer batten, it must be cambered like the beam mold and fair into the true sheerline (Figures 8-20 and 8-21). The clamp is bolted through the frames and planking. It is essential that beams be notched over the clamp, the sure way to prevent panting—athwartships in-and-out movement. So locate the clamp below the sheer the molded dimension of the beams less the depth of the notches. This construction reduces strain on the hold-down bolts running through the clamp. If the plan calls for a few lighter beams of less molded depth, shim sufficiently to raise the beams flush with the camber of the sheer plank. Be sure all beams fit neatly against the inside of the hull. I do not advise screwing or bolting deck beams to bent frames, since any movement could split one or the other member.

You can get the exact length of a beam between notches or sockets by placing the beam upside down over the sockets. Mark at the outboard corner of both sockets. Use your bevel gauge to get the taper of the hull. If you make a gauge as shown in Figure 8-21, you can transfer this cant of the socket to the beam (or to the socket, whichever way you prefer to fit). A little blue chalk on the end of the beam will show you where to shave off the notch. The gauge works well in areas of greater beam but does not make allowance for the flattening of the camber as the hull narrows. Here I would construct the socket, then take off the angle with my bevel gauge.

This kind of construction of side deck framing is satisfactory with glued plywood and plywood underlay decks. It was once standard practice in wooden boats to place tie rods between the carlings and the clamps, because laid decks would not prevent opening of the joints. All dovetailed half laps would provide a similar tie-in. This, however, would be a waste of time with a canted half lap. Figure 8-22 shows the weakening effect of the half lap, then the canted half lap, and then the dovetailed half lap in the carling. It is difficult for me to

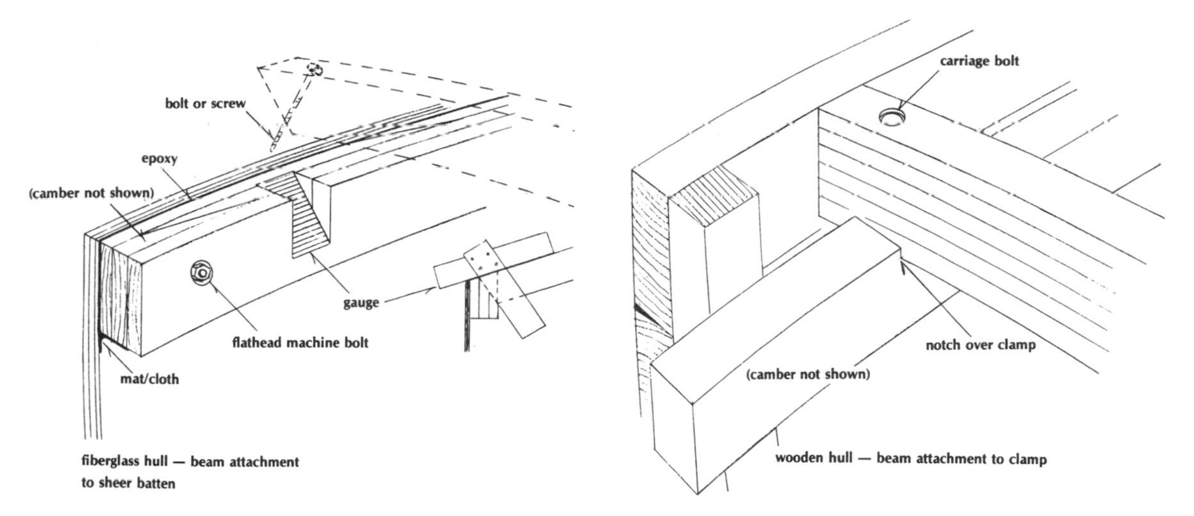

Figure 8-21. *Beam attachments.*

remember that epoxy glue eliminates almost all of the weakness of these joints. It can, however, be a bit disconcerting to lie awake in your berth counting the sloppy joints overhead. And if the joints you can see are bad, how about the hidden ones?

It is interesting at this point to note that L. Francis Herreshoff says to prevent a hull from panting,

Lloyd's Rule . . . calls for the deck beams to be different spacing from the frames so the head of the frames cannot be bolted to the deck beams, which is a very poor arrangement indeed, for the usual single fastening through the clamp does not support the frames or deck beams rigidly.

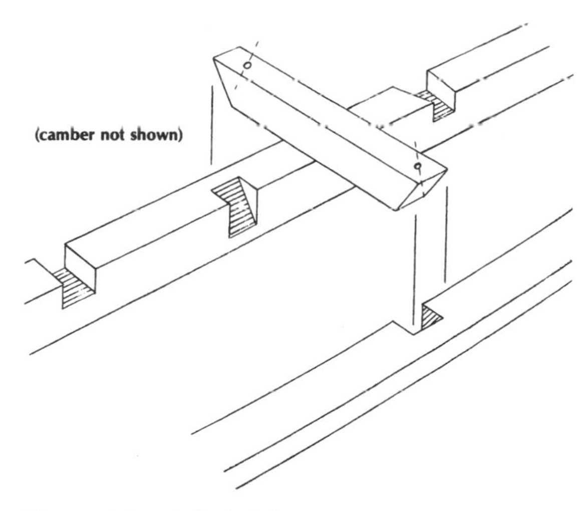

Figure 8-22. *Side deck beam joints—sheer batten to carlin.*

Now, I hesitate to dispute Herreshoff, and believe me, I still hang on his written or spoken words. If I really wanted to split a frame head or a beam end, however, I would stick a bolt or rivet right close to its end and then bounce a heavy load on the deck. The beam loses a bit of its camber for a second, then it stretches, and you have a split in one or both parts. Or possibly a started plank. Notch the beams over the clamp, friends. This resists tension and spreads compression, which occur daily. Herreshoff mentions other things to reduce panting or a change in shape of a section bulkheads and partitions strongly attached to deck beams and/or frames, and cabin sole beams. These have been covered.

A strengthening system down the center of the deck usually is required. It is not uncommon to see a strongback let into the beams of the foredeck. This is not too bad if it is a light piece, say, ½- or ¾-inch plywood let into a 2½-inch beam. It is a crime, however, to cut out one-third or one-half of the strength of five or six beams where it is needed most. It is far preferable to fit blocks (say, 1½ inches thick) between beams, toenailing into one beam and nailing from the other. The tops must be planed off flush with the beams, of course, because there should be no air space between the

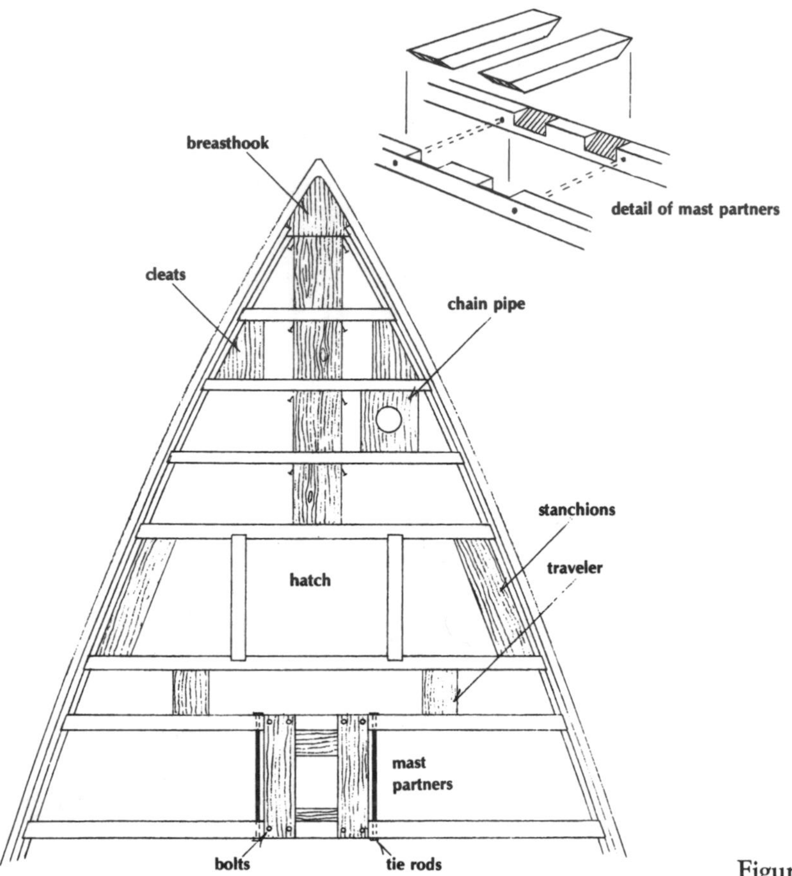

Figure 8-23. *Typical deck blocking.*

blocks and the deck (Figure 8-23). In addition, blocking should be installed all along the deck edge between beams where the covering board joint falls. Even a plywood deck must have blocking for deck hardware so that cleats, padeyes, fairleads, travelers, jib tracks, winches, and so on, can be bolted down securely.

All members that penetrate the deck must be blocked—for example, the rudderpost and cockpit scuppers. Winches, especially those on the cabin trunk, should be blocked to the second beam forward. The blocks should be of attractive hardwood with decorative edges if they will be visible from below. With regard to foam-sandwich decks, I need only remind you that foam cannot take compression or permit shear on the bolts. The answer is blocking in place of the foam in such areas, either by scraping it away before the upper surface is bonded on or by inserting the blocking during lamination. Compression tubes around the bolts instead of blocking would not provide complete resistance to shear, such as from a winch.

The placement of hardware often requires shifting from one location to another. Always install new blocking. This is one of the disadvantages of foam-sandwich decks. The block area has to be routed out; an ugly repair job may result. If your construction plans do not detail blocking for everything, ask your designer's advice.

This concludes my suggestions on basic framing for your vessel. Now you can cover the deck area with plywood, tarps, or plastic for a temporary cover to permit you to work on the interior when it's freezing out and on deck when the sun shines. You might review the construction of some of the joints described in Chapter Seven before starting anything, to make sure you have all the necessary tools and materials.

9 | Decks

The decision on the type of deck most appropriate for your vessel, your tools, and your skills must be made months ahead. Many factors influence deck construction. For example, plywood decks eliminate the need for lodging knees—reinforcements between beams and around openings in the deck (Figure 9-1). If you choose a laid deck, however, many lodging knees must be fabricated and installed meticulously. Also, if there are few plywood structural bulkheads, many hanging knees must be built.

This chapter is devoted primarily to the lower-cost, practical types of deck that many nonprofessional builders seem to prefer. Traditional laid and teak veneer decks are discussed at the chapter's end.

PLYWOOD DECKS

Plywood decks must be built with genuine marine-grade plywood, not exterior grade. Such decks should be laminated of two thicknesses, except on small half-decked craft. It is difficult to lay a deck in a single thickness of ½- or ¾-inch plywood and not have the joints show because of the stiffness of plywood. Even butt blocks between beams may fail to fair out such joints because of the curvature of the deck. Laminated decks prevent this problem by staggering or scattering the joints (Figure 9-2). Moreover, lighter plywood does not resist a slight compound curvature.

Thus, the joints in the top lamination are always backed up by the bottom layer, with both on beams. Scattered joints cannot leak and the flexible sheets form a fair surface and a rigid structure. Two layers of ½-inch fir marine plywood would make a strong deck for about a 40-foot-overall sailing yacht. Combinations of thicknesses can be used to get the total required.

Shift the arrangement of the panels so that no joint falls on a corner of a deck opening. No lodging knees are needed under a plywood deck. If a plywood joint is in a corner of an opening, a stress point is created and working may result. Do try to keep the weight of your deck down to avoid the unhealthy effect of excessive weight above the water line. Massive strength here does not make a superior vessel. Get your designer's advice if this is a potential problem.

A simple plywood deck must be fastened directly into the sheerstrake (or sheerbatten) properly beveled to conform to the crown of the beams. Use galvanized or bronze wood screws or bronze Anchorfast nails. Glue this joint. If the molded glass hull has a flange turned in, I suggest you back this up with a laminated fir batten so the pressure of the through-bolts is spread out. Wooden vessels

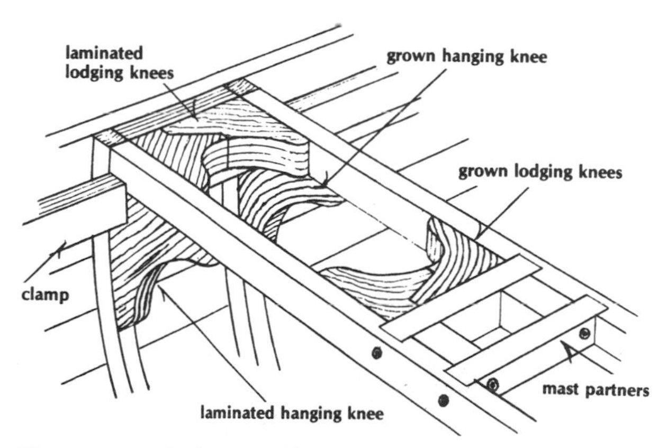

Figure 9-1. *Lodging and hanging knees.*

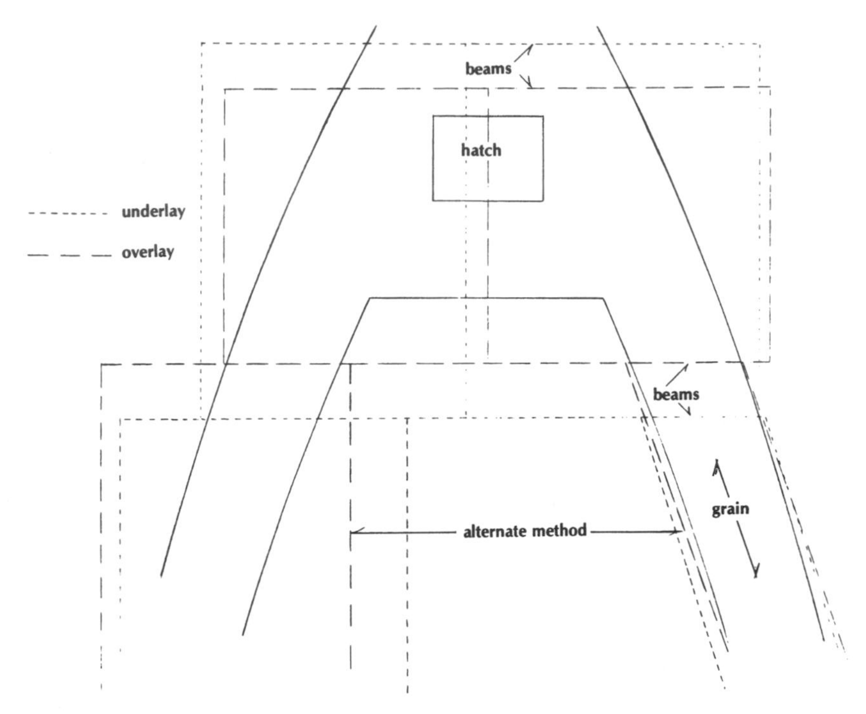

Figure 9-2. *Scattering joints in plywood decking.*

have blocks between the beams and tight against the planking, fastened with screws from outside and toe nails into the beams (Figure 9-3). The top surfaces must be planed off fair with the beams so there are no openings where moisture can enter.

Plan to glue plywood to beams and carlings. Lay down the first layer, fasten temporarily so you can see that all joints match up, mark on the underside where glue must be applied (for Aerolite, T88, or epoxy), remove, and trim close to size. Glue and fasten the first panel, then glue and fasten the second, and so on, being sure all butting edges are adequately coated with glue. Wipe off all excess glue at once, before it sets up.

LAMINATING THE DECK PANELS

There must be no possibility of voids between panels. The adhesive must be watertight or water-repellent and it should be crack filling, that is to say, void filling. Some cracks can be eliminated by using the right adhesive. Aerolite and T88 are crack filling, waterproof, have several days' pot life, and can be used at fairly low temperatures. Aerolite has the advantage of requiring application of the activator to all of one side while the adhesive goes on the other. Chem-Tech T88 may be applied to only one panel. It works at temperatures in the low 50s and loses no strength in gaps up to 1⁄16 inch wide. High clamping pressure is not required with either of these adhesives (see Chapter Six). Weldwood Plastic Resin is not crack filling, but it costs less. In many years of use I have never known it to fail.

When you are ready for the second thickness, spread the heavier adhesives with a laminator's squeegee or a tile-layer's toothed tool. Aerolite may be brushed on; fasten first near the panel center so the air is squeezed out. *Galvanized* finish nails into the beams are acceptable, but use Anchorfast nails or screws around the perimeter. Fore-and-aft joints may require a series of blocks if the panels are too thin to hold screws. Fastened from the underside, they are not offensive. A line of roundhead screws between beams would be good insurance if the top layer is heavy enough. You will need weights everywhere. Place building blocks on battens laid fore and aft, or big plastic sandbags used on pool cov-

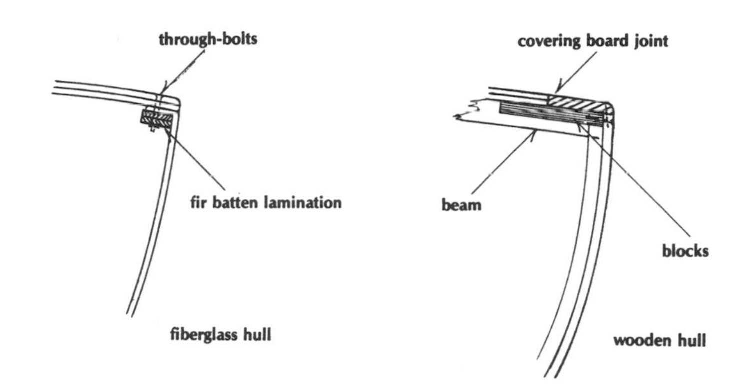

Figure 9-3. *Deck-edge construction.*

ers. If you feel a springy spot, immediately drill a couple of 1/16-inch holes through the outer layer (only!), then tread on the area to force the air out.

COVERING THE DECKS

You have laid the simplest deck. But it is not complete until it has been protected from weather and wear and tear. The mania for glassing everything has resulted in books and articles describing how to use polyester resin. However, such a deck is inadequate even when painted in attractive colors. If there is a bruise, in a couple years moisture enters, delamination begins, and dry rot is not far behind. Remember, too, that polyester is water-permeable and that the panels may work microscopically; so moisture would pass through in time. The only answer is epoxy, such as that produced by the Gougeon Brothers, Chem-Tech (L26), Smith & Co., System Three, and others. Epoxy will not let go and it is nonpermeable, but costly.

The best reinforced plastic deck fabric that I know of is Versatex, a polypropylene, especially when laid in L26 Epoxy Sheathing Resin, because it is highly fluid and designed to penetrate and seal wood and fabrics. With this epoxy, moisture content of wood is not important. It is recommended that L26 be applied to bare plywood and allowed to cure, to increase impact resistance. Versatex is light, stretches slightly, and absorbs less resin, so should be applied to a light L26 first coat to prevent "floating," then a heavier coat rolled, brushed, squeegeed, etc, as described immediately below. Chem-Tech, 4559 Lander Rd., Chagrin Falls, OH 44022, has a 40-page manual you can send for.

Be sure the deck has been swept and vacuumed before you lay out the cloth. Smooth out the wrinkles, stretch taut, and staple around the perimeter, covering the deck-planking joint. Resin may be applied with a large, cheap brush or a paint roller, or poured on and squeegeed all over. Experts swear by the rollers, for they avoid a buildup of excessive resin. The cloth will be transparent when it has fully absorbed the resin. Don't try to fill the fabric surface to a glaze, for this destroys its excellent footing. As soon as you see the grain of the plywood, stop. This means that the surface is well wetted. Watch for the resin to kick off, then trim immediately with razor blades or a sharp knife while the cloth is flexible.

Where it is necessary to join panels, allow for a lap of about two inches. After the cure, sand the under panel to a feather edge to minimize the hump. When the second panel has set up well, sand the joints with 80-grit paper and go over the entire surface lightly. If you insist on a really smooth surface, fill the fabric with a surfacing compound using a hard squeegee or spreader. Chem-Tech has a filler called F-9. It would be satisfactory, however, to give the cloth a coat or two of thinned paint. The barren tennis-court look of your deck will be relieved when you add toerails or bulwarks.

COVERING BOARDS

While covering boards or plank-sheers are essential with any laid deck, they may be used to frame fabric-covered decks too. Covering boards add to your work, but also to the yachtlike appearance and repairability of your vessel (Figure 9-4). I'll show you later how this was accomplished on our old R-class sloop, *Mouette*, with a toerail over the inboard

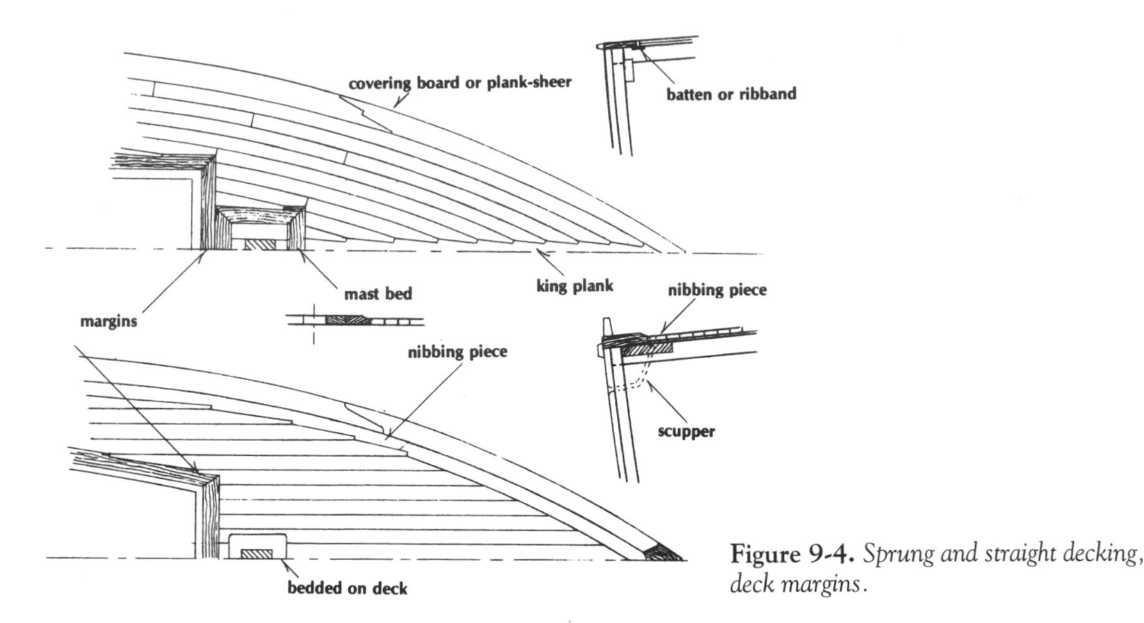

Figure 9-4. *Sprung and straight decking, deck margins.*

edge of the covering board to cover the fabric edge. Of course, what follows has to be done before any actual deck work is begun. *Mouette* was canvas-decked, but the information applies in general to today's materials.

Your first task is to line off the sweep of the covering boards on the deck plan, tapering toward the ends gracefully to about two-thirds the greatest width. Transfer these rough dimensions to the deck beams. Next, spring a ¾- by 1-inch fairing batten by tacking small nails in the marks. Never nail through a batten. Keep shifting the nails slightly until the batten springs fair, then mark on both sides. Don't move the batten.

Here's how to transfer that curve to the deck panels. Weight and brad the panels accurately in place, resting on the batten. Go below and mark along its inner edge. If there will be no king plank, the deck panels can be dressed to the marks, including first and second layers, but do not fasten. If there is to be a king plank, complications arise that I'll try to dispose of a few pages farther on.

Remember that the deck and covering board edges must be supported by a batten or ribband let into the beams, or by blocking, as in Figure 9-4. Locate the ribband notches about 2½ inches wide, ¾ inch deep. Screw and glue the ribband securely, using butt blocks at end joints.

You may, if you wish, provide for eventual replacement of the deck covering. In general, canvas decks are good for 12 to 15 years, but are now rare. Fiberglass lasts for who knows how long (Figure 9-5A). If you go the old *Mouette* route, you fit the covering boards right up against the deck edge; a toerail, its mounting edge well hollowed out, is fastened over the joint and the fabric. In time, the toerail can be removed by digging out the bungs and refinishing the whole thing, or replacing it.

Another yachty style is to insert a ¾-inch sided mahogany spline against the fabric-covered deck edge, with the covering boards fitted against the spline. Well seated in sealant, this spline can be removed carefully when necessary. If the covering boards and deck are very heavy—say, 1 inch plus—the spline may be fitted into a rabbet in the covering board (Figure 9-5B). Carefully plane down to a hairbreadth of the deck, bung over the screws, sand, stain, and fill, and you have a fine joint. But what about covering boards?

Construction of a covering board or plank-sheer is another task requiring skill and patience. Because of the curvature of the hull and deck, these pieces must be made up of short lengths, unless you have on hand a lot of long and wide teak or mahogany. Remember, the procedure above is for a plywood deck. If you go for a laid deck, everything is reversed. The covering board is made up first, with the decking being sprung against it, as I shall describe later. Most covering boards are joined by meticulously fitted scarfs in a half-dozen configu-

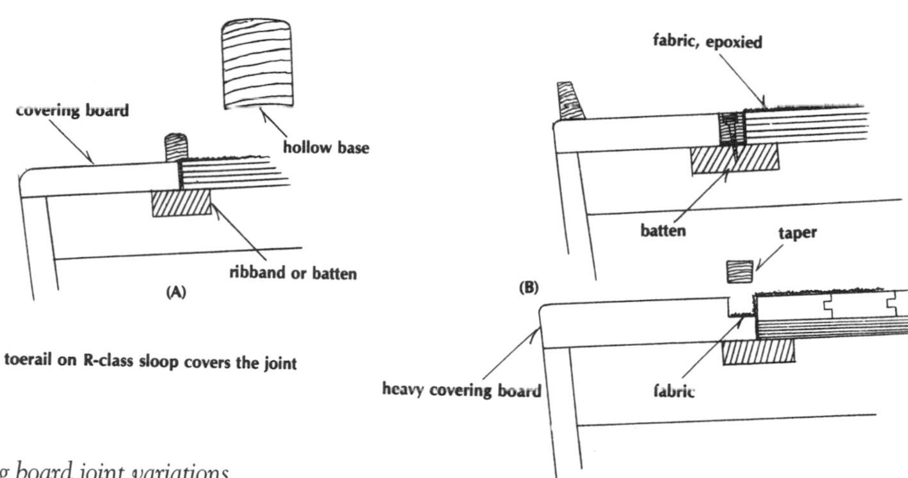

Figure 9-5. *Covering board joint variations.*

rations. The box scarf shown in Figure 9-6 is more than adequate if well done. Make ¼-inch plywood patterns of both sides, A and B, fitted so you can't see daylight. This way all scarfs will be identical, and you won't be heaving large boards all over the vessel. You could do this by a method A, using the patterns only for marking. You should, however, rent a router and save many hours as well as get accurate joints. Use a ⅜-inch combination panel bit. This tool has a shank designed to follow a pattern with precision and dispatch. It is carbide tipped for sharpness and long life.

Always leave the outer edges of the covering boards to the very last, after all scarfs are fitted, so small adjustments can be made. Position piece No. 1 against the stem or the king plank, or to the centerline. Tack pattern A to it so the nib lines up ¾ inch inside the plank-sheer and the center of the joint ribband (Figure 9-6A). Mark the scarf, saw out roughly, tack the pattern to the underside, and rout out. Clean up the corner radius in the nib with a sharp chisel. Bore and fasten No. 1 temporarily in place.

Now lay No. 2 on No. 1. Slip blocks under it to bring it up to the same plane. Use a bent and sharpened pick or small file (Figure 9-6C) to mark the joint underneath. Saw out roughly, tack pattern B in place (on the underside again), and rout to size. Chisel out the nib and try the fit. If the far end of piece No. 2 does not lie over the plank-sheer with ample room for trimming, adjust the position of your pattern, leaving a bit to trim off, and, perhaps, try again (Figure 9-6B). Precise fitting of the length of the scarf may also be completed by trimming a bit off the male end of the nib. Make the plank-sheer far end scarf and bore and fasten down temporarily. Fit the inner and outer edges only after all scarfs are made up for the full length of the vessel.

As you can see, some of this joinery has to be done with a chisel into and across the grain. This takes much control and patience, and very, very sharp chisels. You should have joints that you can't see daylight through. When all the covering boards are fitted, bored for fastening down, and bedded in Thiokol, bore for additional screws, these running edgewise. Some designs call for through-bolts, a possibility if there is no decking in the way of the inboard edge. The last touch is to bore parallel to the nibs and tap in a pine dowel called a stopwater. The covering boards will not swell lengthwise at the nibs, so leakage could occur, followed by discoloration of the fine finish, and possibly dry rot. Of course, use blocking under all scarfs.

It is superior yacht practice to run a covering board above the transom, a fashion piece similar to your rail caps. Support blocking must be constructed to keep the decking and covering board from working (Figure 9-7). The sketch does not show decking running straight out over the transom, but it's not uncommon. This should at least be covered by a continuation of the rail or bulwark cap called the taffrail. The transom covering board may join the side board in a simple miter, but do not allow the sharp ends to extend to the extreme outer edge. A careless helmsman

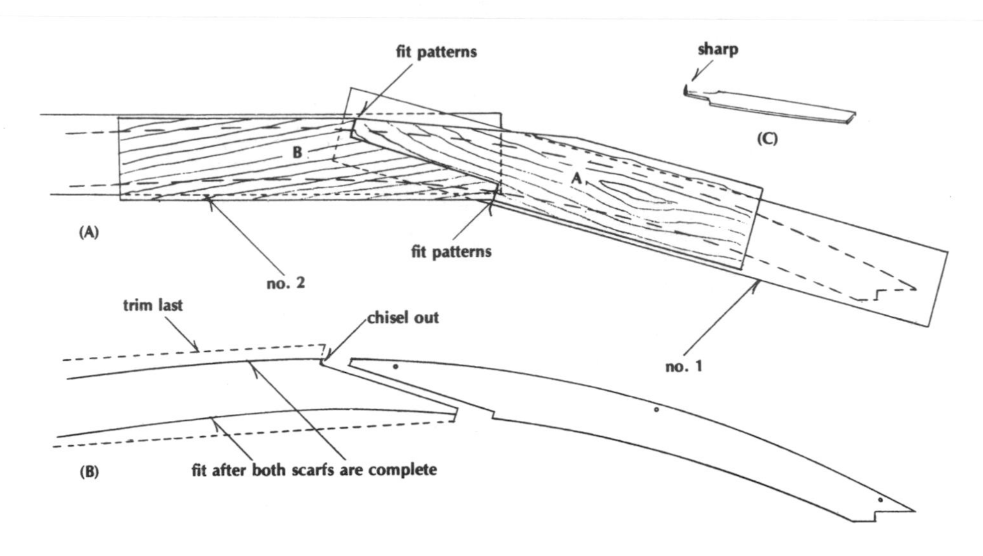

Figure 9-6. *Scarfing with router/shaper patterns.*

might tear off a lot of nice hardwood as he swings away from a pier.

Figure 9-7 shows another style of deck—the herringbone. If any sprung decking meets the fashion piece at a more acute angle than indicated, it could be nibbed in as shown by the dotted lines. Blocking under the deck and fashion piece is also indicated by dotted lines. A great many variations in the quarter knees here are possible, both straight and crooks. All of this joinery must be first class, and all precautions must be taken against leakage. Counter sterns are highly susceptible to rot resulting from frequent doses of rainwater and little fresh-air circulation under the deck.

It is good form and sound construction to build teak or mahogany frames or margins around hatch openings and mast partners, and around the cabin trunk and cockpit well (Figure 9-4). This type of framing and king plank is often quite a bit heavier than the decks, with the step-down usually in the form of a routed cove. Again, such deluxe construction is more often used in conjunction with

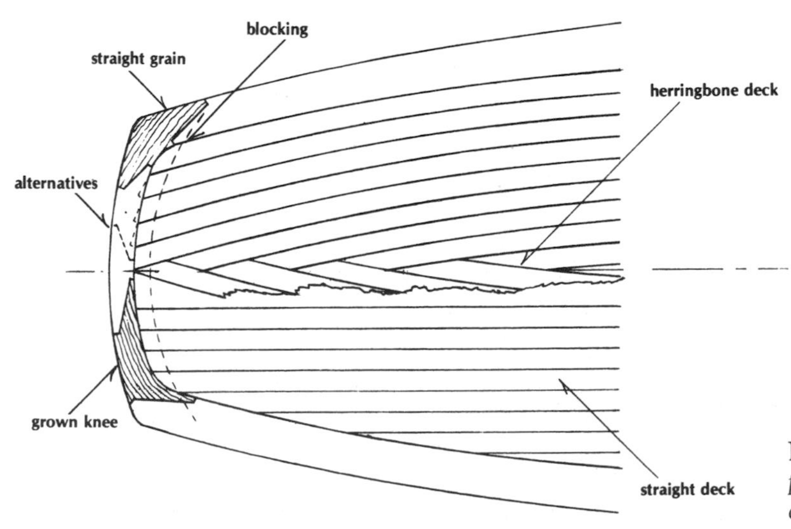

Figure 9-7. *Taffrail or fashion piece and quarter knees in transom covering board.*

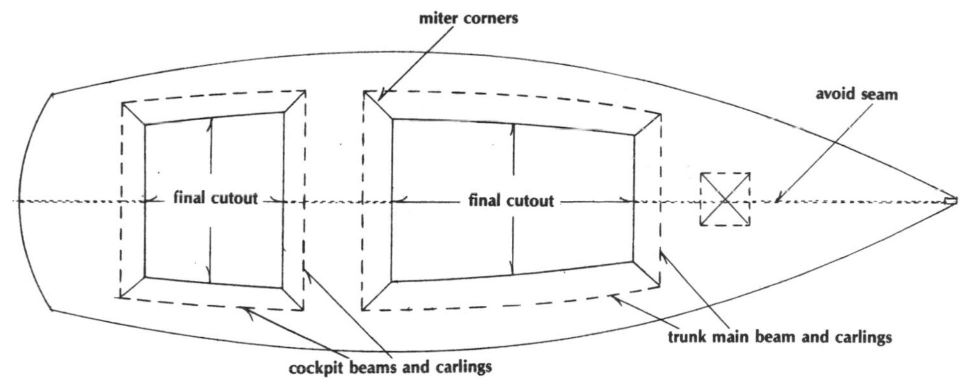

Figure 9-8. *Deck canvas layout.*

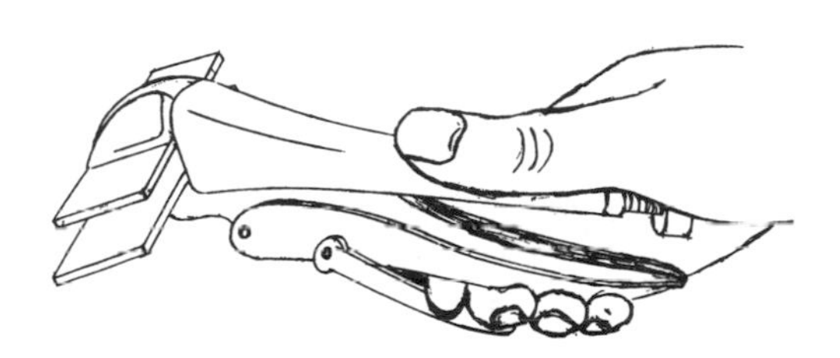

Figure 9-9. *Hand clamp for pulling canvas.*

laid decking, but it can produce an impressive effect with fabric- or fiberglass-covered plywood decks, too. Note the optional nibbing piece, to avoid nibbing the covering board.

CANVASED DECKS

The use of painted canvased decks was revived somewhat, along with the great resurgence of interest in wooden yacht construction in the 1970s and 1980s. Hence, *Practical Yacht Joinery* devoted four illustrated pages to the advantages, disadvantages, problems, and solutions; advantages are low cost and good footing and the relatively simple installation. But there was no feedback, no pros, no cons. Perhaps the great strides made by other materials and techniques outweigh those of the painted fabric. I refer those who might be interested to the old books on wooden boat and yacht construction.

Approaches shown in Figure 9-10 (A, B, C, D) apply in general to fiberglass, Vectra, Versatex, Dynel, and any other fabrics, not just to canvas. Note too that the decision as to which approach you will use must be made long before you build the deck or the trunk cabin. For example, the rabbets in Figure 9-10 (C and D) must be routed out while the sides can be handled.

If the beam of your yacht is somewhat greater than the width of the material, take advantage of the cutout for the cabin and cockpit to spread the canvas perhaps as much as a foot or more (Figure 9-8). Bending it too far, however, may create wrinkles. If the design includes a bridge deck, you may be able to hide joints under the coamings. Just a small piece of surplus would cover the bridge. If you can reach all the way to the transom with one piece of fabric, you may increase the resale value of the vessel by a considerable amount for just the price of a couple of square yards of fabric.

INSTALLING A KING PLANK

In Chapter Eight I described blocking that runs the length of the deck between beams or a strongback let into the beams. This is similar in function to the ribband to which the outboard deck edge and covering board are fastened. Such construction is necessary for any deck that includes a king plank. Also, any margin or framing of trunk, hatches, or other openings must be supported in the same manner, whether the deck is fabric-covered plywood, teak veneer, or laid strakes. Nibbing, especially, must be backed up. Another method, one lacking the strength of the above, is to let in two ribbands to which the king plank and decking are fastened (Figure 9-11A). This saves some weight, increases labor, and decreases strength. I would hesitate to place cleats on this. It might be accepted in a light-

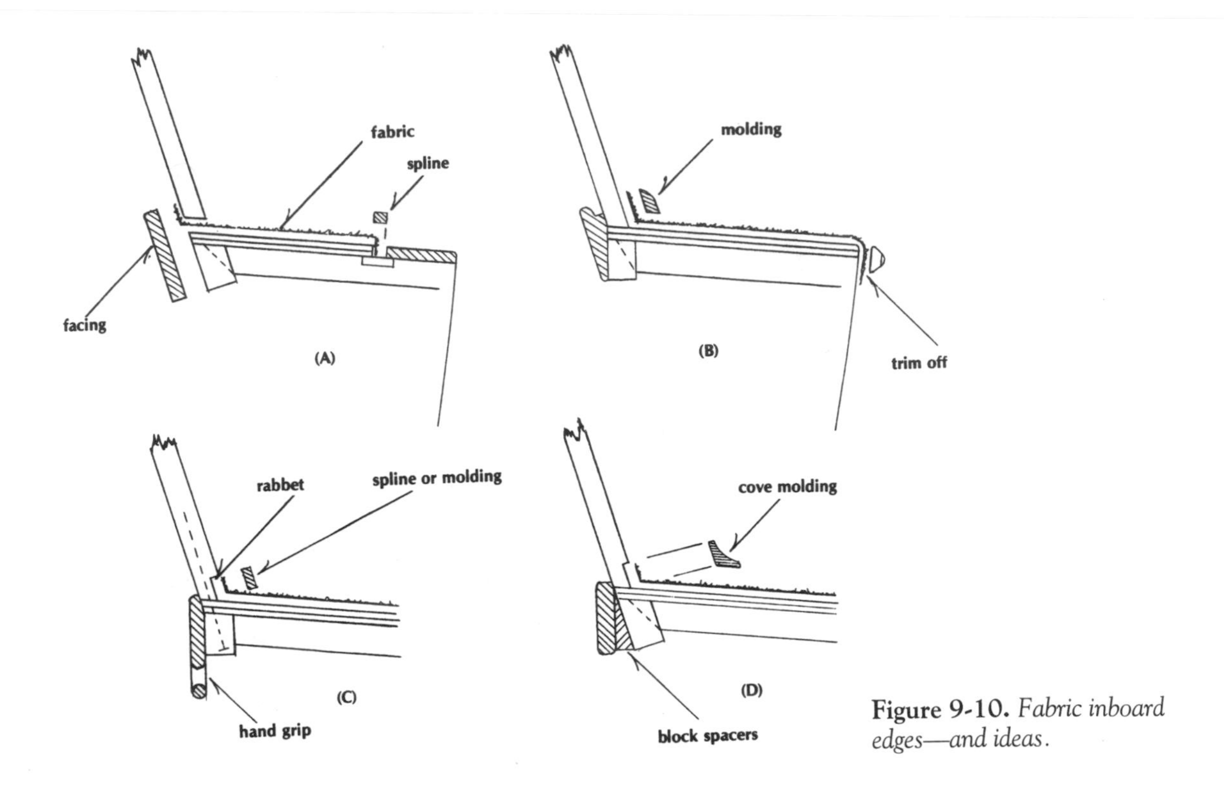

Figure 9-10. *Fabric inboard edges—and ideas.*

displacement sailboat, in which case I would make the ribbands of plywood to prevent splitting.

Here, as at the covering board, the fabric may be made replaceable by covering its edge with a spline pressed into a groove. The splines should be made with a slight taper, as shown in Figure 9-11. Fasten the panels down as described previously, making sure the groove is of precise width to take not only the spline but also the thickness of the fabric. Now it's a simple matter to prestretch, then stretch the fabric hard fore and aft, trim to cover the groove, then remove fabric, wet deck, spread fabric out, and press or tap in the spline. This should pull the fabric into the groove, but don't permit the fabric to bunch up under the spline. Bore the spline for screws and plugs and screw down. Similarly, you must stretch the fabric across the deck to go under a similar spline next to the covering board described above. You *will* need help throughout this job.

I suggest that covering boards, king plank, splines, and so on, be filled, stained, and varnished before the deck fabric is started. This keeps resin, paint, etc. from getting in the pores of the wood. Plastic (compatible) taping is still advisable. It is mandatory if these parts are teak.

LAID DECKS

The process of installing decks of pine, fir, cedar, or teak is covered fully in the boatbuilding books recommended earlier. I included a few words on fastening decks in Chapter Six. Now I'll add some material on the drawings in Figures 9-4 and 9-7. These show the two main classes of decks—sprung and straight laid. The sprung deck, in my opinion, is the most beautiful of all, but a straight-laid deck is also first class! The sprung deck was originally a means of simplifying construction in days when the deckhouse curvature followed the plank-sheer closely, rarely found today. The straight-laid deck matched trunk construction parallel to the centerline, a feature found in many older designs, especially workboats and the yachts of designer Billy Atkin. Both cabin styles are considered obsolete now, but the decking is deluxe.

Either system requires nibbing of the deck

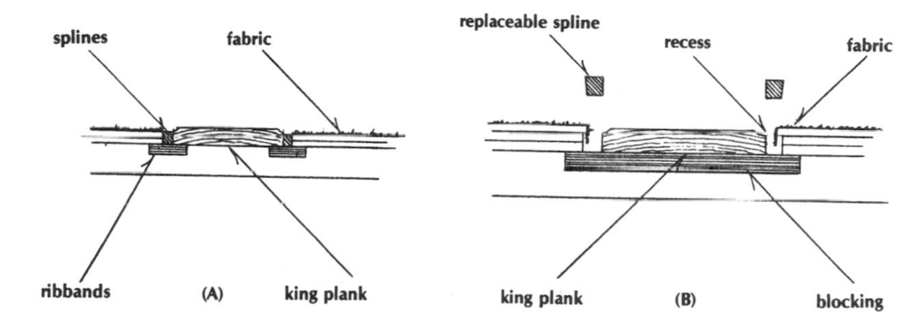

Figure 9-11. *King plank and canvasing system.*

strakes where they run out to a fine end. The sprung deck may be nibbed into the king plank (Figure 9-4) or laid herringbone-fashion (Figure 9-7). The latter is easier to build and very handsome. Nibbing is time consuming because each strake end is squared off, then tapered to its intersection with the covering board or king plank (or to a nibbing piece). It is good practice to outline a margin around all deck openings and the deckhouses (Figure 9-4). This means more nibbing, of course. It is customary to build these margins and king planks of mahogany or teak, perhaps ¼ to ⅜ inch thicker than the deck strakes. All nibs should be of the same depth everywhere, and should match port and starboard. Decking is sometimes nibbed into the transom covering board or fashion piece. It must be if the piece is very rounded and the strakes come in at an acute angle.

Very narrow teak deck strakes might not be nibbed. Also, such narrow decking, if laid herringbone, may be doubled in these staggered joints. This deck may be laid up without a caulking seam. Of course, all joints and butts must be properly sealed with Thiokol, or epoxy, or their equivalent. The bevel required on all caulked seams was described earlier.

One more point. The traditional laid deck is nearly square in section so the sprung deck can be clamped or wedged to the covering board without more than the usual sweat. Moreover, most of the remarks above apply to a deck laid over a plywood underlay (the only way to go!), except as follows. The thinner deck material (say, ½ to ¾ inch by 2 inches) will try to pop up into your face when you spring it the hard way. This is very dangerous. One solution is to have assistants stand on the strake while you draw up the bar clamps and get a few fastenings into the beams. I have seen a half-dozen 20-pound barbell weights perform this duty. The ones I saw in use were covered with plastic. This eliminates bruises. Or, for weights, the old reliable 35-pound lead pigs ballast are cheap, non-bruising, and can be added to trimming ballast.

After the seams have been payed, scrape off as much as you can of the messy compound before it sets up hard. You can sand with a regular belt floor-sanding machine, then finish the job with a portable belt sander and a sanding block. Use 120-grit open-coat aluminum oxide or garnet paper.

An Improved Teak Overlay Deck System

I call this an improved system because the teak veneer deck fastening is very different from that I described ten years ago. It is based on my own and Gougeon Brothers experience plus the feedback

Figure 9-12. *A WEST System teak deck laid over plywood. This handsome deck provides excellent footing and is easily maintained. (Gougeon Brothers photo)*

Figure 9-13. *Saw teak strips from 2-inch flat-grain stock; mark the bonding area and reference points on strips and deck (Gougeon Brothers)*

from builders of my boats over the years. I am once again paraphrasing that fine book *Gougeon Brothers on Boat Construction* (100 Patterson Ave., Bay City MI 48706). I am still impressed by the original book but additionally by their recent *Fiberglass Boat Repair & Maintenance*, Section 5, "Installing A Teak Veneer Deck." It calls for sawing flat grain teak rough into edge-grain strips preferably ⅛ inch (3mm) thick, but not more than 3⁄16 inch (4. 5mm), laid over a minimum of ½-inch (doubled ¼-inch) plywood or heavier deck. The Gougeons say that you may use heavier teak up to ¼ inch but dimensional changes can be limited by using the lightest and that it will provide you with years of service in high-traffic areas. The job will require two wet-outs of WEST System epoxy resin on the bare sanded plywood, plus an epoxy thickened with 404 filler and blackened with 423 Graphite powder. More detail on all that follows.

Here is the sequence:

1. Saw teak strips from 2-inch flat-grain stock or 2 by 2 inches so strips are edge-grain, 1½ inches to 2 inches (38mm to 50mm) wide. You may clamp them together to plane edges straight and smooth; leave both surfaces rough-sawn as that texture improves mechanical bonding characteristics. The remaining marks on upper surfaces will be sanded out after deck is completed. Square the ends precisely. Lay strips aside.

2. Prepare the bonding surfaces. If the plywood is smooth, sand with 50-grit sandpaper (and/or grind off uneven joints and non-skid areas, if any). Vacuum or wipe off dust. Wipe with wax and silicone remover or solvent and dry with a paper towel.

3. Determine your first bonding area conservatively, as epoxy open-time will vary. Lay a few strips in position using 3⁄16-inch wood scraps for spacing, fitting ends with care. Better, use ½-inch # 10 Rd. Hd screws 24 inches apart Mark screw location on the strips since later #10 sheet-metal screws and washers will use the same holes plus others on 8-inch centers to flatten down the strips during bonding. Mark the bonding area and reference points on strips and deck. When fits are perfect, remove the temporary screws and strips.

Figure 9-14. *Play it safe by using ¾-inch screws and washers so penetration of the lowest ply would be next to impossible. Remove the lubricated screws and washers within 24 hours.* (Gougeon Brothers)

4. Prepare for bonding by coating screws and washers with cooking non-stick spray or dipping in melted paraffin. If your plywood is only ½-inch thick, play safe by using ¾-inch sheet-metal screws and two washers so penetration of the lowest ply would be next to impossible. The bonding area of the strips must be wiped with acetone or lacquer-thinner thirty minutes *before* you are ready to spread the resin. Mask off the deck bonding area and cover the surrounding covering boards, rails, etc., with polyethylene tape and sheet to protect from spills. Wet out the bonding surfaces of the first set of strips with unthickened epoxy, also the deck area you masked off. Apply a heavy layer of thickened epoxy to the wet-out deck surface; use 404 high density filler mixed to a mayonnaise consistency, then add enough 423 Graphite powder to make an opaque black. An 808 Plastic Squeegee with serrations notched along one edge works well to apply an even layer on that deck area. Estimate enough of the mixture so it will squeeze up between the strips and bridge the ³⁄₁₆-inch gaps.

5. With care, place the first strip and find the previous screw locations; a couple of ice picks are good tools for this job, but do not punch through the deck. Use the reference marks as necessary. Start the point of the pick where the rest of the screws will be driven 8-inch c. c. Do not drive the screws down until the second strip is snug against them and its own line of screws are holding so it can't slip. Then lay down the rest of the strips in that first bonding area, and go back and clamp down the screws and washers, being sure that each strip is hard against its spacing screws, thus insuring that the seams are beautifully uniform. Tightening down will force the black mixture to ooze up into the seams. If portions are not amply filled, add more of the epoxy/404/graphite mixture so the seam is smoothed flush. Scrape off any excess before the epoxy begins to gel.

6. Repeat the entire process, two, three, or four strips at a time if your epoxy timing is conservative. You may have to adjust the number of strips or size of batch of epoxy. You might be able to clean up some of the mess in the saw marks with a paper towel dampened (not wet) with acetone, but catch it before the mix gels. Later sanding will be necessary, however.

7. Remove the screws and washers within 24 hours. Tighten the screws lightly (5 degrees) before backing them out. If you have difficulty removing a screw, heat the head with a soldering gun's cutter tip; while the screw is still hot, try it again until you are successful.

8. Fill the screw holes with epoxy-graphite mixture. A syringe loaded with the mix will speed the process; be sure you do not trap air under the epoxy, and check frequently for shrinkage. If the screws penetrated a panel, seal the holes with duct tape before filling the holes.

9. Sand to level the surface and remove saw marks. Use a belt or disc sander with 50-grit for the initial sanding. If you have no experience, practice so that you learn to leave no disc

marks and always keep the belt sander moving and in an oval or figure-eight motion. For large decks, a commercial sander works well. Finish with 80-grit, then with 180-grit.

All this would apply if you go for a beautiful sprung teak deck. But bending the hard way is more difficult, so you will be 100 percent dependent on the screws and perhaps on a half dozen barbell weights. Be sure your strips are under control or you could experience a split lip or worse. There is more on this above on solid laid decks.

In spite of the picky work and patience required, either system is worth it if the average amateur builder wants his boat to become a yacht. A good teak deck could increase the value of the yacht by thousands of dollars and attract potential buyers regardless of its other characteristics. The task I have described will pay off in pride when you hear someone say, "Now *that's* what I call a yacht!"

10 | Cabin Trunk Construction

The design of your cabin trunk or house is subject to many variations. It may be of the old style, with sides inclining inward from ½ to ¾ inch per foot of height. Look at the superb creations of such designers as Philip L. Rhodes, S.S. Crocker, Charles D. Mower, Charles G. MacGregor, Murray Peterson, and others. Their deckhouses usually show a moderate camber. The beams are more often sawed than laminated. Forward corners are slightly rounded, unless cost was no problem. In other words, their cabin trunks are simply constructed, straightforward, conservatively proportioned, and always handsome. In many other ways, too, the older designs are totally satisfactory, and they should be considered by any of you making *design* decisions.

In contrast, the deckhouses of modern yachts always tumble home greatly (three inches per foot of height is considered moderate). Many have cambers of more than one inch per foot of breadth, and they have not only extremely rounded forward corners, but also raking forward ends. Combined with a high cambered deck, these features (in wooden construction, cold-molded, etc.) mean increased material and labor costs, shaped carlings, very wide cabin sides, laminated beams, intricate corner posts, laminated or steam-bent forward ends, and so on. No wonder the modern wooden yacht is costly. Because of the complexity of the deck plug and mold, plus the quantities of resin and glass required, the production fiberglass vessel is also costly.

In deference to those who are convinced they must be "modern," I shall discuss a few procedures that apply to contemporary designs, as well as to older vessels. Either way, tumblehome and smoothly rounded forms, especially the blister-shaped structure, tend to reduce windage created by height. High cambers pay off in increased headroom in small boats, so it's not all bad. The Bingham 25 and the Travlr 27.8, for example, have 6 feet 3 inch clearances.

TUMBLEHOME

Your first job after completing the deck is to set up a structure around which the cabin trunk sides can be bent, thus securing them at the designed tumblehome. I'll refer to this as a form and the separate cross sections as strongbacks (Figure 10-1). Strongbacks can also be used to provide a rigid support for laminating a beamless cabintop, if you choose to go that way. This form is mainly 1 by 4 or 1 by 6 construction, with the curved top piece of ¾-inch plywood or lumber. The several radii of the strongbacks may be swung with a wire or with a camber scriber (Figure 5-41). I said "several" because there is a slight sheer to the cabin side when it is viewed after bending (this must be spelled out in detail by your designer). Thus, if you desire a straight centerline profile (not all designers do, nor do I), each strongback must be laid out and supported so as to reach a strong, straight board clamped on the centerline.

It is assumed that your main bulkhead aft and another one forward have been proportioned correctly so you can run strong battens from these over the two or three (or more) strongbacks and along the sides so the trunk sides and cabintop will be fair. A level with a wedge taped to it makes the sides fall to the same tumblehome (Figure 8-20). You can simplify this by making the strongback all one piece. Then you can screw legs to it, eliminating the 1 by 4 cross frame. Don't forget notches for the cabintop carlings, and the many notches for stringers for forming a beamless top, or swing the

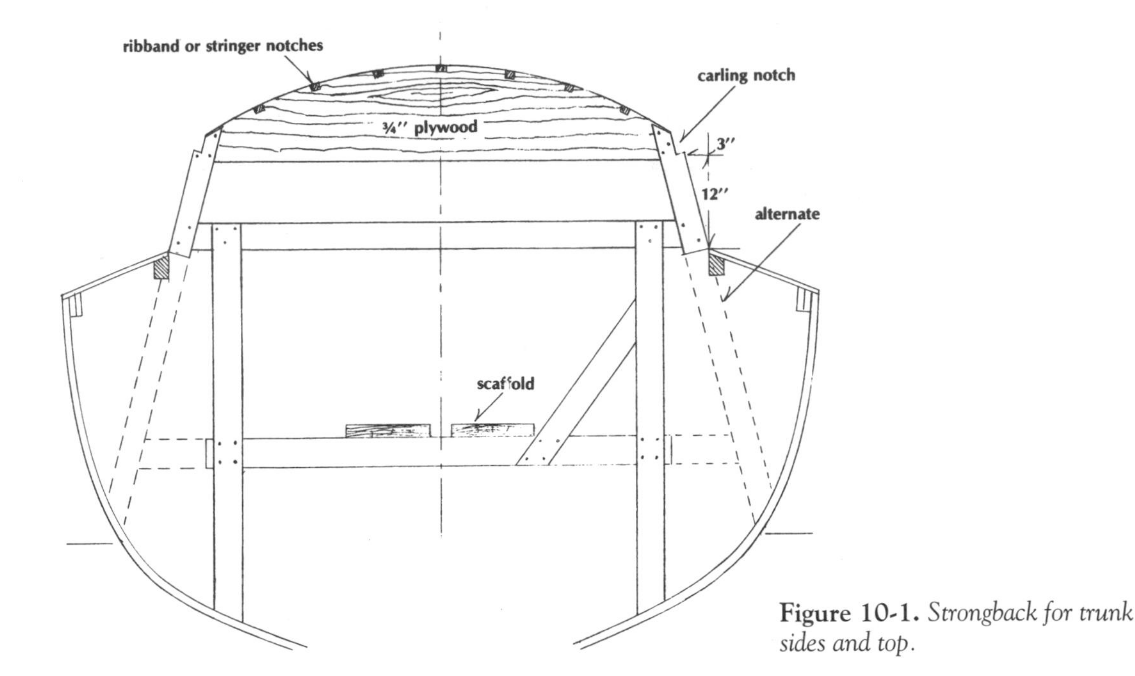

Figure 10-1. *Strongback for trunk sides and top.*

radii adjusted for stringers and avoid the notch job. (If there are to be beams, ignore the cambered strongback and laminate the beams, as described in Chapter Nine, or laminate right on the stringers, also "adjusted.")

A simplified trunk tumblehome form is shown in Figure 10-2. This construction is not strong enough to support the weight of a couple of men laying a laminated top. You can use scrap plywood instead of the open framework. The strongbacks must be tied together to maintain their vertical alignment as the sides are being bent and fitted.

A TRUNK SIDE PATTERN

You will need a pattern of the cabin side to make sure both sides are identical and to show up differences in the sweep of the carlings, and so on. This can be pieces of ¼-inch plywood nailed together with butts between the strongbacks (so they don't interfere on the opposite side). Heights to the sheer of the side can be taken from the lofting or from a projected axis reference line in the design. This sheer, when finished, is almost invisible, unless it is wrong. In that case, it stands out like a three-legged man at a dance. Give this your careful attention, as insufficient sheer could make the side appear to be hogged because of the tumblehome. Notch a stiff batten or two into each strongback, full length. If the side is to be laminated of two thicknesses of plywood, make allowances for the bevels at the deck and at the juncture of side and top. Ports should be located parallel to a line that is safely parallel to the trunk sheer or centered between trunk sheer and deck at all stations. Lay out the ports and window shapes with black tape, and the locations and incline of forward corner posts. Do the same if the main aft bulkhead rakes forward (a nice backrest, but a more difficult piece of construction, of course). With the pattern tacked around the strongbacks and a couple of light battens laid fore and aft, you should now be able to get a pretty fair picture of what your cabin trunk is going to look like.

CORNER POSTS

Now we come to a hazardous and frustrating problem—the forward corners. A simple solution would be to butt the sides against the end with a cleat in the corner. This is lubberly construction. It is unsightly, prone to rot, and seen on cheap boats.

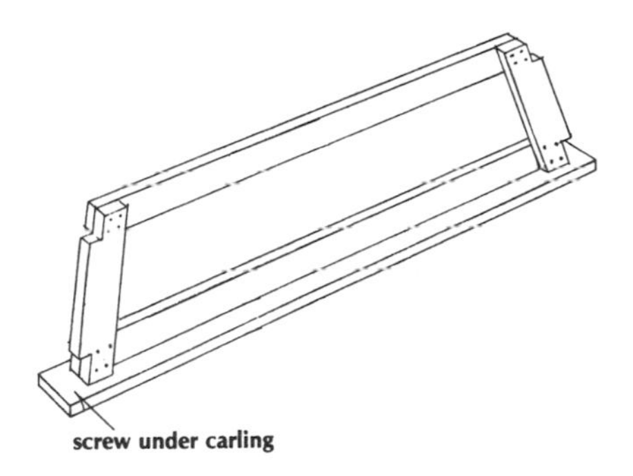

Figure 10-2. *Simple trunk tumblehome form.*

The proper method is to construct a pair of rabbeted corner posts to receive the cabin sides and the front piece (Figure 10-3). The main problem is the tumblehome of sides meeting the rake at the forward end. This is compounded by the need to fasten the post to the beam and/or deck (I design them perpendicular to the deck centerline, but you may not be so lucky). I hope your designer has detailed it well. Herreshoff, Chapelle, Steward, and other authorities have not paid attention to this problem. I have not found a source to which to refer you, so here goes.

To start, build up a laminated timber of mahogany or teak at least 4 inches wide and 6 inches deep, and long enough to make considerably more than two corner posts or blocks. This will be worked into a parallelogram (Figure 10-3B) when fitted, so allow ample stock. The final section will look like the sketch because of the sloping side and front. To get this peculiar shape, clamp the plywood pattern to the cabin strongbacks. Erect another piece athwartships temporarily supported at the angle of the forward end of the trunk. The side pattern was previously marked to the desired angle of the front and should be sawed off (Figure 10-3A). Tack them together to produce a mock-up of the corner. Check the tumblehome and rake with your wedge-level. Place the back of a bevel gauge flat against the side, the blade against the front piece. This gives you the true angle of the vital corner of the parallelogram. Now you can saw out that angle on your table saw in two passes and plane it carefully to fit inside the mock-up. Dress the two inner sides so they are parallel, to be bevelled or rounded later.

Assuming your cabin side and end pieces require rabbets 1 inch by 1½ inches wide, use the table saw or router (Figure 10-4). The block is held securely against an auxiliary fence by a hold-down (comb or feather) raised on a block.

Fitting the lower end against the carling and beam is a fussy job. Set up the cabin side pattern so the corner block sits with its end standing on the deck above the heavy beam and deck (Figure 10-5). Use a block a bit higher than the total thickness of deck and beam as a gauge to mark both the front and the side above the deck. This indicates

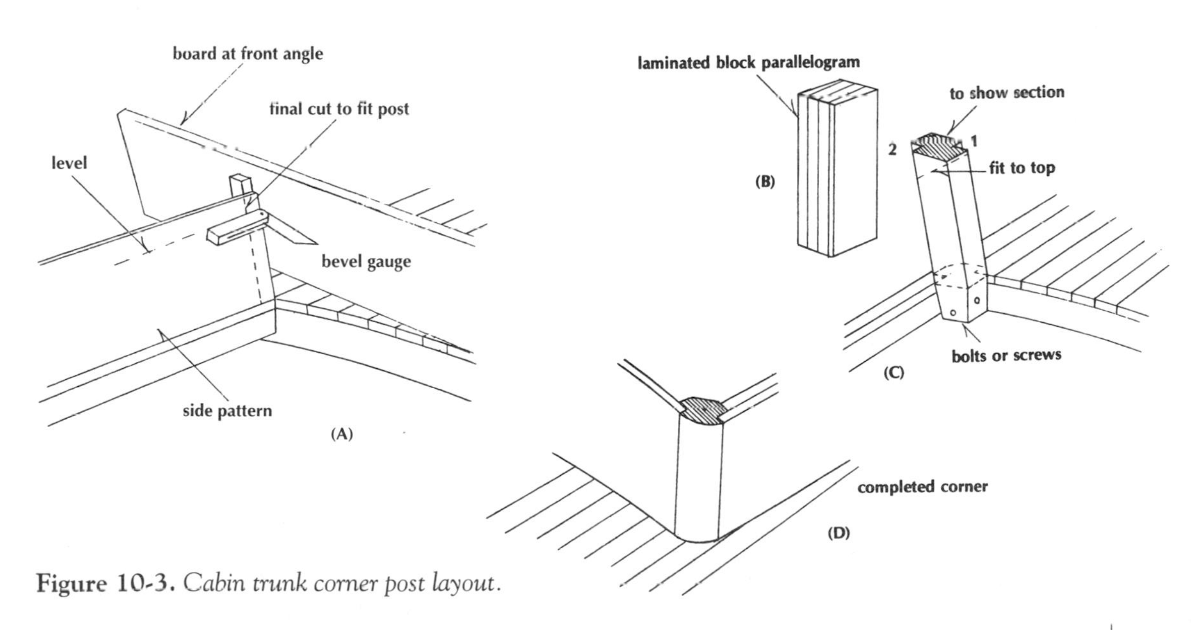

Figure 10-3. *Cabin trunk corner post layout.*

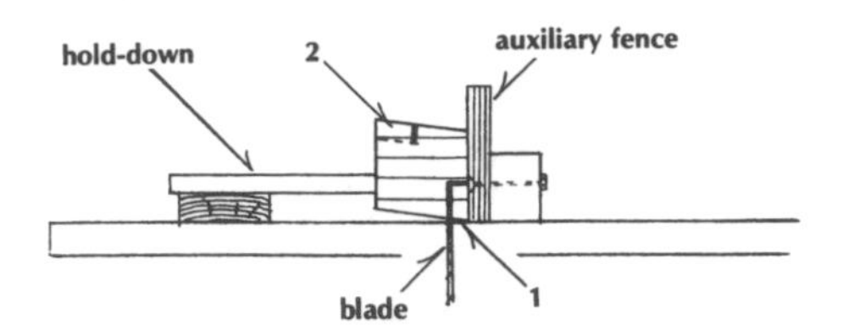

Figure 10-4. *Rabbeting the corner block.*

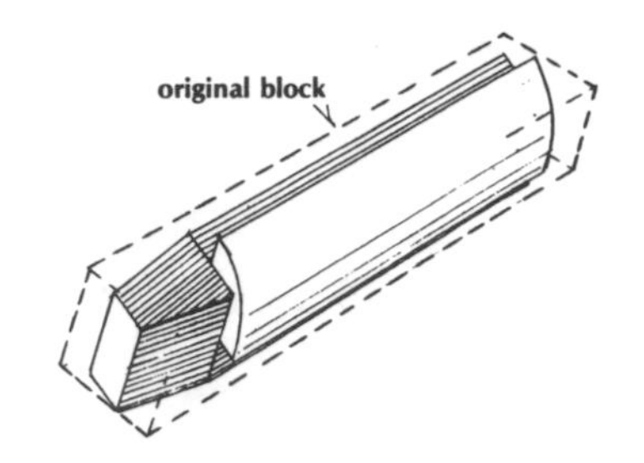

Figure 10-6. *Corner post completed.*

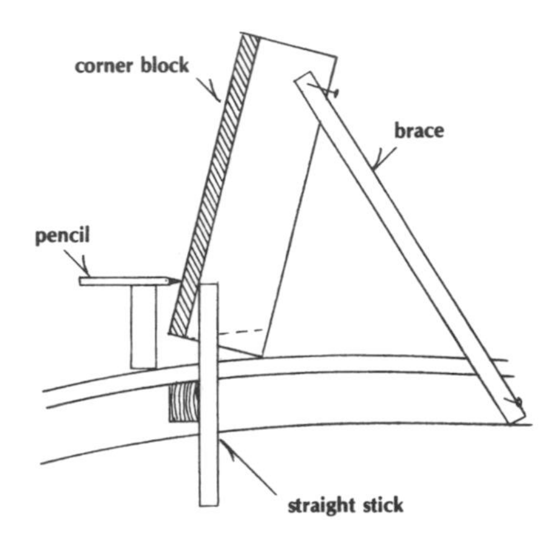

Figure 10-5. *Scribing to deck and beams.*

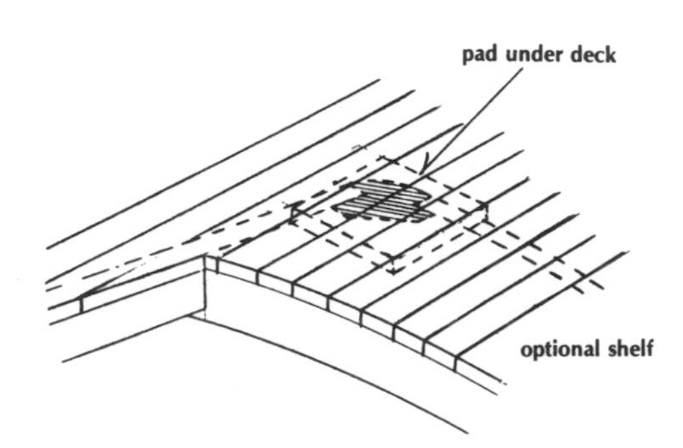

Figure 10-7. *Position of corner post.*

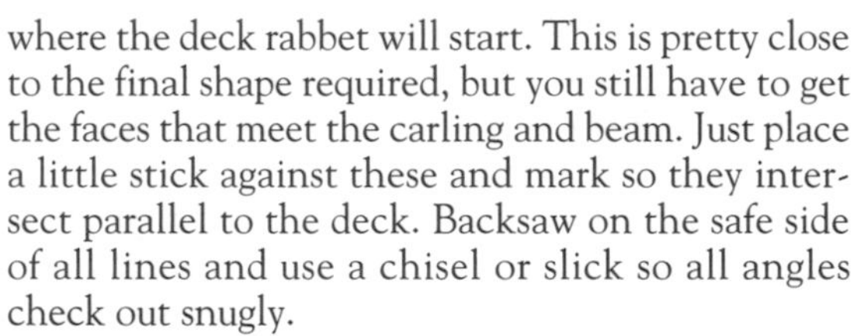

where the deck rabbet will start. This is pretty close to the final shape required, but you still have to get the faces that meet the carling and beam. Just place a little stick against these and mark so they intersect parallel to the deck. Backsaw on the safe side of all lines and use a chisel or slick so all angles check out snugly.

It might be easier to fit the side of the block against the carling, then the front against the beam. A carling canted to parallel the trunk side (see Figures 10-9 and 10-10) might simplify *this* operation, but that, too, is a trade-off. Trim off the lower end to conform to the facing piece to come. Bevel the inside corner or round it off as you please. Finally, the outside corner must be rounded to blend into side and front, as shown in Figures 10-3D and 10-6. The upper end must be trimmed to meet the camber of the cabintop after the final assembly—or bandsaw approximately, then plane off. When all is perfect, counterbore for bungs over bolts through beam and carling. Leave this for now, but plan to use epoxy, T88, or Aerolite in the final assembly. Bed the deck rabbet joints in Thiokol or an equivalent to seal.

SIMPLE CORNERS

There are several easier ways to construct cabin trunk corners. This same one-piece rabbeted post can just sit comfortably on the deck 6 or 8 inches forward of the heavy beam, forming a convenient shelf (Figure 10-7). But first the post has to be set up rigidly so that both angles are correct. Then you scribe with a block or compass around all four sides, as in Figure 10-5. One or two bronze lags driven from the underside is the usual way this is fastened. Blocking under the deck is required. Bed the post in Thiokol or other luting. Once the sides and end are glued and screwed into the rabbets, this will be a strong structure. The forward trunk end must be epoxy glued and fastened from under the deck with long lags, screws, or through-bolts. Nice cabinets on each side of an opening port make a fine yachty touch.

Another system used frequently is the two-piece post. This requires taking angles from sides and end as above. The style shown in Figure 10-8A is a little more difficult than the one shown in Figure 10-8B, and the larger radius is handsome. I would

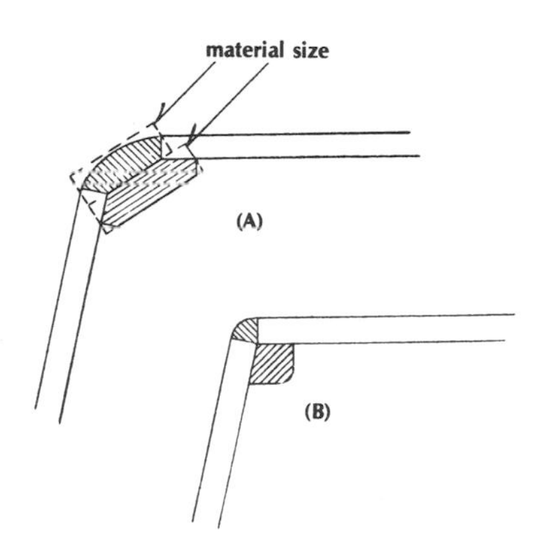

Figure 10-8. *Two-piece corner posts.*

not try to bolt this inside the carling and beam, however, without a filler piece in the corner. The rounded outer corner is fitted after everything is set up, and fastened from the inside.

The trunk aft end has corners that may rake forward 10 degrees (this makes a good backrest on a bridge deck). The answers are the same—tough! More frequently, however, the end is vertical, so construction is easy. Some small sailboats with low trunks continue the side into the cockpit coaming. If the coaming is lower, some of the cabin-side end grain above may be exposed. I can't say this is bad, but I would prefer to see a corner something like the one shown in Figure 10-8B. Or plan for a coaming fairing block to mate with the sloping coamings (this will be detailed in Chapter Thirteen).

Bulkheads meeting the cabin sides must be cleated attractively to provide an ample surface to which to fasten the trunk sides. The side cleat should be mitered to the overhead cleat, which is really a short beam. Run your router around the corners to form a radius, chamfer, ogee, or whatever pleases you.

TRUNK-SIDE CONSTRUCTION SYSTEMS

Sometime prior to this point, you must determine the kind of trunk side that is best for your boat. Your choice is quite broad. Each type has its good and bad points:

1. For a painted trunk, single-thickness plywood, scarfed or butted
2. The same, but laminated of plywood, two or more thicknesses glued
3. Solid mahogany or teak for a bright finish, probably jointed, splined, tongue and grooved, or rabbeted
4. Also for a bright finish, strip-built, small sections sprung around the strongbacks nailed and glued to each other

PLYWOOD TRUNK CABINS

A single thickness of ¾-inch plywood 15 inches or more in width is not "impossible" to bend. At a total length of 10 to 12 feet, this might be possible, although quite destructive. The other difficulty is the fastening from beneath into the plywood edge, passing long lags, or whatever, through the carling and deck with accuracy. Years ago I built a number of small plywood yachts designed by Charles G. MacGregor. He was a genius at simplification combined with excellent engineering. His 27-foot *Threesome* (my name for her) specified a cabin trunk with ½-inch plywood sides. His design is similar to the sketch in Figure 10-9A, except that there is no carling in the accepted sense. Instead, what I label a cleat or sill is sprung down to the necessary sweep of the deck and closely screwed and glued from beneath, with no carling. Long screws go through the beams into the sill. We bent the ½-inch sides around a couple of strongbacks, spiled them to the deck sweep, and screwed and glued them to the sills. Years later, this proved to be very strong construction in spite of its simplicity. There were no signs of opening or movement along the long joint. My construction is massive by comparison. It is further strengthened by the facing piece fastened to both the carling and the sill. Of course, the upper edge had its conventional batten or clamp to take the cabintop beams. This clamp was sprung in, glued, and screwed from the outside. The point of my drawing is to show that the sill (when 1 inch or more) is strong enough to withstand the stresses of bending ¾-inch plywood if the sides are pulled in by using ropes from opposite sides. Since *Practical Yacht Joinery*, I discovered that L. Francis Herreshoff had used a similar system in fine yachts.

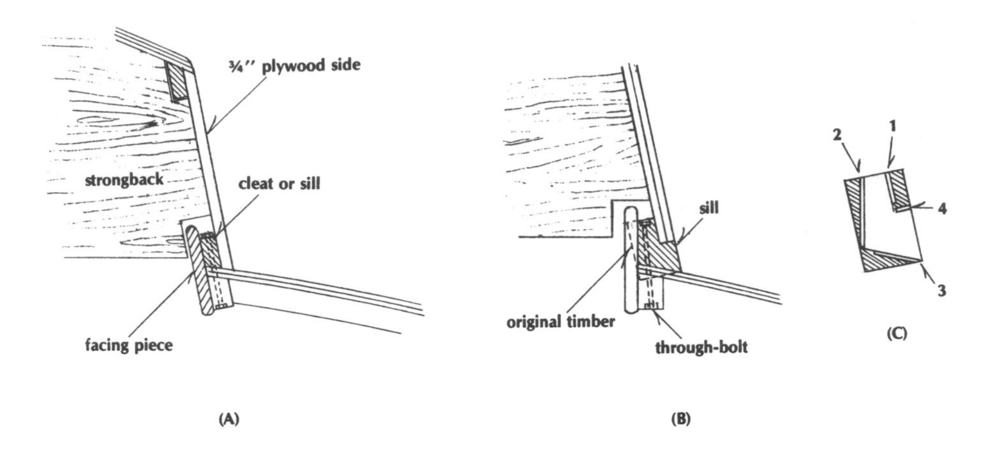

Figure 10-9. *Cabin trunk side cleat or sill construction.*

CANTED CARLINGS

Note that Figure 10-9A shows carling and sill inclined to match the tumblehome. A variation for plywood sides is the strong rabbeted sill shown in Figures 10-9B and 10-9C. This has the inboard surfaces of the carling and sill vertical, but the outboard surface of the sill tumbles home and is rabbeted for the side. This sill is not as hard to make as it appears. I show the sequence of saw cuts you need to shape this piece. The carling may well be redundant.

There are several systems you should consider, even though I do not approve (Figure 10-10). The first is the cabin side fastened inside the canted carling and deck. The second is the cabin side screwed directly to the facing piece (or vice versa). This is seen in some older wooden racing machines. If the facing piece is very heavy, say, 1¼ to 1½ inches, with a shaped hand grip formed in it, it should be substantial enough. Bolts, screws, and glue are advisable. The last, with a wedge-shaped filler full length, beefs up the construction even more. This could be one piece, of course.

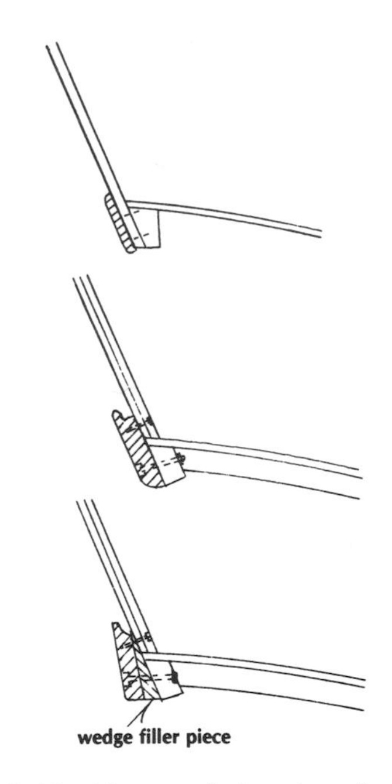

Figure 10-10. *Alternate facing pieces for light yachts.*

LAMINATED TRUNK SIDES

Plywood in two thicknesses is easy to bend. The two sheets may be a combination of ⅜ inch with ½ inch, both ½ inch, or whatever. The first lamination had better be stiffened and faired along its upper edge before the second is glued to it, especially if you see any hard spots over the strongbacks. But battens in the strongbacks would fair the sides nicely. Do not cut out any openings at this time. Your pattern should be useful in fitting this piece into rabbets fore and aft and to the deck or sill. Fasten here with glue and screws.

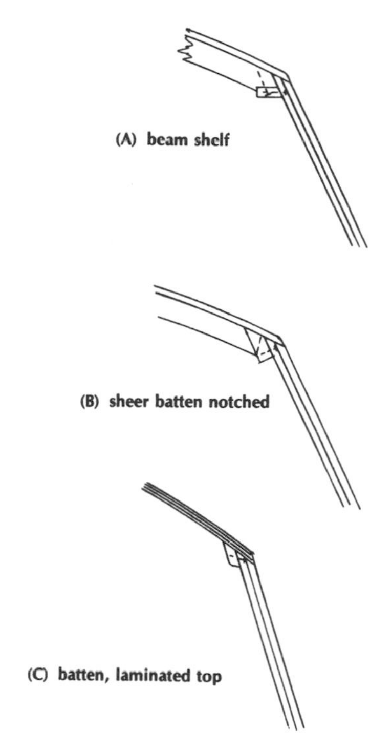

Figure 10-11. *Upper reinforcements.*

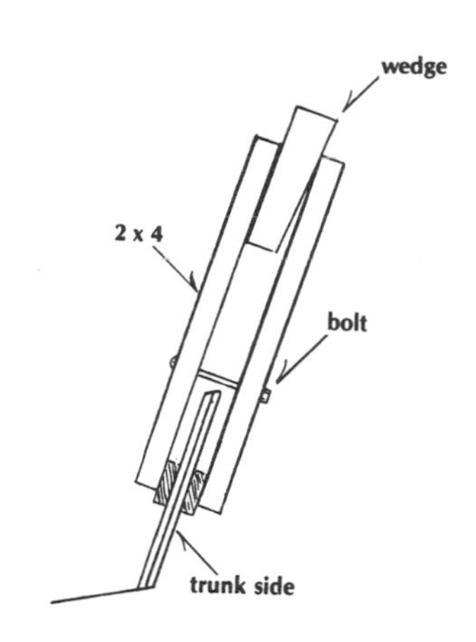

Figure 10-12. *Clamp for laminating trunk side.*

The installation of shelf, clamp, or sheer batten also should fair the side (Figure 10-11). If you are using laminated beam construction, your choices are a shelf (Figure 10-11A) or a notched batten (Figure 10-11B). The former is strong and simple, just a matter of sawing the bevel so the shelf sits level. You can fasten the beams with screws and glue from above or below. If A were laid flat against the side, it would have to be somewhat heavier so fastenings would be more effective. In B there is no mechanical fastening of any real strength. The joint is fully dependent on epoxy glue. The batten (Figure 10-11C) provides a nailing and gluing area for a laminated beamless top (also see Figure 10-15). Because of the great camber, the tops in A and B should be two thicknesses of ¼-inch plywood (not shown). Be sure the angle of the upper surface of these battens is sawed accurately for close fitting of the top material.

Fit the second ply carefully all around, then coat both surfaces with epoxy or plastic resin glue. Four or five large cabinetmaker's screw clamps (the wooden kind) would help prevent air pockets. It's best to place 1 by 4's inside and out, under the clamps. Even with these, however, additional temporary clamping might be required. Try a dozen small nails through small plywood blocks inside and out and also into strongbacks and battens to draw the plies together. Use the Hidden Edge Gauge (Figure 10-15) along the top. Screws or serrated nails and glue in the end rabbets fore and aft should hold well. Here's a tip: If you run into heads of screws in the inner ply when fitting this bevel, use a disc sander very carefully.

Figure 10-12 shows a very useful clamp for gluing up trunk sides or other wide assemblies. You might need four or more of these at a time. Place the 1 by 4's mentioned above under the clamps to distribute the force. A length of 30 to 36 inches should be ample. Tap the wedges in gently. Before the glue sets up, go all over the side with a hammer and a block. If you find any hollow spots, drill a $\frac{1}{16}$-inch hole to allow the trapped air to escape, then place a clamp over the area or run small-diameter screws through blocks. Lay scraps of plastic between the blocks and the plywood, as glue might ooze out of the hole. The hole can be patched up later.

Perhaps you have already thought of applying an attractive overlay of mahogany, ash, or birch doorskin to the inner surfaces. This can be bonded without mechanical fastenings using lots of clamps and blocks. You may use contact cement if you are experienced with the stuff. Follow instructions carefully, as once the parts touch, they can't be moved. This extra lamination adds enormously to the strength of the side. So it permits you to reduce the thicknesses of the first two sheets accordingly.

As you know, weight reduction is vital, especially high above the waterline. Do not overbuild.

One of the side laminations could be a finer plywood, say, mahogany or teak (but check the glue). This would give you a more attractive trunk, either inside or out. Teak plywood is costly, but it is much less than solid timber. The corner posts, of course, may match whatever you use, so this is something to be planned well in advance.

TRUNK SIDES OF SOLID TIMBER

Solid timber is the traditional approach for trunk sides. It is an excellent method. For one-piece construction, however, it requires widths not easily found. You will have to run drifts or through-bolts all the way through the carlings to prevent warping of the wide plank. (More on this below.) The sides can be set into rabbeted sills, but you still need the rods for stiffening. I would consider 1⅛ inches as the minimum thickness unless you have the skill to bore edgewise through 1-inch stock; a boring guide for this task is described in Chapter Five. I think the best way is to clamp the side flat to sawhorses, bore from the top edge all the way, then clamp the side in position and run your bit down through deck and carling. This is for through-bolts. Bore slightly undersize (1/64 to 1/32 inch should do) for galvanized ⅜-inch drifts. Have someone on top of the timber with a heavy weight while you are driving the drifts from below. Otherwise, you might knock everything loose. Slip a washer on the rod first, for the end will upset as you drive. There should be a counterbore for the washer, for looks. Place a drift every 16 to 24 inches, but watch out for openings. You can use rods purely as stiffeners, the actual holding being done with lags screwed up through the carling. This is not, however, the first-class way to go in a vessel greater than 35 feet on deck.

An alternate method is to thread the lower end of the drift so it can be pulled up in a counterbore in the carling. The bore in the side should be just undersize (say, 1/64 to 1/32 inch) and the drift driven from above with your partner supporting the carling. Let the thread appear about 1 inch, place the washer and nut, and pull it up.

The chances are good that you will not need a nut on the upper end of a properly fitting drift. If, however, your cabin side timbers are well over 1 inch thick, you can add this insurance (a through bolt) and ease the drive fit, too. Fill the open counterbores above with thickened epoxy.

If full-width material is impractical, use the splined joint described earlier. The lower piece is fitted, then the upper piece to it, daylight-proof. Mark alignment with pencil. Play safe, fit and clamp up a "dry run" before gluing final assembly. The grooves can be cut using methods in Chapter Seven. Be sure the spline width is about ⅛ inch less than the total depth of the two grooves so there's space for air and glue. With it upright on the bench, brush Aerolite glue in the groove and tap the spline in after wetting it with Aerolite catalyst. Brush glue in the mating groove and joint. Clamp the upper half loosely on the forms and start tapping it down so the spline enters the groove, checking the marks and alignment at the ends as you go. When satisfied, drive the piece down and pull together with pipe clamps. Glue and fasten all around.

STRIP-PLANKED TRUNK SIDES

Strip planking is a kind of construction often used on hulls. Strip planking, in general, is easy, monotonous to apply, very strong, and beautiful if done right (Figure 10-13). If built from ribbon-grain mahogany or teak sawed into strips about 1¼ inches wide, glued with epoxy adhesive, and sanded thoroughly to remove slight irregularities, strip planking the side would be hard to beat. You can't use metal fastenings in the cut-out areas, but ⅛- to ¼-inch dowels work fine. Alignment of each strip (Figure 10-14) with its mate is fussy work.

The first step is to lay off with a batten the sweep of the upper edge of the side. Or you can use your pattern, if it is perfect. Mark each strongback. This is the location of the last strip to go on, making allowance for the upper bevel, and so on. Multiply the height of the strips (I suggest not over 1¼ inches per strip) by the number required to come within about 1½ inches above the deck forward. Mark this dimension on all the strongbacks and the corner posts at both ends. This is where your first full-length strip will go on. (It is No. 1 in Figure 10-13.) Because all additional strips sweep the same as No. 1, you wind up with the beveled edge exactly as designed. But there has to be something on which to nail No. 1.

I show a fairly intricate piece that I call a sill (for want of a better name) into which two short strips

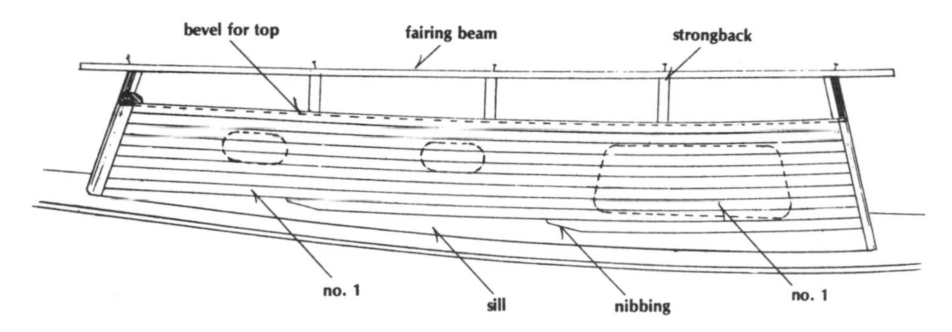

Figure 10-13. *Strip-planked cabin trunk with opening layout.*

or stealers are nibbed. If you insist on avoiding this work, the sill can be made full width to the marks for strip No. 1. Do *not* make the sill with its upper surface following the sweep of the deck, for your strips will run out to feather edges, a bad practice. Whatever the width of the sill, it must be through-bolted or strongly screwed to the carling and bedded on the deck. Each fit must be "dry" to rabbets with care.

The strips may be nailed with galvanized finish nails or bronze serrated nails on about 6-inch centers. Guide your drill so there is no chance that a lead hole or a nail will run out, a real tragedy. In the areas of the openings, drill easy fits for ⅛-inch dowels. Coat with slow-curing epoxy glue or plastic resin. This is strong construction. The long joints must be painted liberally with epoxy glue. Clamp the strips to the forms and line them up, one on top of the other, with little guides, as shown in Figure 10-14. This alignment is vital, as otherwise you will have to dress down both surfaces. Do not permit this tool to smear glue over the sides of the strips. For this reason, you might consider using Aerolite glue; it can be washed out with a wet rag for a few hours after application. It is crystal clear, too. Fasten your top strip with small dowels so the bevel can be planed.

If jointed trunk sides show slight misalignment and unsightly seams, plane them out easily on the exterior with a very sharp plane set fine. The concave inner surface, however, is another matter. Plane diagonally across the grain, then go over the surface with a belt sander, a cabinet scraper, and finally a block sander by hand. It pays to watch the alignment of the strips while nailing.

No doubt there are other ways to build a trunk cabin side. Your designer is *the* authority. All deck

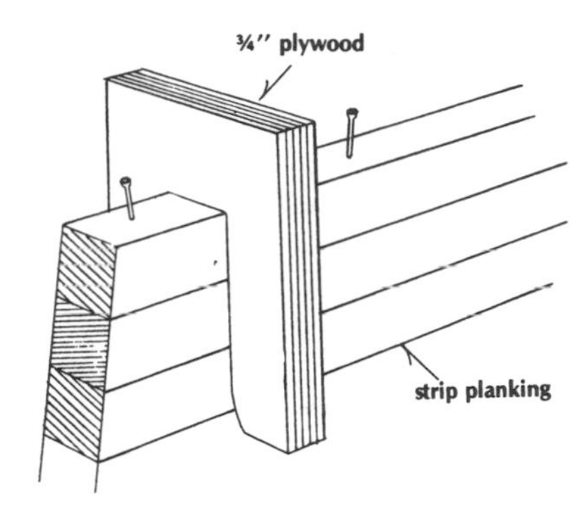

Figure 10-14. *Strip alignment tool.*

structures—trunks, coamings, hatch runners, and so on—must be bedded in an elastic compound such as Thiokol or in a sealant-adhesive that is flexible and crack-filling. The traditional way is to lay a string of cotton caulking well soaked with white lead under such structures before fastening. Never caulk such a joint.

CABINTOPS

Let's start with the accepted type of cabin deck—laminated beams with plywood over. Sawed beams are out, so be prepared to form each beam individually on the floor or bench. Spruce is ideal, or a combination of spruce or ash with mahogany will do nicely. Traditionally, such beams are capped with a piece of mahogany about ¼ inch wider than the beam siding. Plies should be at least ¼ inch thick. Because of various widths of the cabin trunk, no two beams will have the same camber in order to maintain the slight dip in the centerline (should be in plans, such as *1 inch in 10 feet*). Clamp a strong full-length piece with that curve along its lower edge and take your cambers from it (measured above the sheer of the side). A *slight* difference near amidships, say 1/8 inch in height, can be adjusted by cutting the beam a bit long to raise it, a bit short to lower the camber, under the center strongback; so perhaps several beams can be ripped out of a wider laminate. The first ply will stabilize those beams. But fasten with long thin screws into the shelf (and glue). And/or fit blocks along the trunk sheer between beams for ample gluing area there. Also, you may soon want 3/4 by 6-inch blocks where fore and aft joints may be needed. See below. These beam ends and blocking cry out for excellent crack-filling glues such as Aerolite and Chem-Tech T88. Now check all over the beams with a long stiff batten for high or low spots, and correct them. Especially check the corner posts as the upper ends must fit to the decking.

Decking is usually two thicknesses of 1/4-inch marine ply (minimum) glued and nailed to beams and blocks, and amply glued between plies. Run the grain either way, but the conventional way is fore and aft. If it is to run athwartships, the panels must join over a beam, but never two plies on one beam. If the length of 96 inches is insufficient, install blocks equidistant from the centerline (give the corners an attractive cove, ogee, round, or whatever). For fore and aft, join on the centerline over blocks, and again over blocks if decking is more than 96 inches of span. Or a second panel centered (24 inches each way from centerline), joined over blocks. Tack in position, mark all around, using the Hidden Edge Gauge (Figure 10-15) at sheer so a cut there can be close, easy to trim later without striking hidden nails. Don't pull all the tacking nails as these will help relocate the panels after the glue is applied. Brush epoxy in the marks for beams, blocks, and perimeter, and lay panel slowly with great care. Start with 3/4- to 1-inch bronze serrated nails near the center; work outward from center area to avoid buckling, then nail around perimeter on about 3- to 4-inch centers and into beams on 6-inch centers.

Repeat all that with other panels of the first ply. Clean off any smeared glue underneath as soon as possible. Actually, I taped off the beams and blocks of my boats so the surface could be varnished (mahogany underlay). You do a lot of work to avoid a lot of work—unless appearance is secondary. You may trim off the excess material now, if you hate that ragged look.

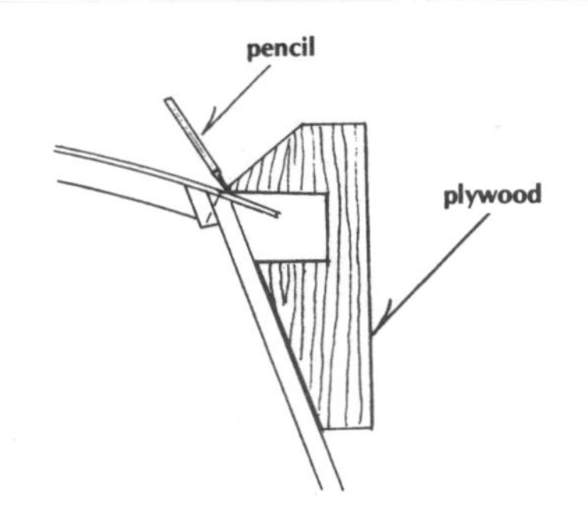

Figure 10-15. *Hidden edge gauge.*

Assuming the second ply has been pre-fitted so joints may be scattered two or three inches away from the first joints, yet be nailed into the same blocks, roll and brush on a generous coat of plastic resin (or epoxy if the cost is acceptable). Coat only the area of one panel at a time unless you have help. Fastest system is to pour copiously from a quart can, then squeegee or roll and brush it out evenly, slap the panel in place, tack it quickly to prevent slithering around, then start nailing outward from the center area into the beams, perimeter, blocks, etc. Serrated 1 1/4-inch nails are recommended for panel thickness up to 7/8-inch; hot-dipped galvanized finish nails are third best (stainless steel screws are second, but slower).

Quickly tap all over for the sound of trapped air; mark any such spot and drill 1/16-inch holes through the *last ply only*, and immediately lay 35-pound lead pigs or barbell weights over. The job on hatch framing, carlings, etc. was done previously as in main deck construction. Saw off excess but stay just clear of the lines so there is no danger of marring the ends, openings, etc., especially the sides where the angle is very obtuse. After planing and sanding down to the nitty-gritty, route a nice radius (1 inch or so) along both ends, then plane and disc sand a generous round over the cabin sides. After sanding off little high spots here and there, apply a penetrating coat of Chem-Tech L26, let cure, lay Versatex (polypropylene) in a wet coat as per the manual. Paint and non-skid to your taste.

A BEAMLESS CABINTOP

After I shipped my *Bay Bird* to California, I found in 1960 that dry rot had had its way with the 20-year-

Figure 10-16. *These hefty forms, or strongbacks, easily supported two men. Each layer was nailed into these, the nails nipped off, and the holes covered by mahogany flat moldings on 16-inch centers.*

Figure 10-18. *Side view of dog house under construction.*

old trunk. Also, the then popular doghouse fad offered a possibility of 5 feet 9 inch headroom where it counted most. Briefly, I scrapped the entire trunk, scarfed ¾-inch fir ply into the new configuration of the sides with pleasing fixed windows and ports, secured bulkheads fore and aft, on deck, installed ample clamps inside the upper side, etc.

Now for the beamless top: Pete, my boatbuilder-helper, and I scrounged some decent 2-inch for high-cambered forms (or strongbacks) over which to laminate three thicknesses of ¼-inch exterior fir ply (Figure 10-16). We needed a broad positive nailing surface, surer than ¾-inch plywood.

Figure 10-19. *Pete setting up husky form supports.*

Figure 10-17. *Bay Bird under sail in the early 1950s with her original cabin trunk.*

Figure 10-20. *New cabin trunk nearly complete. Big windows and later molding reduced its apparent height.*

We set up a slightly sway-backed center piece and took height from this to cabin sides at each station. In a few minutes, we used a pair of beam-scribers (Figure 5-41) and an hour later had four cambers marked for bandsawing (you might want to space closer than 16 inches, but that worked well for us). The end bulkheads had been scribed and trimmed off previously, including openings.

We fitted the forms, cutting notches to clear the side clamp or sheer batten, scabbing 2 by 4s to the lower edges for clamping legs strong enough to stand the weight of two men and lots of nailing. To trim the sheer battens to match the camber, we sprang flexible battens diagonally over the forms and sheer, after chopping it off very roughly with the power saw. Finally we checked all over with a stiff 1 by 2 for bumps and hollows.

Our first laminate was ¼-inch ash, but don't use it unless you are positive the glue is Rescorcinol or other weather-resistant adhesive. Doorskins (only 3/32-inch) of mahogany, birch, teak or other light-colored wood would be fine; you would need another thin laminate to compensate, however. Another alternative would be MDO, plastic-coated plywood, exterior, ¼ inch.

We ran the grain fore and aft for appearance and notched the forms on the centerline for a 1 by 3 support into which to nail. We tacked a panel lightly along the centerline, bent it down, marked form end locations, and all around the perimeter, leaving comfortable 2½-inch overhangs. The opposing panel was tacked up to the first, marked, etc., until the entire top was loosely covered, then removed and sawed to marks. During this, we trimmed as necessary because of the slight compounding. We spread Weldwood, dipping from a half-gallon can, brushing evenly all around the perimeter. Then we stuck the original nails into their holes near the centerline (to prevent messing up the nice wood), and drove ¾-inch wire *finish* nails into the forms on 4- by 6-inch centers, but around the perimeter with bronze serrated nails or galvanized 3-penny finish on 3- or 4-inch centers.

We immediately repeated this with the second laminate, grain running athwartships, and the third. We poured generously from the can, brushing out the Weldwood rapidly and evenly, and again setting finish nails through into the forms, bolder nails around the perimeters. Our aim was to make the curing throughout almost simultaneous. You would probably use epoxy, pre-wet the under side and lay it in the wet-out previous panel. Generous, slow-curing, positive bond, and checked for air pockets as described earlier. Today, I think I would prefer Aerolite glue as it is mixed with water, slow curing, can be adjusted during the first minutes, has long pot-life, and fills small gaps with full strength. There is little waste with Aerolite as the catalyst effect does not begin until both faces contact.

The last job on the laminate was to clamp all around the perimeter. We checked for trapped air by tapping all over, drilled and placed heavy weights over the soft areas, as described before. We allowed the usual 24 hours for curing, removed weights, clamps, etc. At last the "fun job," marking with the Hidden Edge Gauge, sawing close to it, disc sanding a nice radius over the trunk side, and around the ends of the perimeter.

Going below, we dismantled the forms and supports, then found rows of deadly brads protruding, but pulled and nipped them easily with pliers. These marks were covered with false beams of varnished mahogany about ⅜ by 1¼ inch. We then wished we had varnished the ash panel first. I have to admit that this made a very handsome interior. Later we saturated the outer surface with a fir sealer, covered it with heavy sheeting laid in paint, added non-skid. Today, you would probably use L26 and Versatex. Strange, I never sailed *Bay Bird* with her new cabin trunk.

PORTLIGHTS, DEADLIGHTS, AND WINDOWS

There are many types of hardware designed to enclose cabin trunk apertures. In addition, installation is complicated by the different types of trunk construction. It would take many pages to cover all possibilities. I will cover three situations only:

1. Traditional opening ports of cast metal and molded plastic
2. Fixed (nonopening) deadlights, framed and frameless
3. Opening windows and shaped deadlights

First, a few definitions. A portlight swings open. A deadlight may be of similar round, rectangular, or oval shape, but it does not open. I hate to use the word window on a vessel, but what else can one call the large areas of plastic or glass so common

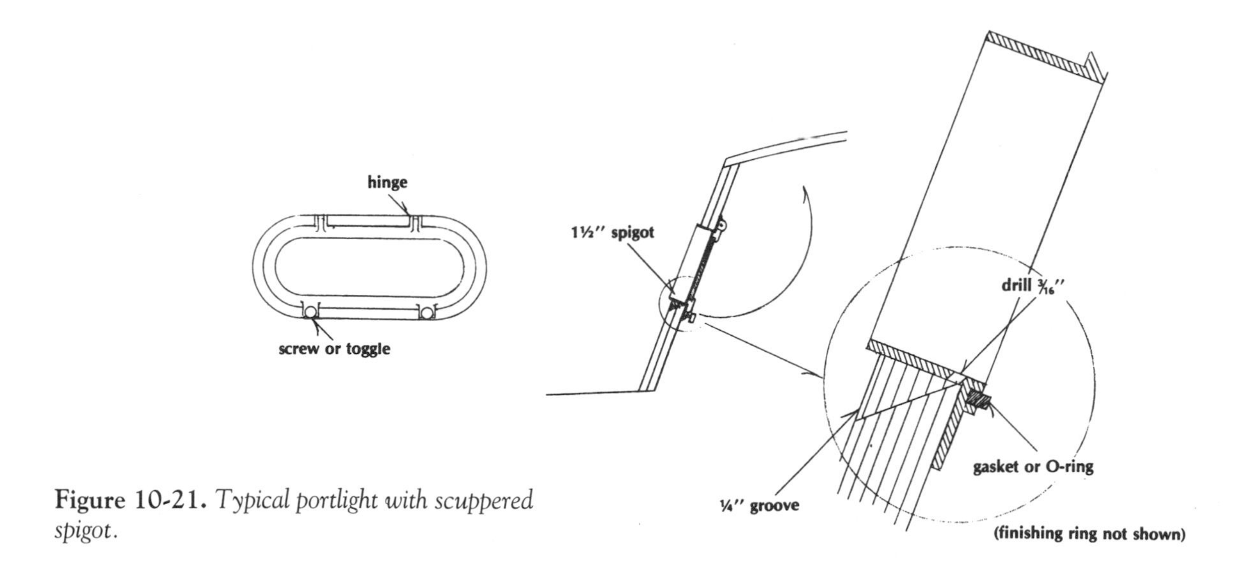

Figure 10-21. *Typical portlight with scuppered spigot.*

today? If it opens, I'd call it a window. If it is fixed, I would call it a deadlight, no matter what its size and shape.

Whether of cast bronze, stainless steel, nylon, or molded plastic, most portlights are designed with a collar or spigot that fits into the opening (Figure 10-21). A frame or bezel may be bedded and screwed to the trunk side to cover the joint and bolt heads, if any. Many prefer to omit the bezel, for it emphasizes the size of the light. Some portlight bezels are through-bolted. This type is probably best able to withstand the force of solid water. The "glass" may be Plexiglas, Lexan, or plate or shatterproof glass. I suggest Lexan because it resists scratching and is very strong yet flexible. Screens are available.

Bronze or steel cast portlights and deadlights are too heavy for most modern yachts. In addition, the great tumblehome of the sides allows water to collect in the spigot. This means the unwary sleeper below gets a shower when the port is opened. I have seen no completely satisfactory method to cure this fault, but Figure 10-21 offers one possible solution.

Installation is no great problem if you have a sabersaw or the larger reciprocating saw. If you have a number of openings to cut, take the time to make a router pattern (or use the side pattern discussed earlier). Use the spigot itself for marking locations, and exercise extreme care that both position and alignment are to the plans. Set the spigot in the hole with heavy epoxy glue or 3M 5200 under the flange and on the edges of the opening.

Some ports and deadlights are manufactured with spigots less than ½ inch deep. Also, you could saw off most of the spigot on a cast bronze port. Then, if the lower edge of the opening were beveled, it would drain. Another possibility is hacksawing and filing a couple of little scuppers in the spigot to minimize water collection. The edge of the wood would have to be saturated with epoxy resin. I show another scuppering method in Figure 10-22. Drill ³⁄₁₆- or ¼-inch holes at the ends of the spigot. Mark the spots, then saw out a groove ¼ inch wide to meet the holes. Sand this smooth and saturate with epoxy resin (use three applications). Keep these scuppers clean.

DEADLIGHTS

Manufactured deadlights that match portlights are no problem, except for their weight and cost. If through-bolted, they probably are as strong as anything available. In large sizes, there is a hazard. The strength of the fitting is almost entirely dependent on screws into wood. Bedding in thickened epoxy glue would be inexpensive insurance, of course.

For light weight and low cost, consider deadlights of plastic screwed directly to the cabin trunk side. See Figure 10-22. For maximum strength at the expense of appearance, these can be screwed or

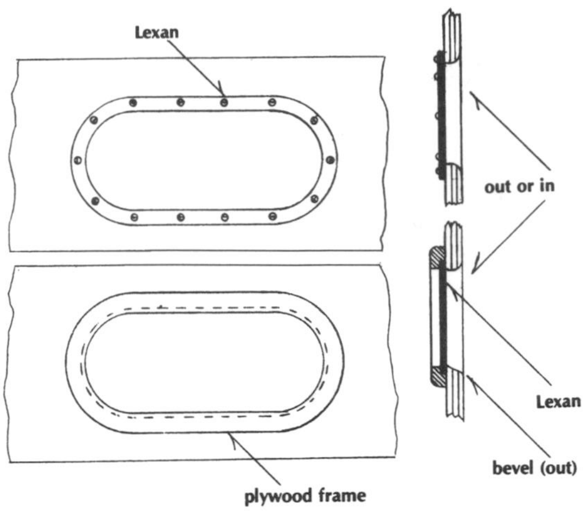

Figure 10-22. *Plastic deadlights direct to side.*

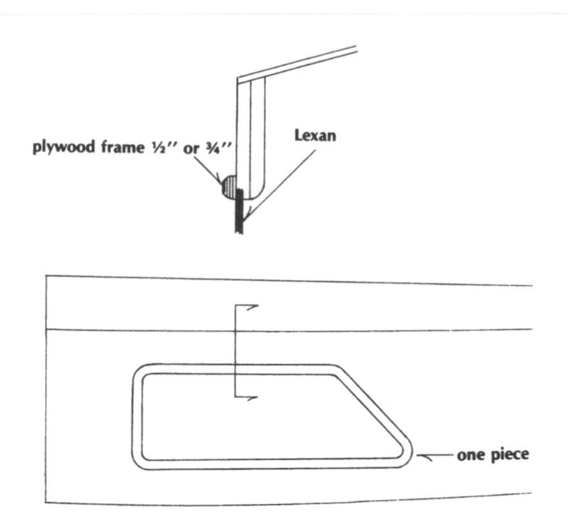

Figure 10-24. *Shaped "window" frame or deadlight.*

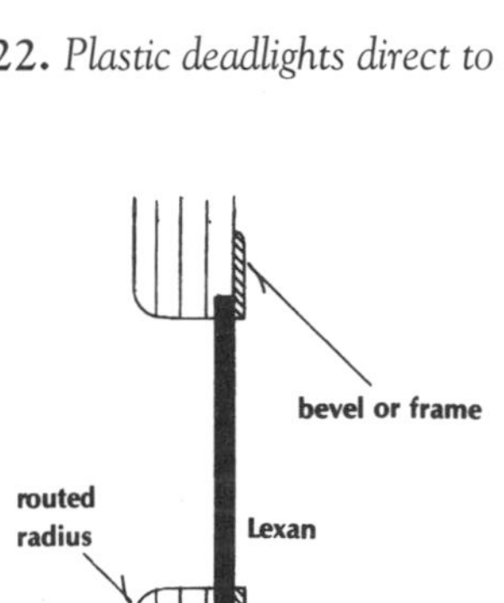

Figure 10-23. *Rabbeted trunk side.*

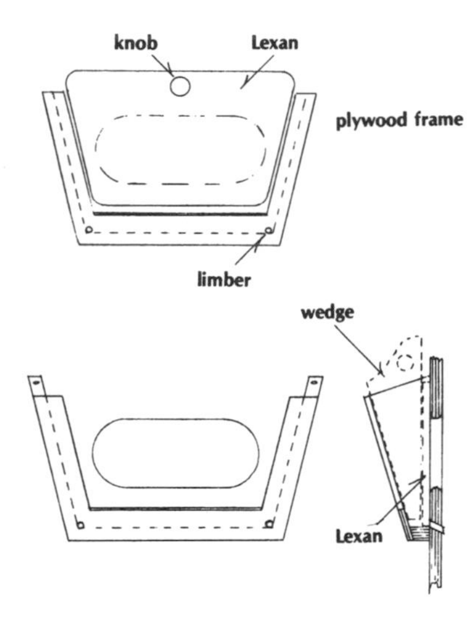

Figure 10-25. *Gougeon and Root portlights.*

bolted directly to the exterior or interior surface, and bedded in a flexible 3M compound to compensate for the different expansion and contraction rates of plastic and wood. The bolt holes should be drilled oversize for the same reason. Do not remove the paper on the plastic until your craft is completed. The paper in the area of contact should be removed, and that plastic surface sanded for satisfactory adhesion. Roundhead screws or bolts with plastic washers against plastic allow the slight movement that will occur and seal out moisture.

Installation on the inside of the trunk side makes an attractive deadlight. The opening outside should be routed to a pleasing radius for drainage and the exposed surfaces saturated with epoxy resin. If you have trunk sides at least ⅞ inch thick, the plastic may be fitted into a rabbet around the opening (Figure 10-23), on the inside or outside, the latter being the stronger of the two. The joint can be covered with a stock frame or bezel, both bedded in flexible compound.

An alternative to the stock frame, especially for larger openings of unusual shape, is a rabbeted frame of hardwood plywood or wood to retain the plastic (Figures 10-22 and 10-24). This can be sawed out of ¼- to ½-inch mahogany plywood and routed to a

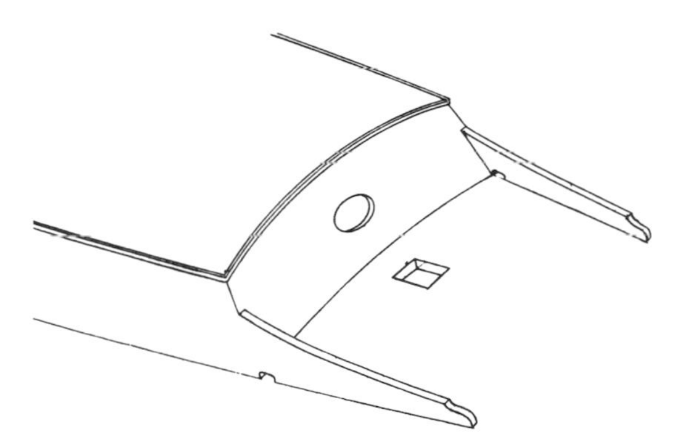

Figure 10-26. *An "old-fashioned" cabin trunk.*

nice radius on the inner corners with a rabbet to receive the plastic. Of course, such a frame can be built up of solid wood with mitered corners—a much more laborious task. I have installed such a frame on the exterior, but it is unwise to emphasize such openings, regardless of size.

OPENING WINDOWS AND SHAPED PORTS

Figure 10-25 shows an idea borrowed from *The Gougeon Brothers on Boat Construction.* The frame should be routed from a piece of 1/4-inch plywood to avoid joints and leaks. The limbers should be fitted with copper tubing of about 1/4-inch inside diameter. Brush epoxy glue in the holes with a pipe cleaner on the tube. I would not expect this window to be impervious to *driving* rain or solid water, but it allows circulation normally. Keep on hand a few little wedges to be inserted inside.

Those remind me of the "Root" ports I once built for our sharpie. Invented by a famous yachtsman of the 1930s, their unique design allows them to be open at the top so that there is ventilation in a downpour, but not in a *driving* rain and wind storm. Wedges are needed to press the panes against the side in bad conditions. These ports always leak, but the water runs freely through the scuppers. Neither the Gougeon nor the Root port would be satisfactory for serious offshore cruising.

11 Building the Yacht Interior

Let's consider the interior fittings of a yacht as furniture, the things that make a barren hull a home. You'll want to furnish your yacht with settees, a companionway ladder, berths, bookshelves, a galley, hanging lockers, doors and drawers, dressers, and so on. In addition, you'll want to ceil your hull for warmth and protection against sweating, and for appearance. This chapter describes ways to provide these furnishings that almost any fair craftsman can follow.

SOME STANDARDS

There is convincing evidence that the human body is larger than it once was. There are certain space minimums we must provide so that the body will function efficiently. The standards shown below come from reliable sources. Some may have to be expanded.

Bruce Bingham wrote how difficult it is to evaluate space by studying the accommodation plan. He recommends here and there placing a figure of a man drawn to scale. You might try stick figures. It's amazing how rapidly some yachts shrink when this is done. "Ample seating" can turn out to be no larger than a phone booth for two. Or a quarter berth may be just wide enough for your shoulders, and the cockpit beams too close over the mattress for your feet.

I know I am encroaching on yacht design with these suggestions, but I hope you will do something similar before making a final design choice. On the other hand, if you are past that stage and already own a big, beautiful hull, I hope you will undertake another experiment—one that takes some time but little money. Rough in the bare essentials of settees, sea berths, backrests, galley, and toilet enclosure. A mock-up engine made of crummy plywood might expose an uncomfortable situation.

ROUGHING IT IN

For example, knock together a plywood box 16 inches square and 15 inches high. That's a portable toilet. Set it on that narrow stepped-up floor and try it for size. Elbow room OK? Do you have to unscrew your feet to relax? Headroom OK? Or is this a "sitter" only? You have seen designs showing a quarter berth extending under, alongside, or both, the cockpit well. Many of these narrow down substantially at the foot. Raising a berth or settee only an inch may widen it considerably, depending on the hull section. This is especially true in the forepeak, where raising berths not only widens them but creates large storage spaces as well. Many designers (including me) raise the feet ends 1½ inches to broaden the width (and to compensate for the two bodies in the bow). Perhaps a double there instead of V-berths would permit building hanging lockers outboard of the pillow end. But perhaps your crew will hate climbing over the head of this berth.

Try it before you build it. Stretch out on the rough boards; don't just rely on the blueprint. If you're over 6 feet tall, you'll have to fudge here and there to work in a 6-foot 6-inch berth if the bulkheads are already in. Can you cut an aperture and extend the berth? Try it for size!

BERTHS IN GENERAL

Berths can vary greatly in otherwise similar yachts in terms of their usage. Think carefully before you

SOME BASIC DIMENSIONS

Item	Minimum	Comments
Berth width, at head	22 inches	28 inches in port, 22 inches at sea
Berth width, at foot	15 inches	18 inches is luxury
Berth length	6 feet 3 inches	from 3 inches to 4 inches over the sleeper's height
Settee width, if low	23 inches	22 inches if cushion is 14 inches wide
Settee width, if high	18 inches	19 to 20 inches if cushion is 16 or 17 inches
Settee clear, cushion to carling	36 inches	(more if decks are wide)
Berth clear, butt to shoulder	22 inches	(more if access is difficult)
Headroom, full	6 feet	6 feet 2 inches under beams
Hanging locker, rod length	9 inches	never enough
Hanging locker, depth	17½ inches	3 inches more than hanger length; additional for hooks
Hatch, escape and vent	18 by 18 inches	24 inches square, especially if used for sailhandling
Hatch, sliding	24 by 24 inches	larger, if over galley
Galley height	33 inches	36 inches with 3-inch kick space under
Passage width, forepeak	18 inches	OK if shaped for ease and ventilation
Passage with door	22 inches	OK (also double hinged, sliding, and accordion folding)
Clear above stove	30 inches	cover woodwork with asbestos and stainless steel or aluminum
Toilet, 15-inch height	headroom 36 inches above seat	learn to sit, regardless, for sanitation and safety
Portable toilet (build up to 15-inch height)	headroom 36 inches above seat	handy size may mean small capacity
Enclosed head, width	24 inches	allow 27 inches elbow room if possible, also foot room
Cockpit well, width	24 inches forward, 18 inches aft	just wide enough to brace feet against
Cockpit seats, cushioned	15 inches wide, 17 inches to cushion	helmsman must see over trunk comfortably
Side deck catwalks	12 inches	you can't get them too wide

act. Don't build broad, comfortable nests if you're likely to go bluewater cruising. And vice versa. In coastal cruising where 90 percent of the nights are spent in quiet anchorages, you'll probably want lots of roomy comfort. In a seaway, however, these berths would roll you back and forth—not the best setup for adequate rest. In such circumstances, you need narrower bunks you can chock yourself into with bunkboards or other devices. Consider berths that can be tilted when the boat is on the same tack for many hours, days, or weeks. When the occasional guest or extra youngster comes aboard for a couple of nights, certain space-saver types can be used. More about these later.

It is customary for a sleeping berth to be built with a hardwood face and raised portions at head and foot to keep pillows and bedding in place. A section used for sitting has a raised face piece generally much less than the mattress thickness. A berth used exclusively for sleeping, however, should have a face about 10 inches high, but with a lower center for ease in climbing in and out. Settees do not need more than about a 2-inch face or fiddle to keep the cushions in place.

Double berths are not for true offshore cruising. I have used double berths (with a minimum of 48 inches at the head) and admit that they offer great comfort and other benefits. If you must have a double berth, you'd be wise to include a "bundling board" between the mattress pads, probably with pins fitted into sockets in the berth top. Otherwise, both sleepers end up in the lee. No further comment.

Dinette berths are definitely for quiet anchorages only. In rough conditions, they are too wide for one sleeper; two sleepers end up on the sole. Their construction has been described in dozens of books and articles.

TRANSOMS, SETTEES, AND BERTHS

A transom berth is a settee that is also used for sleeping. There is no way you can sit with comfort on a settee wide enough (28 inches) for sleeping. The wider it is, the more uncomfortable the sitting is. This discomfort is compounded if the backrest is vertical (10 degrees is recommended), and the seat is not likely to be angled as chair seats are. Designers, of course, have come up with different ways to cope with this problem.

Extension berths are one answer, even in a yacht of modest beam, say, 8 feet (Figures 11-1 and 11-2). The sloping backrest may be hinged for access to the space behind it, and it may be upholstered, or you can use loose cushions or pillows. Blankets stuffed into attractive zippered covers make fine bolsters. The berth top is built right back to the hull, extending into a recess beneath the storage compartment. This gives you a normal seat 16 inches wide, but more if the seat is more than 15 inches above the sole. The extension can extend to within 6 inches of the centerline of the sole and still leave a narrow space for passage. Some berths extend to join at the middle. This makes a huge playpen for a litter of cubs, or friendly folk.

A proven arrangement is the high fixed berth known as a pilot or sea berth, possibly because you

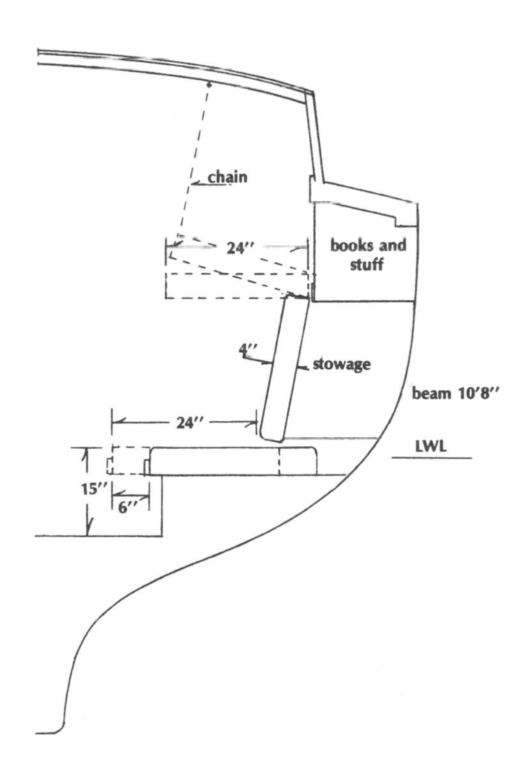

Figure 11-1. *Extension transom and sea berth.*

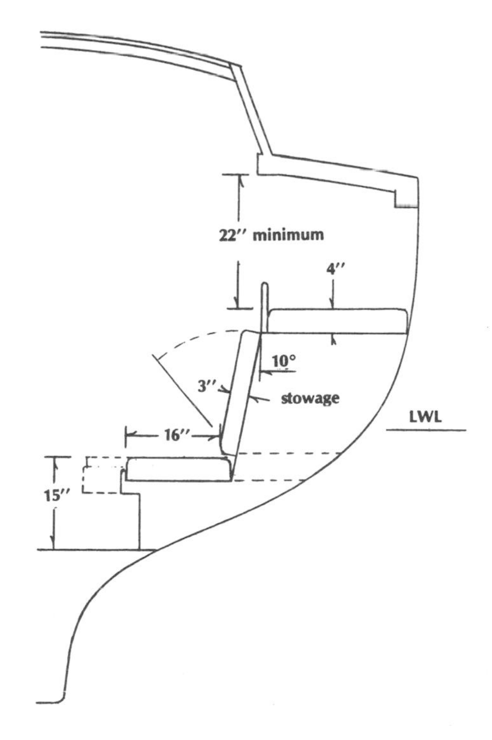

Figure 11-2. *Extension transom and upper berth.*

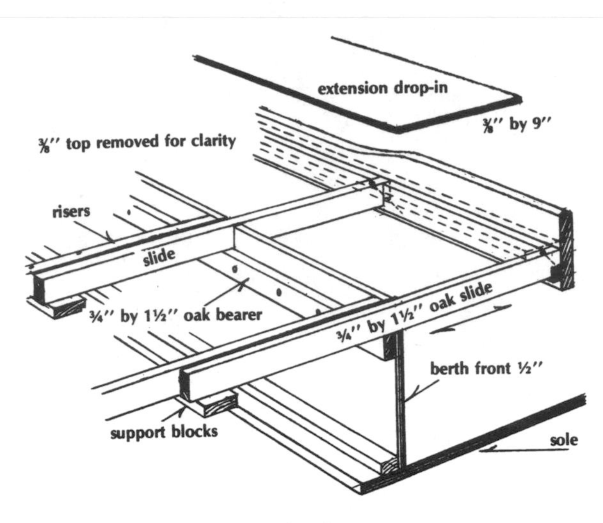

Figure 11-3. *Extension berth construction.*

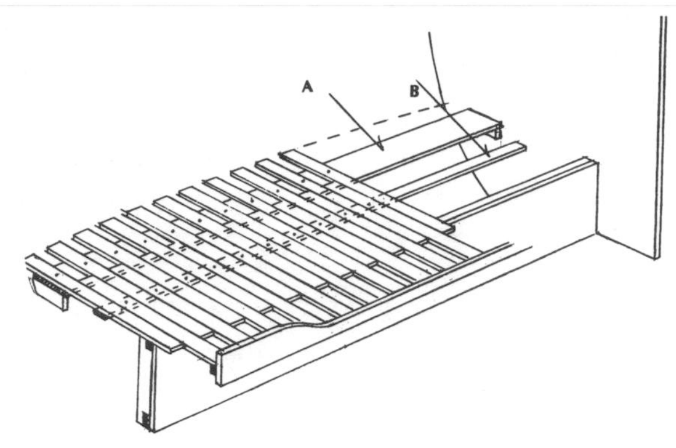

Figure 11-4. *Extension berth—slat construction.*

can chock yourself in with generous boards for comfort and safety in any seaway (Figure 11-1). Today each member of the crew has his own sleeping bag, so he tosses out the warm one and climbs into his own. A sea berth requires a pretty fair-sized vessel to get minimum hip and shoulder room of 22 inches. Width, however, should not be more than a big guy's shoulders. A quarter berth is for the off-watch while the settees are being used by others.

Another possibility, where there is ample beam and height under the carlings, is to swing up the backrest, supporting it with chains hooked onto padeyes or eyebolts overhead (Figures 11-1 and 11-2). This is good for occasional use, although I suppose I could live with it. You must have a minimum of 22 inches vertically between both berths and the carling, so wide side decks might prohibit this. This upper needs little in the way of safety boards unless you are tacking frequently, as the chain may be adjusted to the angle of heel. Such a backrest berth could be a pipe berth with laced canvas, covered with perhaps a 2-inch foam pad.

The sliding framework of the extension transom is ¾- by 1½- to 3-inch stock, riding on an oak bearer, so there is no reduction of storage space below (Figure 11-3). The drop-in filler piece is ⅜-inch plywood. It stows under the mattress pad back out of sight.

Another extension that is quite easy to make, but adds weight, is shown in Figure 11-4. For this one, you set up the berth front in the usual way, fastened to the slats (on a cleat). The outboard support board (A) may be 6 to 8 inches wide, fastened to cleats at each end and supported by a couple of ¾- by 2½-inch legs bonded secured to the hull to take the weight of sitters and sleepers. For the slats, the lazy way is to have all 22 pieces of 3½-inch fir or plywood cut to one length. If there is a lot of hull curvature, you'll want to cut overlength, fit each one against the hull, then nail the alternate ones in position properly spaced. Place the sliders in between, out to the hull. Clamp a straightedge end to end for a saw guide and cut them off. Glue and screw the sliding slats to the cleat on berth face. Slide the whole thing back against the hull, clamp the batten (B) against the board (A) to locate it, and screw about six of the slider slats to the batten (B). Do not glue, as this must be disassembled if you should want to remove the extension, so cut nice, large openings in the berth front. Be sure the batten (B) is in place before you fasten the fixed slats.

The sea berth in Figure 11-1 has a hardwood face or coaming (see Figure 11-5 A) to keep you and the pad in place. In extreme weather, however, you'll need a bunkboard. The one shown at the top in Figure 11-5 is exaggerated in thickness to clarify the detail of the brass pins. Use the berth face as a pattern for the addition (B), and make the board about 1 inch deeper than the opening to avoid feather edges. Bandsaw out and fit with care. C-clamp in place between pieces of scrap and bore two end holes down through the addition, plus another above the receiving hole in the center of A (see Chapter 7 for more on dowell centers). The pointed brad system would work well here. Round the upper edges to match the berth face. When the piece is not in use, stow it against the hull.

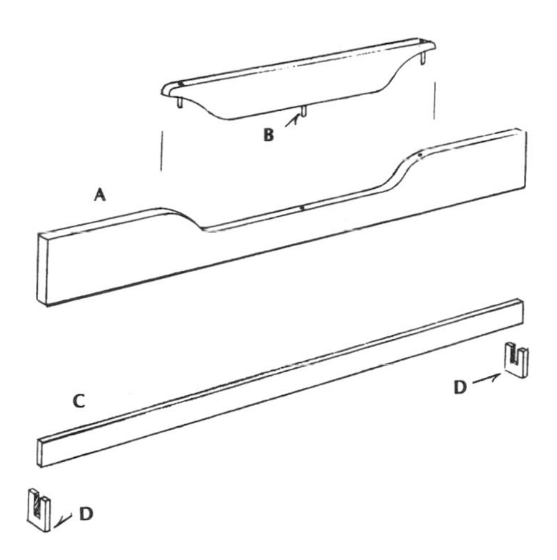

Figure 11-5. *Bunkboards.*

The simple bunkboard (C at the bottom in Figure 11-5) is self-explanatory. The chocks (D), of ¾-inch marine plywood or laminated wood, are glued and screwed to the bulkheads.

SPACE-SAVER BERTHS

One of the most practical, comfortable, and underrated berths is the pipe berth. This isn't joinery, but the discussion of berths would be incomplete if this equipment were omitted (Figure 11-6). As I show it, it's not the handsomest addition to a fine yacht, but it can be improved by covering it with attractive material. With a 2-inch foam pad, it might make a fine backrest with bedding stowed behind. You don't need bunkboards of any kind, for the sag forms a natural nest for the body. In extreme angles of heel, the hanging chains can be adjusted. On the 18-day reach from Catalina to Diamond Head, I would choose this berth. The structure is 1- to 1¼-inch galvanized pipe. The inside dimensions of the pipe frame are 24 to 26 inches by 6 feet 6 inches. Grommets are placed all around on 4- to 6-inch centers, and, of course, the canvaswork must be strong. The brackets may be hook-shaped so the berth can be removed.

THE ROOT BERTH

The Root berth dates back to the 1930s. It is a stretcherlike arrangement that was dreamed up by Elihu Root, an imaginative cruising sailor (Figure 11-7). I made a pair of these berths using 1½-inch aluminum tubing with 1¼-inch fir dowels inside, for *Mouette*. Epoxy poured into the tubes and coated on the dowels would *greatly* increase stiffness. The canvas is a simple sleeve (A) strongly stitched and hemmed across the ends. This could be of striped awning cloth. Set the bars in chocks something like the one shown as B.

If you want to provide more sag for really bad weather, make a second set to go inside the first. Clamp boring so the expansion bit does not split the hardwood. Use 1- to 1 1/4-inch stock or doubled plywood to prevent the bars from coming out of the chocks. Glue and screw, then plug over. If there are no bulkheads there, rig spreaders (C) supported by rope lanyards or chains to padeyes. It takes five minutes to stow these. Or, when folded back and stuffed with blankets, they make passable

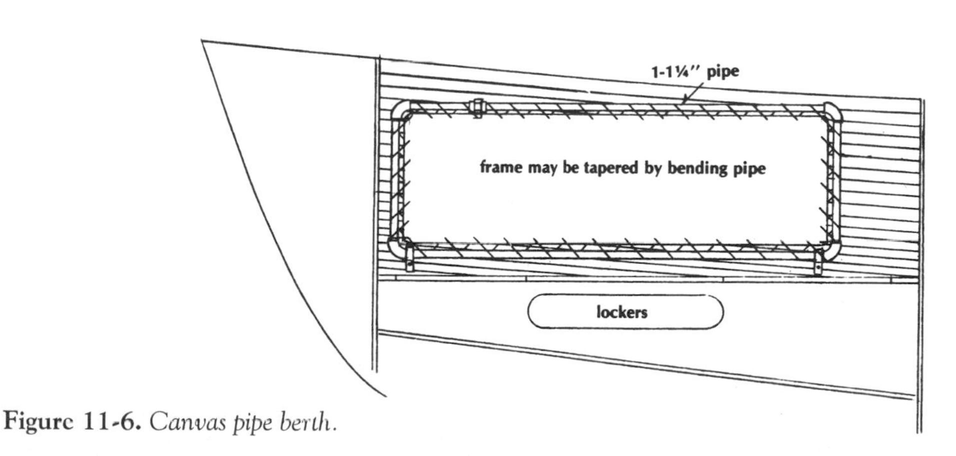

Figure 11-6. *Canvas pipe berth.*

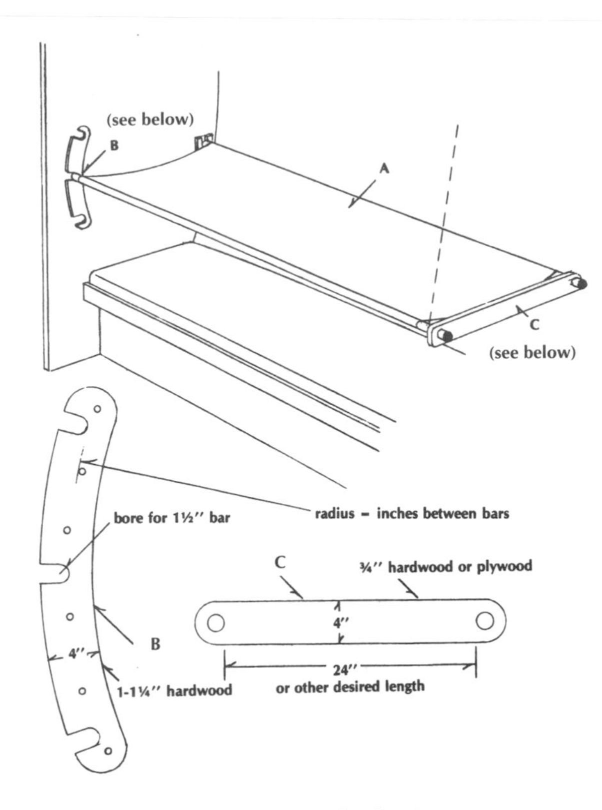

Figure 11-7. *Root-type stretcher berth.*

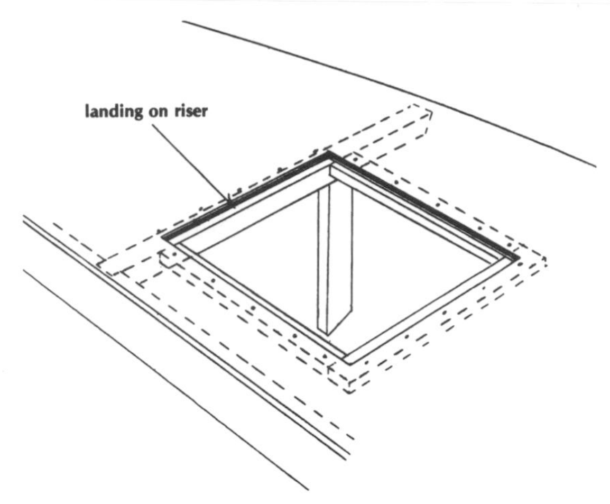

Figure 11-8. *Framing berth top opening.*

backrests. In the forepeak, a Root berth makes a fair bin for sails when not being slept in. The cost and the weight of this contraption are minimal.

BERTH TOPS AND FRONTS

In all construction, keep the weight down. Every pound you add is a pound less ballast and less sail that the boat will carry. Berth tops do not have to be heavier than ⅜-inch plywood—even ¼ inch will work if it is properly framed. Openings in settee or berth tops can be closed with the cutout if a circular saw is used to start the cuts and there are no bored holes (Figure 11-8). The cuts to the corners can be done with a sabersaw, handsaw, or even a hacksaw blade. There must be a strong frame inside the opening to support the cutout, about ¾ inch by 1½ inches, laid on the flat so there is a solid landing for the cutout and a full inch for screwing and gluing to the underside of the berth top. Berth tops take a beating from fat behinds and jumping juniors, so place screws on 4-inch centers with T-88 epoxy, Weldwood, or Aerolite. If your openings can be arranged to fall on the risers, the supports running from front to back, this is even better. To keep the structure light yet tremendously strong, place one or more legs under risers, and bond them to the hull.

You should have generous access openings in the berth fronts also. These can be fitted with hinged or sliding ventilated doors, and you may elect to trim around the opening with hardwood. Berth fronts need not be over ½ inch thick if they are plywood. Even ⅜-inch plywood would be satisfactory. For this lighter construction, Figure 11-9 shows a method of trimming and reinforcing often used in superlight dinghies and racing craft. The frame may be of ¼- to ½-inch mahogany plywood, with edges routed or shaped and sanded, then glued to the plywood front. The plies will show, but this can be made very attractive when filled, stained, and varnished. Solid hardwood, too, may be used for the framing.

BACKRESTS

The simplest backrest is a plywood panel hinged along the lower edge for easy access to the storage space (Figure 11-3). Some backrests require considerable expense and work. The smooth surface with snap-on cushions, however, is most comfortable and easy to construct. For ultra-simplicity, use colorful pillows or zippered bolsters for folded blan-

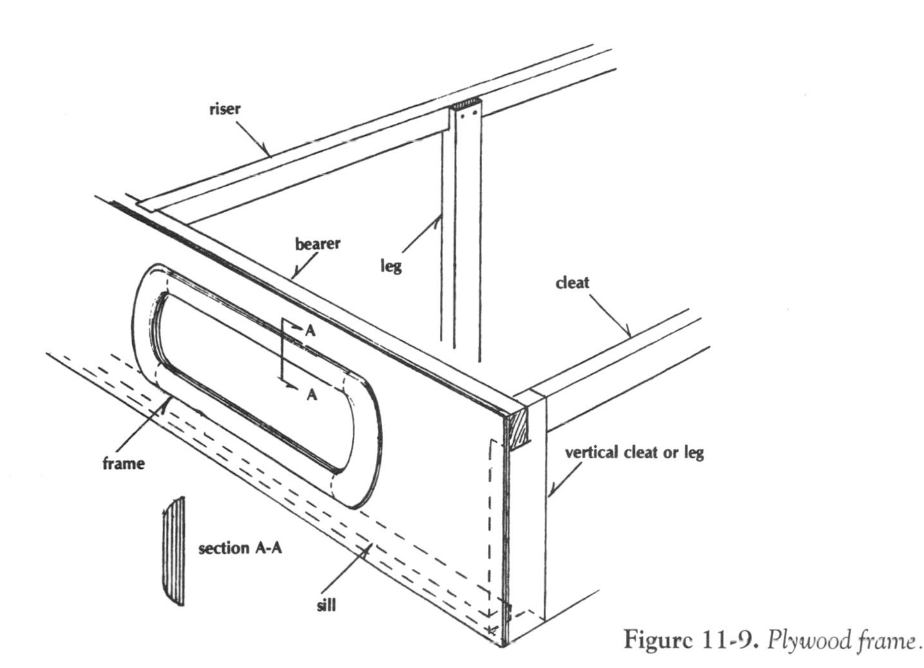

Figure 11-9. *Plywood frame.*

kets. Cloth or vinyl pads can be upholstered right on the panel or frame at low cost. Any visible frame should have a decorative edging of mahogany or teak. Descriptions of doors and jambs follow.

Backrests can be built-in cabinets with three or four flush, paneled, or sliding doors. Remember, backs should slope about 10 degrees. The seat, too, should slope, but if it's to be used as a berth, this is not practical.

Half of the port backrest in my *Allegra* design Plan One does double duty as a dining and chart table (Figure 11-10). This expedient saves space and is quite practical, ¾-inch plywood trimmed all around with ¼-inch mahogany. Fiddles are on the inside plastic-laminate surface. The table is held in place with loose-pin hinges and a turnbutton. To set it up, you screw the pipe leg into the 1-inch pipe flange under the table. The pipe is then passed down through the plate in the sole and into a socket bonded to the hull or to a floor timber. The forward end of the table rests on a block, held there by a brass door bolt fitted into a brass plate. Another method is to mount two brass angles on the bulkhead with a pin passing through them and the plate. Or the table can be stood on two pipe legs independent of the bulkhead. A clever builder could devise a drop leaf for the hinge side to allow the diner on the starboard transom to eat in more comfort.

FRAMING BASICS

Whether you build a berth, dresser, galley, locker, or whatever, the principle is just about the same. If the structure is to rely on vertical supports (legs) to carry weight, the front can be very light indeed. It is common to screw and glue two vertical cleats or legs to the end bulkheads, and a sill lengthwise on the bottom (maybe on the sole) (Figure 11-9). Screw and glue the ⅜- or ½-inch front to these. Add a bearer or stiffener that carries the risers full length along the top—this may be set down the height of the risers so they sit on it. In this case, insert spacers between the risers so you have something solid to fasten the top to. If you use glue everywhere, very few screws are needed, as finish nails are ample to hold everything in position under pressure until the glue sets up, especially if it's epoxy. The framing material may be 1 by 2 clear pine, fir, Philippine mahogany, or oak. Watch the weight, however, and remember that oak does not take glue well. Rely on screws if you use oak.

THE GALLEY

Decide what equipment will be in your galley. You may have to juggle things around to make the best

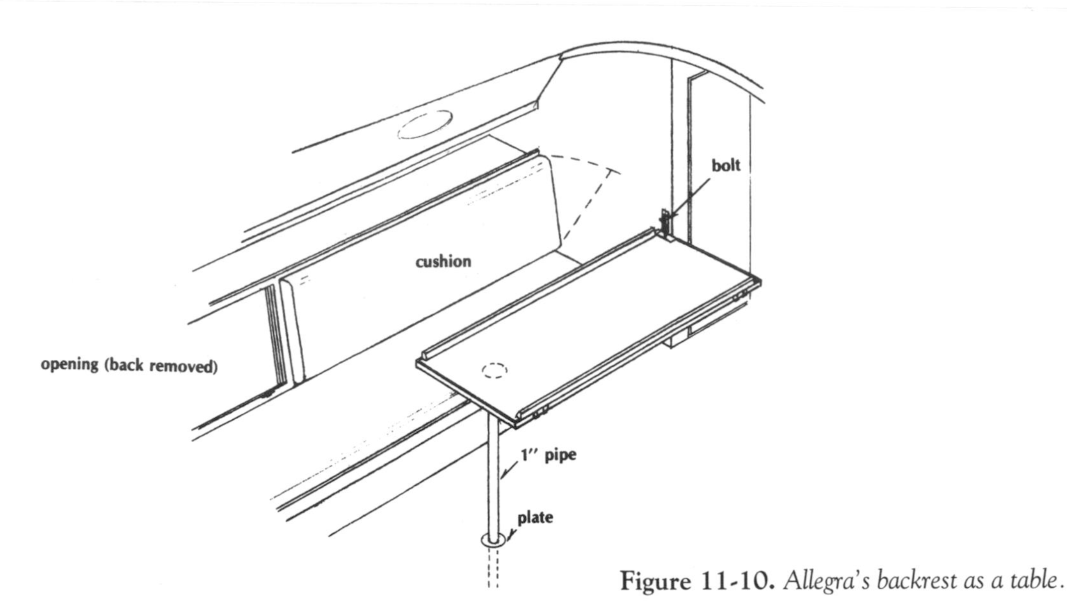

Figure 11-10. *Allegra's backrest as a table.*

use of a small amount of space. If you go for simplicity, you'll choose a drop-in two-burner stove that needs only a cutout. A gimbaled range, however, must swing in a recess built to order. You can settle for a plastic dishpan set into an opening, or select a double sink of stainless steel complete with pressure water. Will it be an ice chest or a refrigerator? Everything hinges on space available and your pocketbook, and perhaps the influence of the person using the galley most.

Figures 11-11 and 11-12 show the framing, front, and top of *Allegra's* galley. This is just about a minimum, being only 52 inches long. However, it does provide for a gimbaled range that takes up a lot of space. Study this quite complex framing. Little that you build will be more involved.

To start, fasten cleats (A and B) vertically and horizontally on bulkheads. Allow at least 4-inch clearance for the length of the range recess. This opening has to be deep enough so the range clears the base as it swings. Measure from the gimbal to the lower corner of the range; add 1 inch for clearance. This locates the height of the horizontal notched into the center vertical leg (E). Swing this dimension outboard to be sure the range also clears framing or any other obstruction (such as a bin). The frame from the sole to the underside of the top is 35¼ inches. Notice the kick space (D). This eases reaching into the ice chest, washing dishes, and getting up close when the boat is heeling. If this piece (D) can't be nailed on top of the sole, it must be spiled to the hull and bonded. Make it 3 inches by 3 inches clear. Framing of ¾-by-1½-inch stock is typical.

Make allowances for the thicknesses of the front and top as well as the range enclosure, but leave the back of the stove recess open into the bin. Most of the framing at C and E could be assembled on ¼-inch plywood, working on a floor, a rigid unit. Clamp to a straightedge across the top and bond to the hull. Do not install the framing around the ice chest (dotted lines) until the chest is complete and chocked in place. There should be a light plywood bottom to keep articles from becoming wet from occasional bilge water. The bin behind the stove also should have a bottom. Doorskins are fine for this because of their low cost and light weight.

The Galley Front and Top

The one-piece galley front is sawed out of ⅜- or ½-inch plywood to fit snugly between the bulkheads. The openings can be marked from the inside of the final framing so you have a clean saw cut on the exposed side. I prefer shelf space to drawers, as drawers are space wasters. You can, however, install two drawers under the sink and one under the range. Details on drawer construction follow. For now, put the front aside until the ice chest has been installed.

The countertop should be ½- or ¾-inch plywood to which a laminate will be applied. Fit it by tacking thin plywood rippings or scrap into a template

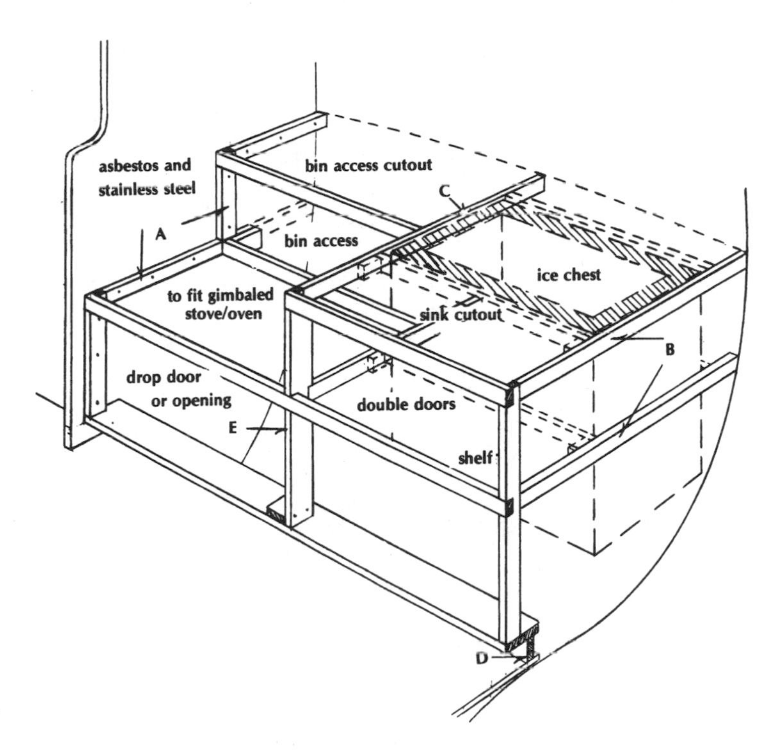

Figure 11-11. *Allegra's galley frame.*

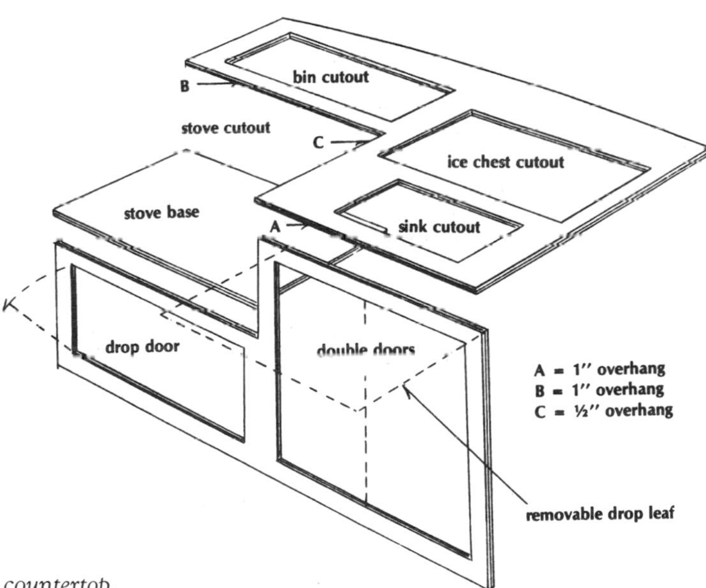

Figure 11-12. *Allegra's galley front and countertop.*

so it conforms to hull curvature. Apply a plastic laminate following instructions on the can of contact cement. Again, mark the openings on the underside. The cutout for the sink is 10 by 14 inches for the larger single sink. Let the front edge project ½ to ¾ inch past the plywood front. There should be small allowances around the range recess, too, for looks.

Saw to the lines by clamping on a thick straightedge for a saw guide, except on the ice-chest opening. The remnants can be used for covers. Note the removable drop leaf in front of the sink. Cover these with laminate and give them a nice hardwood edging of mahogany or teak. There must be generous fiddles around all sides of the galley counter.

I think it would be handy to make the galley top fairly easy to remove. You can fasten small angle brackets to the frame at four or five locations accessible from the front or through the openings. Be sure the screws do not reach the laminate. It would also be possible to have the top slide out by applying moldings at each side on the bulkheads. A removable top makes it easy to reach around the ice chest or to pull it out, if necessary.

Working surfaces are severely limited in small yachts. One possible solution (an expedient, really) is to make a cover for the galley in two sections. Fit these inside the fiddles. If there is no water pump, you have a large area suitable for a chart table. It might be possible to have half the cover in place with the range in use. Make the cover sections light enough so they can be stowed under a berth pad.

DRESSERS AND LOCKERS

Framing for dressers and lockers is similar to galley construction. But fronts may be very slim. There are many ways to construct a dresser front. Let's discuss these first, as they affect the framing procedure. The simplest front is of good ⅜- or ½-inch plywood (Figure 11-13). Mark on back side and saw out as described above. The second front, shown in Figure 11-14, is assembled right on the vertical cleats or legs. This will give you the look of solid mahogany or teak. The joints must be absolutely flush and the sanding done with extreme care to avoid unsightly scratches across the grain.

The rails (A) and stiles (B) must be cut very accurately with a fine-toothed blade, so clamp a stop on your table saw. Position it just this side of the blade to prevent jamming between the blade and the fence, or use the sliding auxiliary table (SLAT) described in Chapter Five.

A less conventional way to build a dresser front of hardwood is shown in detail from the reverse side in Figure 11-15. It's best to assemble this on a floor or bench on a sheet of light plastic. Tack blocks around three sides of the perimeter to the exact dimensions of the front. Make the stiles (B) full length, in conventional cabinetmaking style. Cut the rails (A) very accurately and cleanly to make a snug fit between the stiles. Now carefully mark the exact position of the rails on the stiles. Glue the mating surfaces of the stiles and rails. This is a good place to use Aerolite, because it is strong and sets up crystal clear. But put the resin on the end grain

Figure 11-13. *Dresser front (one piece).*

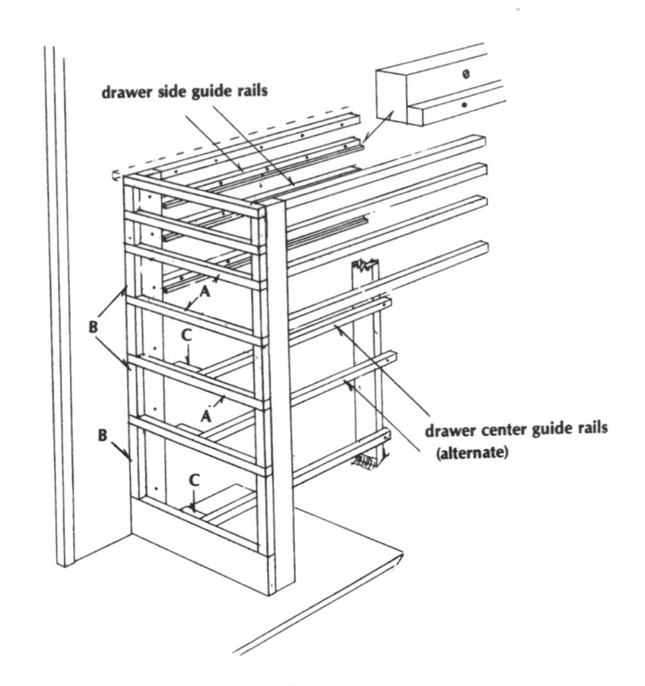

Figure 11-14. *Dresser frame.*

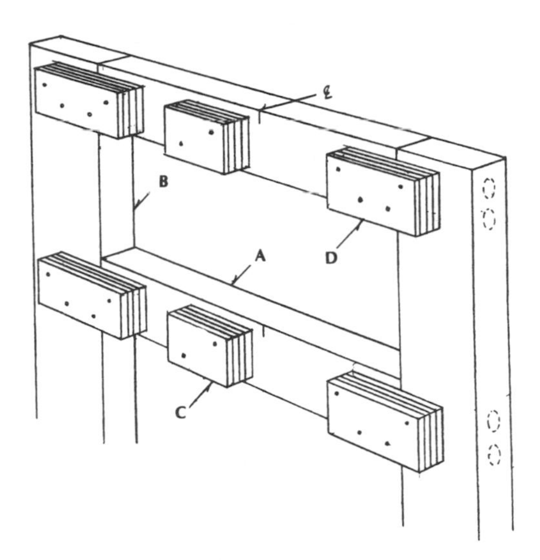

Figure 11-15. *Built-up dresser front, glue-block style (inside view).*

generously, and the activator on the stile. Place all the parts in position, then tack down the holding blocks on the remaining side of the jig. The glue blocks (C) don't have to be cut accurately, but a similar setup is useful when the quantity makes it worthwhile. Place the glue blocks as shown (the small finish nails should have been driven in previously).

Clamp the whole thing with light pressure, using four or five bar or pipe clamps. Then drive the nails well in until the joints appear to be flush. Give the clamps a turn, but do not crush. If you don't have clamps, tap small wedges between the stiles and the jig blocks. In about four hours, if you handle the assembly very carefully, you can lay it on its back and wash off any Aerolite glue that may have squeezed out; water will dissolve Aerolite before it sets up permanently, but once it's hard, there's no hope. The only disadvantage to this kind of front is that it requires wider stiles than the other built-up front. Thus, it wastes a bit of space and you can't have side guide rails.

This assembly could be put together by doweling the joints, as previously described, using a boring jig (or by boring from the outside of the stiles). You could fasten with long screws, but screws do not hold as well in end grain.

There's still another method, one that is really yacht style. The rails are halved into the stiles (Figure 7-11). This is picky work requiring fine, sharp tools, preferably a dado on the table saw. In ½-inch mahogany or teak it would be a nice challenge. But where is it written that stiles can't be of ¾-inch material? I guarantee you'd have an exceedingly strong front, and if you make fine, clean joints, you'll have something to enjoy as long as you own the boat.

Drawer Guide Rails

Which comes first, the framing or the drawers? You can't proceed with one unless you know how you are going to build the other. The trickiest part of this job is installation of the drawer guide rails (Figure 11-14). These keep the drawers level, square to the front both ways, and prevent the drawers from swinging from side to side. In dry-land cabinetry this is accomplished efficiently by using roller devices or plastic glides that ride on grooved rails. You can use these, but often they require more than 1 inch between drawers. This is a lot of wasted space, but your drawer rails don't need to be much larger than ¾ inch by ¾ inch because they support very little weight.

First, let's talk about the rails at the top half of the illustration. These are simply screwed and glued to the bulkheads. The configuration shown may be two-piece or rabbeted out of heavier stock. However, there's an important detail: drawers tend to slide out when the vessel heels, so they are often notched to lock onto the dresser front. Thus, in this case the guide rails must be located ⅛ to 3/16 inch below the front pieces (A). (Also see the drawer construction shown in Figure 11-16.) The guide rails must also be dead level athwartships and across from one side to the other. You'll need a short torpedo level. While drawers may be of several lengths because of the shape of the hull, the rails don't have to reach full length if on a bulkhead.

The center guide rail may be mandatory because of the width of the dresser front stiles (B). It's complicated. This guide rail must level up with the front where it is attached to the glued block (C). It must also be dead center between the bulkheads so the drawer front matches the dresser front. Even a very small misalignment will be obvious. Prepare a vertical piece from 1 by 3 stock to take the outboard ends of the center rails. Clamp this for a moment to the dresser front and transfer the spacing to it. Now, when this vertical is erected in its final position and level from any front rail to any

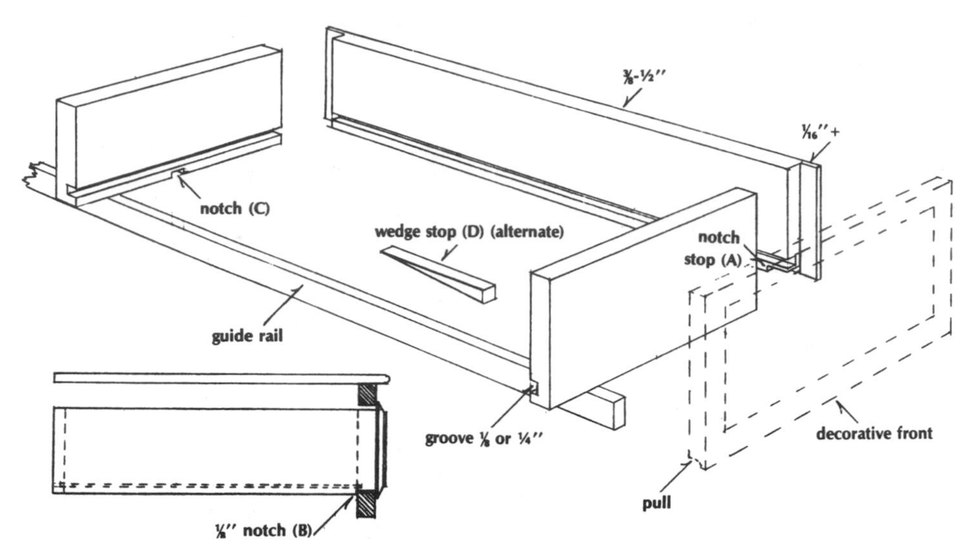

Figure 11-16. *Typical drawer construction.*

corresponding mark, all drawer guides will be level. The drawers will fit.

How to fasten this vertical depends on the type of boat. My guess would be that bonding with fiberglass cloth and epoxy resin would be the best, as this allows flexibility for adjustments side to side and up and down until the resin sets. Or it can be screwed and glued to horizontal pieces at the top and bottom of the cabinet. There may have to be two verticals because of different drawer lengths. Don't screw the rails in until the drawers are completed. Shimming or paring may be required if the vertical is not perfect. Rub a bit of paraffin on the upper surfaces of the guide rails for lubrication. I would use oak or ash for these rails.

Drawer Construction

There are many ways to build drawers. Start with the premise that every drawer, fully loaded, must withstand the shock of being *slammed* shut a thousand times. In addition, there are certain desirable characteristics. Two examples are a double or rabbeted front that acts as a stop and a bottom fitted into grooved sides and ends. And I like the appearance of closely fitted flush fronts, but the alignment must be perfect and the spacing all around must be close and uniform for all drawers in a bank. Any of the overlapping or rabbeted fronts cover the opening completely, so your small errors are hidden. Figure 11-20 shows four ways to make the insides of drawer fronts.

The bottom panel fitted into grooves is the strongest possible drawer construction. It squares up the drawer assembly almost automatically (Figure 11-16). Bottom panels may be of ¼-inch marine or exterior plywood or ³⁄₃₂-inch mahogany doorskins if you want space, appearance, economy, and adequate strength in a drawer less than 24 inches square, not for tools. Of course, bottom panels can be nailed and glued on the sides and ends to save space and time, but this is hardly yacht-style workmanship.

I prefer the deeply rabbeted corners shown in Figure 11-16. This structure creates more gluing surface than any other except the dadoed or dovetailed joint. And the appearance of sides and ends is good, although I'm assuming you will fit a decorative front. Dovetails are ideal, and they can be done quite easily with a dovetail router template (Chapter Seven). Another strong joint using a single dovetail is only a bit more difficult than the rabbeted corner (Figure 11-17). Rather careful layout with a knife is recommended. Make up a couple of trial joints using scrap identical in thickness to the drawer sides and ends. Make the front first. Cut the dado as shown in Step 1, ⁷⁄₁₆ inch wide for a ½-inch side, ⅜ inch deep. Set the blade to 15 degrees when all rabbets are cut, and make this single saw cut in all fronts. This is Step 2. For Step 3, clamp up a

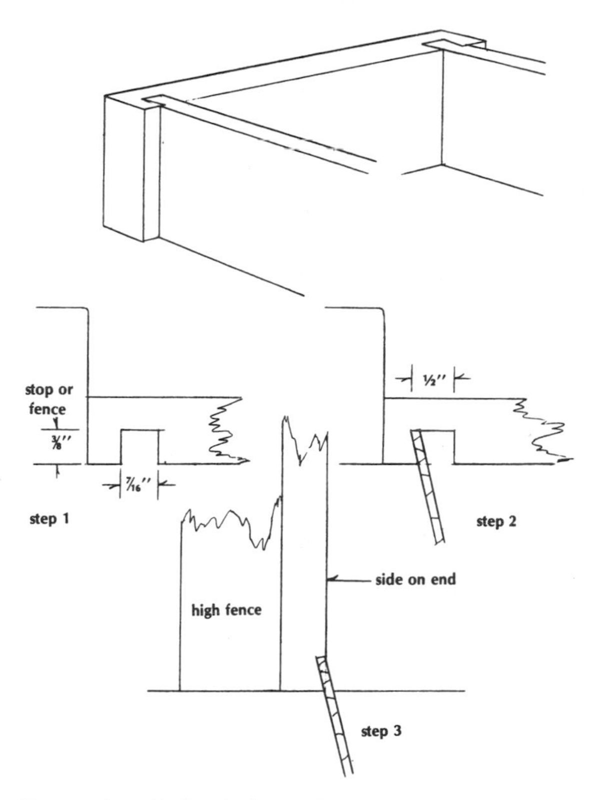

Figure 11-17. *Single dovetail.*

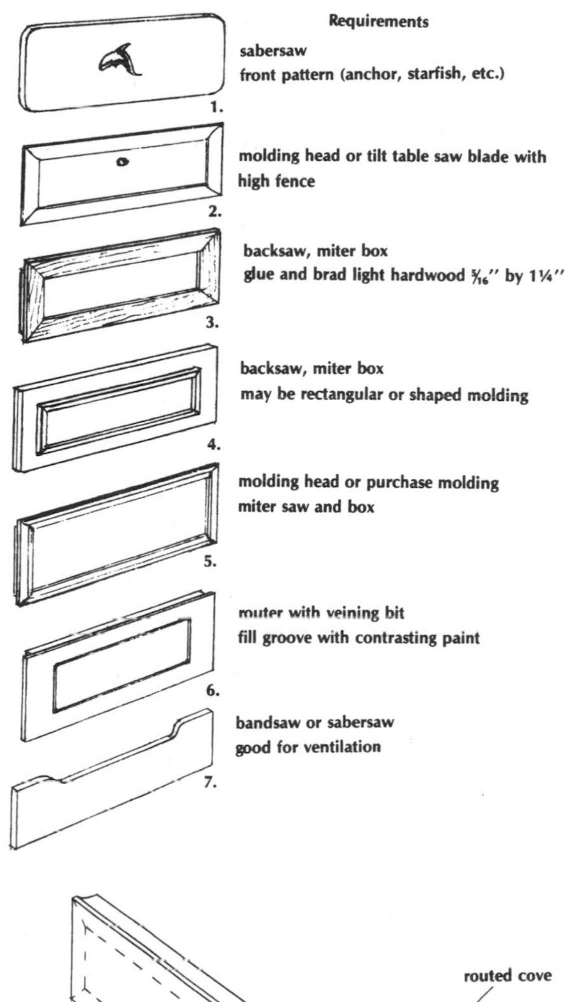

Figure 11-18. *Drawer fronts.*

high fence to keep the pieces vertical. Form the corresponding angle in the side. Groove the parts for the bottom panel, as above. Assemble with Aerolite or T88 glue. No fastenings are required. Incidentally, use an ordinary rabbet joint for the back end. This drawer will take any type of added front over the single dovetailed front.

Drawer Fronts

I show only a few drawer front styles in Figure 11-18. Read some of the fine cabinet and furniture construction books available in your public library. A great deal depends on your skill, tools, and patience. Number 1 in Figure 11-18 is just about self-explanatory. The fancy cutouts can be routed, sabersawed, or scroll sawed. If you want to repeat the design in other lockers, berth fronts, or what-have-you, make a template for a fine router bit. Keep the design simple. Sharp corners must be filed or sawed out. Use your imagination. Anchors and sea gulls are a bit hackneyed. How about sharks, dolphins, swordfish, ducks, eagles, seals, whales, stars, burgees, and so on?

Number 2 is a raised panel, very handsome if made with a molding head so the surface is slightly concave. I used a single cutter head from Sears. I ground the bevel to the shape, an easy task, for there was but one blade. Another method is to set

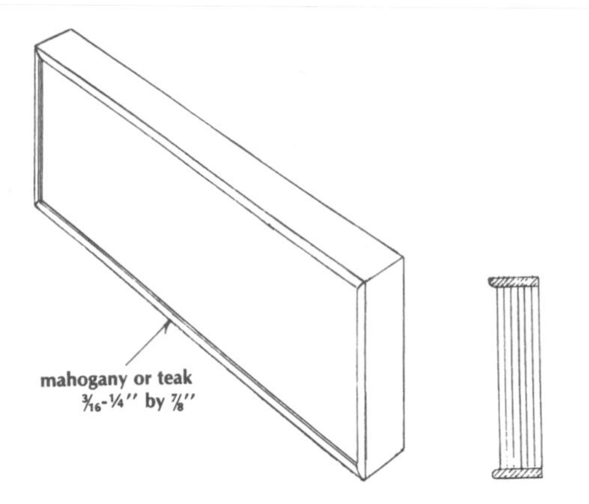

Figure 11-19. *"Yacht style" door and drawer trim.*

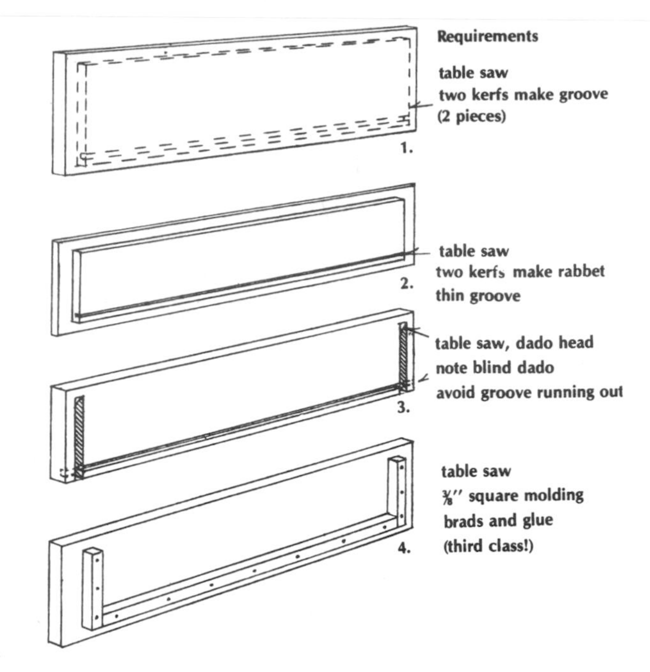

Figure 11-20. *Inside drawer fronts.*

up a high fence and incline your saw blade. The edge is not quite as pretty and requires a lot of sanding, but I have made hundreds this way. A fine touch is to make the area to be molded of hardwood. It is then mitered and glued to the center panel. No nails should show, of course. Glue will do the job. Finish the molding bright against a painted center—beautiful!

Number 3 is a simulated panel effect. I call it a drop center. It looks best on drawers and doors if the mitered frame is rather delicate—say, 1/4 inch by 1 to 1 1/4 inches. The center panel may be painted or bright if of the same material. Glue thoroughly and brad. This job requires a good miter box or a SLAT. If you have confidence in your table-saw miter gauge, use it.

Number 4 also can be done with hardwood moldings, perhaps a tiny half round or a beaded molding. These are seldom found commercially in hardwood, so see what you can make with router bits, with the router mounted upside down in a little table. Even pine, painted a contrasting color or antiqued, looks great. All of this applies to Number 5 as well. In this drawer front, the edge might be molded with an applied molding added.

Number 6 shows a pattern routed in with a veining bit. This requires a guide template of 1/4-inch plywood. Number 7 allows lots of ventilation and is easily made with a sabersaw or bandsaw. A router or shaper straight cutter can turn these out in minutes if you make a template. If made of plywood, this drawer front should be painted. In addition, it must be rabbeted, for it cannot be attached to an inside front.

Number 8 is a modern finger-pull front. The edge is undercut either by sawing the angle or by forming with a router, shaper, or molding head. A first-class job would be a hardwood mitered edge glued on, but it can be done in 3/4-inch plywood, well sanded, and painted.

One drawer treatment I have not shown probably should have been Number 1, as it is pure yacht style and easy to make (Figure 11-19). If you go for the one-piece plywood dresser front, the cutouts can be used as drawer fronts. Simply edge the front piece with hardwood about 1/4 inch by 7/8 inch, miter the corners, and have the strips project 1/8 inch or less. Bullnose the corners and finish bright. This treatment is excellent for both cabinet and passage doors.

The inside views of drawer fronts need little comment (Figure 11-20). The two rabbeted styles can be done with the table saw alone, or they can be dadoed. The third requires a dado set and great care to see that the bottom groove does not run out and ruin the appearance. The fourth is simple. You'll need only enough tools to make the light square molding. Rely on glue for strength.

Drawer Problems

One problem in any sailing yacht is drawer movement when the vessel heels. Designers, of course, are well aware of this problem. Steps can be taken to solve it. Look at Figure 11-16. You'll see notch stop (A) at the upper right-hand side, front. It is shown again in side elevation at the lower left (B). The notch should be just deep enough to let the drawer drop down over the rail of the front construction. Also note the clearance above the drawer side so it can be lifted up out of the notch. There is a difference. Notch (A) is cut out of the side only. Thus, it permits the drawer to slide forward ¾ inch, then it stops. Notch (B) requires the inner front to be reduced to the depth of the notch, perhaps by taking off a saw kerf before the decorative front goes on. Now go back a page or two to dresser framing. You will see that the side guide rails were located ⅛ inch lower than the front rails to accommodate the drawer dropdown. If you use the center guide rail, there must be a notch (C) in the back panel of the drawer. The alternative, the wedge stop (D), eliminates front-end notching. It automatically raises the drawer ⅛ inch, then allows it to drop down and lock. The back notch is not necessary if two guide blocks are glued on to ride on each side of the center rail.

If you use a center guide you must adjust the position of the vertical support for the guide rails. It's all right if you measure from a bulkhead at both ends to make sure the guide rail will be parallel. You'll have to shift the vertical support accordingly, or pare or shim it. If you install the drawer and then clamp or wedge it flush with the front of the dresser, you can mark the position of the guide rail at both ends. If they are not exactly alike, you'll have to move the vertical support as above.

The second problem is connected to the first. Guide rails must be square to the dresser front or an impossible jamb will result. If you use the side guide rail construction, you merely have to see that the bulkheads either are absolutely parallel or allow so much slop on either side of the drawers that they square up when closed against the dresser front. This is a poor solution. Perhaps you could shim out the side rails from the bulkhead to line up the drawers.

There's still another way to line up a drawer square to the dresser front, regardless of the kind of guide rails used. Fashion quite long wedges to be glued to the bulkheads on each side. Place them in or out so they move the drawer sideways until it squares up. Brad and glue them in this position. The drawer will still be sloppy when it moves in and out, but perfect when fully closed. This is similar to the idea of shimming the side guide rails mentioned above.

A final word on drawers. Get used to shelves instead. With little fiddles on shelves, they are almost ideal because you get full use of the space, even if you have to pull everything out to find what's in back. Shelves save a lot of time and material.

SMALL DOORS

Let's cover cabinet and locker doors first, then get into passage doors. Some of the moldings for drawers already described would be attractive on small doors (Figure 11-18). The last one described, a simple hardwood trim strip, would be practical and attractive for passage doors as well. Number 2, called the raised panel, gives a rich effect for a full overlapping door, that is, a door not rabbeted. Number 3, the fake panel drop center with mitered hardwood trim, also has a traditional look about it. This would be good for stiffening a light door less than ¾ inch thick, but I would not recommend it for a passage door. Any door large enough to warp should be weighted down on a dead flat surface while being glued. The molding pattern, Number 4, could be used on any door, but Numbers 5 and 6 are suitable for cabinet doors only. The vent in Number 7 makes it good for a hanging locker, especially for storm gear. In this case, I would repeat the cutout at the bottom of the door. The more ventilation here, the better.

Most doors block air circulation, of course, so here are a couple of ideas for ventilating panels you can make (Figure 11-21). This job looks much more difficult and time-consuming than it really is because of the inadequacy of the drawing. All you do is set up your dado width, say, ⅜ to ½ inch, to cut to a depth slightly more than one-half the thickness of the material. The drawing shows 7/16 inch for 13/16-inch mahogany. Cut the dadoes on one side, then flip over and rotate the piece so the grooves intersect. This will leave a square of daylight where they cross. Screw and glue a substantial rail on the end-grain edge, assuming the grain runs horizontally. Use Aerolite or another good epoxy or Weldwood Plastic Resin. You may go all around with rails and stiles, or with a simple flat hardwood trim well fastened. Knock the sharp corners off the grooves with a sanding block.

How do you space the grooves and dadoes? In

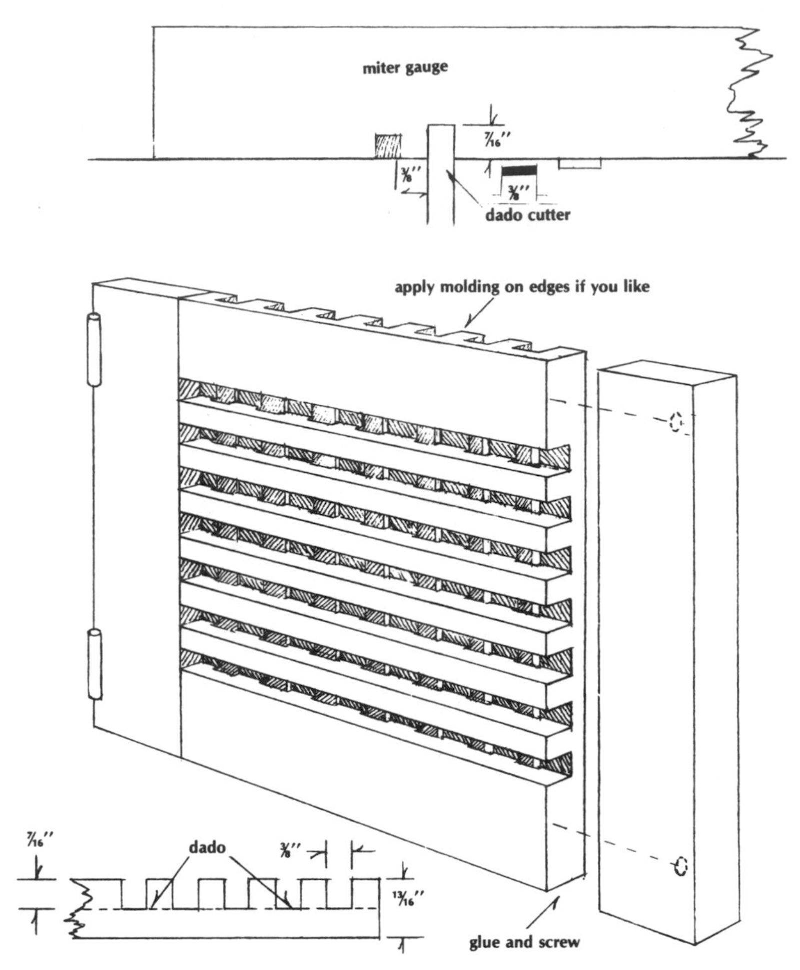

Figure 11-21. *Vented (dadoed) door.*

Chapter Five, I described a method of making box or finger corner joints. This is identical. Fashion a tiny block 3/8 inch wide, then glue this to your miter gauge. Place your piece against this block and cut your first dado. Now place that dado over the block, and run the second cut. Then just keep repeating to the width you want. If you like the heavier frame effect shown in the drawing, make the first cut, then fix the block in place. If this block is a nuisance, drive in a small brad, but remember to press the piece against it on each pass. If you have a sliding auxiliary table (SLAT) for your saw, the task will be easier and the result more accurate.

Plywood doors are made more attractive by cutting out ventilation apertures. A series of slots is a simple design. Another design has holes bored in a pattern, using a template so all doors will be identical. Three-quarter-inch doors may be sawed out to leave stiles and rails for paneling, as in Figure 11-22. The openings may be filled with leaded glass held in place by a rabbet in the back and small moldings. To avoid such unusual weight, use doorskins, 1/4-inch plywood, or even molded plastic designs. Not yachty enough? Correct, but they are inexpensive, almost indestructible, light in weight, and no finish is required. Tempered Masonite is available in a variety of styles. This is true, too, for textured polystyrene and, of course, the ubiquitous wicker (preferably natural, brass, or aluminum, but not steel). If none of these will do, you can fashion very

Figure 11-22. *These ash plywood doors were sawed out in one piece. Rabbeted in the back, they are strong enough to carry leaded glass.*

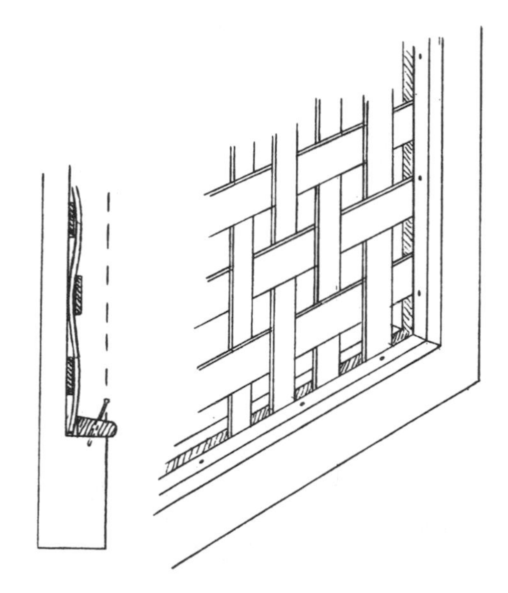

Figure 11-23. *Woven door panel.*

attractive panels by weaving thin strips of mahogany or teak 1/16 inch by 1/2 to 3/4 inch (Figure 11-23). This is quite a task, but appearance and practicality make it well worthwhile.

Woven Door Panels

Cut the strips to 1/16 inch or less, making them about 1/2 inch longer than the rabbeted opening dimensions. Sand well. With the door frame rabbeted side up on a bench, square off the desired spacing, marking on the rabbets so they won't show later. Spacing equal to the width of the strips is about right. Your moldings have already been made and mitered. Weave the strips together on the bench to get started, then space them out over the marks on the rabbets. Excess lengths can be nipped off with a sharp chisel. Clamp a couple of battens over the strips so they won't move around and start bradding in the moldings. It's easy if the brass brads (#18 or #20) are driven well in beforehand and slightly canted. Press the moldings down hard with small C-clamps, being sure they are straight. Do the same on the opposite side, remove the temporary battens, and apply the third and fourth moldings. Glue under the moldings might help, but it's not vital. Hang this door either way. If it's hung molding side out, make the moldings wider so they project slightly, as in the sketch. A hardwood quarter round can be used. Miter the moldings, perhaps finish bright, with painted lattice.

SIMPLE PANEL DOORS

I have not mentioned traditional paneled doors because of space limitations. The procedure for building them is described in dozens of public-library sources. The drawing in Figure 11-24 shows a simple panel door with optional half-lap construction and without milled rabbets. Instead, rectangular strips are used to make a simple built-up rabbet. Quarter-round trim is then used. Figure 11-25 shows a door assembly made by dadoing the rabbets. A 3/8-inch groove in the stiles and the ends of the rails takes the splines of 3/8-inch plywood glued in. The rabbet corners have to be chiseled out. Moldings cover the panel and the exposed edges of the rabbet. When it is glued up, the door must be held down on a dead flat surface. Bar clamps or wedging are required. This makes a very strong door for cabinets, lockers, or passage.

A method of stiffening a 3/4-inch plywood door is shown in Figure 11-26. The frame is heavier than the plywood because it helps to counteract warping. It looks good, too. Like the door in Figure 11-25, this door depends on 1/4-inch plywood-strip splines, but they run full length. This is a simple job

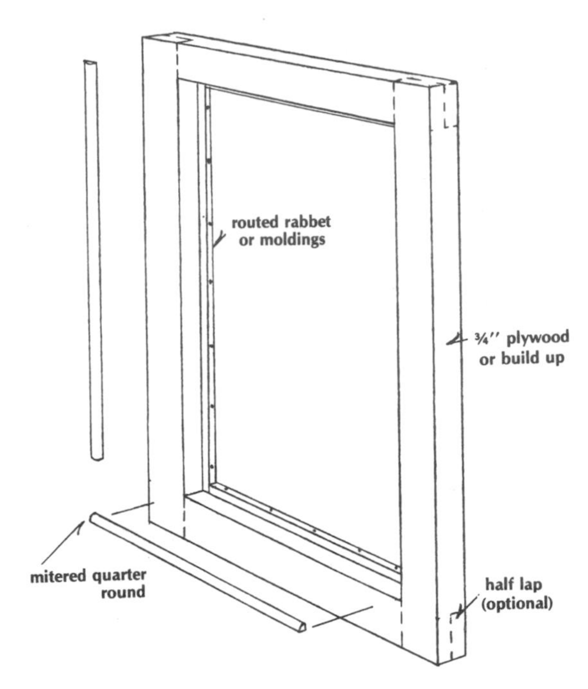

Figure 11-24. *Simple frame for paneled door.*

Figure 11-25. A *built-up door frame with splined corner joints and routed or dadoed rabbets. Assembly must be done on a flat surface and door must be weighted down while glue cures.*

with a dado set, or less so with repeated saw kerfs. Set the fence back 1/16 to 1/8 inch when grooving the frame so it projects equally on both sides of the plywood. The mitered frame is like any other. The splines do not have to be mitered or even meet in the corners. Plastic resin painted into the grooves would work well here.

PASSAGES AND DOORS

Every bulkhead passage opening must be edge-trimmed. In addition, door jambs must provide a very solid stop. The one-piece construction shown in Figure 11-27A is best, because while it is light, it is also strong. Quite intricate jambs can be built up, as in Figures 11-27B and C, to minimize waste. The section in (A) is not difficult. It just requires time. Plan ahead. Jot down how much door jamb or simple strips will be required. Set up a dado for the groove in (A) or the rabbet in (B), then run all pieces through. If you elect to make these with repeated saw kerfs, repeat the cut in all pieces, then reset the fence, and so on. Don't finish one piece, then set up all over again for the following pieces.

The drawings give suggested dimensions. For a ¾-inch bulkhead, you might want to make the piece in (A) from 1¼-inch mahogany or teak so the stop would be heavier. I suggest the following sequence.

1. Joint 1½-inch dressed mahogany plank edge.
2. Rip off ⅞ to 1 inch.
3. Joint this (plane) to ⅞ inch or more by 1¼ or 1⅜ inches.
4. Dado, shape, rout, or saw groove (bulkhead thickness).
5. Dado, rout, shape, or saw rabbet. The last pass takes out the dotted area.
6. Sand all surfaces except bulkhead groove, including the rounding of corners. The form in (B) is similar but it comes out of full ¾-inch stock, so it saves a lot of material. The built-up form in (C) is all strips. You can get two 5/16-inch dressed pieces from ¾-inch stock if you use a fine carbide-tooth blade.

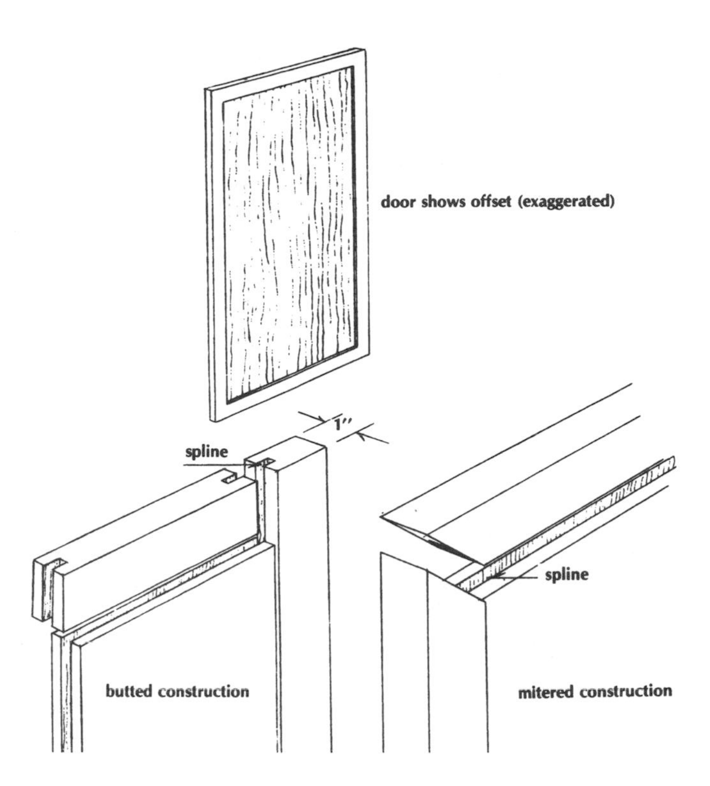

Figure 11-26. *Stiffening a ¾-inch plywood door.*

Need I remind you that hold-downs are absolutely essential when working with small stuff like this? Safety as well as consistency are at stake. Use glue and 1-inch brass brads or galvanized finish nails to assemble. The corners are mitered, of course.

The door edge trim shown in Figure 11-27 is self-explanatory.

One of the most beautiful door designs has a 4- to 6-inch radius in the corners, both top and bottom (Figure 11-28). A passageway with a perfect arch top and bottom is even lovelier. These are a lot of work. They can be botched if you lack skill and patience, but they can add thousands of dollars to the value of your yacht if well done. If you follow the procedures below for the construction

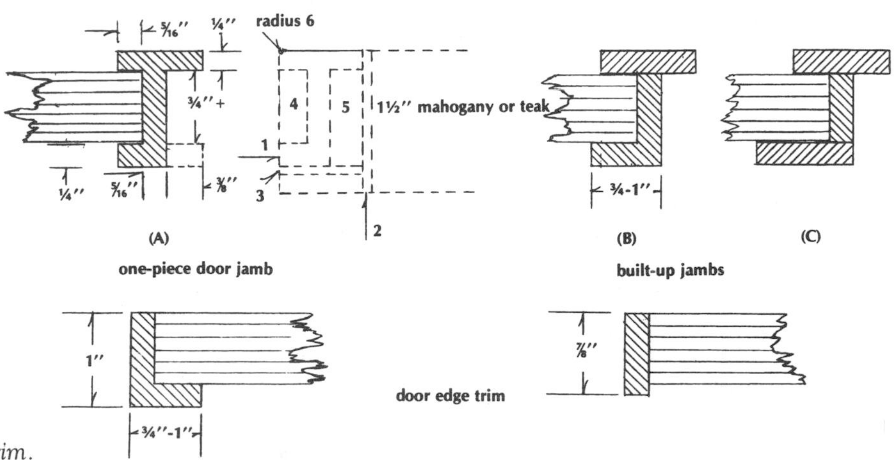

Figure 11-27. *Door trim.*

Figure 11-28. *"Yacht style" passage.*

of corner trim for cutaway bulkheads as well as for passage doors, you will be halfway home. You'll create a beautiful interior. A word of warning. Don't start this involved work unless you are prepared to carry it from forepeak to aft bulkhead. The scheme must be coordinated and have integrity.

First of all, this work requires a shaper or a router set up as a shaper, with collars on the cutters for forming grooves in inside and outside curves. Second, some of this work is dangerous. You'll be handling small pieces up close to a cutter turning 28,000 r.p.m. Often you'll be cutting against the grain. To avoid having small pieces thrown about like missiles, always try to do the risky cutting while the piece is still in the plank. Don't bandsaw it out and then try to shape it. The plank's weight is your protection.

CORNER TRIM

First consider an outside corner trim for a door or bulkhead. For trimming ¾-inch you need mahogany or teak dressed to 1¼ inches thick, for ½-inch trim you need ⅞ inch dressed. See section (A-A) in Figure 11-29. The width of the piece is 6 to 8 inches, depending on the desired radius. Let's do a 4-inch radius for doors and partial bulkheads using a shaper or router-shaper. Radius 1 in the drawing is 4 inches to the bottom of the groove to fit the bulkhead radius. Thus, radius 2 is the inside radius—3¾ inches. Make your shaper pattern to cut that radius. Space out on your plank the number of pieces required, marking around the pattern for approximate location along both jointed edges.

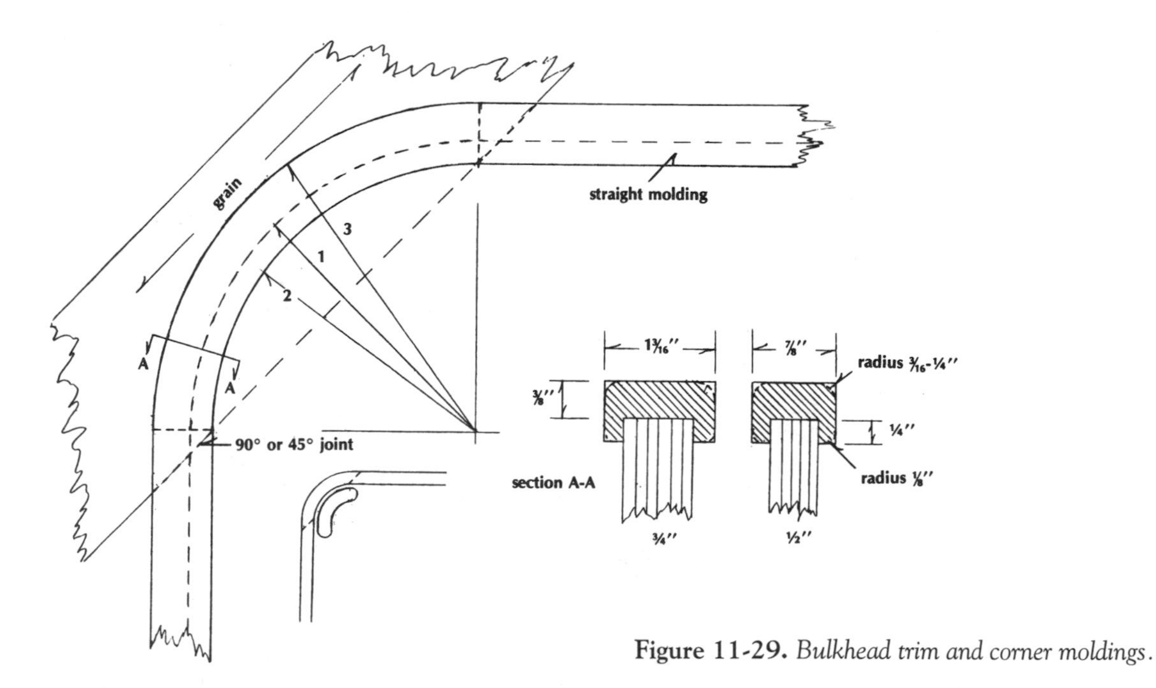

Figure 11-29. *Bulkhead trim and corner moldings.*

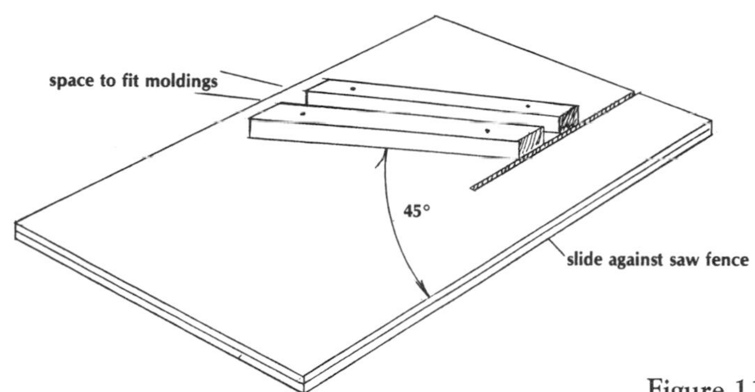

Figure 11-30. *Molding mitering jig.*

Allow plenty for the piece to radius 4⅜ inches, plus some for dressing. Make a pattern for later bandsawing of this outside curve.

The next step is bandsawing or sabersawing around the inside radius. Tack the shaper pattern in place. If you have a shaper cutter 1½ inches high, set it up with a collar to follow the pattern. If you are using a 1-inch cutter, make two passes; with a router, use a trimmer but ride on the pattern. This should give you a surface that needs only a few strokes with sandpaper, or on the end of the belt sander. Do the same for all parts.

Next, we need a groove ¼ inch by ¾ inch or ¼ inch by ½ inch, depending on bulkhead or door thickness. Cutters for rabbeting ¾ and ⅜ inch are available, the latter requiring two passes, of course. Set up with a collar ½ inch less than the cutter diameter, and adjust the height so the groove is dead center. Now run the radius against the collar. In making these shaper cuts, feed the work in very gingerly until you acquire the feel and control it needs. Remember, half of this pass will be against the grain and the shaper will try to fight you. One or two shallow freehand cuts are advisable until you see whether or not the reverse grain is going to split. Don't rush, but don't feed so slowly that the material scorches, as this will take the temper out of the tool. Always start the feed against the cutter rotation. Repeat this process for all the parts and the job is more than half done.

The next step is easy. Bandsaw outside of the radius mark, and use your bench disc sander to take the sawed surface down to the mark. After the corner has been connected to the straight trim, it will have to be sanded well with the grain to remove the cross-grain scratches. Or the edge can be planed after the piece has been installed, and then sanded. This molding is a little scant for plugging over screws. I suggest 1-inch galvanized or brass finish nails, set in well and filled.

The joint with the straight trim is up to you. I feel that the 45-degree joint permits closer fitting. Use the 45-degree sawing jig shown in Figure 11-30. Whatever you do, however, tack all the edge moldings in place before gluing any. This way, precise pressure joints can be made. If you use the 90-degree joint and you cut a straight molding $\frac{1}{32}$ inch too short, that piece will have to be used elsewhere. After fastening, take fine cuts on the sides with a block plane or a sanding block. The last step is to rout a radius all around on both inside and outside. This should be generous on the outside, possibly as much as ½ inch. On the inside, ⅛ inch is plenty. This means you will have to grind off a good part of the pilot on the bit. If you prefer the easy way, rout the radii before assembly. Be careful, though. They look bad if they do not match up.

Figure 11-29 shows a hand grip just inside the molding. The hole should be about 1¼ inches in diameter. Sabersaw the hole out before moldings are installed.

Corner moldings of this type can be purchased in teak and mahogany for both ¾- and ½-inch bulkheads. The radius offered is 2½ inches, standard in my designs. The manufacturer is H & L Marine Woodwork, Inc. The moldings are sold at most marine supply dealers.

MORE CORNER TRIM

The process of making inside corners by the method just described is about the same, but in reverse. There will be more waste because the

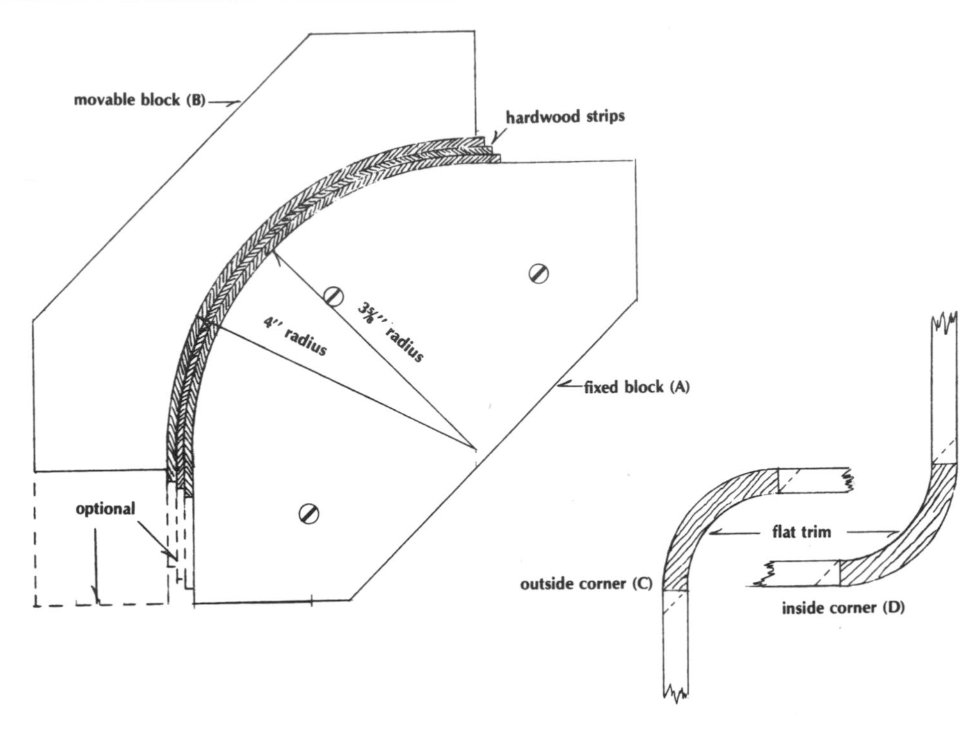

Figure 11-31. *Bending jig for laminated bulkhead and door trim.*

pieces can't be nested together as well. There is a laminating system for forming both inside and outside corner trim (Figure 11-31). I know that this will draw snorts of derision from some because it is not difficult and few tools are required. You will need two forms inside and outside, as the radii are different by the thickness of the lamination. The bent segments are covered after installation by flat quadrants of whatever dimensions you choose. One definite advantage of this trim is that the flat quadrant can be dimensioned to make a stop for a door.

Make the forming jig of 1½- or 2-inch lumber or plywood. Make the strips ⅛ inch by about 1¾ inches. Lay down a piece of plastic and screw block A on top. Apply glue to the contacting surfaces, press block B against the strips, and clamp loosely. Tap the strips down against the plastic, make any necessary endwise adjustments, and clamp up snugly. When the glue has set, rip the lamination to the width wanted and cut the ends square or to a 45-degree angle, as you wish. Dress to match the bulkhead or door thickness. These can be fastened with glue and finish nails.

The flat quadrants are self-explanatory, except that the door jamb needs a stop. Lay out with a shorter inside radius so the door finds a rabbet to bear against. The straight trim can be shaped from one piece or built up as above. After all pieces match up, sand the surfaces of the laminate to a perfect blend. I think the door molding should be bullnosed, but it's optional for the door jamb and casing (as they are called ashore).

Some will think, "There's got to be an easier way!" There is, but it doesn't have as much class. It's a compromise using 45-degree angles instead of rounded corners (Figure 11-32). I've seen this a few times and it's not all that bad. If you go this way, all of the molding can be dadoed out in one step. The pieces all meet at 22½ degrees, so all you have to do is add two guide pieces to the 45-degree jig above. Round all the corners slightly to protect your ribs, and rout corner radii.

SHELVES

Bookshelves are simple structures, and designers don't give them much attention. In a wooden hull, no one expects shelves to do anything but carry a little weight. In a fiberglass vessel, however, they

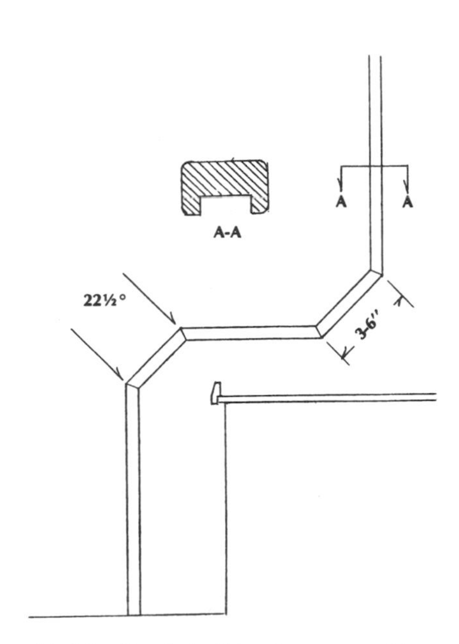

Figure 11-32. *Simple bulkhead corner moldings.*

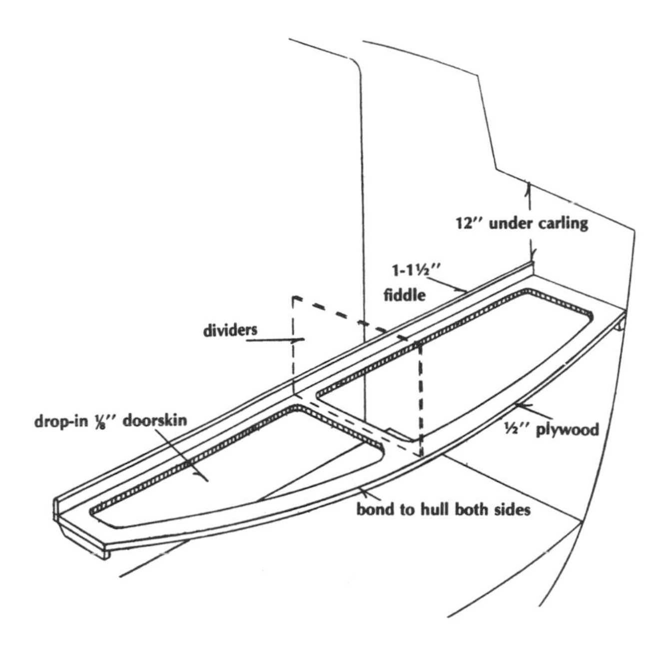

Figure 11-33. *Lightweight bookshelf.*

sometimes act as light but important structural members. For example, they form a T beam when bonded to the hull. Thus, they must fit well and then be glassed in with cloth tape over a fillet of Cab-O-Sil or pulverized limestone. This applies to berth tops against the hull as well. Weight, too, is important. Bruce Bingham reworked all the shelves in his manufactured *Flicka* with a sabersaw. He left flanges all around 1½ inches wide along the hull and 1 inch at the front (Figure 11-33). Then he dropped in scraps of plastic laminate (I would use doorskin). This let him throw out many pounds of plywood trash. If too many dividers interfere with this job in your older yacht, do it the hard way, on your back.

Plan your shelving so you have at least one good length with more than 12 inches of vertical space. Many marine books are being published in larger sizes. Depth, too, must be adequate. Finally, fiddles that are too high make it awkward to get books in and out. I would go for 1-inch-high fiddles and a removable batten set in chocks about 6 to 8 inches above the shelf. A good trick is to stretch shockcord between the ends or dividers. Shelves are frequently enclosed. See the medicine cabinet in Chapter Twelve for an example.

COMPANIONWAY LADDERS

Ladders are great space wasters. Thus, Figure 11-34C shows an attempt to make some small use of one. Likewise, we can't afford to make ladders comfortable to use, so my treads are much shallower than the standard, 6 to 8 inches. You can't use much more than the ball of your foot going up and your heel going down. The treads and rails (sides) should be full 1-inch stock. The rise must be 10 inches or more. Nonskid material must be cemented to the treads, bent under the front bullnose, and nailed underneath. All of this applies if space is at a premium. If your yacht is larger, the companionway ladder can be a stair, even a winding staircase.

Ladder A is straightforward. It has 1-inch treads dadoed ¼ inch into 1-inch rails. Bullnose the treads to ½-inch radius with a router or shaper. The plywood on the sole is one way to keep the ladder from slipping. It's a U-shaped chock that the foot of the ladder fits into. A wing bolt tapped into a plate holds the head. Ladder B is curved to clear an engine. Notch the foot of this one to fit over a block or batten. Ladder C has a back of ¼-inch plywood and little compartments under the treads. These are

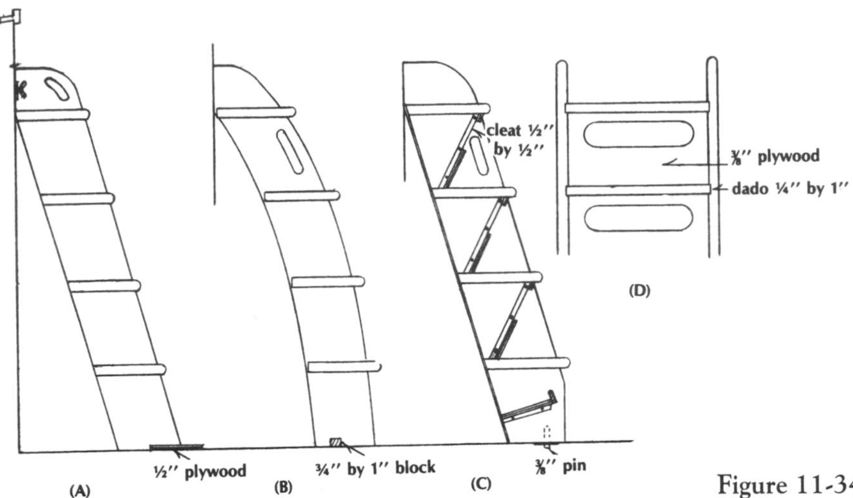

Figure 11-34. *Companionway ladders.*

accessible through the apertures shown or through the back. Small as they may be, little compartments like these are valuable on a cruising yacht, for light tools, twine, wire, a first-aid kit, an extra winch handle—even a few drops of schnapps. The shelf under the bottom step is easier to build. Of oiled teak or mahogany filled and stained, and with five or six coats of varnish, any of these ladders would be a handsome piece of furniture (Figure 11-35).

ICE CHESTS

The costliest refrigerator is only as good as its insulation. This technology is changing so rapidly that my knowledge may be obsolete. However, here's an ice chest that will preserve food for two weeks, less if you open it every hour for a drink. Boxes with side-opening doors should be avoided, as they spill out precious cold air at each opening. As for foam insulation, I suggest you check with your designer and local supplier for the latest information. Buy the best and the thickest one that you can crowd in.

Allegra's ice chest is too deep for total convenience—about 24 inches. It's 20 inches fore and aft by 17 inches across, inside dimensions (Figures 11-36 and 11-37). To determine your size, deduct about 6 inches from each dimension of space available. This is ample allowance (3 inches) for insulation. There are a number of ways to build this

Figure 11-35. *Skilled amateur Steve Soltysik's beautiful and strong companionway ladder is laminated pine and mahogany scrap. Note exquisite desk and bulkhead molded trim.*

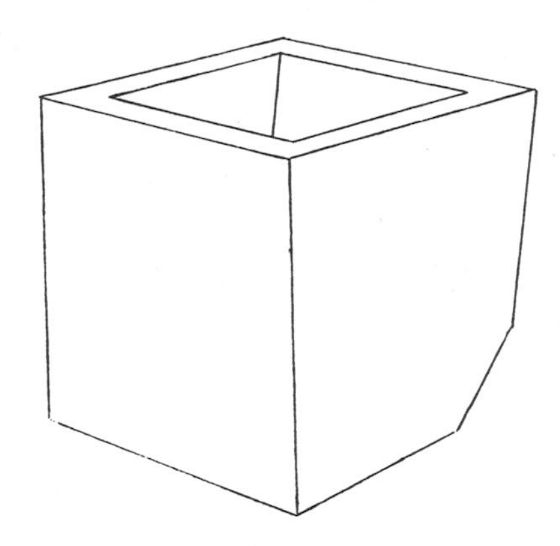

Figure 11-36. *Ice chest shaped to hull.*

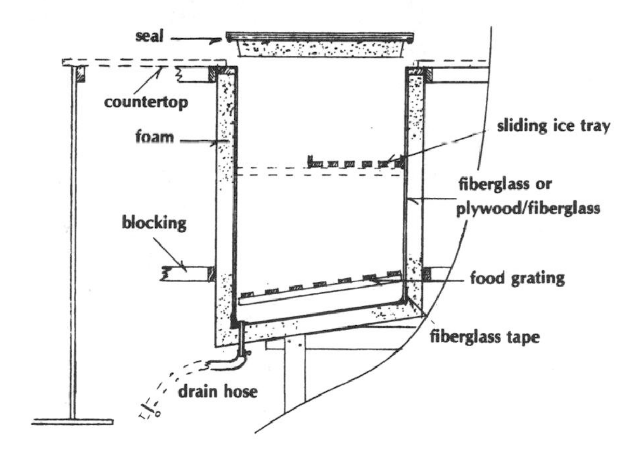

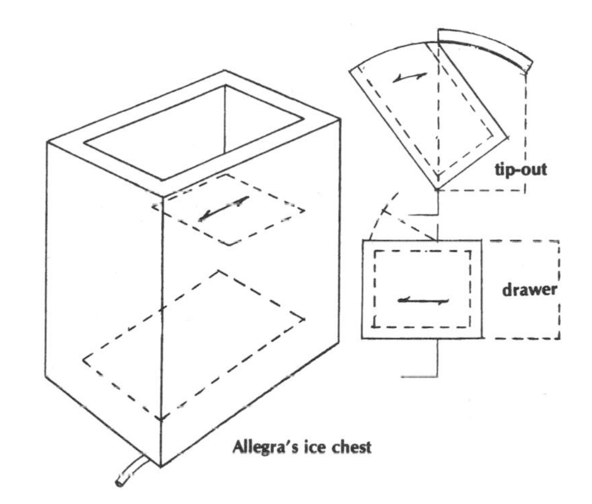

Figure 11-37. *Ice chest ideas.*

box. It can be a plywood shell, lined with fiberglass and covered with foam. It can be a foam shell, lined with fiberglass. Or you can start with a plywood *male* mold and lay up a fiberglass shell. Foam on the shell completes this last job. Details of construction are not the province of this book.

CEILING

Ceiling adds a lot of class to any yacht, and it's always good insulation. It helps reduce the sweating common in fiberglass hulls, and it makes an attractive traditional covering for insulating foam or other materials.

Wooden hulls pose no problem, as ceiling may be screwed lightly to the frames. There's precious little to screw into in a fiberglass hull, however, unless it is well over ½ inch thick. I'm haunted by a vision of hundreds of lead holes going right on through, of little points sticking out all over like five o'clock shadow. Let's see what can be done to avoid such problems. The drawing in Figure 11-38 is supposed to represent almost vertical battens sprung inside a hull. They lie naturally without forcing. The battens can be spruce, pine, fir, or strips of exterior plywood about ½ inch by 1½ inches, thicker for heavier insulation material. To get them to bend without cracking, kerf them about two-thirds of the thickness, spacing the cuts closer for sharper bends.

To me, it's best to put in these strips with contact cement, plus an occasional self-tapping screw if your hull is heavy. Read the label on the cement can and follow instructions rigidly. Locate the battens about 12 inches apart. If you want to have foam insulation between the battens, rip the foam sheets into convenient widths and lengths, say, 12 by 30 inches. If these refuse to press to the hull slash or dart where needed. Coat the foam and the hull with cement and allow it to set up. In about 10 minutes, when the cement is set enough to permit paper to just cling to it, but still pull away easily, cover the area with newspaper. Locate the foam sheet with one hand over the newspaper, raise the sheet at one end, and pull out part of the paper. Press the foam onto that area, lift the other end, pull out the rest of the paper, and let the foam fall back onto the cement. This stuff grabs instantly. There can be no second tries.

Go all over the foam with a small block and a hammer. Your next batten should locate against the foam you just cemented, and so on. I think

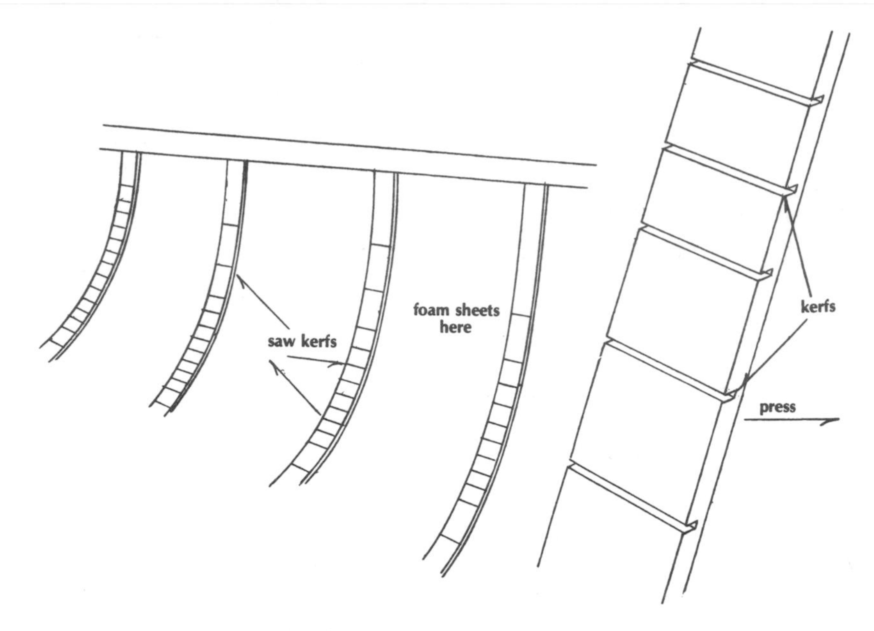

Figure 11-38. *Installing insulation in hull.*

½-inch foam should rectify any sweating problem. If, on the other hand, you want to live aboard through a Maine winter, you'll need heavier foam, and the battens, too, will have to be thicker, perhaps ¾ inch. My old friend Steve Soltysik ceiled his *Talofa* with ¾-inch battens of exterior fir ply, saw-kerfed for the bends. He screwed these to the ¾-inch hull with self-tapping stainless steel screws. Then he placed fiberglass batts, insulation you can buy at any building supply store, between the frames. He tucked them in with the foil side against the hull. No cement was needed. His ceiling is Philippine mahogany, 7⁄16 inch by 2½ inches, fastened with oval-head brass screws. Each strip is rounded off slightly.

Spacing the ceiling looks good, but this probably reduces the insulation value of tight ceiling alone. The back of the ceiling should be sealed with Firzite, Rez, varnish, paint, or best of all, epoxy. Installation could be a two-man job. Use a couple of spacing blocks of ⅛- or ¼-inch scrap. Keep moving these along, following the sheer as you drive the screws.

Mastic with fiberglass batts stuck to it might be a good system for metal hulls, or for any hull material but wood. I don't know how you would fasten ceiling to a metal hull, unless you bolted wood cleats to the sides of the frames. Hulls built frameless with full-length stringers suggest clips needed for wooden verticals.

I have heard about insulating with two thicknesses of indoor-outdoor carpeting epoxy-glued with ceiling over. I once repaired a 24-foot production sailboat lined entirely with one layer of nylon loop carpeting. The pieces seemed to conform easily to the hull with little cementing. Apparently the panels had been a rather snug fit, and the friction alone was almost enough to hold them. I must say that the carpeting looked very neat, but it wasn't seamanlike. I'm sure, however, that it insulated somewhat and muted the noisiness of her hull.

Ceiling a hull is a lot of work. Materials are costly and you lose some space. Again, however, you just might add thousands to the value of your yacht. The next chapter deals with a few ideas that will add to your comfort and convenience and enhance the desirability of the vessel should it ever be placed on the market.

12 Niceties and Necessities Below Decks

To many, a toilet in an enclosed compartment is a necessity, not a nicety. This is especially the case when cruising with another couple or with children. Unfortunately, however, working such a facility into a small cruising sailboat sometimes forces the designer to use some of the most desirable space in the yacht, at or near amidships. Thus, the enclosed head chops the accommodations into two parts, with two berths far forward and two far aft. As a result, the space needed to seat six or eight people is lost. The answer? An unobstructed cabin plan with a hidden marine pump toilet, the popular Porta-Potti, or even Herreshoff's wooden bucket. Hiding this can be accomplished in several ways that I will describe later. Let's talk about "the head" first.

THE ENCLOSED HEAD

The head enclosure must be at least 24 inches between bulkheads. The minimum sole (standing room) athwartships is about 15 inches, raised above the normal sole to compensate for the angle of the hull (Figure 12-1). It may be necessary to raise the portable or fixed stool on another floor 4 to 6 inches higher to move it farther outboard. The broken lines try to indicate the hull taper between bulkheads. This means that one's feet may dangle or straddle the stool; it's a bit like riding a wide bicycle seat. If you locate it farther outboard and up, you're too close to the carling and side deck (minimum 36 inches clearance). Extending the bulkheads farther toward the centerline solves the problem, but what does that do to the main sole and ease of passage? Design problems, not joinery. But which comes first, chicken or . . . ? In the drawing of *Allegra's* cubicle I show padding on the structures over and behind the pot. In my opinion, all crowded heads should be sitters, regardless, for sanitation.

A head under a seat between V-berths is well-known in the small-boat world, but a dreaded expedient. A true offshore sailyacht might place it in the forepeak, even if opposite a pipe berth or built-in. An open plan shown in Figure 12-2 (my Travlr 27.8 Plan 3) has the head in an area least conducive to lounging, a step above the sole, 5½ feet aft of the first chain locker bulkhead, with space for toilet and basin, general stowage, shelves, bins, etc., forward. Move the pot farther aft to get bowing-headroom under the cabin top, put the basin under the locker door. Note the curtain, dresser, open hanging locker and shelves and bins. Chain slides under the stowage bin. Hint: install shock-cord or ties to keep swaying clothing from being worn away to buttons. Joinery? Well hardly.

WASH BASINS

In toilet rooms as small as those discussed here, a wash basin is often dispensed with. In a beamier yacht, a small enclosed vanity can be mounted outboard of the head, perhaps with a small linen cabinet beneath. You could install a stainless steel basin with pump or pressure water, draining into a holding tank, encouraging waste. But all of this is unnecessary if you just use the galley sink. If you must have privacy, it would be simple to rig up a shelf covered with plastic laminate, hinged on a bulkhead or against the hull. Saw a hole in the shelf for a plastic basin. This eliminates plumbing, holding tanks, and holes in the hull. Or fit a cover in the basin and hold it all together with small brass turnbuttons. Or simply stow the basin when it's not in use, and use the cutout to close the hole flush with the top.

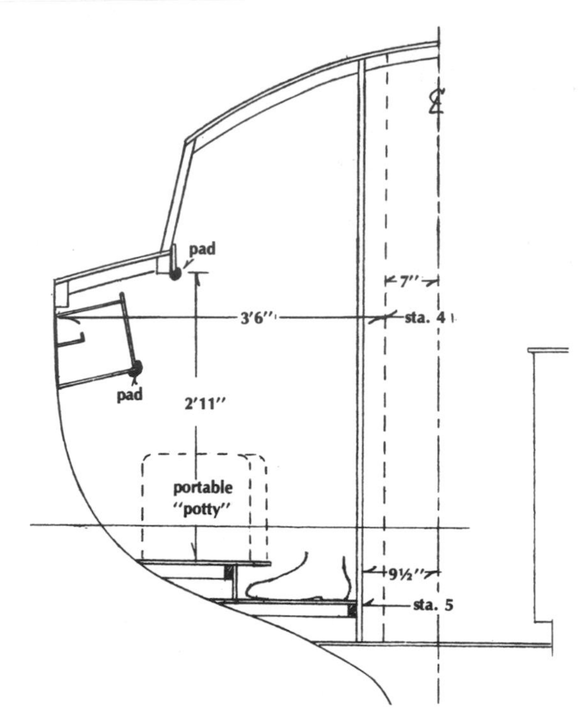

Figure 12-1. *Allegra's "cubicle."*

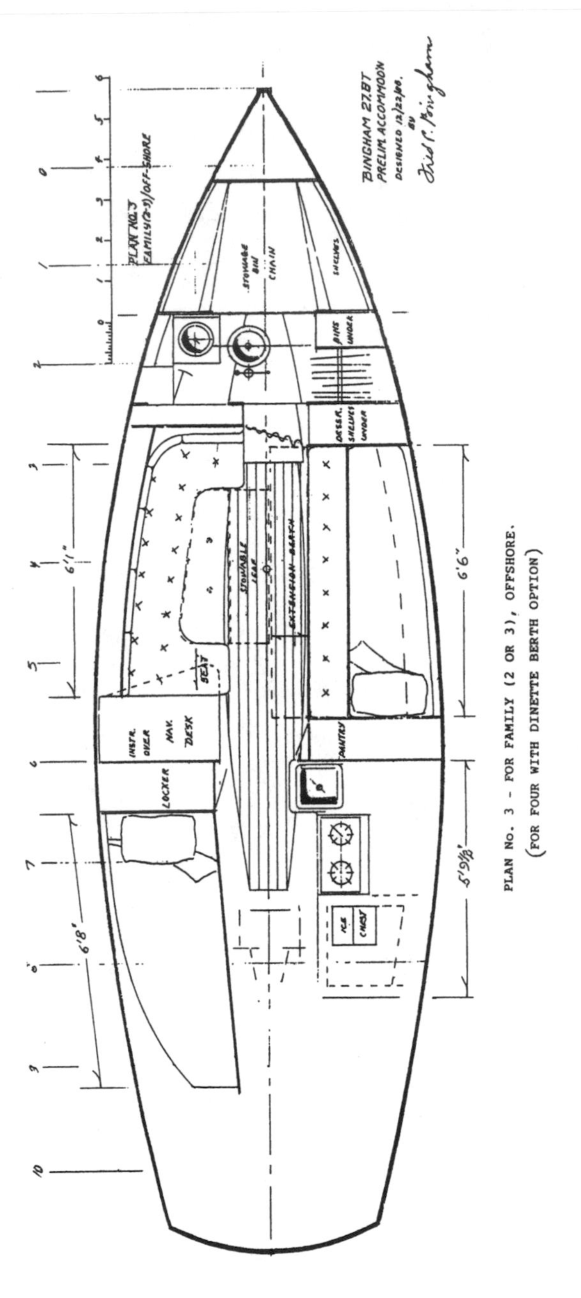

Figure 12-2. *My Travlr 27.8 has the head in an area least conducive to lounging, a step above the sole, 5½ feet aft of the first chain locker bulkhead, with space for toilet and basin, general stowage, shelves, bins, etc., forward.*

A Folding Basin

At one time you could find folding wash basins in lovely hardwood cabinets, ready to be hooked up to the plumbing (see Figure 12-3). With a little patience, you might build this one in marine mahogany or teak plywood. The cabinet could be ⅜- or ½-inch plywood glued and finish-nailed together. No other dimensions are suggested, as much depends on the space available, the type of basin, and so on. For pressure water, a valve could be inserted in the small brass pipe. The 90-degree swinging bent pipe or L directs the flow into the basin; the basin tilts to empty and latches closed.

The interior of the cabinet can be coated liberally with epoxy resin, fiberglassed, or both. The stop block allows the top to recess so it rests vertically. Be careful in locating the bolts or screws that pivot the cover. A brass chain supports the cover when it's open. If you have a marine head discharging overboard, you can lead the drain tube to it. Flatten the end of the tube so it fits under the seat. Otherwise, lead the drain tube to a holding tank or a large jug in the bilge or behind the head.

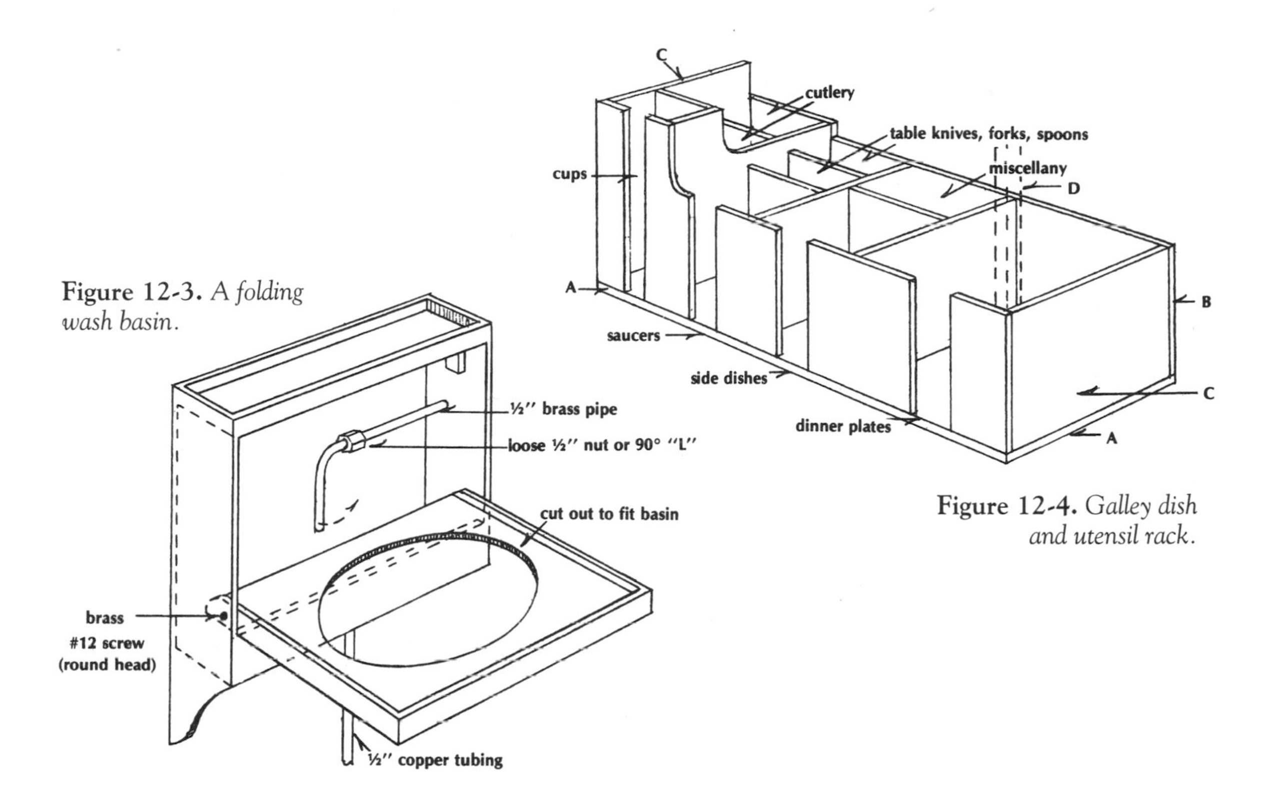

Figure 12-3. *A folding wash basin.*

Figure 12-4. *Galley dish and utensil rack.*

DISH RACK IDEAS

The china and utensil rack shown in Figure 12-4 can be built with ease if all parts are cut with precision from ⅜- to ½-inch plywood. Check the vertical space available. Allow for the rack to be at least 6 to 8 inches *above* the galley surface so there is minimum encroachment on that space. Be sure, however, that you can still get cutlery into its slot. First mark the divisions on the base board (A). Transfer the athwartships divisions to the back piece (B). Saw all dividers and ends (C) to exactly identical lengths with allowance for the front panels. Saw to their various heights. Measure fore-and-aft spacing—allow ¼ inch total clearance around the china—and mark square. Allow cup handles to protrude.

Now drive brads along lines through the base board so their points show. Pull these and drive from the underside. Glue and nail back, ends, and all dividers with Aerolite, T88, Gougeon Brothers, or other epoxy glue. Lay out the front panels of mahogany or teak plywood so slots are about 2 inches wide with corners and edges nicely rounded (not shown). Nail and glue from the underside and front into dividers. The exposed edge of the base (A) may be covered with a light molding. Rest the rack on cleats on each bulkhead or hang from the overhead on posts (D).

TABLES

The problem with tables is that they are very difficult to stow when not in use. Earlier I showed *Allegra's* backrest/table, an expedient. If your vessel is large enough to allow a fixed table, there's no problem—unless you like to strip her for racing. The drop-leaf table shown in Figure 12-5 and the tubular-legged one shown in Figure 12-6 are awkward to stow, but they can be left ashore. A table mounted on the bulkhead is almost ideal if the bulkhead is wide enough (Figure 12-9).

The drop-leaf table is quite straightforward. It could be improved if you joined with matching drop-leaf molded edges. A good piano hinge will work well. The swinging supports for the leaves might interfere with seating. One in the center, however, might do the job. The trough marked "storage" would be great for books, bottles, condiments, and so on. Incidentally, the base can be much shorter than the top, perhaps 10 inches at

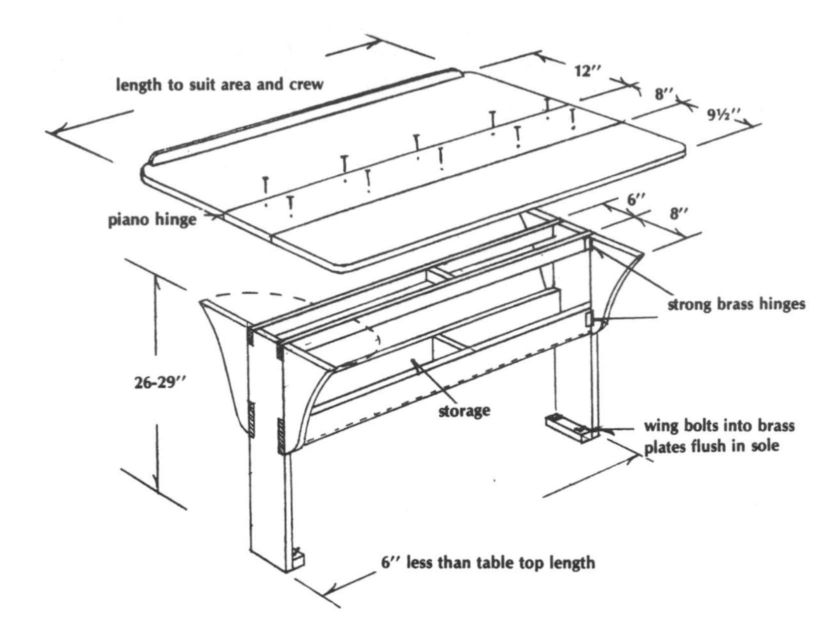

Figure 12-5. *Drop-leaf table.*

Figure 12-6. *Jim Burks's folding table is a fine production job. It is removable in moments. The tubular legs could be made of solid teak by an ambitious amateur builder. And the fiddles could be removable for harbor dining.*

each end. I have not indicated fiddles throughout the table drawings, but there definitely should be high fiddles around the center section and 1 inch on outer edges.

The base legs should be of 1-inch solid mahogany or teak. The top may be hardwood plywood with rounded edge trim or solid wood. Later I'll describe how such tables should be secured to the sole.

The tubular-legged table in Figure 12-6 is from *Prime Time*, Jim Burks's IOR yacht, where weight is a consideration. It straddles and is clamped to the mast. The whole thing can be removed in five minutes. The same design could use solid wood legs, of course. Steve's beautiful creation in Figure 12-7 is a permanent piece of furniture, but fiddles are too high.

The gimbaled table shown in Figure 12-8 looks rather spartan. With nicely rounded corners and gracefully shaped legs, it would be quite handsome. The pendulum trough marked "weighted" can be loaded with lead pigs or cement covered with ¼-inch plywood. If the latter, the trough can be used for storage. The pivot brackets may be of oak (two pieces each), bronze castings, or brass brazings bored to take ⅜- or ½-inch bronze bolts. They must be stout. A pin locks the table at normal or other

Figure 12-7. *Steve Soltysik's huge table has a large drop leaf. Lift panels cover tableware storage. Note the precise fits and excellent fiddles.*

angles of heel. There must be generous fiddles all around. The drop leaves can be supported by sliding cleats. Again, piano hinges would be satisfactory.

The table folding up against a bulkhead shown in Figure 12-9 has possibilities. Try to find a long aircraft hinge with a removable wire-like pin. With this at the bulkhead end, the table can be removed. The hinge in the center joint is under great tension, so it should be installed flat on the underside of the 3- to 4-inch oak cleats. The two table halves must be closed up snug when the hinge is screwed down or the table will sag. As I show in the drawing, there could be another off-center leg hinged near the cleats, to lie alongside the leg shown. Without it, the fiddles take quite a strain, so they must be well glued and screwed.

If your table tops are to be built up of solid hardwood, form joints as in Chapter Seven. Table tops should be about 11 to 12 inches above the seat cushion, one of the disadvantages of a gimbaled table.

Because of the many stresses, a table must be fastened down securely. You can't use less than 3/4-inch plywood for the cabin sole. Locate the legs adjacent to a sole beam, if possible. Figure 12-10 shows an oak cleat or pad under the sole, counterbored to hold the nuts. The best way is to bore down through the foot cleat, the sole, and the temporarily located oak cleat simultaneously. Just let the bit point come through, then remove the cleat and counterbore for the nut. Wax the bolts thoroughly, then run the bolts into the nuts. Daub each counterbore with thick epoxy glue. Back off the bolts a bit so the threads have no glue, or take them out for cleaning. When the glue has set, tighten the bolts. Although these nuts will never turn, you can do a nice job with a tapped brass plate let in flush or on the underside of the sole. This must be through-bolted, of course. You can also use brass angles instead of wood leg foot cleats.

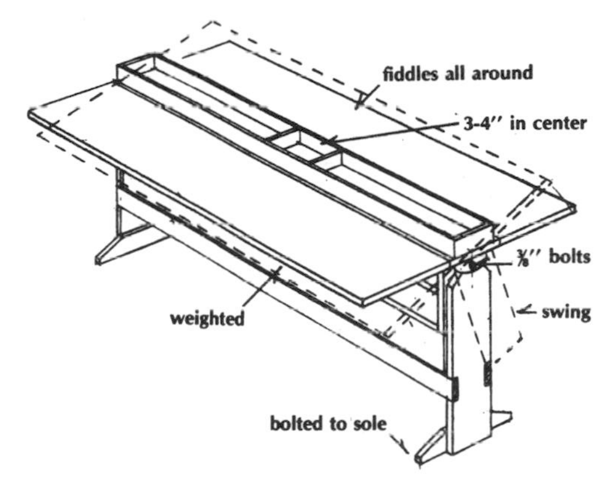

Figure 12-8. *Plain gimbaled table.*

A MEDICINE CABINET

To fit a cabinet between bulkheads—say, a galley rack or medicine cabinet—you must check

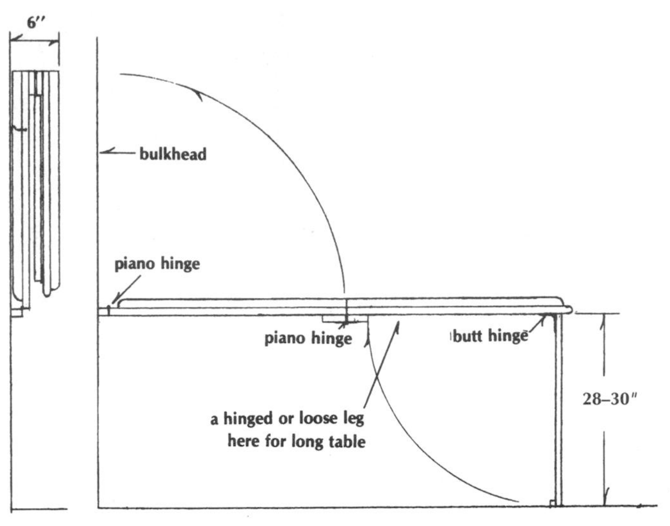

Figure 12-9. *Bruce Bingham's folding table.*

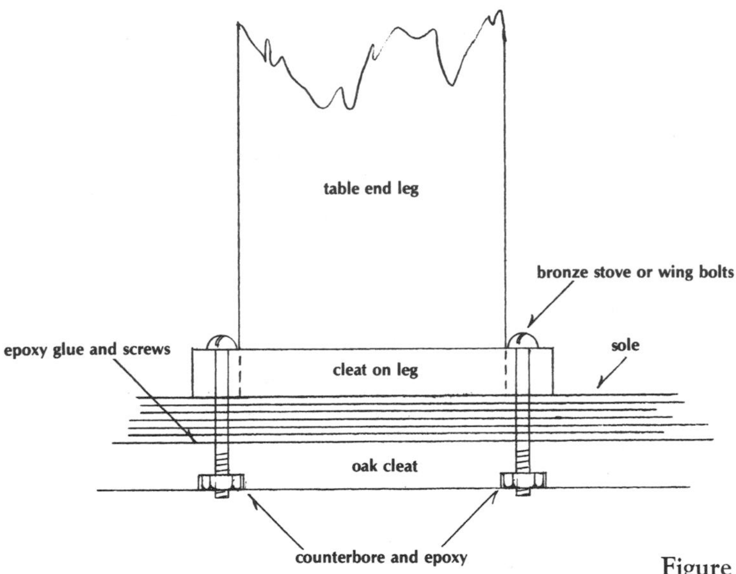

Figure 12-10. *Removable table leg.*

measurements with care. If the bulkheads are already trimmed with overhanging moldings, you may have to spring them apart to slip the structure in. If the cabinet must fit the shape of the hull, however, you have a different problem. The cabinet front must be longer, as it is at an angle from bulkhead to bulkhead. *Allegra*'s stations are 25 inches center to center. The head enclosure is designed to be 27 inches in between. Thus, this medicine cabinet's front (Figure 12-11) is 28 inches, longer than the box itself. Only a little trimming at the low angle is needed to make a nice joint against the bulkheads. I suggest fastening to ¾- by 1-inch cleats set back at about 70 degrees so things can't dive out when the vessel is heeling. The cabinet should be far enough beneath the deck so its top can be used for small articles.

The front panel is ⅜- or ½-inch mahogany or teak plywood. Mark on the inside and saw so all splintering occurs on the inside. Then trim with a

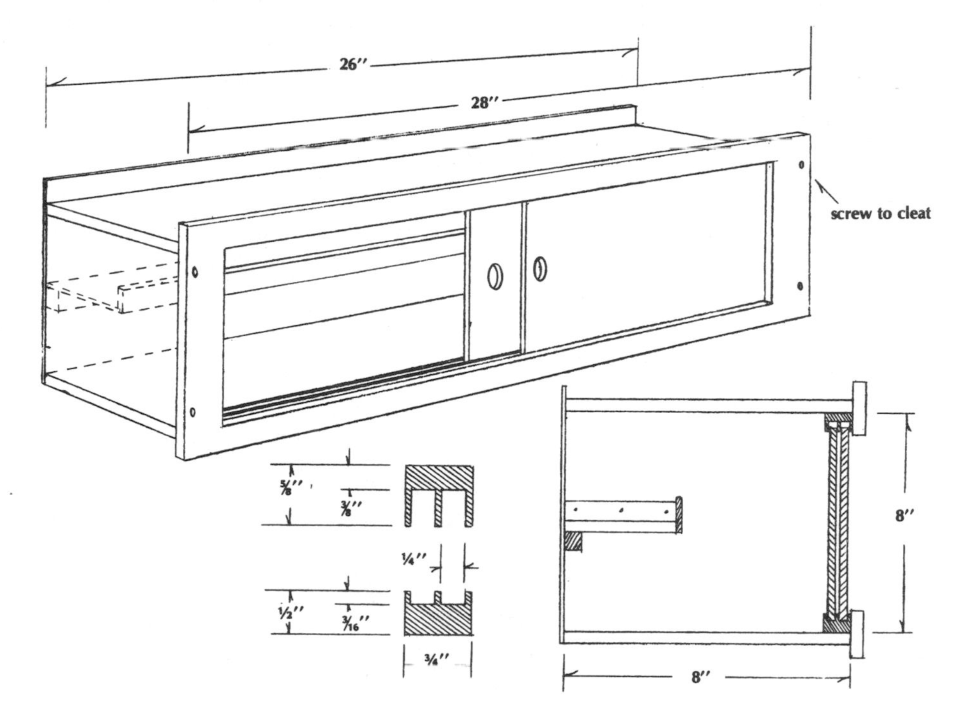

Figure 12-11. *Allegra's medicine chest.*

fine 1/8-inch edging of hardwood (not shown). The ends are 1/2 by 8 by 8 inches, or dimensions you choose. Brad and glue the small shelf cleats. Trim the shelf with a hardwood fiddle and fit it to a cleat on the doorskin back panel. Lay aside. The door tracks are quite delicate. They are easier dadoed, but repeated saw cuts work. Use hold-downs on your table saw for accuracy. Glue and brad these to the bottom and top, then assemble the ends, top, and bottom. Install the back, clamping and gluing the shelf to the side cleats previously fastened (you won't be able to drive brads). Finally, fasten the front to the box with glue and finish nails set in.

The sliding doors can be 1/4-inch hardwood plywood, glass, or even mirrors. Carefully make the doors 1/8 inch higher than the distance from the bottom of the lower groove to the upper track. Paraffin rubbed on the edges of the doors makes them slide nicely. To install each door, pass it *into* the cabinet, insert the top edge well up into the deeper groove, then drop the lower edge into its groove. A towel rack could be screwed to the bottom of the cabinet.

FIDDLES

A fiddle is a necessity that can be a decorative nicety (Figure 12-12). Style A is strong enough to keep pots from sliding off the galley top. It looks best with the ends slanted back about 60 to 70 degrees, or ogeed, leaving the corners open for cleaning. Style B is popular with a graceful ogee in the ends and well rounded all over. To make it removable, insert 5/16-inch brass pins or dowels (see Chapter Seven). It's advisable to have removable fiddles on the galley counter at several points to keep everything in its place. The front one, too, might be made removable if you use the countertop as a chart table. Style C is usually seen on shelves. The dowel holes can be bored accurately if you clamp upper and lower pieces together and bore simultaneously, bottom up, being careful that the bit point does not emerge. Rout 1/4-inch radii as shown. To install, glue and finish nail the lower rail in place, then glue and clamp the dowels and top rail. Marvelous turned posts are available at lumber yards.

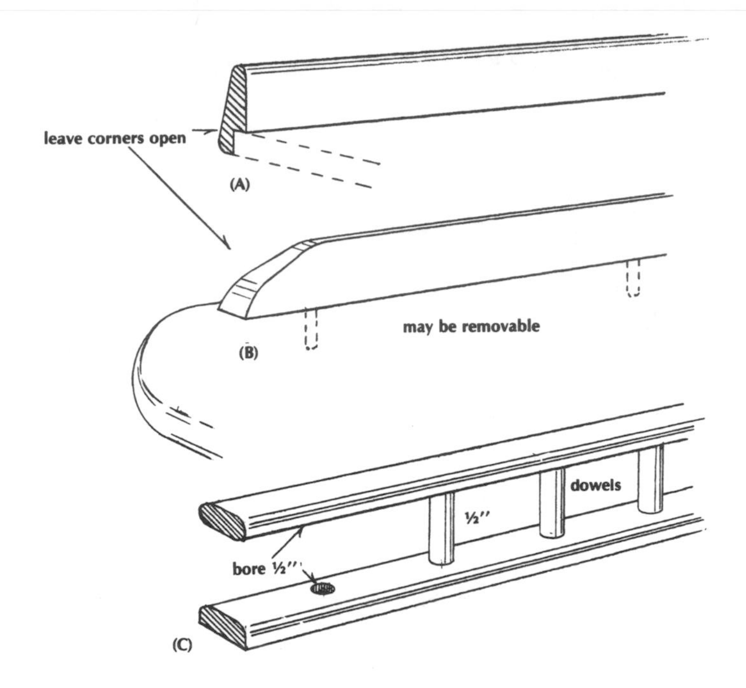

Figure 12-12. *Some fiddle styles.*

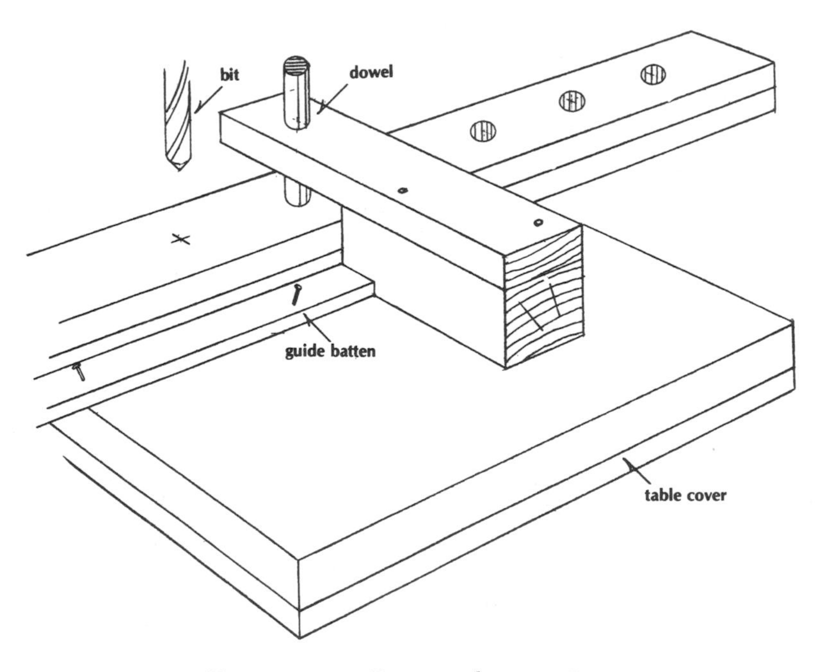

Figure 12-13. *Boring and spacing jig.*

Figure 12-14. *Mike Lafayurie's unfinished stove cover of exotic woods. This is a durable, easily cleaned, and handsome piece of work.*

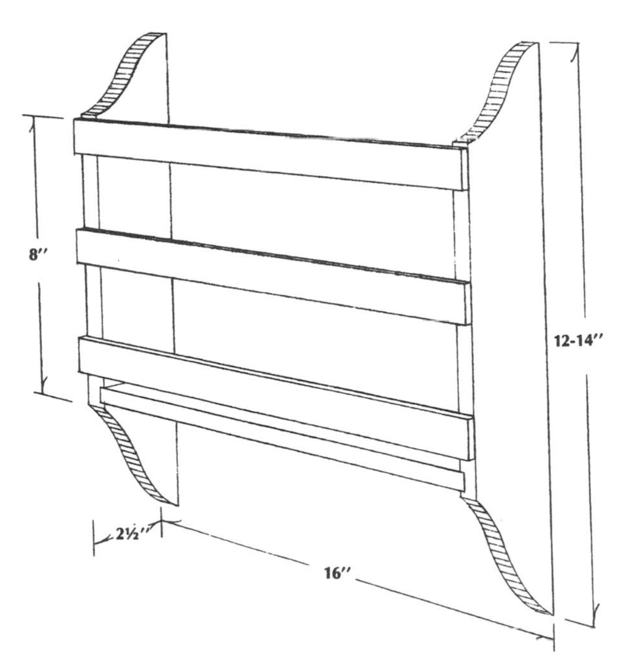

Figure 12-15. *Typical magazine rack.*

If you want precise spacing on many holes, build the boring jig shown in Figure 12-13. The dowel must be loose enough for easy insertion into each hole as it is bored. Set the drill press depth stop accurately. Use a Forstner bit rather than a twist drill bit. Your drill-press table should have a wooden cover for all woodwork.

STOVE COVER

Galley space is at such a premium it's necessary to make double use of it. Figure 12-14 shows construction of a handsome cover for a four-burner range. The strips are from ¾ inch to 1⅛ inches in width by ¾ inch thick. Zebrawood, walnut, Honduras mahogany, teak, and maple are alternated. The glue can be any crystal-clear epoxy. Three pipe clamps were used to prevent distortion.

Figure 12-16. *Finished bulkhead racks.*

The cover was dressed down with a belt sander. If a cutting board, it should be left bare and oiled periodically with mineral oil. Otherwise, four or five coats of polyurethane varnish make a beautiful and durable finish. The cover can be trimmed with light hardwood molding. A complete galley counter made this way would be an interesting and rewarding project.

BULKHEAD RACKS

The racks shown in Figures 12-15 and 12-16 can be made from scraps of teak or mahogany by resawing full 1-inch stock, then planing down to about 3/8 inch. The slats can be resawed from lighter material to about 3/16 inch or 1/4 inch by 1 to 1 1/4 inches. I laid out the two ogees on a 1/4-inch plywood pattern for router or shaper. From this I easily made matching racks for magazines, navigation books, instruments, first-aid kit, binoculars, and so on. Note that the shelf piece is dadoed in, and that a slot is left for cleaning. This lets you see pencils, but they can't fall out. Vary the dimensions and add vertical dividers to suit your needs. To mount, drill through the thin end of the ogees and drive roundhead screws at both top and bottom. I rubbed in a finish of varnish thinned with turpentine. If you use teak, rub in teak oil.

That's it for interior furnishings. Now let's go up on deck.

13 Things on Deck

It isn't possible in this space to discuss everything on deck, so what follows is a mixture of major projects and simple jobs. Let's start with the most involved task on deck, the main hatch.

COMPANIONWAY HATCHES

There must be as many varieties of hatch construction as there are builders. As I sketched the four shown in Figure 13-1, four others came to mind. I have yet to see the perfect hatch. This I define as one that never leaks, that you can walk on, is beautiful, requires no maintenance, and is burglar-proof. I should add that the perfect hatch also should be easy to build. With regard to leaks, the only sure way to keep a hatch from leaking under severe conditions is to leave it ashore. Wind-driven rain and spray invariably find the narrowest crevice, usually directly over the chef's neck. The forward end where the hatch beam meets the coaming is particularly vulnerable. I'm happy to see that this problem is solved on some modern boats with a cover husky enough to be walked on and to carry the mainsheet track as well (Figure 13-2). Let's look at the four hatch covers in Figure 13-1 before we get into hatch runners, which also are called slide logs and coamings.

Hatch A has a traditional feel, perhaps because it is straightforward, strong, and not too difficult to build. A rabbeted corner is used more often than the box or finger joint shown, but the latter looks good and would be very strong, especially if you put a dowel down through it. I do not show a separate side half round or other molding, as the top is supposed to extend past the runner and is nicely rounded to match the ends. End moldings often require steaming to shape. An external beam with molded hand grip is a distinctive treatment (Figure 13-2). Construction may be tongue and groove, spline, laminated veneers, or triple plywood. A brass or stainless steel 1⁄16-inch strip running in a wide saw kerf has proven excellent for keeping water out (Figure 13-1A); an offset rachet screwdriver is the only way, the hatch being now forward of the coaming. All hatch runners must be spiled carefully to the slight sag of the top so the metal does not bind in the kerf. The unseen beam in the forward end must be formed to just clear the cabintop. Of course, for looks, the after beam lower edge may be cambered to match the drop boards and the hatch top.

Style B is handsome and may be removed easily because the screws are accessible, but there is the tricky matter of fitting the split tube under the strip into a shallow rabbet. I suggest routing or shaping the two outer pieces of the top to fit the tube, before jointing and grooving for splines, then make up the other splined pieces in the top. Careful width measurements are needed; install the two outer pieces to fit to the strips. Tap the last two splines in (with glue) after all the pieces are screwed down and plugged. This top would have to be not less than 3⁄4 inch thick. The tubing and strip thicknesses are exaggerated in 13-1B; 1⁄16-inch brass should do for the tube, 1⁄8 inch for the strip. More follows this.

Style C is heavy, allowing a large radius on four sides (see Figure 13-3 also). Its Lexan top could be fitted into any hatch heavy enough to take a 1⁄8-inch rabbet. Holes for screws must be oversized to allow for expansion and contraction of the plastic. Because of this movement the panel must be bedded in a flexible compound or rubber tape. The brass runner strip with the crimped edge could be quite difficult for an amateur metalworker. Most

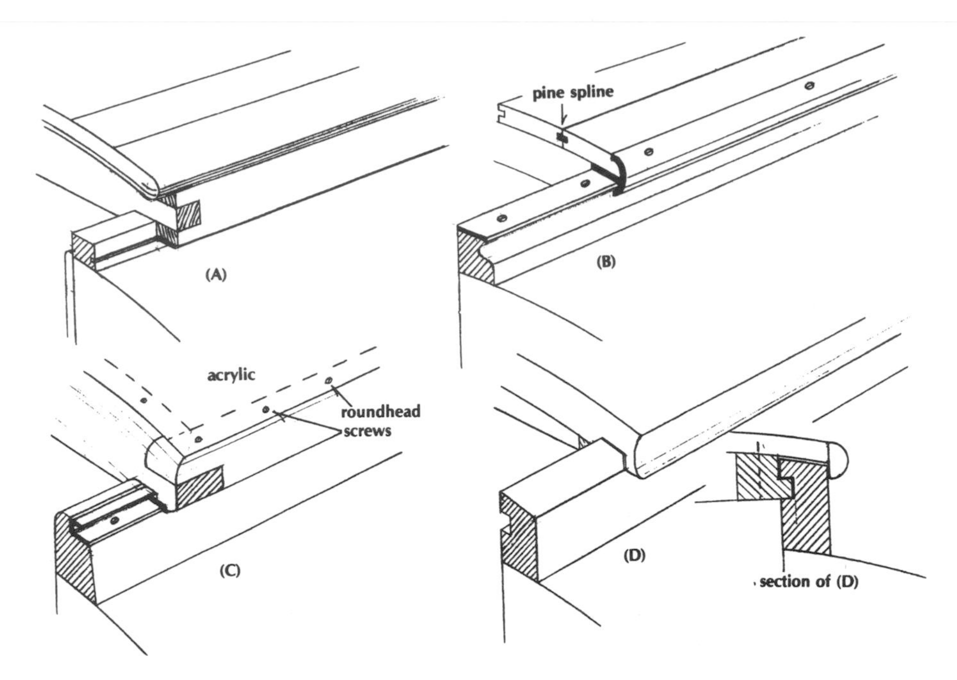

Figure 13-1. *Sliding hatches.*

Figure 13-2. *This hatch cover prevents leaks, avoids damage, and incorporates an efficient traveler. Note the external hatch beam molded into a hand grip.*

sheet-metal shops could form it of 1/16-inch brass or stainless steel in a bending brake. The brass that rides in the crimp should be full length. This hatch, too, can be installed easily, as the screws are accessible. To foil thieves, fill screw slots with epoxy resin paste.

Style D needs no metalwork. The hatch framing must be heavy enough to permit formation of the 3/8- by 5/8-inch projection. Don't use oak, however, for it does not take glue well. The beams are shown rabbeted into the sides. These could be half lapped, box jointed, or even mortised if set back a bit from the end. Also, you're not limited to half-round trim. A rectangular shape rounded over and covering the end joint would be acceptable.

HOW TO SPLIT A BRASS TUBE

Tubing has a tendency to rotate as it is being sliced. The jig in Figure 13-4 is intended to prevent this. Pick a piece of plywood (A) large enough to clamp to your bandsaw. Rip it for half its length. From the split, mark off the outside radius of the tube to locate the fence (B). Clamp this assembly to the table. Carefully saw the tubing about 3/4 inch deep and check for centering. Drive a small nail (C) through block D, insert its point in the saw cut, and nail the block to the fence. The nail must be free in the kerf but not sloppy. It's best to use a new metal-cutting blade, for wandering may occur if the set of the teeth is off. Blade

Figure 13-3. *The large radius on the hatch requires heavy material and looks great. The acrylic panel should be protected with hardwood strips.*

tension must be maximum. Do not rush. Belt-sand the rough edges after cutting.

HATCH RUNNERS, COAMINGS, OR SLIDE LOGS

The timbers on which the hatch slides must be substantial, although some think it looks sharp if both hatch and runners are very low. It is customary also to pitch down slightly toward the forward end. Height depends entirely on style, as shown in Figure 13-1. The runner need not be more than 2 inches high by 1½ to 1¾ inches thick. This should allow the hatch side member to clear the cabintop by a fraction of an inch when forward. The runner in B may be about the same dimensions. The groove may be simplified by shaping a simple rabbet and rounding off the lower corner. The tubing then rides against the brass strip alone, although I would prefer to see it ride the groove, too, for a better seal. Style C could be simplified but not improved by eliminating the rabbet entirely. Looks would suffer, and the crimp could be damaged without the protection of the rabbet. Note the side taper here. This is optional on all runners.

Style D shows a rather tall runner, more properly a coaming about 1¾ by 2½ to 3 inches. The hatch slide, however, is very low, so the overall effect is excellent. Note that this hatch would be strengthened greatly by doubling the *after* beam and also fitting the end of the doubler into the groove.

Runners are fastened with long screws into the carlings and beams. Avoid drilling into the slide grooves. The runners are usually about twice the hatch length, terminating in a graceful ogee or bullnose. It is vital that the runners fit the trunk top perfectly before fastening down, for the slightest distortion will cause your grooves to bind the brass strips. Spiling or scribing is necessary. My experience with three-ply molded trunk tops indicates almost no flexing (but tops are usually "swaybacked"). I would rather feel confident that my heavy hatch runners were contributing rigidity, too.

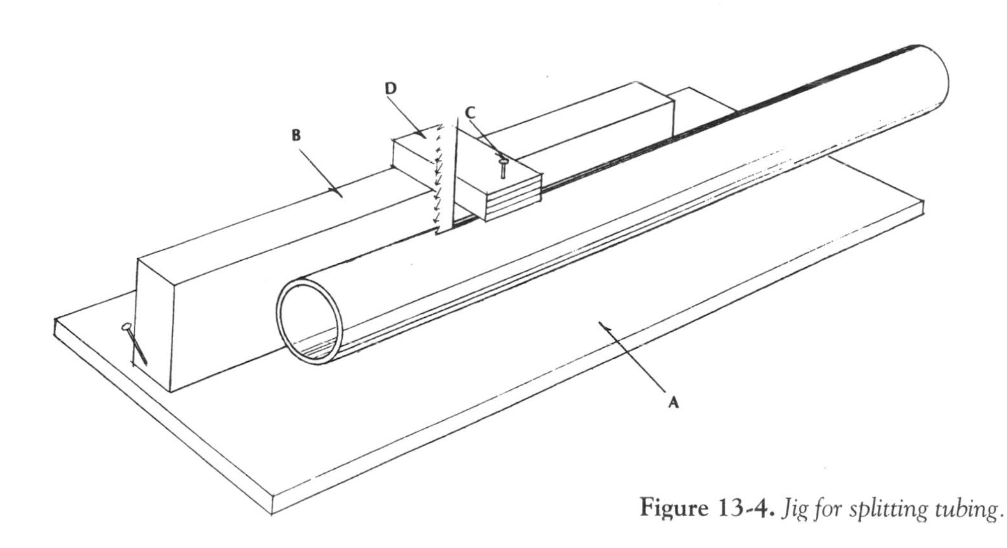

Figure 13-4. *Jig for splitting tubing.*

LEAKPROOFING THE FORWARD COAMINGS

When it's blowing hard and rain and spray are beating against the hatches, you'll pray that the forward coaming is doing its job. The simplest construction is a hatch beam that just clears the deck crown and presses against the forward (removable) coaming. You can put weatherstripping on the beam and on the coaming, but it may still leak. You can compromise by making the forward beam and the coaming shallow enough so they pass, but you need a removable strong stop covering the opening. One answer to this problem is a molded or laminated hatch cover, so you can do away with the after beam altogether, adding an external beam for strength insurance and for pulling the hatch closed. Your forward beam and replaceable coaming can be full depth. This should give you a weathertight hatch. But see below.

A DOUBLE-COAMING/DOUBLE-BEAM HATCH

The hatch coaming shown in Figure 13-5 is one that I guarantee to be leakproof. It's a bit more work, however. The heart of this hatch is the double coaming matching double beams in the hatch cover (shown before the top is on). As you can see, water can be blasted through the joint, but it will run out through the second scupper. The stop piece is screwed on. This allows removal of the hatch easily, if necessary. Or remove the brass strip. If you make the side facings ⅞ to 1 inch thick, they may be well rounded to blend into the hatch top.

That's enough on sliding hatches, even though many questions are left unanswered. When you build a hatch like the one shown in Figure 13-6, you'll have found the answers.

DROP BOARDS

Look at several yachts for variations in drop boards or slides. Vents are needed to keep your vessel dry and sweet smelling, the vents draining aft so water cannot creep in. Holes bored canted in interesting patterns and slots are very common. Most boards are rabbeted into each other and into the sill, or beveled to keep water out. Figure 13-7 shows a hatch cover overlapping the boards about 1½ inches. The drawing shows pieces let into the ends to prevent warping and to cover end grain. Straight end trim with glue and screws would be quite strong, but not as seamanlike in appearance. A rabbet or spline joint would be an improvement.

LOUVERED DOORS AND DROP BOARDS

Don't hesitate to tackle a louvered drop board or door. If you have a table saw and a dado set, it's easy. Build or saw out the frame as described previ-

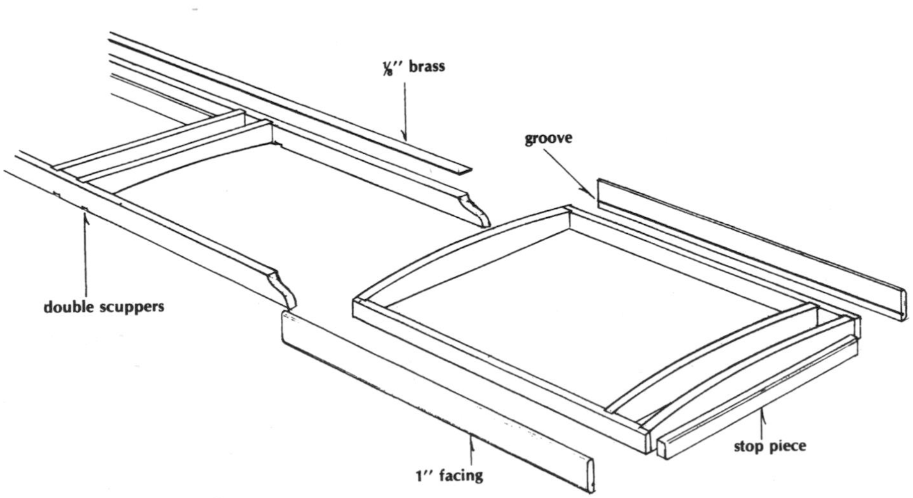

Figure 13-5. *Double-coaming sliding hatch.*

Figure 13-6. *Note the contrasting plugs in this hatch. This fine workmanship helped transform the vessel into an outstanding yacht. (Bruce Bingham photo)*

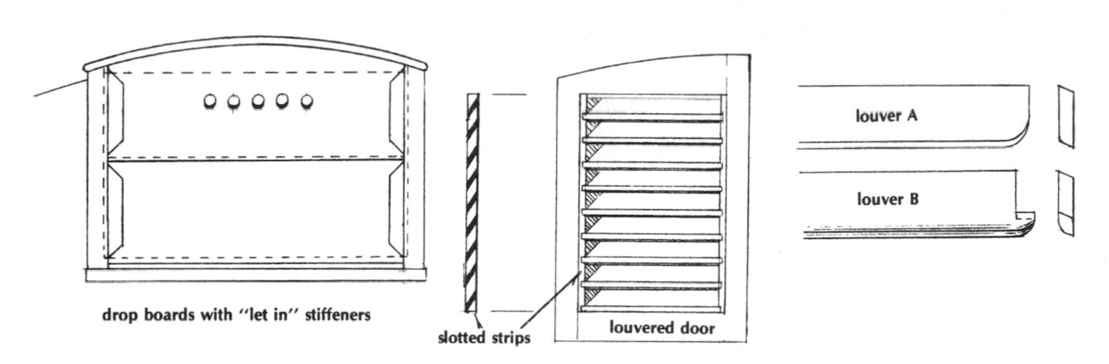

Figure 13-7. *Drop boards and louvered door.*

ously. Then make two strips to fit each side of the opening. These should be about ¾ inch square. Divide the strips according to how many louvers you want. I show the louver slots at about 35 degrees from the perpendicular. A higher angle would be more watertight, but admit less air. The amount of overlapping of the louvers is more important.

Now go back to Chapter Five, where I described a jig for mitering moldings. The louver jig is similar except that you can't run against the fence of the table saw. Instead, just clamp a batten to the table for the base to slide along. Also see SLAT, Chapter Five. Set your dado cutters to saw out the thickness of the louver material—¼ to 5/16 inch. Don't go over 5/16 inch deep. Use a piece of scrap to adjust correctly. Place it between guide pieces and cut a slot. Check for accuracy. Now whittle a bit of louver material into a tiny batten the width of the slot. Move the strip forward so the next slot will be at exactly the desired spacing. Make the cut, stop the saw, but hold the strip there while you insert the tiny batten into the first slot. Tack or glue it to the base or SLAT. This makes a stop so that all spacing will be exactly even. If this takes more than 15 minutes, I am not communicating. Another five minutes is needed to run each strip. Then glue and finish nail the strips in the opening just flush with the front.

Figure 13-7 shows two shapes for the louver ends. Louver A has a rounded corner. Louver B is cut to form a projection that covers the slot. Various shapes are common, and often the strips are covered by a small molding or frame all around (Figure 13-8). The lower edges of the louvers should be left sharp so water will drop free. After the louvers are glued in with epoxy, plane off the inside edges flush to the door. Fit a piece of screen over the opening and cover the raw edges with a neat mitered frame.

Figure 13-8. *Traditional louvered companionway doors have hinges permitting instant unshipping. Note the modern plastic hatch top.*

Incidentally, the bottom of the door should fit into a rabbet in the door sill, which itself should be covered with a formed brass or copper sheet (or at least a brass half oval where wear occurs).

Doors should swing on the type of hinge that permits lifting the door for removal. If you plan to use your yacht in cool weather, make plywood or doorskin inserts to fit inside the screen frames. Turnbuttons will hold them quite well.

LIFT HATCHES

There are almost as many options in lift hatches as there are in sliding hatches. The four shown in Figure 13-9 will be described in detail. Hatch A is as simple as they come, B is hinged to lie flat on deck, C is streamlined to match companionway hatch C, and D has a system of double coamings. Elements of any one could be combined with another.

Hatch A is a traditional hatch. There's no guarantee it will be completely watertight under all conditions. This is considered workboat style by some, but we had one like it on one of our boats and I don't remember that it ever leaked a drop. The top can be a single thickness of ½-inch plywood if flat, canvased or glassed. If there is a lot of camber, as in a cabin trunk, build the hatch with a matching frame and a top of doubled ¼-inch plywood. It is customary to cover the canvas or fiberglass edge with a half-round trim. The frame and coaming can be rabbeted. This isn't the best way to go, but it's quite simple and, with modern epoxy glues, should last a long time.

Scribe the ends of the coaming to the camber before assembly (then the upper crowned edges, if any, to match). Let the sides extend slightly above these so they can be planed off after installation. Do not build your hatch coamings inside the carlings and beams. This is an invitation to leaks and rot. The ones shown are all fastened through the deck down into the carlings and beams with long screws plugged over. Assuming the cover top and coaming have matching cambers, leave a 3⁄16-inch space between so you can apply weatherstripping inside the cover for insurance. A facing is mitered inside to cover the beams and carlings.

Hatch B has several interesting features. As shown, the facing just described extends to about ¾ inch above the coaming, forming a water stop. An improvement would be a heavier coaming with a rabbet sawed or dadoed in. This would make the half-lapped corners correspondingly heavier than indicated. Either way, such laps would be very strong if you put a dowel or a long screw down through the joint. Any hatch frame or coaming is a good candidate for the box or finger joint described in detail in Chapter Seven.

To hinge a cover so it lies flat on the deck, the height of the hinge pin must be more than the thickness of the hatch cover at its center. This controls the dimensions of the coamings. Figure 13-9B shows a piano hinge, but marine stores carry hinges made for this purpose (Figure 13-10). Butt hinges are also acceptable. Personally, I am not enthusiastic about hinged covers that lie flat, since there is always a temptation to step on them. My comments about weatherstripping apply here, too, if the facing or flange extends to a near contact with the hatch top.

Hatch C is shown well in Figure 13-3. This one is fitted outside the coaming rather than on the coaming. Also, the plastic (perhaps Lexan) is not rabbeted fully into the cover. The hinges are not the special type mentioned above; they force the hatch cover to stand upright. The construction is

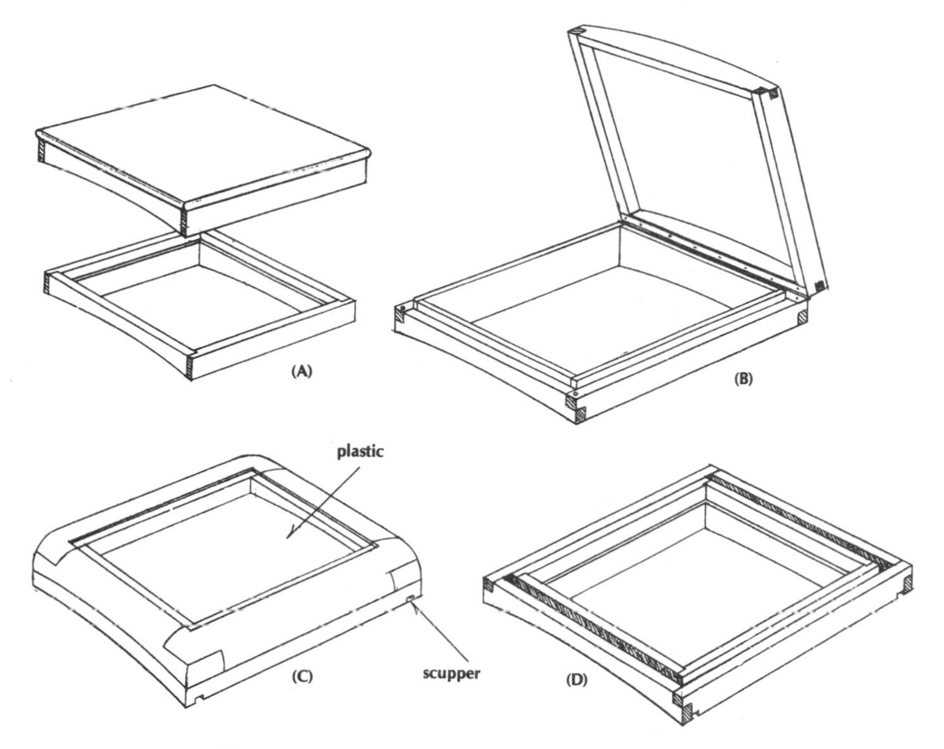

Figure 13-9. *Four hatches—simple to complex.*

mitered or half-lapped and mitered (a very strong joint). This hatch looks massive—probably a 3- by 3-inch frame. Because of the generous radius, it does not take kindly to plywood or solid topping. These plastic panels should have a number of teak battens or light beams extending across the plastic to prevent scratching the surface. Note that the screws are not countersunk. There may be fiber washers under the heads, to allow microscopic expansion and contraction.

Figure 13-10. *These hinges allow the hatch to lie flat on deck. The hinge pins are removable. (Bruce Bingham photo)*

Figure 13-11. *A typical Dorade ventilator doing double duty as a running-light screen.*

Hatch D shows the coaming I would suggest for C. Water blown through the joint runs out of the scuppers in the corners. It can't climb the inner coaming because the scuppers relieve the air pressure. I assumed this coaming would be used with a hinged cover. A lift cover, however, also would work well if you installed small blocks or molding inside the cover to line it up with the coaming. Better yet, you could rabbet the cover into the coaming as in hatch B.

The drawings show the hatches with sharp corners. Both covers and coamings, however, can be rounded generously on the vertical exterior corners. Take a look at the fine modern yachts in your harbor. The top corners of covers are all well rounded. All other corners carry a slight radius so varnish will last. Sharp corners wear fast.

DORADE VENTS

Back in the 1930s, the "Boy Genius" Olin Stephens designed *Dorade*, a yacht with real class. She cleaned up on everything in ocean racing and for years was world-famous for her speed. Then her star finally dimmed and died. But wait! It's not so dead after all. Stephens utilized several clever ideas. One is her ever-flowing ventilators that let nary a drop of water below. To this day, they are called Dorades (Figures 13-11 and 13-12). The photograph shows the ventilator turned aft. Note that it carries a running light. Others have additions such as a built-in box for winch handles. Strong construction is mandatory, but the canted forward end is optional. Now look at the drawing. You'll see "dam optional" dotted in. This could be easier for you than the plastic pipe glued into the cabintop. Be sure to screen the opening and provide a sliding door for controlling the ventilation. Flexible plastic ventilators are a fine idea, because they reduce the fouling of lines and sails and barked shins. The Dorade's box should be scuppered at several places along its length.

STORAGE BOXES

The beautiful mahogany box in Figure 13-13 was built by Frank Stapleman to enclose the LP gas tanks on his yacht, *Samurai*. The corner treatment is a fine example of what I call a box or finger joint. Frank ran a dowel down through the joints—a tremendously strong construction method. You might like to vary this design in an interesting and practical way by crowning or sloping the cover to shed spray and dew. The sides are spaced about ¾ inch above the deck, and there are vent holes bored in the back and bottom so air will circulate. Since LP gas should be kept cool, I suggest lining the box with foil, reflective side out, or fiberglass batts. Such a box also would be great for fenders, spare line, sail stops, or what-have-you.

A fine variation is Katy Burke's generator cover (Figure 13-14). The box is mahogany marine plywood with rabbeted corners. Apparently it has an underlay cover of lighter plywood, with a mitered

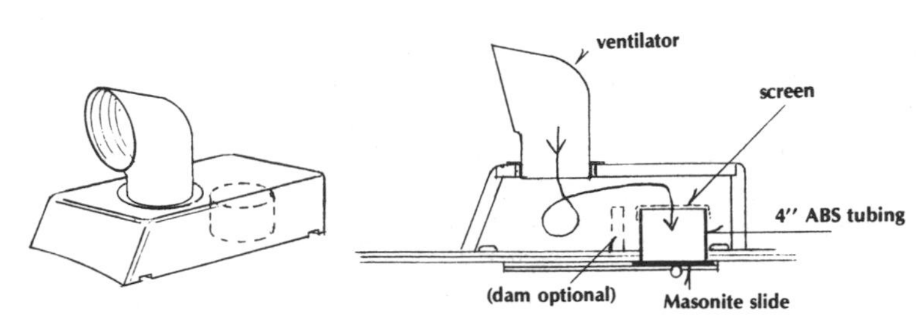

Figure 13-12. *Dorade ventilator.*

Figure 13-13. *Frank Stapleman covers his propane tanks on Samurai with this nicely jointed mahogany box. Note the box joints—a dowel runs down through for strength.*

Figure 13-14. *Katy Burke concentrates on fine workmanship in this generator cover. (Bruce Bingham photo)*

frame of ⅜-inch mahogany. The V-grooved planking is being laid on a bed of epoxy glue. Because of this light construction, I suspect that the planking is weighted down without fastenings, with contrasting plugs put in later for looks. A small half-round molding covers the exposed edges of the cover. The generator gets its air supply through the open bottom.

MAKING HANDRAILS

Grabrails or handrails should not be thought of solely as deck features. You can be thrown about unmercifully in a cabin, or simply be unsteady on your feet, so it's worth putting in a hand grip here and there. There should be a short one near or over the head and on the galley bulkhead. Long ones running fore and aft under the cabin overhead are great. Spread these about 30 inches from the centerline so they do not interfere with headroom, even when heeling. On some yachts, they are on the facing covering the carlings. I have seen the facing itself formed into a grabrail.

Figure 13-15 shows a method of turning out grabrails of any length with a minimum of waste. Don't waste your time making them of ¾- or 13⁄16-inch stock, unless they are very short. They'll break just when you need them most. For short ones, 1 inch is all right. I would not consider less than 1¼-inch mahogany, teak, or oak on deck. You can vary some of the dimensions specified in the drawing. The radius of the bored holes can be reduced from 1¾ inches if you wish a lower silhouette. Keeping in mind that you have to dress the sawed edges, however, don't go under 1¼ inches. Leave space for fat fingers. Spacing, of course, can be varied, perhaps to match beams (if any).

The first step is to joint both edges, then lay out the centerline for locating borings. Place clamps across the piece when you bore it to prevent splitting. Use an expansion bit at low speed in a drill press or a solid ½-inch drill or a brace. Then rip end to end. Or to save material and heavy boring, bore at A and B, then bandsaw or sabersaw the rest to save the nice chunks of hardwood. Dress all surfaces and rout a ½-inch radius all around. When fitting any type of rail, whether to a highly crowned trunk or not, hollow the base by running the piece diagonally across a saw blade. Then bed it well. Use No.

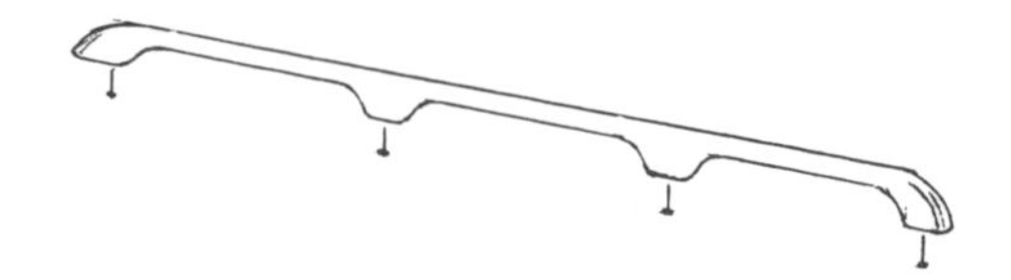

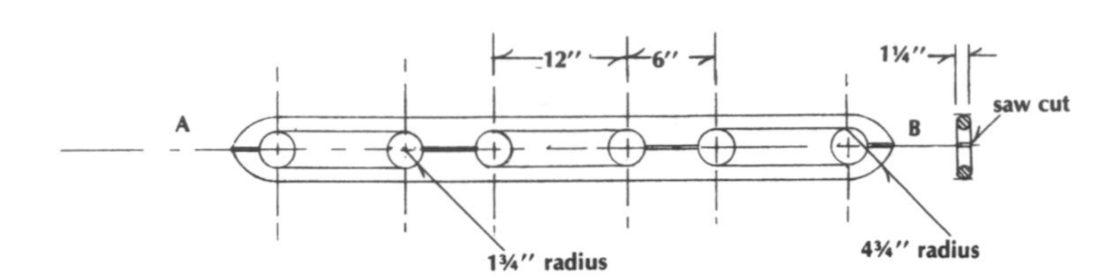

Figure 13-15. *Layout of typical hand grip (grabrail) (not to scale).*

12 bronze screws or through-bolts. Figure 13-16 shows a grabrail that terminates in a graceful ogee. To spring long rails easily, install two or three good fasteners at the wider ends, then gradually pull the forward ends inward, using several turns of light line. Better to hold with oval head bolts through cabintop or #12 wood screws into beams, since this minimizes splitting. Avoid screwing from below.

ANCHOR CHOCKS

The anchor chocks in Figure 13-17 were made by Katy Burke. Before bandsawing the pieces out of a full 1-inch plank, she drew in the shapes of the recesses, then routed them out freehand. (Of course, she practiced on scrap first.) You'll find that a straight ¼-inch router bit can be maneuvered along a curved line reasonably easily. At such a slow feed, the bit will scorch, but that surface has to be pared down anyway. You might laminate two or three thicknesses of mahogany or teak with the grains running diagonally to keep these delicate forms from splitting during use. If you have several sets to make, use plywood router patterns.

Figure 13-16. *A nicely shaped and finished mahogany handrail. (Bruce Bingham photo)*

LIGHT SCREENS

Figure 13-18 shows a nice pair of light screens. These are wired for electricity, but there is another set for oil lamps standing by for voyaging (now illegal, incidentally). Both were made to be installed in the shrouds, but they could be mounted on the cabin trunk.

COCKPIT COAMINGS

Many fine yachts of the past had cockpit coamings that were more or less continuations of the cabin trunk sides. If the inboard surface of the coaming was not doubled to provide a backrest, or the coaming canted back in a separate piece above the deck, the man at the helm sat in a virtual torture chamber (Figure 13-19A). I show several ways to avoid a kidney-killing coaming. Coamings are seldom high enough, or the seats low enough, to match the ease of a good old straight-back chair. Unfortunately, too, any canted coaming decreases the usable cockpit space and seat width, and base for winches. Sloping seats would add comfort, but they collect water unless drained by a rather involved system of channels and scuppers.

Figure 13-17. *Mahogany anchor brackets made by Katy Burke. To prevent splitting, these should be built up of two or three laminates. (Bruce Bingham photo)*

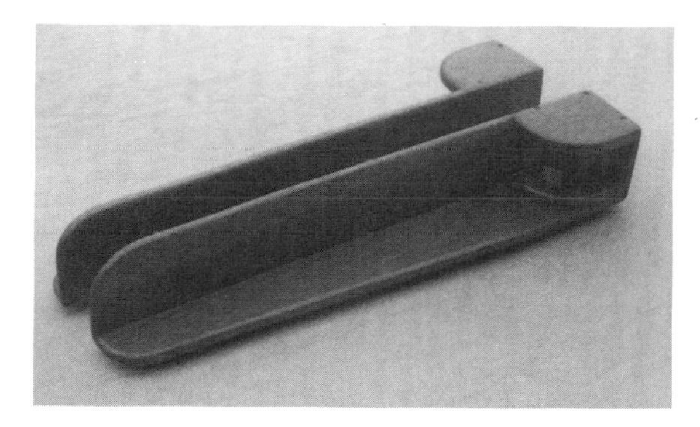

Figure 13-18. *A pair of running-light boards or screens for mounting on lower shrouds. (Bruce Bingham photo)*

Sketch B shows what happens when the trunk sides tumble home drastically, as they do in most current fiberglass designs (dotted line). The only escape is to go far outboard, and never mind the loss of foot room on the catwalk, as in C. The only good thing in such a design is that you have the beginning of a fine winch base. If you build C, you get passable comfort but a more difficult woodworking job where the coaming meets the opposing angle of the trunk. If you go to D, you get more complicated construction but an easy and neat transition into the trunk side. Both B and D allow small storage spaces, but you should bore holes in the deck of D for ventilation. These coamings may be of plywood with a nice mahogany or teak cap, finished bright. I am not suggesting that you select one of these; I merely hope they stimulate your thinking.

COAMING BLOCKS

A discussion of cockpit coamings leads naturally to the joining of coamings and cabin trunks. Perhaps you have seen some of the awesome structures found on wooden yachts. There is no easy way. The three shown in Figure 13-20 are laminations. The block in A may be built up parallel to the coaming or to the side, or it may be built up starting flat on the deck. If there is no bridge deck, the block has to be installed forward so the joint with the coaming is over the deck. If there is curvature in the coaming, it will have to be steam-bent or built up of double plywood—this requires a carling separate from that of the trunk. Style B is

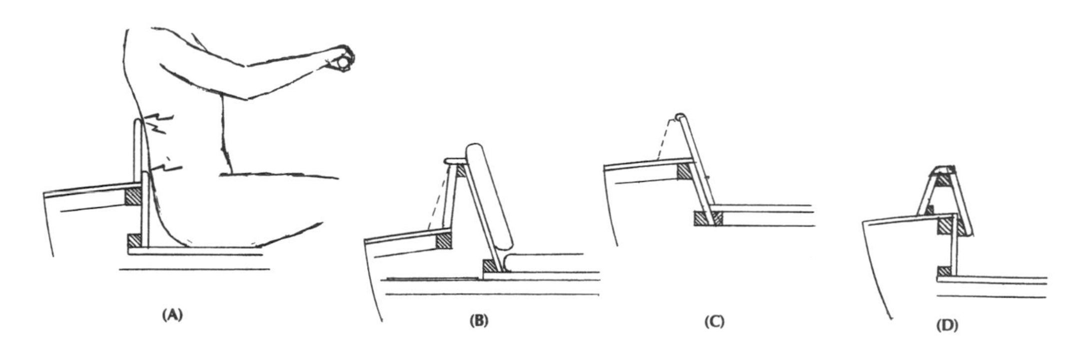

Figure 13-19. *Cockpit coaming construction.*

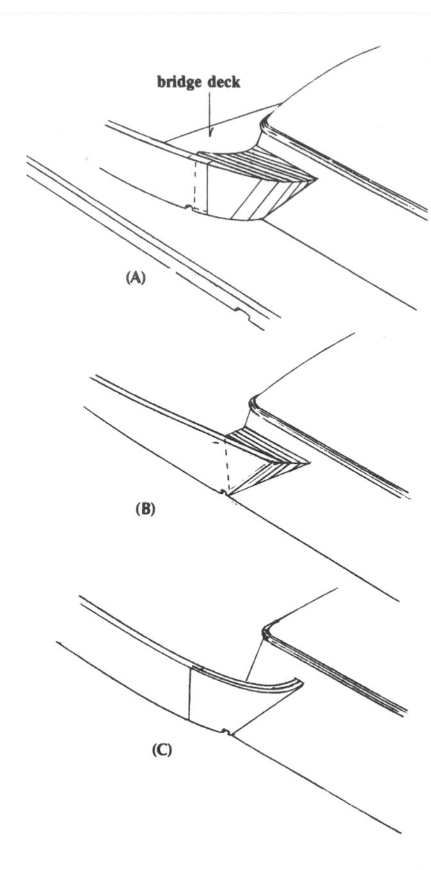

Figure 13-20. *Fairing coaming to trunk.*

Figure 13-21. *The beginning of rot can be seen in this nicely molded coaming block. It is formed of several laminations of teak plywood with a neat half-round cap.*

Figure 13-22. *This molded corner is rabbeted to receive the coaming side rabbet.*

fastened to a continuation of the trunk carling (from beneath). The angular block ought to be much easier to construct. Just round off the outside corner of the coaming piece to blend it in. Style C can't be done unless you bend light plywood (or doorskins) over a form, as has been done in Figure 13-21. Notice that the builders have worked a small half round along the top of the bent coaming—a nice touch. Too bad they forgot a scupper. Otherwise, this is a nice job. As shown in Figure 13-22, the aft corners of the cockpit coamings match this construction. Both corners are rabbeted to the coamings.

COCKPIT SEAT HATCH SCUPPERS

I could do a chapter on cockpit design, but let's concentrate here on watertightness. Typical hatch framing is shown in Figure 13-23. Inside the opening, the channels (A) are screwed and glued to the beam risers (B) and beams (C) with epoxy to make the connection strong and waterproof. These channels or scuppers can be routed out from a pattern with a straight ¾-inch bit in two or three passes down to about ¾ inch deep. This will leave rounded corners, as I have tried to show. Or the channels can be run off with a dado or saw, if you miter the corners. Note, however, that some

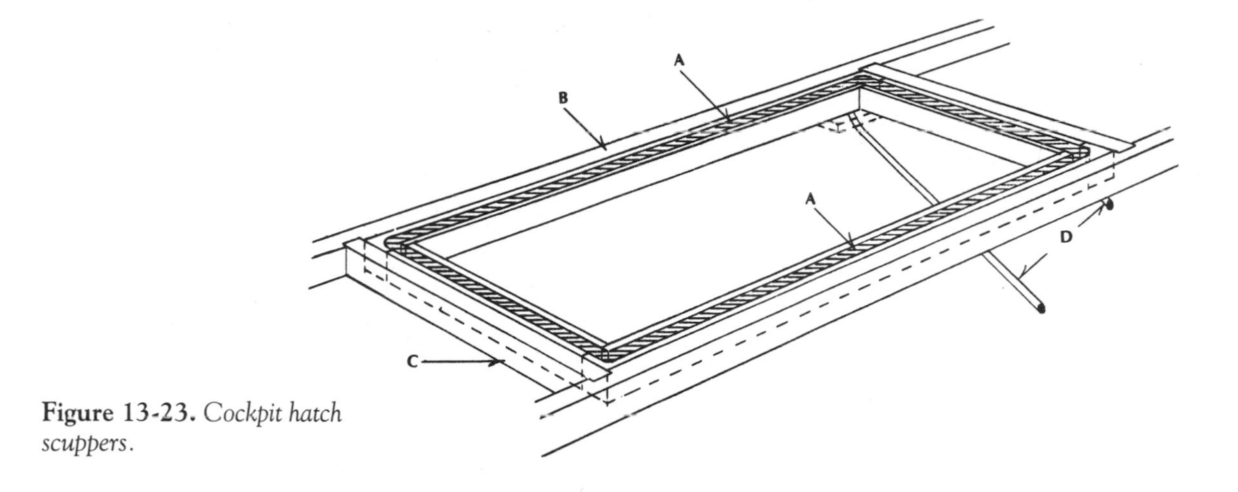

Figure 13-23. *Cockpit hatch scuppers.*

hatches are not rectangular, so mitering is a bit more difficult. A rabbet would be strong. These joints must be filled with epoxy paste glue. Bore for ½-inch copper tubing (D) near both the inboard and the outboard corners. Note that the lower corners as the vessel is at anchor may not be lower when she heels, so you may have to install three tubes. It would be wise to stabilize these tubes by gluing a block under the point where you bore. Bed the tubes in epoxy. Brush three penetrating coats in the channels. You might use short stubs and connect with plastic tubing. The scuppers should extend about ⅜ inch past the plywood front of the footwell.

COCKPIT SEAT HATCH COVERS

Figure 13-24 shows a handsome seat and hatches with natural teak decking and framing. Construction here is similar to that shown in Figure 13-25. It is basically ½- or ⅝-inch plywood to which ⅜- by 1¾-inch teak decking is glued by weighting down. No fastenings are used, but they could be if the decking were ½ inch thick. Since water travels along the underside of flat surfaces, you must encourage it to drip where you want it to—into the channels. The drip strip is 1/16-inch aluminum or brass set into a groove approximately ⅜ inch deep. Allow the metal to protrude ¼ inch (the drawing is somewhat exaggerated). You can make this fine groove with a veining bit or fine saw blade or continue the saw cuts right out and fill the kerfs with glue and pulverized limestone or Cab-O-Sil. If you use a fine saw blade, some sharp knife work will be needed near the corners. Work epoxy glue into the groove and press the strips in. Don't worry about the groove being wider than the strip—the epoxy will fill it. Put a couple of drops in the corners also. They must be located so the water drips directly into the channels. The surrounding deck and/or frame should overhang the channels very slightly and the joint underneath must be watertight. I show a piano hinge because this helps to keep water out. Butt hinges would be satisfactory.

Figure 13-24. *The yacht shown in Figures 13-21 and 13-22 has handsome teak hatch covers built over a plywood underlay. Scuppers (if any) under joints are not visible.*

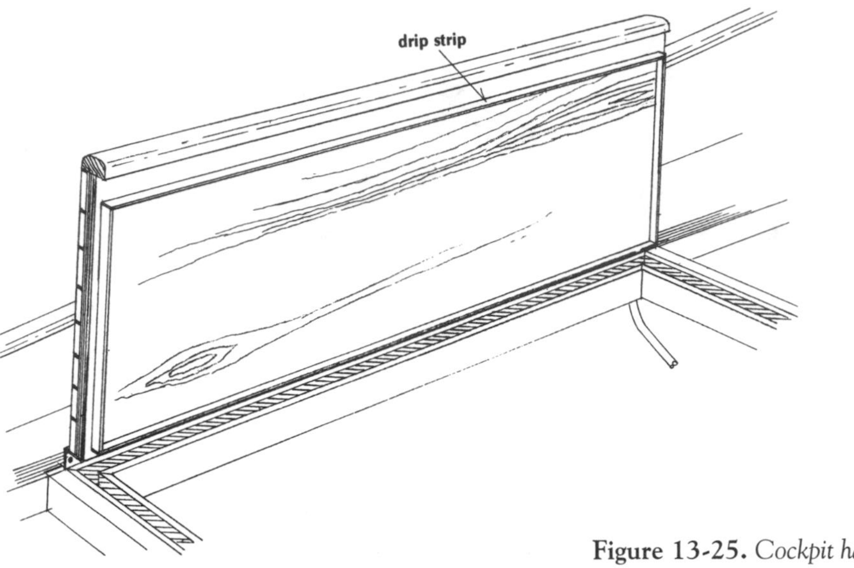

Figure 13-25. *Cockpit hatch cover.*

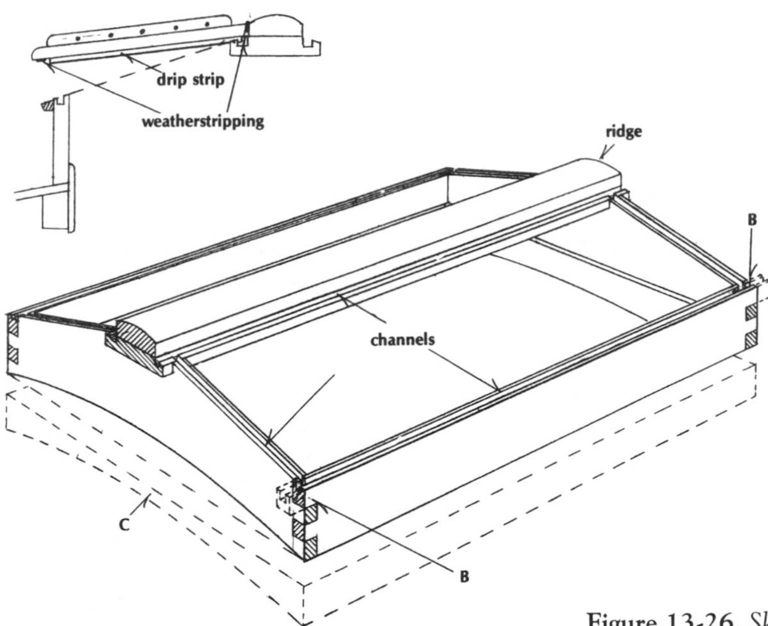

Figure 13-26. *Skylight frame construction.*

SKYLIGHTS

Skylights are closely linked to hatches, because that's what they are. Part of the construction of the one shown in Figures 13-26 and 13-27 is similar to that of the cockpit seat hatch just shown. Like most hatches, skylights are prone to leaking, only more so, primarily because of the two hinged joints along the ridge. The outer edges are also subject to leaking. You can apply moldings and hope to cover some joints, and you can fill some joints with weatherstripping, which, of course, deteriorates.

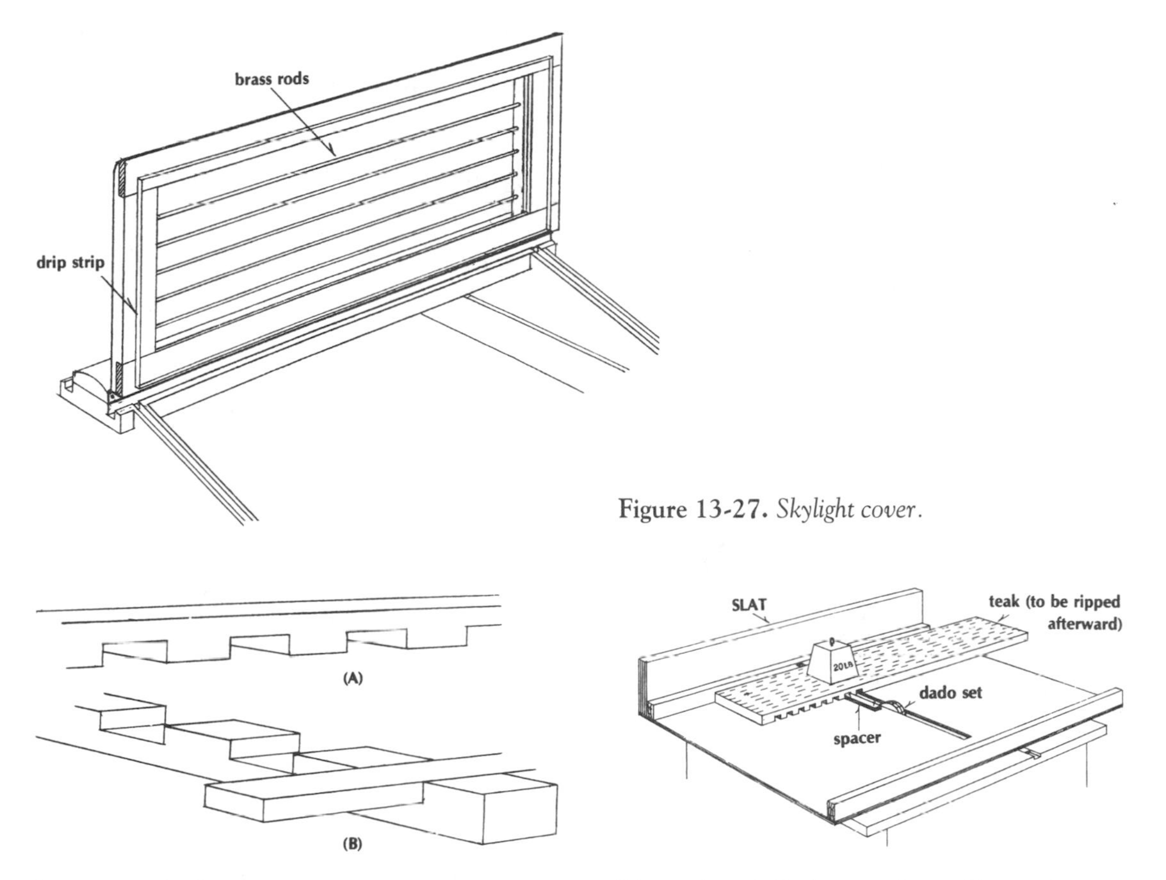

Figure 13-27. *Skylight cover.*

Figure 13-28. *Gratings.*

Figure 13-29. *Dado setup for making gratings.*

After these fail, just direct the water into a scupper. Let the moldings and weatherstripping be insurance. Use them to *slow down* the entrance of water and driven air.

The upper portion of the corners (B) in Figure 13-26 should extend to act as spouts, or you could bore for a small drain tube to keep water from running down the frame. The ridge extension of about 1 inch shows this clearly. Put a weatherstrip inside the piano hinge. Do not put this hinge on flat. The dotted lines (C) indicate an optional separate coaming if you want a removable skylight. If you build this perfectly square, you have a skylight that you can turn to catch the best breeze. Of course, you must have hatch hold-downs to secure the coaming, and the skylight covers must have dependable locking devices and lifting mechanisms.

The corner construction on this project is the box joint described earlier. To simplify matters, the ridge is constructed from two pieces. The lower part carries the grooves, channels, or scuppers that drain off the water. I suggest you dado or saw these out of the full piece, then saw the bevel. This applies to the outboard frame member also. Of course, the frame is bedded to the trunk, and fastened with long screws from beneath. The cover should be half-lapped, and of material not less than 1 inch finished. When gluing this up, be sure it is held down with weights or clamps to a dead-flat surface. Aerolite glue would be ideal. There is about a 1-inch overhang of the cover so that moldings could be included.

Use Lexan plastic or wire glass or you could build the cover solid with two or three large deadlights, good for serious offshore cruising. The rods should have an additional support. Moreover, an opening this long would need another cross member in the center.

Figure 13-30. *A teak cockpit grating before construction of its frame.*

Again, for other ideas, look at yachts in your marina and sales lots and take photographs or make sketches—with permission, of course. Ask about performance under extreme conditions, then go build the perfect skylight.

GRATINGS

To make a grating my way, you must have a sharp dado cutter, preferably with carbide teeth. And a fine-toothed carbide blade on your table saw, too. Again, I urge you to equip your saw with the sliding auxiliary table (SLAT) described in Chapter Five.

There are at least two good ways to make your grating. The first is better, being all ¾-inch or better teak (Figure 13-28A). The second, (B), uses heavier teak only in the long members, the shorter pieces being strips ⅜ inch thick. There is no difference in strength or utility if the grating is fully supported. Spacing uniform dadoes precisely is done on the principle of the box or finger joint jig. But this time let's use the SLAT instead of the less rigid miter gauge (Figure 13-29).

Make a 2- or 3-inch spacer piece the exact width of the desired dado—say, ¾ inch wide by ¼ inch thick. Select enough wide pieces of ¾ -inch teak to produce all the cross members (21 in Figure 13-30). Joint one edge. Practice on scrap. Hold this firmly (a 20-pound weight leaves both hands free) on the SLAT while you run a dado. Check to see that the spacer is a snug fit. Again place the trial piece against the dado blade, using the spacer as a gauge. Cut a dado and check the spacing against the spacer. They must be identical. When you feel confident, tack the spacer in place as in Figure 13-29. Run three or four dadoes across your scrap, placing each dado, as it is cut, over the spacer. Then do the same with three or four narrow pieces about like your grating will be, say, ¾ inch wide. These should press together into a dummy grating, flush on top.

If the spaces are too wide, too shallow, or too narrow, adjust your dado cutter carefully. Get it right! When you're finally satisfied, glue and tack the spacer, weight your wide teak down on the SLAT, run a dado, jump over to the spacer, run a dado, and so on.

What do you have now? Three or four pieces of teak with dadoes running across them, right? Now these have to be ripped into pieces that will fit into

Figure 13-31. *A neat oval cockpit grating. A fine production job a skilled amateur can match.*

Figure 13-32. *This rounded helmsman's seat on Hoku Kea offers slightly more comfort as the yacht heels. It doubles as a lazarette hatch cover.*

the dadoes, and snugly. Try with scrap. Once you achieve this, rip all your teak; clamp on hold-downs both ways for precision. Don't worry about dressing over the saw cuts—if you used a fine-toothed blade, the surface is all right. Put a dab of glue in each intersection, then press it all together with weights.

You'll need a frame all around the grating, the dimensions of the cockpit sole, allowing ¼ to ½ inch total clearance. If you have a rudderport in the cockpit, the grating must be in two sections. Also, if you like to rig up bunks in the cockpit, two lengthwise sections are handy. To simplify, however, my instructions are for a single grating.

Make a tacked-strip pattern of the cockpit sole with ¼-inch clearance, including corner radii, all around. These dimensions are the outside of a teak frame, say, ¾ (at least) by 3 inches, with a rabbet ⅜ by ¾ inch around the inside. The best-looking corner is mitered half-lapped, but a straight half-lap would look good with the rounded corners, too. The rabbet can be run in the stock with a dado set or a router, or after assembly with a router, leaving a small radius in each corner, an interesting effect.

During these steps, your rough grating has been pressed together with glue in the many joints, turned over and nailed with ¾-inch brass brads (epoxy is not reliable on teak unless the surfaces have been de-greased with acetone), about every fourth joint. If you use Aerolite, brads are not necessary.

Clamp the rough edge grating on the frame, square and centered perfectly, then mark very *slightly outside* of the rabbet in the frame, using a sharp knife, because the grating has to be a fine fit into the rabbet. Saw free-hand close to the line so you can take off the surplus with a long coarse sanding block, then fine grit until it fits the frame. Next step is to saw, dado, or rout a rabbet out of the lower edges of the strips to match that in the frame: depth must be perfect. Apply Aerolite or epoxy glue to the rabbet back and lower surface for maximum strength. If it appears that pressure has bowed the frame outward, place two to four bar clamps before starting the nails. Flip over and drive brass brads along the rabbet, say, at every other batten, then pile blocks on top over a dead flat surface and let cure 24 hours. Finally, sand the entire grating surface to remove any irregularity, using about 250 grit and avoiding crossing the grain into the frame.

Figure 13-28B shows a simpler grating but one just as strong and good looking. The difference is that the strips running athwartships are just half as thick as the ones running fore and aft, so there are no dadoes, other than those in the long pieces. The frame is identical, and the attachment of the grating into the frame rabbets is much easier.

I must warn you about using brads. Try not to set the heads below the surface, as the points could show. Easiest answer might be to drive brads at an angle, or to not drive flush with the surface, then either pull them out later or grind them off. Of course, you could have the mill dress your teak 9⁄16 inch or heavier. Even without fastenings, Aerolite glue or epoxy will hold.

Here is another interesting and practical project for almost any cockpit (Figure 13-32). This helmsman's seat is also the cover on the lazarette hatch. As the vessel heels, the man on the wheel has a seat that is far more comfortable than the usual sloping seat so hard on the spine. The margin shown adds a touch of elegance, probably laminated teak. I'll leave the rest of the construction to your imagination and skill.

14 Log-rails and Toerails

LOG-RAILS AND TOERAILS

Unless they are specifically designed for extensive offshore cruising, most modern yachts do not require bulwarks. Instead, metal stanchions and lifelines or liferails that, theoretically, prevent crews from falling or being swept overboard, are added. Owners and designers are reluctant to increase windage, of course, but there is a mysterious feeling of security that comes from the sight of a generous rail. Part of this is pure illusion, however, for a 4- or 5-inch rail will do little to prevent a body from being washed overboard. A 2-inch toerail will keep you from skidding off the deck and save the life of many a tool, but that's it.

The distinction between a log-rail and a toerail is not precise. A log-rail generally follows the contour of the topsides (flared, straight, or tumblehome), may be made of two or more pieces of lumber on edge, and often, but not always, has a cap. Log-rails taper down in height toward the stern also. Toerails are predominantly one piece of lumber (or they are laminated to look like one, and to make good use of costly lumber). The outboard faces are vertical (with very few exceptions), while the inboard faces always cant outboard at a uniform angle. Toerails may be tapered or of uniform height, and they may be as low as ¾ inch on small boats, but the best style tapers aft. All rails must be set back a minimum of ⅛ to ½ inch from the edge of the plank-sheer or fiberglass hull.

Figure 14-1A shows a log-rail about 6 inches high. The lower log is drifted (rods driven down) into the deck beams or the sheer batten (especially if the hull is fiberglass, as in C). The second log can be screwed or drifted or nailed into the lower log. The cap should be grooved to prevent movement and damage and to ensure a fair line. On Chesapeake Bay they once used an open log-rail consisting of a heavy cap resting on short pipe stanchions. Drifts went through the cap and the pipes into the lower log. This is a practical workboat style that could be adapted to some so-called "character boats," replacing expensive turned wooden stanchions. A cap can be screwed to a narrower single log-rail also (B), with the latter drifted, spiked, or screwed into the sheerstrake or sheer batten.

The forward ends of lower logs and toerails are always wider for a foot or more from the stem to provide for bow chocks, which must be screwed securely through the rail and preferably let in so the screws are relieved of shear stresses (Figure 14-2). The screws should be long enough to hold in the plank-sheer. Chocks can also be fitted into a mortise in the rail if you can find the right dimensions. The greater width can be carried aft, tapering down at about a fourth of the total length. Where chainplates penetrate the rail, it is customary to swell the rail out to double its thickness elsewhere.

Toerails may be of one timber scarfed to make the length or strip-built of two or three laminations. In the latter, it's all right to use plain tapered scarfs 1 by 8—or butted if yours is a "plain" boat. If the scarfed lengths can be handled, make them up on a bench. But if the length is such that it might break of its own weight in handling, the scarfs can be made on the plank-sheer or a lower lamination. Care must be taken to make the joint so fair that it is invisible from above.

The safest way is to do the fitting on a bench (see the scarfing jig in Chapter Five) and tack in a couple of small finish nails so the relationship is easy to repeat. Lay the two on the plank-sheer, fasten one, apply glue, tap the nails into the original holes, and clamp the *sides* of the scarf between two long pieces of scrap. This will hold the scarf so that

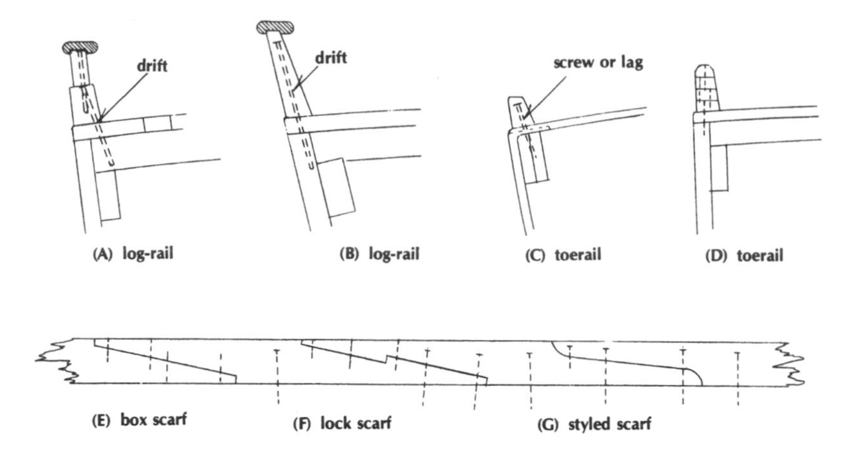

Figure 14-1. *Rails and rail scarfs.*

it will bend fair while you screw down the rest of the rail or lamination. Fasten down through the scarf with screws and finish nails. Let it set for at least 24 hours before releasing the clamps. Scatter such joints in laminations as far apart as possible.

If the toerail is higher than wide, don't use a common scarf, especially if it is 3 or 4 inches tall. Use a box, locked, or styled scarf, as shown in Figure 14-1E, F, and G. Side clamping is even more important here, and it should be predrilled for long screws. If you can get several in from below while the assembly is on a bench, do so, and give it *time* to cure.

All rails must be set in bedding, since you may have to remove them. Scarfs are epoxy glued, of course.

The toerail shown in Figure 14-1D has four laminations. This rail, if one piece, would be difficult to bend and fasten. I suggest sawing all the bevels (the inboard surface) and dressing carefully so they need only sanding after installation. The outboard edges can be sawed to the decreasing widths as the rail is built up, planed as accurately as possible before installation, then planed and sanded after the rail is completed. This convex surface is easy to work, whereas the inner concave side would be difficult. Where scuppers are cut through the lower piece, a hard spot is likely to show when the piece is bent to the curve. Clamp a two- or three-foot length of stiff scrap inside the piece so that the hard spot is faired out, then glue and screw to the deck. Use your eye.

The types of scarf in Figure 14-1E, F, and G are the same for caps. Strong rails such as in A should be able to take a cap sprung around the curve. This is preferable to cutting it out of a wide plank. However, if B has any deflection because of its height, its cap will have to be sawed from a pattern. Scarfing from a router pattern is described in Chapter Nine.

Here is an expedient for you who want the appearance of a cap rail (or "rail cap," as I prefer). First, I warn you that a rail as low as two inches might look ridiculous with a cap. But if you have, say, a naked toerail or log-rail, from 2½ inches up and perhaps 4 or 5 inches in height, this idea might make sense. The easiest way is to nail or screw and glue half-rounds to both sides of the existing rail. If its corners are now rounded slightly, you would have to set the moldings below the radii, then plane off the excess. But first check the security of

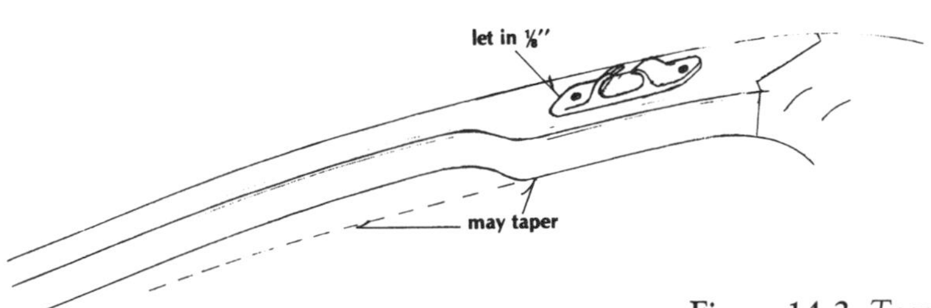

Figure 14-2. *Toerail at bow.*

existing wood plugs, to avoid leaving such shallow seating that they might come loose. Here is a second system: fasten two quarter-rounds flush with the rail's top, then cover the total with a solid piece—1 by 4 (or whatever fits)—on which you have shaped or routed radii to match the moldings. Plane the surface to catch any high spots first, for first-class appearance.

Finally, the best-looking and certainly the strongest construction would be a solid cap with a groove dadoed, shaped (two passes) or sawn (many passes and chiseled) to fit the toerail or log-rail. Regardless of which way you decide to follow, the treatment of the ends will be your problem. Perhaps a simple bull-nose or ogee will do. Use your imagination.

15 Spars

I have enough material on the design, layout, and construction of wooden spars to fill a fair-sized book. On the other hand, what I know about aluminum spars I can write on my cuff. I'm attempting to steer you toward wooden spars because the non-professional boatbuilder can't build an aluminum spar. A properly engineered aluminum mast is an excellent replacement for a wooden mast, so go to a recognized company for a metal mast if you can afford one.

Unfortunately, the best wood—Sitka spruce—is so costly that you may not come out ahead. One major reason for this is that the metal mast comes with tangs and other attachments neatly and strongly built on, if you choose, and with a groove for the sail. The chances are the whole rig may be significantly lighter, too, which might pay off in sail-carrying ability.

Am I trying to talk you out of building your spars? Yes, if you have the least doubt about your capabilities. Because this is where careless workmanship, or less-than-perfect material, or faulty assembly will almost always result in a dismasting, an injury, or worse!

But hold on. There are good reasons why you should choose wooden spars. Only the costliest tapered aluminum spar even remotely approaches the beauty of a gracefully shaped spruce mast gleaming like pale gold under many coats of varnish. The ugliest rig in aluminum is the gaff rig, unless it is viewed from a half-mile away. Look at the massive 9-inch mast on a Cape Cod catboat. Note how it fights the thrust of a mighty gaff yet slims down rapidly to its head, topped off with a neat truck. And remember that wooden spars still stand up to their work, just as they have for thousands of years throughout the world.

In keeping with the purpose of this book, I shall show you simple ways to construct spars. Note that much of what follows has already been covered by giants such as L. Francis Herreshoff, Howard I. Chapelle, Robert M. Steward, and others. I have used most of these methods in my small shops with inadequate tools under conditions quite similar to yours—if you are the average amateur boatbuilder. I have never hewn a mast from a sapling or tree—from a timber, yes. But these subjects are rarely needed.

HOLLOW BOX SPARS

Round masts, solid or hollow, are completely obsolete except for small boats and traditional rigs such as sprit rigs, Cape Cod cats, gaff rig schooners, etc. In these cases you build exactly to the spar plans provided by the designer. Also, round solid masts are required for gaff rig, since the sails need rings or lacing, except in rare cases. Not that a round or oval spar is not *occasionally* built with a hollow core. If you are concerned with round spars, you will find adequate information below.

The familiar box mast is the only way to go for an amateur builder, except for the mast of a dinghy or very small boat. The box is stronger than a solid timber of the same section, and uses less material. Thus, it costs less. A box mast requires a lot of time, but not really hard work. The layout must be precise and the workmanship equally accurate. The assembly must be held rigidly in a jig or bench of some sort. You'll need many clamps, as well as a pair of assembly (gluing) helpers about every eight or 10 feet of length. Other than that, the job is straightforward.

HOLLOW-SPAR LUMBER

Spruce is the most desirable lumber for any spar because of its stiffness and strength combined with light weight. Spruce weighs in at 26 to 28 pounds per cubic foot. Its appearance, a light creamy tan, is beautiful. Try to find the type of Sitka spruce known as "aircraft quality." Inspect this or any other lumber for spars before buying it. Look for almost perfectly straight vertical grain, which should not run out of the board at more than a 1-degree angle, if at all. If the grain is straight, but runs at a greater angle across the board without bending, you may have to settle for a much wider board. The waste can be used later to make small sticks, tillers, and so on. Do not accept lumber with knots, pitch pockets, checks, or other blemishes. But do expect short checking at the ends.

Sitka spruce is difficult to locate and extremely costly. Because it comes from the Northwest, you can only inspect it personally if you live up that way. If you need a fair quantity of it, try contacting a lumberyard near the coast for a recommendation. It's possible your small order (100 board feet and up) can be shipped into your area with a large load of other lumber. I am confident you can trust a reputable yard to supply your exact needs—if the material is available to them. Sometimes they, too, have to wait. And there may be others in line before you. Perhaps a boatbuilder in your area could take care of these details as your agent—for a "small" fee, of course.

Some dealers handle rare, exotic, cabinetmaking and boatbuilding woods. Be sure to state in writing what you need, and for what purpose. Last but not least, be sure to consult a local boatshop operator about woods from your part of the world that might serve nearly as well as Sitka spruce.

If you are unable to acquire the right kind of spruce, go to Douglas fir vertical-grain lumber, available in lengths to 30 feet. Its weight runs as high as 40 pounds per cubic foot, but it is stronger. Most designers say you can reduce the thickness of the staves proportionately. Usually the outside dimensions of the box remain the same, with only the wall thicknesses being reduced from, say, ¾ to ⅝ inch. The desirability of Douglas fir, also called Oregon pine, is well established, as it was used almost exclusively by L. Francis Herreshoff (Figure 15-1D). His spar for the R-boat *Yankee* had a section 5½ by 7 inches (his standard proportions), with side staves only ½ inch thick and ends ⅝ inch thick. Don't think of just copying this design. The R-boat had an intricate double-spreader rig that kept the stick in place. The mast scantlings alone did not do the job.

MAST AND SPAR SECTIONS

Figure 15-1 shows mast sections designed to a ratio of about 5 inches in width to 7 inches in length (fore and aft). Herreshoff used 5½ to 7, but note that today's modern foretriangles, tremendous headsails, powerful winches, and adjustable backstays place stresses on spars that he never dreamed of. Many modern masts are occasionally under loads that equal or exceed the vessel's total displacement.

Figure 15-1A shows a simple box section. This is suitable for small boats with under 250 square feet of working sail area. The dotted line indicates an optional thickening of the aft stave if there is

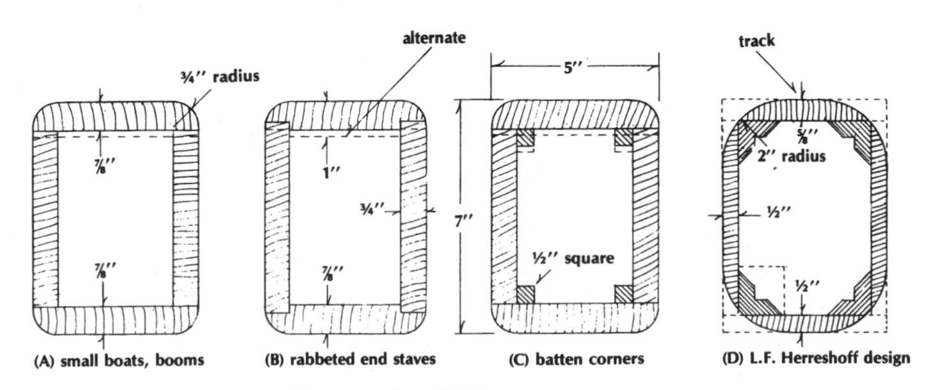

Figure 15-1. *Hollow spar sections.*

any question about the holding power of the sail-track screws. The usual batten under the track also increases the mast's rigidity. The corner in B is stronger because of the much-increased gluing area. The rabbet makes it easier to glue, as it aligns the four staves. This is a major problem with configuration A.

I like the style in Figure 15-1C, as it can be made easily with a table saw. The battens can be lined up precisely with a little spacing jig and fastened quickly with glue and small nails. Even style B can be rabbeted on a table saw if you are careful with the blade settings. None of these corners needs sanding if you use a cabinetmaker's combination or planer blade, as glues such as Aerolite and T-88 need a slightly toothy surface and very little pressure. Style C makes a slightly heavier spar, of course. You can refine it by making the battens triangular instead of square or rectangular.

SPAR LAYOUT

Spar dimensions are shown in the designs of all responsible naval architects and designers. You must not trifle with these. However, I have seen designs showing mast tapers as a series of straight lines, some with a straight taper from about midpoint to head. It is up to you to see that these deficiencies are corrected, by going to a qualified designer or sparbuilder, who will charge you for his time. Sections and walls proportional to those I have included here would not be far off, but rigs do vary. The drawing in Figure 15-2 is of the mast of *Allegra*, a 24-foot heavy-displacement cruising cutter design.

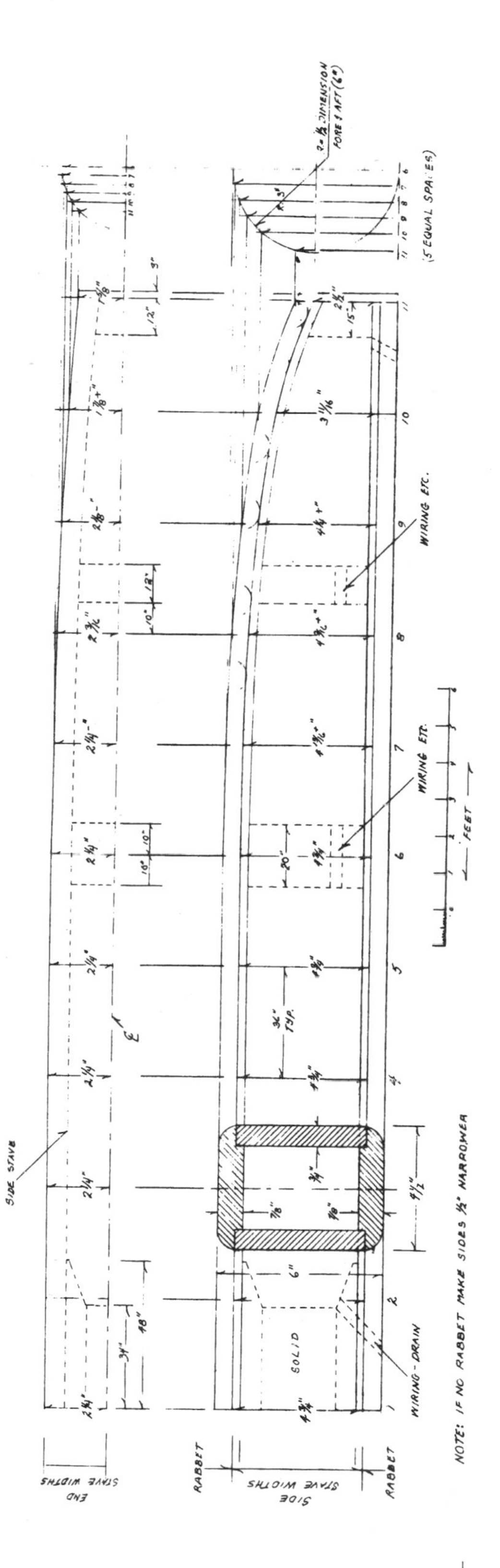

Figure 15-2. *I drew the section to a scale of 3/8 inch equals 1 foot for better reproduction in this book. The longitudinal scale, however, is shortened, for the curves are more easily faired in a short bend. My scale is 1/2 inch equals 1 foot. (My working drawings for amateur builders have a longitudinal scale of 3/4 inch equals 1 foot.) See Howard Chapelle's* Yacht Designing *and* Planning *(W.W. Norton) for traditional spar design mast.*

LAYOUT OF FORE-AND-AFT (GAFF-RIG) SPARS

Later I'll describe briefly how to construct solid round spars for use with gaff rig, so here are the essential proportions that Chapelle specifies for such masts.

Catboats with an unsupported mast: Diameters: at deck, .02 of total length, heel to truck; at gaff jaws, .90 of deck diameter (a gentle curve on forward side); a fast taper to peak halyard block, .65 of deck diameter.

Sloops and cutters, normally rigged: Diameters: at deck, .02 of overall mast length; at head or peak halyard eye, .70 of deck diameter; at butt or step, .50 of deck diameter. If heavily rigged, deduct about 10 percent of diameters.

Schooners: Mainmast diameters: at deck, .023 to .029 of the length from deck to hounds; at hounds, .85 of diameter at deck; at masthead, .80 of diameter at hounds. The foremast should be larger than the mainmast by about 10 percent. *Gaffs*: Greatest diameter .015 to .018 of total length; diameter at jaws, .90 of greatest diameter; at head, .72 of greatest diameter. *Solid booms, round*: Diameter at mainsheet blocks, .015 of total length; at jaws, .80 of greatest diameter; at aft end, .70 of greatest diameter. *Schooner fore boom*: At sheet blocks, .022 to .026 of its length; at jaws, .90 of greatest diameter; outer end, .80 of greatest diameter.

This is only a portion of the information Chapelle provides in his famous book. I have tried to boil it down to essentials. I used these proportions and others for unsupported masts in designing the rig for our 36-foot *Bay Bird*, known as a three-sail bateau in Marylandese. We called it a modified sharpie with a modern bugeye rig. Her 36-foot sticks whipping in a breeze gave many a midlander heart attacks.

SPAR BENCHES AND ALIGNMENT

Let's talk about ways to keep spars aligned properly during construction and gluing. They must be aligned precisely whether they are hollow-box (most common) construction or worked out of a massive timber or log. Most of you will be building box spars (Figures 15-1 and 15-2), so we'll postpone solid-spar support systems until later.

On all spars, masts, booms, and gaffs, the side to which the sail is attached must always be dead straight. No sailmaker can construct a proper sail for a crooked stick. Thus, some means of holding the spar components in alignment must be incorporated in the spar bench. As you see, the mast consists of four staves. The two side staves are identical, each being tapered toward the head on its forward edge only. The forward and aft staves taper in toward their centerlines, so they are identical in shape, but they may vary somewhat in thickness, as shown. The other components are blocking at points where stresses occur, such as at the bury from deck to step, the gooseneck, the area of the spreaders and lower shroud attachment, the forestay intersection, and the head. See Figure 15-3.

The supports for your spar bench must be solidly immovable or so heavy that movement will be unlikely. Figure 15-4 shows four types of bucks for a spar bench. Many others could be devised. If you are working outside or in a rough building with an earthen floor, No. 1 would do fine. No. 2 shows the bench frame attached to building studs (which are never in alignment). No. 3 represents a series of sawhorses built of very heavy timbers to resist movement. Note that these are toenailed to a wooden floor. If the floor is concrete, you'll count on the weight of the timber to be sufficient. No. 4 has a built-up frame nailed to a wooden floor, ideally, or lagged down to a concrete floor, less than ideal.

Make each buck not less than two feet long and space them from four to eight feet apart (the latter for large masts); see Figure 15-5. Set up the end bucks level, then stretch a wire or 50-pound nylon monofilament as tight as you can. A line level would help to level these approximately. Tack ¼-inch blocks under each end of the line so it is spaced ¼ inch above the actual top of the bucks. Now the bucks in between may be shimmed or raised or lowered by using a ¼-inch block as a gauge. If your spacing is so long that sag is visible in the line, you can work from a center buck out to the ends. Failing this, you may have to use a long level or a level on a long straightedge. Being truly level is not critical, but having a dead straight line from end to end is.

The principal bench-top board should be of straight ¾-inch stock, carefully butt-blocked with glue and screws. Its width should be such that it cannot interfere with clamping. Make up the backboard width less than the spar's greatest dimension, joining with butts. Nail the backboard to the bench-top board to form a continuous girder. Then this entire assembly must be lined up with a chalk-

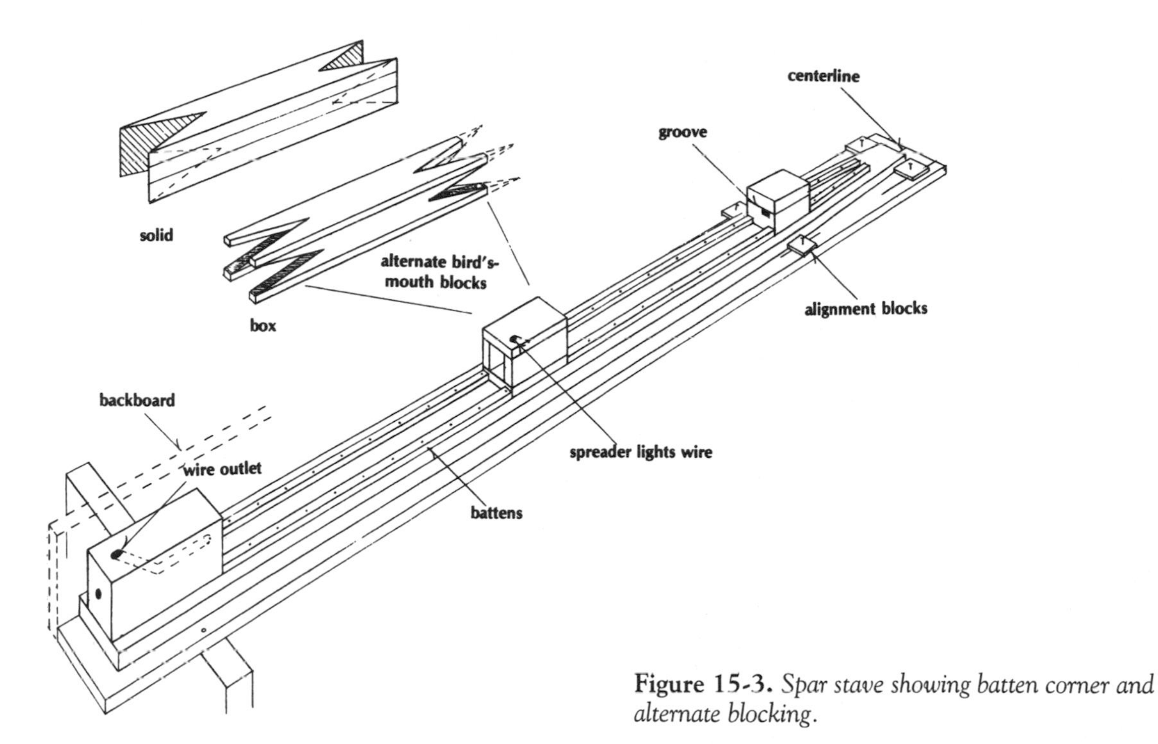

Figure 15-3. *Spar stave showing batten corner and alternate blocking.*

line and fastened. The backboard must be vertical, so shims or blocks behind it may be necessary. If the bucks are not perfectly level, you ask for trouble. You can't have any humps or valleys.

An alternate bench called a trough is shown in Figure 15-4. It, too, can be built with the opening in the side, but either way the sides must be rigid. The trough could eliminate the need for clamps entirely if the ties were to be nailed securely with wedges under them as well as inside the trough

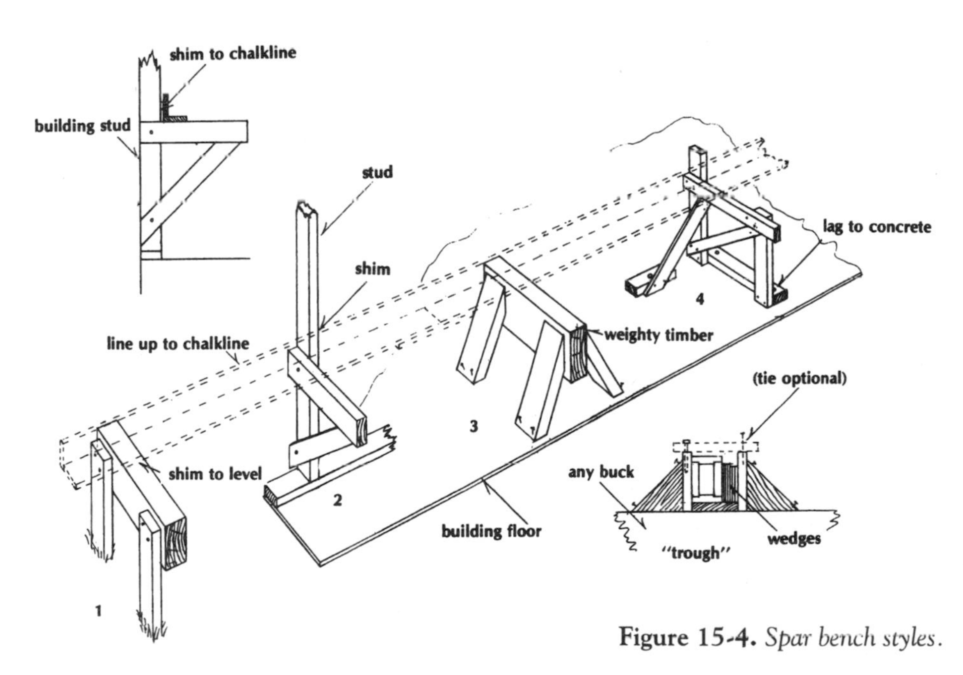

Figure 15-4. *Spar bench styles.*

Figure 15-5. *Sixty-two-foot rectangular mainmast built by skilled amateur Leland Cass. Note the perfect scarf and threaded rod clamps on glued-up mizzenmast. These were Cass's first spars.*

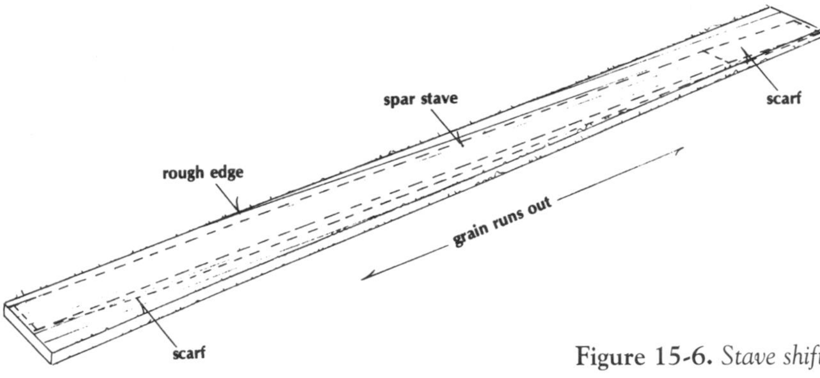

Figure 15-6. *Stave shifted to follow grain.*

itself. This device is often built on the wall studs or on a wooden floor.

SCARFING AND ASSEMBLY OF STAVES

You may have to use spar lumber considerably wider than the principal width of the staves because of grain run-out. In this case, the grain will not be perfectly parallel to the edge of the boards, although the grain itself should run perfectly straight (Figure 15-6). Unfortunately, this creates some scrap. You may have to piece together three or four short lengths to get a complete stave. This involves work, of course, but do not be nervous about the strength of a properly constructed 12-to-1 scarf. (A scarfing jig was described in Chapter Five. Its use was then discussed in Chapter Seven.) Plan to join continuous lengths before assembly of the box. Do not expect to join the scarfs during the box glue-up. Also, unless you have been lucky enough to find spruce or fir in which the grain runs dead straight in the boards, you can't just scarf them together willy-nilly, then lay out the widths and tapers.

Your first step is to snap a chalkline on each board parallel to the grain. Then tack a stiff, straight batten or board on the scrap side of the line so that this straightedge can guide the saw, whether you use a table saw or a power handsaw. All your pieces now have a sawed straight edge, but two will need a considerable taper on the opposite edge, and two will taper on both edges. Remember that the scarfs must be scattered, that is, no joints opposite or close together. Take the lengths, allowing for the checking at the ends, and lay these out in pencil on a sketch of the spar plan. Number the pieces of lumber accordingly. Lay these out on the spar bench in their relative positions, allowing for the scarf overlaps, and mark off locations of the ordinates (Figure 15-7). At these points lay out the approximate widths of the tapers and the general widths of the lower spar.

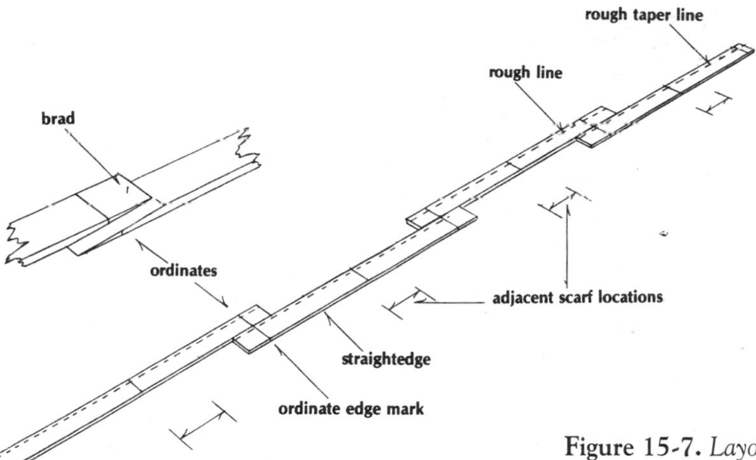

Figure 15-7. *Layout of joints and ordinates.*

Let's see why this is done. First of all, consider these questions. Do you rabbet the pieces (1) before scarfing, (2) after the scarfs are fashioned but before gluing, or (3) after the full-length assembly is glued up?

Let's go through these step by step so you can make your choice intelligently. Let's agree that rabbeting, dressing, or any operation is easier to do when the components are short and easily handled. The rabbeting can be done with a router, with a table saw, with a router-shaper, or with a shaper. The steps for No. 1 are to lay out excess widths on the temporary ordinates so you can leave at least ⅜ inch more than the desired width, because some lumber takes a set or spring when a portion is ripped off. The excess must be enough to allow corrections, to go back to a straightedge or chalkline and still have enough stock for the final opposite edge. Once this edge is sawed straight to your satisfaction—and it might be from the first cut—dress it accurately to the line. Mark the widths on the ordinates, connect with a batten, and saw close to that line. Plane down, then form the rabbet on both edges with the best tools you have. Now you can go ahead with machining the scarfs (see Chapter Five), being careful to locate them so the ordinates also marked vertically on the *edges* match up perfectly.

No. 2 is just the opposite. The scarfs are located by the ordinates—the marks on the edges—then fitted, and the pieces are bradded to prevent shifting. Now take them apart and plane the straight edges, from which you mark the stave widths on the ordinates. Saw to these lines, dress, and rabbet both edges.

Whether you followed No. 1 or No. 2, the pieces are ready to be glued into full-length staves. If you have one or two helpers, you can follow No. 3. When the pieces have satisfactory straight edges, rout or plane the scarfs, glue all together (details follow), and let cure. Lay off the dimension on the ordinates and dress to the line. It should be easy now to rout the rabbets without help. But if you

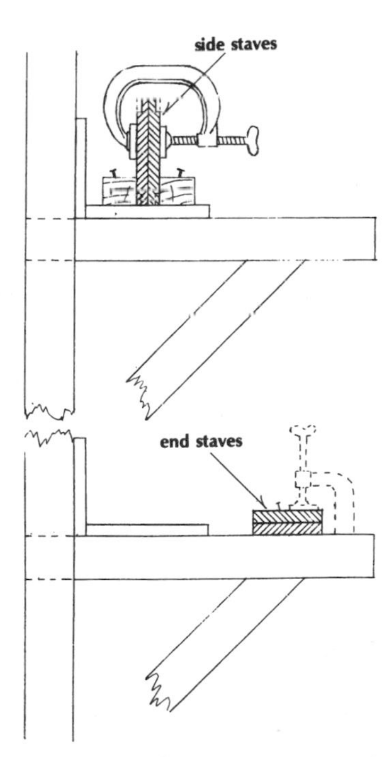

Figure 15-8. *Matching staves on spar bench.*

want to cut the rabbets on a saw or shaper, you'll need help to handle the stave's long, limp weight. Be sure you set up ample hold-downs on the saw or shaper table. In addition, (to saw the rabbets) you'll need a high fence on your table saw and two strong hold-downs so the rabbet comes out precisely square to its matching part.

GLUING THE SCARFS

Butt-block several good pieces of lumber to make a gluing surface on your spar bench. Cover this with vinyl or waxed paper. I prefer Aerolite or T88 adhesive, but Plastic Resin Weldwood has stood up for many years, and it does cost less, and now there is WEST System epoxy. Lay the scarfs in position dry against a line of blocks tacked on a chalkline. Drive in a brad or small nail to prevent shifting and leave the head out for pulling later. Take apart a joint at a time, apply the glue, get the nail back in the same hole, and line up against the blocks (or the bench backboard) with a few C-clamps so nothing can slide around. Cover the area with plastic sheet or waxed paper, place blocks, and clamp up firmly but do not crush.

Observe whether the joint appears to be closed uniformly. Plastic resin requires pressure, the others do not. Wipe away the glue that will extrude from the joint for several hours, more or less. When all the scarfs in each stave are glued, let the assembly cure for not less than 24 hours. Then block-sand the joint surfaces for appearance and a fair surface. Be sure there are no globs or runs of glue in the rabbets. Of course, if you have decided to build with corner battens (Figures 15-1C and 15-11A) instead of rabbets, this will be a future step. My only advice on this is to use galvanized 3d nails or brads. And make a little gauge for locating the battens uniformly so the outer surfaces meet smoothly everywhere. But first, match the staves for width.

Obviously, both pairs of staves must be exactly alike when finished. Even if the rabbets were made in the two side staves before you scarfed them to full length, try to make them match perfectly. Set these two up on their straight back edges between blocks on the spar bench (see Figure 15-8) or against the backboard. Press them down against the bench and clamp here and there. Be sure the ordinate marks coincide. Now run your hand along the top edge of the pair. If you feel or see the slightest difference, shave this off. Replace any pencil marks that were planed off. Now go back and rework the rabbets accordingly. If you elected to rout the rabbets after scarfing, or if you plan to use the batten glue system, you have no problem. Just dress the staves to match.

The fore-and-aft staves may not be so easy to match. Lay one on top of the other to check and clamp together here and there. Now turn the pair on edge on the open part of the bench and shave one edge. Then turn and shave the other edge. Replace any ordinate marks planed off.

To maintain the vital side tapers and squareness when the box is being clamped to the backboard, I suggest a series of graduated spacers. Position the aft stave with the centerline up, the lower straight edge blocked firmly against the backboard. Soft clamping will hold it down. Rip pieces of scrap into various thicknesses to insert every 12 inches or less along that curve. Simply measure from the backboard to the centerline of the stave in three or four locations so it is held perfectly straight, while you tack a mess of spacers vertically to the backboard. Now clamps cannot pull the assembly out of line.

Figure 15-9 may be confusing because it does not show the backboard between both types of clamps and the spar (see Figures 15-3 and 15-4). Also, remember that the backboard should be less than the box height at the masthead so it does not interfere with clamping. Make a dry run, putting the parts loosely in position so there will be no last-minute foul-ups.

I show C-clamps for this matching job in Figure 15-8. If your mast is to be painted, it would be easier and quite all right to use small nails, 18 and 16 gauge. If you are in such a hurry that you decide to skip this matching operation, your spar will probably hold together, but don't expect perfect or almost invisible joints.

GLUING UP THE BOX

I have glued up several conventional box spars without help, using Weldwood. But I had many sleepless nights afterward. Did I get the 40 or 50 clamps on before the glue started to set? Were the joints fully closed and under proper pressure everywhere? Was the sun too hot that day? You can avoid these problems by getting four or five people to help for an hour or so.

First of all, there's the clamp problem. You should have C-clamps of ample span, and enough

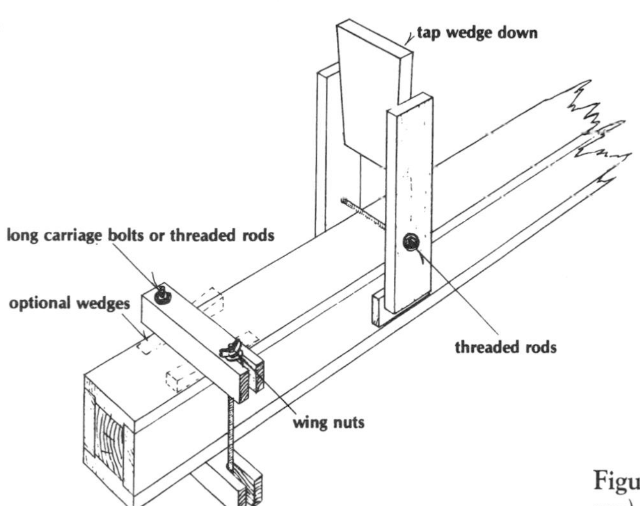

Figure 15-9. *"Make-do" spar clamps (for occasional use).*

of them to place them 12 inches apart. It's unlikely that the average amateur builder will have that many clamps on hand. Borrow or rent. If you plan the assembly for a weekend, you might be able to rent clamps from a couple of local cabinet shops. Or a welder might have some. Professional boatbuilders, of course, will have plenty.

If you can't round up enough clamps, you'll have to build your own. The most common type of make-do clamp does an excellent job (Figure 15-9). The length of the bolts could be a problem if your mast is large. Ask a hardware dealer if he will let you use 100 of his longest carriage bolts (diameter is unimportant), nuts, and washers for a couple of days. Return them in good condition and pay a rental fee much smaller than their selling price. Or buy them and share with another sparbuilder. Threaded rod works just fine, too. The two-bolt clamp shown can be made from any 2 by 2 stock, or use lighter stock for smaller masts. Allow for the bench-top board beneath and long scrap and/or wedges on top. The slots can be bored and then bandsawed or table sawed.

The fast clamps may not be required closer than every two feet, set up in advance of the hold-downs just tight enough to press the sides against the back of the rabbet (or batten) gently, and especially against the blocking. After the end staves are seated well in the rabbets, these one-bolt clamps can be given a tap and then all clamps are pulled down. If you see only a small amount of glue oozing out, tighten down. This assumes that you spread the glue evenly.

A fast crew could probably glue your spar in one step, after the locating and gluing of inside blocking as shown in Figure 15-3. That is, if you apply glue to all four corners and staves simultaneously. This enables the box to rest on its back, with one side against the bench backboard and taper spacing blocks. There's no reason to move anything. Experience here would be very valuable. Two stages might be easier for you. The only real problem is that often the sides have a tendency to go out of square when under pressure. The solution is to cut a dozen plywood spacers to fit temporarily inside. The spacers must be varied to take care of both tapers. Wrap plastic around them to prevent bonding. Also, don't forget that the side against the backboard can be clamped square.

Another way to guarantee squareness is to use the front stave as a spacer only, without glue. Then, when this three-sided box has cured at least 24 hours, remove the front stave, spacers, and so on. Remember to paint the interior with at least two coats of epoxy resin to prevent rot caused by sweating. (Of course, hollow blocking should have been so treated before assembly.) Any bored holes should be stopped at one end while resin is poured into the holes or applied with a soaked pipe cleaner or toothpick. The necessary wiring should be ready or installed for lights, coaxial cable, conduit, or what-have-you. Also, the front stave must be given two coats of epoxy resin, in advance. You can unclamp your spar in 24 to 36 hours. Then plane and sand it clean.

Make the masthead crane (Figure 15-12). Rout

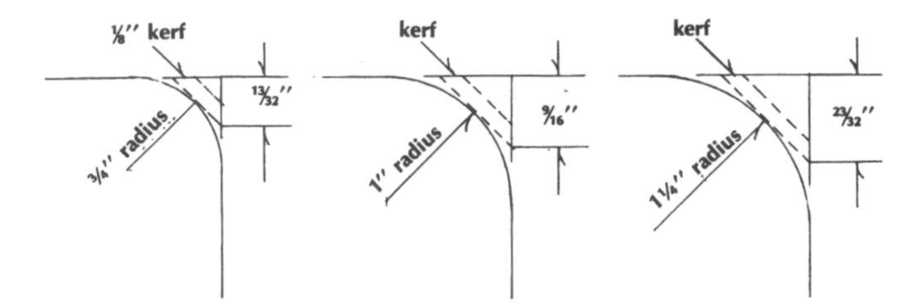

Figure 15-10. *Saw settings to produce spar radii.*

the mast radius if your bit is large enough, but skip the head area. Small 1- to 1¼-h.p. routers usually handle no more than a ½-inch radius. You may be able to rent a model that takes a ¾-inch radius bit, or larger for a heavier mast. A moderately heavy shaper could do this job, but you would need a helper or two, several strong hold-downs, and some skill. Other than these possibilities, you'll have to plane or saw a 45-degree angle, then knock off the 22½-degree flats with a plane, then the high spots, and end up with a hand scraper blade or pieces of broken glass.

The major job, the 45-degree angle, can be made easier by sawing it off using your power handsaw—if the guard permits close adjustments. The sketches in Figure 15-10 show radii of ¾ inch, 1 inch, and 1¼ inches and the saw setting required to leave just a smidgen for planing. If you use a table saw, you'll need a temporary wooden face screwed to the fence or clamped to the table. And a helper. Luckily, in this job a slight movement takes off less material, so no damage results. When you get ready to sand these rounded corners, remember that uniformity is a must. Irregularities will mark it as a backyard job. Make your sanding blocks of tooling foam or plaster of paris formed on a board to fit the radius.

Figure 15-11 summarizes spar construction in simplified form.

MASTHEADS

Almost every marconi mast of any size has an enlargement at its head. This is usually called the crane (Figure 15-12). The masthead crane is formed from spruce blocks bandsawed to the

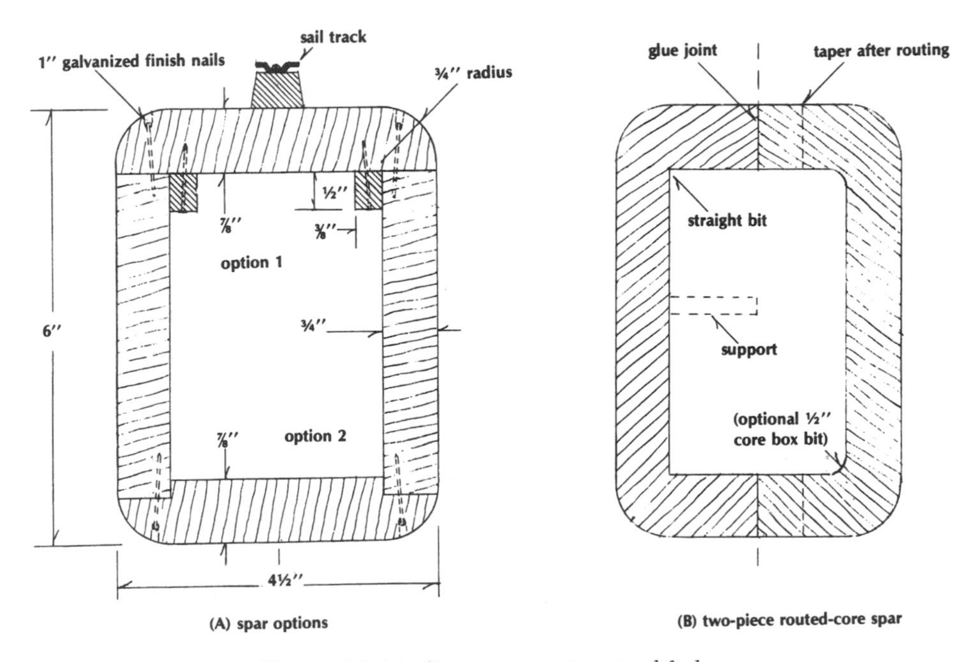

Figure 15-11. *Spar construction simplified.*

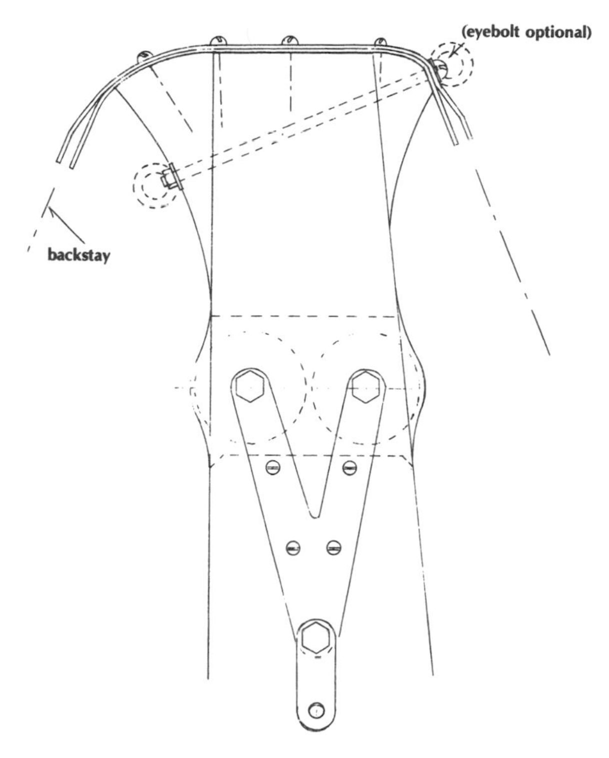

Figure 15-12. *Masthead crane and approximate tangs.*

desired profile the full width of the masthead. These are then glued and faired in before the mast radii are routed. The radius is formed all the way around the new profile, leaving a flat for the tangs bent over the head. The halyard sheave (or sheaves) is also covered with spruce blocks to prevent a wire halyard from jumping over the score in the sheave. These, too, are formed along with the radii, to match.

The slot for the sheave(s) is usually bored out undersized, then cleaned up with a sharp slick and chisel to form a deep mortise. A copper lining in the mortise is no longer considered good practice, because it wears and fouls the halyard. I have seen several light stainless steel boxes, bored for the sheave pins, installed neatly in this slot. To get the bolt (or pin) holes to line up, clamp the box to the side and use it as a guide for the bit. Insert the box, line it up with a pin, then bore clear through. Incidentally, you can save significant weight in rigging by using hollow bolts of bronze or Monel, or threaded stainless steel tubing or pipe.

Figure 15-12, a hypothetical masthead, shows a long diagonal bolt. This is an old Herreshoff system designed to resist the stresses of large headsails. I think this would have to be a special bolt. I suggest instead a rod of stainless steel or bronze, threaded at both ends to take eye nuts and washers. Or use stainless steel or bronze threaded rod from a marine supply house. These may still be available from good hardware manufacturers. If not, they could be fabricated by a clever welder. The forward eye could be used for the spinnaker halyard block. The one aft would be perfect for the topping lift.

The tangs in the drawing are only approximations of what your designer may specify. All screws should be roundheaded. There are several good reasons to avoid flathead countersunk screws. First, working of the tangs has been known to wedge screw heads out of the countersinks. Also, the knife edge of hard, thin tangs can act like a shear, literally slicing off screw heads. Bolts running through tangs and the spar should be encased in compression tubes to prevent crushing the staves and loosening up. Do not scale the drawing in Figure 15-12—it is just a sketch.

GAFFS AND BOOMS

Gaffs in large sizes should be hollow, if possible. Saving weight in this big spar is extremely important. It makes the difference between a stiff vessel able to carry her sail and a miserably tender tub never up to showing its inherent speed. In addition, because the gaff is under severe bending strains, these loads should be spread by the use of proper bridles and many blocks. Avoid eyebolts except at the ends. Study the sail plans of fine large yachts and old fishing vessels for workable rigging schemes.

There is usually little advantage in trying to save weight in a boom. Weight here acts something like a vang, helping to keep the sail down flat, and it often reduces slatting around in light air. Have a member of the crew stand on the cabin trunk and sit on the boom. This is much faster and safer than a bothersome preventer. I designed a T-boom for *Allegra* (Figure 15-13) because it is the lowest-cost rigid spar and the easiest for an amateur builder to put together, but I recommend staves of full 1-inch Douglas fir to add weight. Clear stock called stair treads comes up to 1⅛ inches thick and is ideal for this purpose. Warning: The mainsheet bail on a T-boom exerts a great twisting strain that can damage

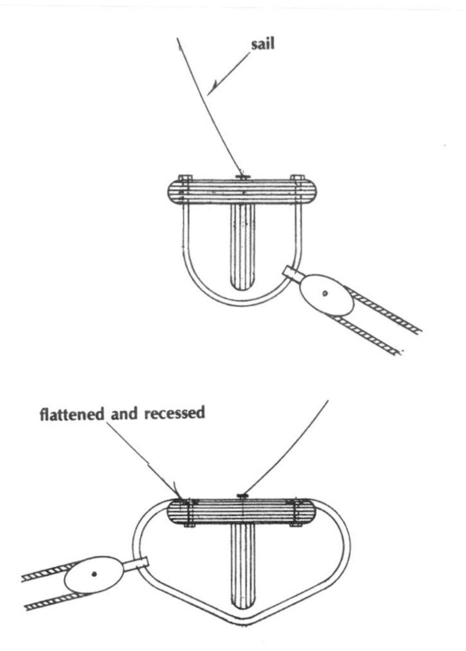

Figure 15-13. *T-boom mainsheet bails.*

the gooseneck or split the boom. Provide some sort of swivel or lots of play in the gooseneck, and install a bail that permits the sheet blocks to travel.

SIMPLIFIED MAST ASSEMBLY

For years I have been an admirer of yacht designers, boatbuilders, and sailors from New Zealand and Australia. In the 1960s these folks had the guts to risk all by diving headlong into ferrocement construction. They had the intelligence, skill, and perseverance needed to turn out a large number of successful ferrocement yachts while we belatedly and sloppily overbuilt ours, creating revolting failures or mere disappointments. A better reason for my admiration is that sailors Down Under are so accustomed to their *awesome* weather that they often sally forth for a casual sail or a race in gales that would send many of us into hiding. They design and build their yachts accordingly.

Consequently, when R.T. Hartley, N.A., described a method of nailing spars instead of clamping, I listened and learned.

The use of nails has long been taboo because of rumors of added weight. The effect of a pound of nails distributed over the length of a mast, however, cannot be measured by any instrument I know of. On the other hand, pressure is not distributed as equally as required by glues such as Resorcinol and Weldwood. Because of this latter argument, I never used or advocated the use of nails instead of clamps. I did not have the courage. But now we have powerful adhesives with crack-filling qualities that give even a tyro's 1⁄16-inch gaps 100 percent strength. That's why I am including a description of how to secure with nails until the glue cures.

Here's the procedure (Figure 15-11). Locate the blocking very accurately on the aft stave, and nail and glue, using 1- to 1½-inch galvanized finish nails according to the thickness of the material. If you have helpers, go right ahead, but if you are shorthanded, let the assembly cure for 24 hours. Next, lay old lumber or blocks on the bench and drive the same size nails the full length of the forward and aft staves on about 6-inch centers. Let the points come through about ⅛ inch, and note that the nails are at an angle. This angle and protruding points will tend to pull the side staves tightly into the rabbet or batten gluing corner, assuming that you have previously nailed and glued the battens accurately.

You'll need several small blocks on the bench to elevate the tapered ends of the side staves so their centerlines are level. Set the two staves on edge, apply the glue, match the aft stave with the various pencil marks, and start driving. Have someone follow immediately to set the heads in about ¼ inch or enough to clear the finished radius. Remove the elevating blocks, flip the *three-sided* box onto the bench, and weight down or clamp the box. Wipe off the oozing glue. Paint the interior with two coats of epoxy resin and let the assembly stand for 24 hours. If there is any indication that a side is not standing square at any point, press or clamp in a small spacer, to be removed later. Install the wiring, conduit, or whatever.

In the meantime, as above, coat the forward stave inside surface with epoxy resin. Drive nails as described, and weight and clamp to the bench to maintain alignment. Complete as for a clamped-up mast.

I expect critics to ask, "Ain't them one-inch nails a leetle small?" No, because even if you could pull all them nails, the spar would still be full strength. After all, do the clamps hold the conventional spar together? Hardly. Now, I would not use nails with Resorcinol or plastic resin, for they require high pressure evenly distributed to produce hairline joints. The new adhesives only have to be held together immovably during curing.

There is one fault with nailing. The final surface is possibly unsatisfactory for varnished spars if you look closely. When they are filled for painting, however, you wouldn't know that nails were under the lovely contours.

A TWO-PIECE MAST

The routed-core spar construction method is suitable for small masts and booms. I built a 36-foot stick about 3⅞ by 5½ inches by routing out the centers of two full 2-inch-thick vertical-grain spruce planks. Of course, this method wastes material, but it is fast and produces a good spar while eliminating most gluing problems. The planks would have to be scarfed to full length.

Once you have a straight edge on each, the tapered curvature on the leading edges must be laid out, sawed, and dressed. Match the two pieces by planing simultaneously with the pieces clamped together. Don't be concerned about the side tapers at this point. Tack pieces of scrap to the planks or staves in the areas to be left solid so there will be a neat termination of the routed hollow. If you don't have a manufactured router guide, screw a guide piece to the router base. This is to follow the outer contour of the spar. If the hollow works out to be wider than the router base, leave a ridge near the center for support, or add a wide plywood base. Rout in two or three passes for depth. In some cases you may have to deepen the cut by pulling the bit out of the collet slightly. *Be sure the collet is locked tightly*! Any diameter of bit from ⅜ to ¾ inch will do. The latter will create a load on the motor if you do not take shallow cuts.

Now for the side tapers. The material must be removed *from the open side*, so lay out the curvature on ordinates in the usual way. This is an easy freehand resawing job for a power handsaw, but not a table saw. If the kerfs do not reach from each side, the remaining bit can be sawed with a bandsaw or even a hand ripsaw. If you do the resawing entirely with a bandsaw, play safe on the line. Plane the two side by side, identically. When the two are dressed down to the line, saw or rout out generous grooves through the solid areas for wiring, or whatever, and provide a scupper in the butt. Also rout the slot for the halyard sheave or sheaves. Now, when you glue up on your straight bench, the sides will be *sprung in* so there is no run-out of grain. It's best to lay out blocks on the bench to locate the tapers equally about the centerline, the spar on its back. Then glue and clamp in the usual way. Complete the mast as described above.

FORE-AND-AFT (GAFF RIG) SPAR TERMS

Here are important words in the ancient language of spars. *Stick length* is measured from the tenon in the heel to the truck. The *tenon* is that reduced square or rectangle that fits in the mast step or socket. *Bury* is measured from the bottom of the socket to the top of the deck or cabintop. *Hounded length* is measured from deck to the *hounds*, the point where the shrouds join the gaff mast, eyesplices resting on a hardwood bolster or shoulder or shackled to an eye band or *wye*. *Deck to pin* is from the deck to the pin in the peak halyard sheave or block. There is usually a shoulder at this point to support an eye band or the splices of the upper shrouds. The *head* is the rapidly tapering portion from that shoulder to the tip or truck, which is also confusingly called the head. The *truck* is a circular hardwood disc mortised over the tip of the head and containing sheaves or slots for the flag halyards. The *pole* is the length from the hounds to the shoulder at pin height. Some of these terms are included in Figure 15-14.

SOLID SPARS FROM A TIMBER OR BAULK

A timber or baulk is a rough-sawed squared piece of lumber, not a sapling or tree. Solid masts are used today only in gaff rig, such as cat boats and some schooners. Again, Sitka spruce is the best, but being soft, it does not take the chafing of gaff and boom jaws well. Douglas fir makes excellent spars, even though it is heavier. Other good woods are white or Norway pine and a spruce in the Northeast that is quite knotty. This makes good large spars if the knots are quite small—the size of a dime and, of course, tight. I don't know whether any is available at this time.

Try to pick a straight timber in which the heart is near the center at both ends. You can check the solidity of the heart clear through by having a pal rap with a hammer on the heart while you place

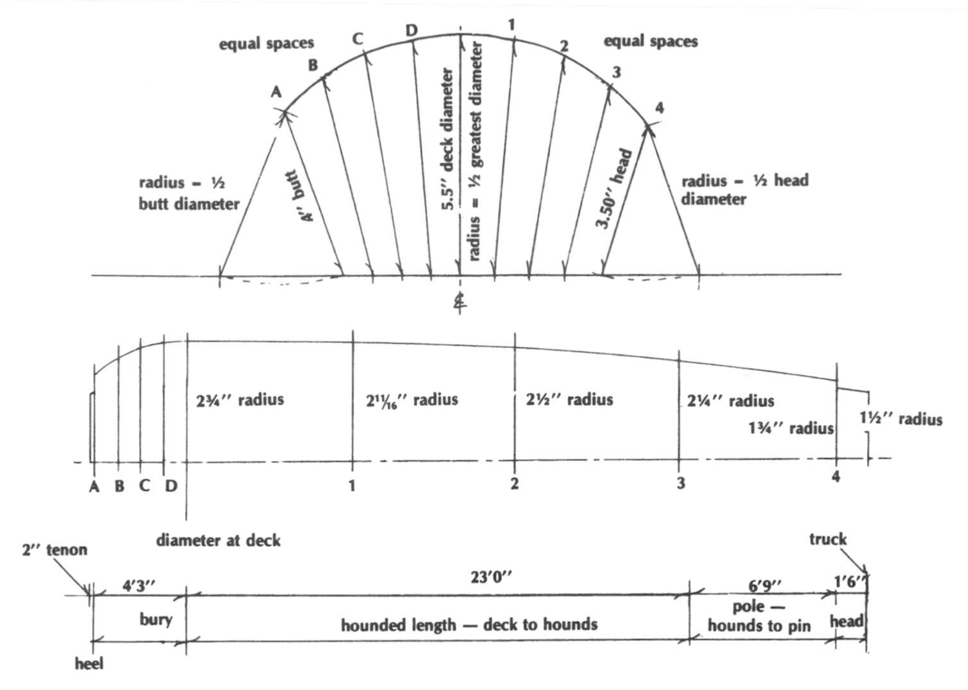

Figure 15-14. *Spar layout, fore-and-aft rig.*

your ear at the other end. A solid heart will conduct this sound readily. A bad heart will transmit a mushy sound. If a good timber otherwise has a slight bend, you may be able to take out the bend by blocking up the timber and placing heavy weights on it for six months or a year. Look out for cross grain (run-out), exposed heart, sap pockets and sapwood, checks anywhere but near the ends, and so on. A spiral grain is very strong. However, you will need three or four feet more timber than the total length of your spar (its *stick* length) because of the tendency to check at the ends. If a good timber shows some wind (twist), you may have to purchase a larger size and work it into shape. This is extra work, but perhaps you can buy such a timber at lower cost. As mentioned, the West Coast is the most likely place to find large timbers, or some southern states.

If your solid spar can come out of a 6 by 6, the rough shaping can be done with a big power saw (or the new portable sawmills—up to 16 inches). Four or five ordinary sawhorses would make a satisfactory temporary bench. If you are an expert with a chain saw, perhaps that's your answer. But be sure you have the skill to saw right on the vertical. Suffice it to say that your horses or blocks must be level across, and long enough to permit rolling the timber 360 degrees.

Your first move is to decide which is to be the straight aft side. If the timber has a slight bend (1 inch in 30 feet) that could be pulled out by the head or springstay, pick this hollow side to face aft. There must be no bend sideways. Lay the timber with the hollow side up. Use a couple of big weights to pull it down to the blocks and a long wooden jointer or jack or an electric plane to take off the sawed surface. Check for fairness with a chalked 1 by 3 on the flat and level across if there is wind (wynd) in the timber. The other sides must be squared to this first surface.

The next step is to snap a chalkline down the center. Is the timber actually relaxed? You'll see that you can shift an end slightly without moving the opposite end. To relax the stick, lift an end so the sag clears all the blocks or horses. Prop it there, then level down to the blocks and mark each location. A series of blocks tacked in these spots will fix the timber while you find the centerline. It doesn't matter if the stick now shows a slight bend. It is relaxed, and your shaping will produce a straight spar. You can see why the rough timber must be somewhat larger than the finished diameter of the mast.

Locate the center of the heart in each end. Plumb a line top to bottom. Hook your chalkline around a nail in the marks and stretch it taut. Stand near the center of the stick and press a thumb firmly on the line. Carefully lift the line far to the right and let it snap, then lift far to the left and snap. You now have a perfect centerline from which all diameters will be laid out. Go over the chalkline with a pencil and straightedge so you don't lose it.

Now you have both sides rough tapered. Plane these surfaces close to the lines, preferably with a wooden jack. Mark the ordinates square to the aft face, then swing the same diameters or merely pick up the measurements from the face. Again saw fairly close to the lines and dress down nice and fair. You now have a clean square timber tapering gracefully to head and butt. There may be just a slight curvature in the aft straight side, with the head bending aft a smidgen.

LAYING OUT A FORE-AND-AFT SPAR

There are different, slightly contradictory methods for laying out spars for the fore-and-aft rig. You can use the proportions from Chapelle given earlier, or you can use a geometrical projection system (see Figure 15-14). I have used both without complaints. I have also arbitrarily picked a diameter at the gaff of 90 percent of the deck diameter, picked two-thirds for the head, sprung a batten from the hounds to head, and—voila—a mast layout! But this was for my own spar. So don't do as I *do*, do as I *say*. Also, there is a nice shape to the geometrical design, and the masts might be nearly identical if one were superimposed on the other.

Start the projection method by swinging an arc equal to the 5½-inch deck diameter on a piece of plywood or paper—a 2¾-inch radius in Figure 15-14. Take 3½ inches (approximately .65) as a rugged masthead (where the peak halyard goes), and step off the radius of ¼ inches along the arc to point 4 in the drawing. Then swing from there to the baseline. Divide the arc geometrically from the vertical centerline to point 4 into four equal parts (or as many as you choose). Do the same along the baseline. Connect as shown. These are the radii at four equally spaced ordinates on the mast from deck to head. Divide the left half of the arc similarly to find the diameters of the bury.

Lay out the ordinates and radii on the timber surface and all other points of your spar length. Drive brads at these points and spring a batten fair to correct small errors. Sawing close to this line will be little problem if you have access to a resaw (bandsaw) large enough to carry the weight and accommodate the thickness. Perhaps a steady helper can aid by pulling moderately as you guide and push the stick. Or use the portable "saw mill" above. Of course, you could do this in one pass of a chain saw, but please allow plenty of room for error.

AN EIGHT-SIDE GAUGE

Now we come to an intriguing part of spar shaping, the octagon. The spar must be eight-sided in the same proportions from butt to head, unless it is to remain square in the bury. Figure 15-15 shows several methods, using gauges and rules, to accomplish the lining off rapidly and accurately. The upper detail is a gauge laid out for spars from just under 8 inches in diameter down to as small as 3 or 4 inches. The lower left-hand sketch dimensions a gauge for 6-inch and smaller spars. It is easier to use. These gauges are simply dragged along with their dowels against the side of the stick, leaving pencil lines in their wakes. At the lower right, I show the use of a rule, making marks at the 7-inch and 17-inch points all along the spar, to give you a series of dots as close as you have time for. Or you could make a gauge on the 24-, 17-, and 7-inch marks.

Set your power handsaw to an accurate 45 degrees and rip off the corners. Stay safely away from the lines and roll the stick if the kerfs do not meet. To rotate a heavy timber, clamp a four-foot 2 by 4 to the butt and heave. The last time I built a solid spar, I knocked off the rough with an electric plane and saved hours and sweat. Then I took a couple of passes on the corners to 16-side the spar to just above the gooseneck. If there had been jaws on the boom, the mast would have been rounded to below the jaw's lowest position and an oak ring would have been fitted to support the boom.

SHAPING A SOLID SPAR

I discussed a drawknife in Chapter Two, but large sizes are needed for spar work. This is the traditional tool for knocking off to 8 and 16 sides, but it is designed for rough work. It is too easy, if you pull the drawknife straight along the stick, to catch

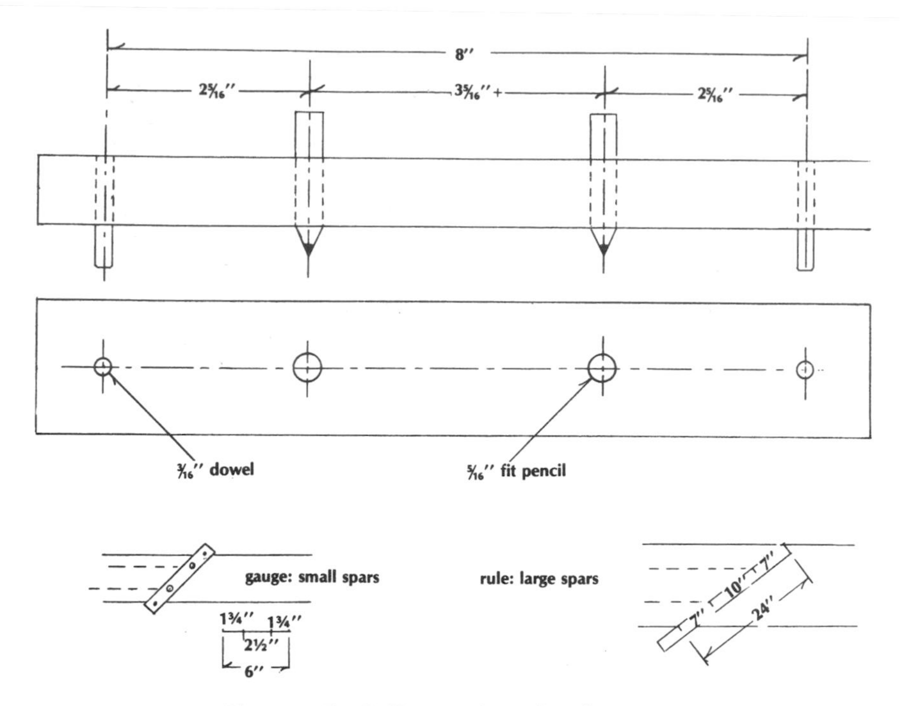

Figure 15-15. *Devices for eight-siding spars.*

in the grain and rip off a splinter as long as your arm. This is wrong. The proper draw is a spiraling motion at about 30 degrees across the stick. Find a drawknife with a blade well over 2 inches deep and about 24 to 26 inches between the handles. If you let the knife bite too deep, mark the area with a crayon so you can avoid planing there and creating a low spot. I made the simple calipers shown in Figure 15-16 many years ago. Set yours to the diameter of the original planed eight sides; then you can gauge the high spots. I prefer to work on a quarter for the full length, rather than rotating the spar. By the time you have worked all four quarters, you will be amazed at the improvement in the fourth quarter. No matter. Go back and shave the high spots until your eye and your calipers tell you that you have a fair spar. You can't get perfection with a drawknife. So, it's time to plane.

I have planed several good-sized spars with small planes. I hope you can buy or borrow a wooden jack plane about 16 inches long. Work end to end, a quarter at a time, then repeat, then rotate for the next quarter, and so on. When you think the spar is perfect, go over it again with a 24-inch jointer set fine. It will find high spots you won't believe are there. Never plane at the slightest angle to the grain. Your chances of finding a hollow wooden smooth plane of the right hollow are remote. So use ordinary window glass broken up so the fragments have a concave sharp edge. Even though these may not match the spar's actual radii, glass is great for taking off fine shavings like fuzz.

Follow with a coarse 50- to 60-grit open-coat garnet sandpaper or aluminum oxide production paper. Do not waste your money with white flint paper. You may prefer to double over the sheet, holding it under your entire palm. The hollow sanding block in Figure 15-17 was made with plaster on a board. You may want several, perhaps foam, to fit the spar reasonably. In general, stroke around the spar at about a 45-degree angle. Next, use about 100- or 120-grit paper and go with the grain on the last half-dozen strokes if you intend to paint. If varnish is your thing, finish with 200-grit folded sheets, but no blocks. If your spars show checks, do not panic. These have no effect on strength as long as they follow the grain. Use an oil can to squirt clear Cuprinol generously into the checks. Most checks will open and close with the weather and humidity. Never fill them with any-

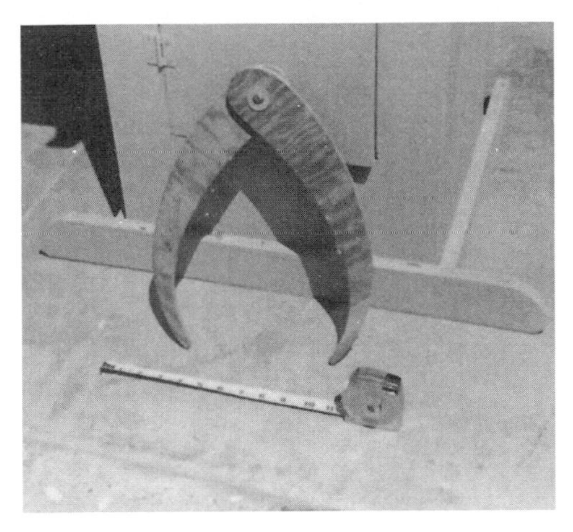

Figure 15-16. *Simple handmade spar calipers.*

Figure 15-17. *Sanding boards of plaster, paper tubing, and shaped wood.*

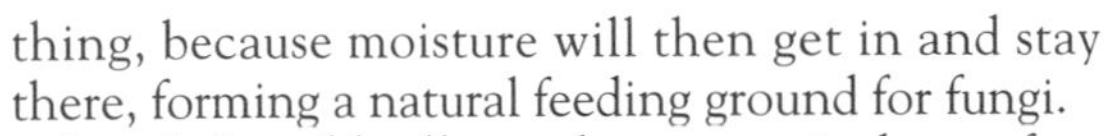

thing, because moisture will then get in and stay there, forming a natural feeding ground for fungi.

I wish I could tell you that epoxy is the perfect sealer for solid spars, but no one seems to be sure of that. I would, however, cheerfully settle for a couple of coats of boiled linseed oil thinned with one-third turpentine (some swear by thinning with kerosene). After this is well dried, scuff-sand lightly and varnish with four coats. Or don't varnish, but give the spar a couple of coats of the oil during the season. I have had good results by soaking the spar with two thinned coats of varnish, followed by three or four of full strength. Sand lightly between coats, of course. Polyurethane varnish sounds good and looks great on some things I have built, but these were things that were not out in the weather. Get competent advice on modern finishes before you use any of them.

If you have remained alert to this point, you already have the essential ingredient (perseverance) of a respected joinerman. If you show that same splendid characteristic in performance of your work, you will not merely succeed, you will excel. And I shall take pride in having contributed something to that excellence.

PART 3

16 Woodworking Projects for the Home

If you are an adequate woodworker now but want to be more than that, or if you think you have two thumbs on each hand, or *know* you are nearly a pro, take my advice: *read* the earlier parts of this book. Why? First, because it is easy reading, not at all like a textbook, with only a little unfamiliar "boat lingo." Second, it is about simple, proven, how-to steps that can lead you to turn out fine woodwork. On every page there is another idea, gimmick, hint, or trick about how to make woodworking projects stronger, prettier, more economical, and more fun! And, hey, do not for a minute assume it's not for you because you know *everything*; many experienced carpenters, boatbuilders, and cabinetmakers have told me they have found solutions to old problems, here and there, that made the original edition of this book easily worth the small investment.

What qualifies me to set myself up as a woodworking expert? In addition to my many years as an amateur and professional boatbuilder, I had another 10 years' experience operating a one-man cabinet shop. I was sort of an advanced amateur cabinetmaker, and only because I had learned a great deal as a boatbuilder and had read and *studied* every book on the subject. My old friend who suggested the cabinet shop was right when he insisted, "For cryin' out loud, you've got a lot of tools and can do anything in wood, so why not build things for ordinary folks, never mind the boat and yacht crowd!" I took that advice, threw together a tiny shop in Santa Barbara, ran an occasional three-line classified ad that began, "RETIREE builds. . . , " followed by cabinets, bookcases, entertainment centers, vanities, kitchens, desks, etc. Work poured in from the ads like water from a faucet because my prices were in line with my overhead and I could draw adequate perspectives of the pieces I intended to build. I gained a vocation doing what I loved (second to boatbuilding) and got paid for it, too.

When I wrote the original *Practical Yacht Joinery* in 1983, I wasn't thinking much about home cabinetry. But soon after, it occurred to me that the publishers and I had missed a good bet: the close relationship of boat tools and joinery to woodworking tools and cabinetry. They are identical but for a few exceptions. For the most part, only the end products differ in appearance and function. So in this new edition, I decided to add a chapter on practical cabinetry. For those turning directly to this section, I again emphasize the value of reading through the entire book. For cabinetmakers, it is a veritable gold mine. Every tool, tip, or technique found there can be applied to the kinds of woodwork you see around you every day and that you can create if you take your time and work carefully.

This chapter shows you how to build a few straightforward contemporary designs, which I have built at one time or another in my old shop. I do not repeat the "how-to" details of each step described and illustrated in the previous chapters. If there is some step you have trouble following, the answer lies within the earlier sections of the book. I do, however, include alternate ways of construction according to available tools, if appropriate and not previously described. See these methods in earlier chapters: method A (hand tools only), method B (powered hand tools), and method C (advanced bench and floor machines).

Here are the projects:

1. Bedside table or telephone stand/bookcase
2. Space-saving bookcase
3. Cedar chest/cocktail table masquerading as a seaman's treasure chest

4. Simple hutch with open bookcase above, storage below, with tips to make it "colonial."

WHAT ABOUT TOOLS NEEDED?

I shall not repeat this advice about tools. An impoverished workman—or a masochist—might actually succeed in building cabinets by method A. Our ancestors did and many of the results are cherished as beautiful antiques 100 or 200 years later. Lumber or plywood certainly can be hand sawn and dressed with a plane, as my father taught me when I was eight or nine. But this book is aimed at slightly more practical levels. You can do accurate work with a power handsaw, but much better by making yourself a portable table saw on a box (see "Happiness Is a $5 Table Saw" in Chapter Five). That is a method B that becomes a method C. It is admittedly a bit short of the myriad advantages of a full-size table saw, but if you equip the saw with a carbide-tipped 7¼-inch blade with twenty or more teeth, or with a planer or cabinetmaker's combination blade (next best), you can make almost polished cuts. As you carry it around or in the trunk of your car, you will wonder how you got along without it.

Do not fail to build a SLAT (shown in Chapter Five) for that portable or full-sized saw. Proved by years, even generations, of use, it always cuts precisely square and makes perfect miters. With a plywood square-guide, you can even use it to do short ripping jobs. Long after writing *Practical Yacht Joinery* I clamped a Skilsaw in a Black & Decker WorkMate, with a SLAT fitted to it, and it is still turning out fine work.

Yes, some of the pieces require the use of man-sized tools; if you read the word "dado" or "groove" or "molded," etc., the text will not go into those operations. You will find the detailed explanations in the earlier chapters.

THE PROJECTS

I believe you will see that most of these projects were designed to answer a specific need—such as a tall, narrow bookcase for a room with a shortage of wall space. Needless to say, I hope these projects are pleasing to the eye. But, for the most part, esthetics are secondary. As Frank Lloyd Wright said, form follows function. All the projects here were designed to do a job, not to decorate a room. I've given basic dimensions, but I encourage you to customize each of these projects to fit your own taste and space. Even the little sailboat, *Trifle*, described in Chapter 17, was designed for an adult education manual-arts course for the easiest possible construction, roomy enough for a couple and their two youngsters (or for two or three big teens), fun to sail, sophisticated enough for teaching your family, unsinkable, requiring no precision frames, mold, or finicky jig, car-toppable, and, of course, low in cost. Anyone who can follow detailed instructions can build it.

PROJECT 1: A SIMPLE BESIDE TABLE OR TELEPHONE STAND/BOOKCASE

I am one of those weirdos who thinks he must have a clock he can read in the darkest night. This little bookcase is just the right height (24 inches) for a lighted-electronic-display clock-radio next to my pillow, along with glasses, my current book, etc., while lower down are my next-to-read books, and a family of mysteries. Additionally, for four years I've used it as a telephone stand with three city phone books on the adjustable shelves. However, if I had designed it for *that* purpose only (as you may do), I would have drawn it, perhaps, to a more convenient 40 inches tall for more comfort. I say this in part to get your attention; all the pieces here are intended to be adaptable to your needs. As you see the bedside table here, the height of sides is 22 inches, the depth is 12¾ inches, while the base and top measure 13¼ × 16¾ inches (less trim). The base sits on 1-inch hardwood balls (available at any self-respecting lumberyard). If you want it taller, make it 40 to 42 inches or so. Add a fixed shelf at the midpoint (dadoed in) with three adjustable shelves.

Since the construction material is ¾-inch mahogany, birch, or oak plywood, the edges are covered by mitered hardwood strips (I used ¼ × ¾ inch). For years I have glued edge trim and held it temporarily with strips of masking tape 3 to 4 inches long pulled down hard and spaced every 3 inches. There is no need for mechanical fasteners. Using a good carpenter's glue, such as Elmer's, I have never known this technique to fail. The miters(and all sawing) must be cut with a fine 10-inch carbide-tipped blade (no less than 40 teeth) on your table saw, and using a SLAT (see Chapter Five) for perfection.

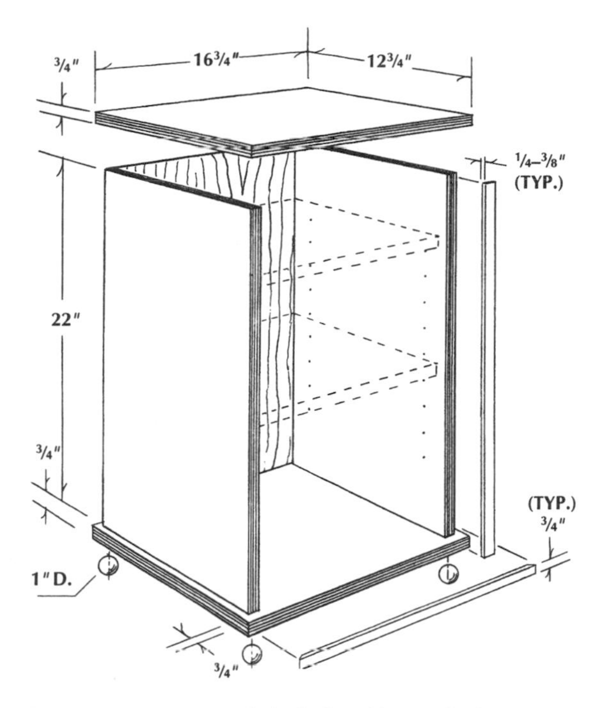

Figure 16-1. *A simple bedside table or telephone stand/bookcase.*

If you prefer a little less contemporary styling, replace the trim strips with moldings made with your router, shaper, or molding head. Whatever shape you choose, mold it on the jointed edge of the solid hardwood lumber, then rip off the molded trim as you see in Figure 4-49.

I fastened the top on the sides with glue and screws plugged over, the grains matching; fasten the bottom as you please. The cabinet you see is made of Philippine mahogany finished with only marine mahogany paste wood filler rubbed in across the grain until so clean it is nearly polished. Someday maybe I'll give it three coats of urethane varnish. I glued on a back of 1/8-inch mahogany doorskin, the edges dressed back to 45 degrees, so it was almost completely hidden.

PROJECT 2: A SPACE-SAVING BOOKCASE

This tall and narrow bookcase is another example of cabinetry created for particular needs. Over the years, my wife and I have lived in dozens of smallish homes packed with hundreds of books, most of them linked to my calling. But others were prized, too, some being oversized encyclopedias, dictionaries, illustrated histories, special National Geographic editions, etc. Short of both floor and wall space, the only answer was to go up! So I went up—83 inches. Just right for one of our narrow corners `tween divan and doorway. For you, perhaps the design will fit into otherwise lost spaces of your home. Certainly, it is a useful design in those small apartments or condos where some of the headroom may be only seven feet. But of course the top is useful under a 96-inch ceiling, too. For

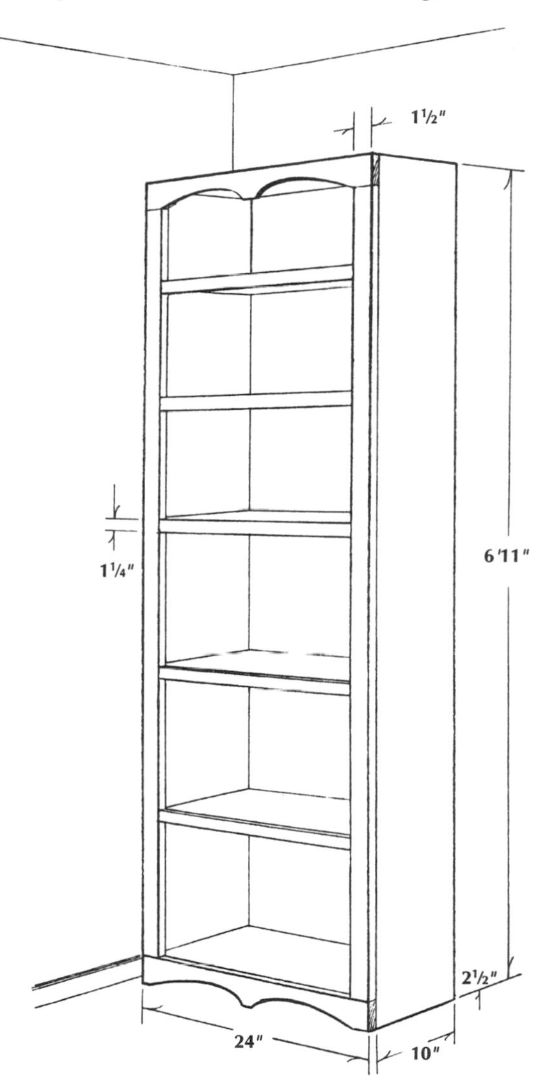

Figure 16-2. *A space-saving bookcase: Basic dimensions.*

us, a 24-inch width was mandatory. But make yours whatever width and height pleases you, allowing one inch to clear the ceiling.

Construction was of cured yellow oak and 3/4-inch oak plywood. I found that, as with many planed hardwoods, the planer corrugations were impossibly difficult and time-consuming to sand out. There are probably half a dozen different shapes of scraper that can remove these, but the adjustable cabinet scraper (Figure 2-29) has a 2½-inch blade mounted in a frame with two handles for pulling or pushing the tool, always at an angle to the corrugations. Rectangular blade scrapers do a similar job, but not as easily. Follow sharpening instructions in Chapter Two. Best to work on the stock about 2¾ inches wide, then rip to joint and finish-sand to 1¼ inches wide for rails and stiles. A belt sander clamped to a bench belt-side-up is acceptable for this light work.

I divided mine into six unequal spaces, the lower three being 14⅜ inches in the clear, between shelves. The upper section I divided into three spaces about 11¼ inches clear. The lowest shelf is 2½ inches above the floor.

Cut your rails precisely (whatever your width may be) between stiles and with the usual fine carbide-tooth blade. The assembly is to be dowelled, flat on a bench with weights to keep it flat, or you may clamp it all together on the carcass (a British term for frame) after the shelves are in place according to your own design. Dowelling procedures are explained in Chapter Seven.

This case is easy to put together with glue in dadoes in the sides ¼ inch deep, and a similar rabbet for the top. Each one is a pass or two with a router or over a dado set or adjustable (wobble type) dado blade, or even multiple saw kerfs if done precisely. The rails will cover any small imperfections in workmanship. If you decide to avoid the work of dadoing for the shelves, you may fasten with screws covered with wood plugs, or, descending to a lower quality, use finish nails and glue; these, however, will almost certainly be visible.

Do not studiously avoid variations. I like some shelves to be adjustable, and perhaps you might consider this convenience, say, for the two shelves of the upper section (so no rails there, of course). This might work well combined with doors for the lower two or three spaces, making a nice storage cabinet. Doors are easiest to make in this order: overlapping, rabbeted (or lip), flush. The overlapping type with a finger-pull recess along the top is very popular; this requires a matching heavier molded edge on four sides. You will find ideas on doors and drawers in Chapter Eleven, including how-to details. An adjustable shelf drill jig is described in Chapter Five and shown in Figure 5-21.

Gather three or four pipe- or bar-clamps before beginning the assembly. To ensure perfect squareness, measure from corner to corner of the back and when diagonal measurements are precisely even, tack on a couple pieces of scrap wood to hold the carcass square. Turn over and check with the dowelled frame, unless you built it on the carcass. I assume you seated the dadoed shelves perfectly into the recesses, but check again with a long straightedge along the sides. If there are no bumps or valleys, clamp the dowelled frame in place; or go the other route, assembly on the carcass. When gluing the joints between stiles and risers, with either method, wet the end grain with glue first, but keep the glue about ⅛ inch behind the front surface, or a dark line may be prominent. When that glue is tacky, brush on a little more in the same way and also on the edge of the stile. This job is made to order for Aerolite, a crystal clear two-part adhesive with a slow kick-off. Now glue and screw stiles and rails (or assembled front) to sides and shelves, with wood plugs over the screws (see Chapter Two).

If you can feel the joints of risers and stiles, stick a piece of plastic tape across whichever feels a smidgen low, then sand with a hard block and 300- to 400-grit production paper. Stop when you start to cut the tape as you want to avoid sanding scratches across the grain. Remove the tape and very carefully sand the high spot a few strokes, keeping the block right in line. You may hold or clamp a scrap on the low spot so the sanding does not encroach into that area. Patience is the password.

Again, I glued and bradded on a back of birch doorskin on which I had brushed two coats of varnish, no filler or stain; I like the books to contrast against the light background. And again, I rubbed in an oak filler stain on the case, just because I do not care for glossy furniture.

PROJECT 3: A "TREASURE CHEST" CONVERSATION PIECE

As you may have surmised, it is difficult for me, a Piscean, to stray far from the sea. However, this modified replica of an old seaman's chest is for dry-land storage (it's even cedar lined) and may be con-

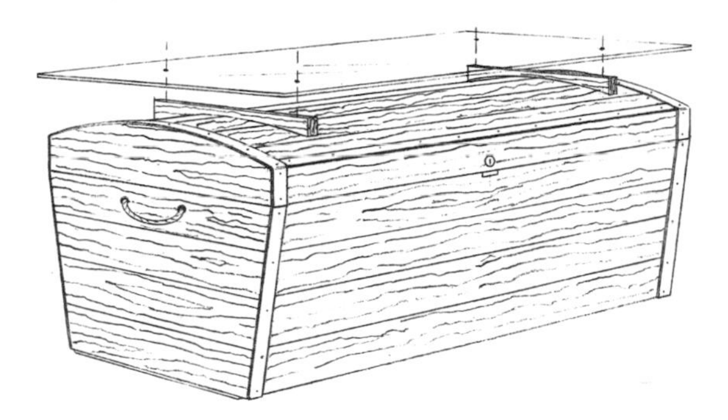

Figure 16-3. A *"treasure chest" conversation piece. It might make an interesting cocktail table.*

verted into a unique cocktail table. Several construction methods are described that require a bit more skill than the preceding projects, and almost the total elimination of fastenings.

The strange shapes of sea chests evolved over centuries. The fo'c's'les (once "forecastles"), compared in volume to a small bedroom today, for many months or years housed from six to twenty or more men. Each seaman kept nearly all his worldly possessions in his sea chest; all space was extremely restricted, leading to shapes that we think are strange. I believe the reasons are these: first, the sides incline, approximating the hull's shape, so foot-room is increased. Second, when sitting on the chest, a seaman could pull his feet back so his toes wouldn't get trampled. Third, that crowned or cambered top (undoubtedly an uncomfortable seat) was to shed the seawater that dribbled or poured below too often, and to prevent the stacking of other heavy chests on top.

Most sea chests were built strongly to withstand years of rough handling and voyaging around the seven seas; some were decorated, real works of art, usually bound with iron or brasswork. I have designed one much lighter in weight, attractive in appearance, versatile, and one unusually constructed with ordinary or exotic materials, including a little metalwork you can do. As I suggested in the first two projects, you certainly may increase or decrease some dimensions, but please do not destroy the chest's character by eliminating the inclining sides or the cambered top.

Here are some options to consider: construction with ash plywood, oak, mahogany, or teak; a cedar lining; one or two trays; a removable glass cocktail table top; and a lock. Metal trim, in my opinion, is not optional.

For the woodwork, you will need a table saw with a SLAT (Chapter Five) for precision work, and a fine-tooth carbide-tipped blade. The brass trim requires only a hacksaw, small drill, and file. Look for a source for about two quarts of epoxy resin adhesive. If you go for teak construction, be prepared to cut tongue and groove on the saw or have tongue-and-groove bits for your router, or own a molding head. For the cambered top, a few minutes with a band saw or sabersaw is all that is necessary.

NOW THE SIMPLER CONSTRUCTION

If this chest were ordinary box construction, the sides would lap over the ends, but that would expose the sides' end-grain. The metal bands, in order to hide the grain, would be much more difficult. Therefore, the ends *overlap* the side material, unless you resort to mitering the corners, which I don't recommend because it leaves them very vulnerable to damage.

Saw both sides to the dimensions 13½ × 35 inches; and saw the ends to 13 inches on the bottom, 18 inches on top, and 13 inches in height. Thus the sides are ¼ inch oversized to provide for dressing to a 9½ degree angle on both top and bottom, and for close fitting when top is ready to be installed. The ends may be planed to finish dimensions now.

The four parts should show a planked effect. Divide each into four spaces 3⅜ inches wide, which should be scored into the plywood; a clamped-on straightedge and a small nail-set or nail will do this nicely. If you elect to go for solid construction with tongue and groove, the exposed surface must be the same 3⅜ inch width. Make the scores or joints deep enough so paint will not fill them years from now; try to match the sides and ends for appearance; and remember that the end pieces cover that end grain.

Now a radical, tricky procedure. Pre-assemble the chest in preparation for gluing with epoxy adhesive. Also put in place the corner and perimeter reinforcements. With the side and end part upside down, hold them together with tape so that they are in accurate relationship while you drive two or three brads through each end, leaving the heads barely out. Carefully roll the chest right-side up and measure corner to opposite corner, then brad a diagonal across for absolute squareness. Place

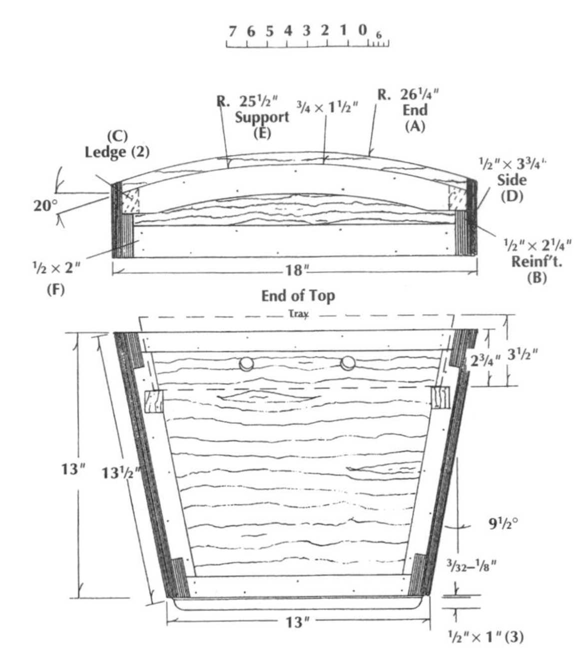

Figure 16-4. *Construction of chest and top ends.*

this "mock-up" on a piece of ordinary doorskin and mark around the bottom. Saw inside of the lines to about 45 degrees, so it will be nearly 100 percent invisible later, strengthened with three ½ × 1 inch cross pieces after gluing, along with corner reinforcements also coming.

Release the pieces of tape, but do not pull the brads since they will aid in reassembly with glue. The best method is to wet the gluing surfaces with unthickened catalyzed epoxy resin first, let it start to kick off, then apply epoxy thickened with a small amount of pulverized limestone (calcium carbonate) to make a syrupy consistency. (You can buy calcium carbonate in an eight-pound bag for around $2.95 at garden centers and hardware stores. Another thickener is wood flour, available in paint stores.) Using the brad points to guide you, reassemble, tape together, replace the diagonal and the tape. Before the glue is hard, pull the brads and let everything cure for twenty-four hours. As is, this structure is quite strong, but slightly flexible, so the next step is to install the bottom doorskin and reinforcements everywhere.

The simple doorskin (or other light plywood) needs only a good carpenter's white glue and small nails. However, as insurance, to support unusual weights, add at least three ½ × 1-inch battens or runners so the chest may be moved easily over rugs and carpeting.

The bottom corners lengthwise call for two ½ × 2 inch, one edge sawed to a 9½-degree angle, epoxy glued and bradded from inside. All inside corners need ½ × 1-inch reinforcements. However, those inside of the top edge must be deeper, front and back, for hinges and lock or latch (if any), so ½ × 2¼ inch is shown. Plywood may be used for all of these strengthening members. The upper ends are stiffened by the smaller size. All the upper perimeter may be clamped for gluing, the brads everywhere being overkill.

Figure 16-4 indicates blocks for supporting a tray; build two trays if you like, but the tapered sides mean a lower set of support blocks. I have seen double trays, a pair one-half width so one can be lifted out, the other slid aside, thus you can dig below easily. In that case you would need a full-width batten instead of the little blocks; ¼ × 1½ inch on edge would be ample support (I always worry about saving space). I feel that the upper tray should be located so it extends well into the rounded top.

WHAT'S THIS ABOUT A CEDAR LINING? AND TRAYS TO MATCH?

Aromatic cedar does protect woolens from attack by moths, they say. Lumberyards, home centers, hardware stores, etc., sell kits or packages of cedar tongue-and-groove boards 4 inches wide, ⅜ inch thick. The smallest pack covers 16 square feet, just enough for the chest and top (about $25, 1992). Just cut to length between corner reinforcements, stick on with panel adhesive (in a tube). No fasteners needed; besides, your heirs may prefer to remove it.

For a tray or two, consider cedar, perhaps calling for the next larger cedar kit. It will cost a few more dollars, but matching trays would add a refined touch to this unique piece of furniture. The cutaway drawing in Figure 16-5 is based on ⅜-inch cedar with corner and perimeter reinforcements (⅜ inch square), small but all glued with epoxy, bradded from inside. You may also build with ¼-inch plywood (AB or BB grade) and reinforcements could be as large as ½ inch square, maybe from leftovers from the chest construction. If you want to exercise your skill, you might tackle compound mitering of the corners, in which case the

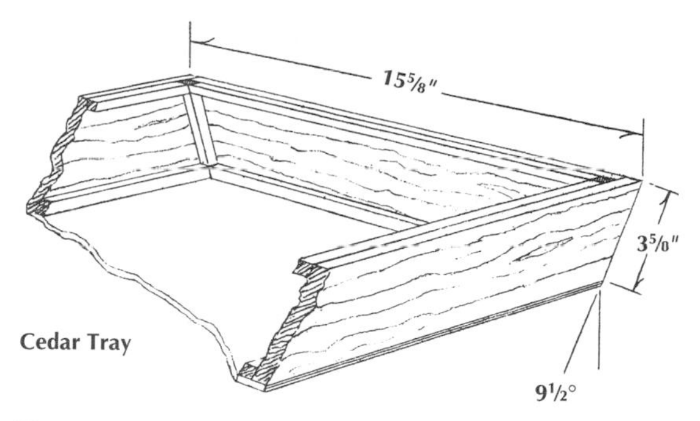

Figure 16-5. *A tray for the treasure chest.*

reinforcements would have to be stronger, not less than ½ inch square. Remember to wet the end-grain gluing surfaces first with unthickened epoxy.

To assemble, start by building up the ends with corner reinforcements glued along the 9½-degree angles, followed by the longer stiffening reinforcements at top and bottom. When these are cured, you'll have no problems adding the sides and their respective stiffeners. However, to avoid misalignment, I advise you to use clamps on the corners, the assembly resting on a flat bench and held with weights until cured. Fit the long stiffeners, plane off excess. Then add some brads and a bottom of doorskin or ⅛-inch plywood, glued, nailed, and weighted down once more. I see no need for handles as fingers under the stiffeners should be more than enough.

THE CAMBERED TOP

To me, and on paper, the top looks best cambered to a radius of 26¼ inches (see Figure 16-4). If you laminate the top of two layers of ¼-inch plywood, plus a final laminant of hardwood doorskin (hopefully ash to match the chest itself), you will have a rigid top ⅝ inch thick. This reduces the radius of the heavy support piece to 25⅝ inches. You must decide now on the materials and consequent thickness of the top so you can build up the end assemblies as shown in the drawing, Figure 16-4. The support piece (E) is sawed to a radius of 25½ inches because that provides for a ¾-inch cover or decking which could be three ¼-inch plywood laminants, or tongue-and-groove "flooring" you make to that thickness. If for the support piece you choose hardwood or fir flooring, both of which should have a thickness less than ¾ inch, the radius would have to be changed accordingly.

The tricky part is that the end assembly has to provide for the support piece (E), the reinforcements (B), and the ledge (C), and this last cannot be installed until the frame is fixed. So I suggest that you follow the same procedure as in the chest itself, the ends and sides taped and glued with epoxy, cured while weighted on a floor or bench (a diagonal holding it to fit the chest top absolutely). This way the support can be fitted right into the corners, and the side reinforcements or stiffeners (B) can come next. That all-important ledge (C) follows; note the tangent of the camber at 20 degrees, so the decking will be flush with the end (A) and sides. There could be a lot of stress on the ledge, so in addition to epoxy glue, I would add three or four screws plugged over, just as long as their points do not penetrate the sides. I hope it is clear that these corners have ample reinforcement, no verticals needed, or possible. And once again I suggest that you try the fit of all parts, a "dry run," right on the chest itself so errors are not graven in stone.

DECKING OPTIONS AND HOW-TO

My decking preference is for three plies, say two ¼ inch decent plywood and the third ⅛-inch (approximately) doorskin to match the chest, including a scored plank effect. The plies, as well as several other decking schemes, will require no less than three solid forms for support during laminating or other construction; these must be sawed to the radius of end support (E) used as a pattern. Finish is not important but material should be ¾-inch scrap. Tape them in any old way. It is O.K. to use good carpenter's white glue or plastic resin. Nail through into the under-supports starting near the centerline, and into the ledge (C) and support (E), using tiny blocks or scrap so 1¼ to 1½-inch #18 wire nails can be replaced later with brads in ends and ledges. See Chapter Ten for beamless laminated construction.

If you should feel a spongy spot or so, drill a tiny hole to let the air out and place weight on the areas during curing. Add more weight to prevent twisting.

Also consider using ½-inch plywood as in the chest sides, or ¾-inch plywood (matching sides) saw-kerfed within ⅛ inch of total thickness, spaced approximately 1 inch apart. Try spacing and depth on scrap; if spacing is too great, flat spots will show. If you fill the saw kerfs after installation with epoxy glue (catalyzed slowly) using a big rubber syringe, your decking will be very strong; or press in a putty of epoxy thickened with pulverized limestone. Sand these seams, of course. If you are nervous about strength, make two small beams, say, ¾ inch wide by 1 inch deep, preferably of plywood, and have these ready to epoxy glue and brad as soon as the seams are filled. All brads from outside should be set in, and filled if too obvious.

In a moment of mental weakness, I mentioned several kinds of tongue-and-groove decking, but you must be a masochist to do it. The backs of each piece must be backed out to fit the radius of the support piece (E), requiring that your jack plane must be fitted with an iron radius ground for that purpose, or that you are lucky enough to find the narrower roughing or scrub plane (see Planes, Chapter One). Other than that, making tongue-and-groove decking is almost fun with a pair of bits in your router or a molding head on your table saw. But such work in teak, oak, or ash is not the same as in fir, pine, or mahogany. Commercial flooring almost always has a recessed underside, so backing out may not be required; do not expect the joints to close up tight unless you relieve the lower corner. The decking will also need the support of two light beams, although filling the joints with epoxy glue will add greatly to their rigidity. I suggest that you drill lead holes for the diagonal nails (1½-inch #6 galvanized finish nails) and that the decking be glued to the beams and all around the perimeter.

Whether scored ash plywood or doorskin, or meticulous tongue-and-groove hardwood, or whatever, lightly fine-sanded where appropriate, under three or four coats of urethane varnish, this top will be beautiful. So now let's get into the metal trim.

BRASS TRIM FOR BEAUTY AND DURABILITY

The metalwork is simply straight flat bars of polished brass, about 3/32 or ⅛ inch thick by 1 or 1⅛ inches wide. You need only a hacksaw, file (or simple grinder), and a drill. No brazing is necessary if you can accept close-fitting joints where cover verticals meet the formed pieces. If you want these corners brazed, consult your metalworking shop, then file or grind to lengths required. The slightest over- or under-length will be a tragedy. The shop will not work on your chest and cover, but it might be useful to have a close pattern of the top handy. Drill holes for escutcheon pins (small round-head brass nails) about every 3⅛ inches; but at the ends of each length, drill for round or countersunk oval-head screws because the pins have inferior holding power. Always bend before drilling the smallest hole in metal.

If you want antique-looking hardware, such as lifting handles instead of the rope I show, try a local custom cabinetmaker. He can either supply you with catalogs or tell you where to find such hardware. I am sorry I have not been able to find my catalog from ten years back. Perhaps the Thomas Register of Industry in your public library will give you a steer, or, of course, your hardware dealer—for locks, latches, and hinges, too.

You may have discovered that I am not an instructor in finishing of cabinetry, furniture, or even boats. Materials, formulations, techniques are evolving so rapidly that my suggestions would become obsolete before the ink were dry. I would not approve of the best fir plywood other than painted (and that is best for this job if antiqued a bit). Ash has a beautiful grain, much like oak, and can be finished with clear varnish for a light golden color, or stained like dark oak. Consult first with a good paint dealer regarding the many materials from which you can make a selection and the different colors and tones available to meet a color scheme.

If you build your treasure chest with good tools, patience, and love, you will have a piece of furniture you will be proud to display in your living room, den, or bedroom. Now let's talk about making it truly unique!

HOW ABOUT A "TREASURE CHEST" COCKTAIL TABLE?

This unique glass-topped table is sure to bring comments from guests at your next occasion. Believe me, it was difficult to draw in perspective, but adapting the glass to a rounded top is a piece of cake. It begins with a sheet of heavy plate glass or mirror, two inches wider and longer than the chest; at 20 by 38 inches it is nicely balanced. Make two hardwood rails about 1⅛ by 1½ by 15 inches spiled (see Chapter Two, Scribing and Spiling Compass, also Ogee) to fit the camber of 26¼-inch radius; a graceful ogee formed on the ends would be pleasingly appropriate. Screw-fasten from inside without glue (you or a later owner might decide to eliminate the glass). Cover the upper surface with a strip of felt, plastic, or rubber. The glass must be drilled and countersunk by the glass shop for four #12 brass or bronze screws. And have the corners nicely rounded and edges ground, perhaps polished, if cost is within reason.

There might be rare occasions when you want to get into the chest. The glass will be heavy, so first install a brass chain to keep the top just past the vertical when opened, and, second, weight the bottom to prevent the chest from flipping over when opened quickly. If you should test the screws or replace them, be very careful that you do not stress the glass or you will be out big bucks!

Thus endeth the lesson on Bingham's Treasure Chest. Enjoy!

A MODERN HUTCH BECOMES AN ANTIQUE

The contemporary hutch in the photo was designed for a lady who wanted a pair of these. They were intended to be installed in her two eight-foot bedroom closets. For this simple purpose, contemporary or"modern" seemed most appropriate. However, she liked them so much she put this one (without drawers) in her living room after I added a durable plastic countertop for use as a bar. She did the finish. It passed as a genuine hand-rubbed oil finish but took only a few hours; details later.

In the photo you see the lauan mahogany plywood appearing to be lighter because of the strong California sunlight. The style is close to Danish: straightforward, the plywood edges nicely disguised by 1/4 by 3/4-inch Philippine mahogany trim glued on with no fastenings (described in Project 1). A slightly more involved creation than the bookcase in Project 2, a whole world away from the treasure chest in Project 3.

However, to give that basic hutch a more substantial, heavier feel and more for you to cut your teeth on, I drew the dimensioned perspective in Figure 16-7. It shows a 1 3/4-inch frame around the bookshelves, mitered joints, a 3/4 by 1-inch trim on the counter edge mitered to a 3/8 by 3/4-inch flat molding on the sides. The mitered frame may be dowelled, half lapped, or splined for strength, all described in Chapter Seven.

Notice that I show the shelves as adjustable (see Chapter Five for a drill jig for this and future shelves).

Perhaps your decor needs less severe treatment, something nostalgic, reminiscent of our five hundred years of distinctive home furnishings design and construction. So if you want to try your cabinetry skills on a more ambitious and complicated hutch, see Figure 16-9, developed on the bare bones of the contemporary piece. I call it "Bogus Colonial" because I don't vouch for the authenticity of any of its details.

FIRST THING'S FIRST: BUILDING THE BASIC HUTCH

Figure 16-6 can't convey the true color of luan mahogany plywood (a type of Philippine mahogany). It was used throughout—sides, countertop, shelves, doors, top, bottom—even the 3 by 3-inch kick. Being a cedar rather than a true mahogany such as African and Honduras, Philippine mahogany has a coarser grain needing a filler-stain, but it also costs a lot less. More later.

First of all, I routed a rabbet 1/8 by 1/2 inch all around the back edges, sides, top, and bottom, for gluing and nailing a doorskin back. Your table saw with a fine blade will also do this in minutes. Although I made my shelves adjustable, thus relying on the back to prevent wracking, it is sound construction to dado the shelves into the sides 1/4 inch deep. A 1/2 by 3/4-inch rabbet is needed to secure the top to be glued and screwed at an angle downward into the rabbet later. Other parts must be cut 1/2 inch longer than the inside width; but if you go for adjustable shelves, their length should be about 1/8 inch less than the inside dimension if you plan on dowels for supports; or 3/8 inch less for most metal or plastic support brackets or clips. The depth of the shelves, too, may vary, as fixed shelves should be tight against back panel and side trim, whereas adjustable shelves are easier to handle if there is a bit of clearance. In the Danish style, trim is flush to the inside. But if your mahogany was planed to 13/16 inch, as mine was, let it remain overhanging inside.

Trim may be applied better after assembly, with miters in the upper corners, using glue and numerous patches of tape pulled down hard. Watch that it does not creep if it extends past flush. There must be trim on the ends of the countertop and across its forward edge, about 3/4 by 1 inch mitered, glued, screwed, and plugged, so it will hang over the three sides approximately 1/2 inch.

Figure 16-6. *Danish mahogany hutch.*

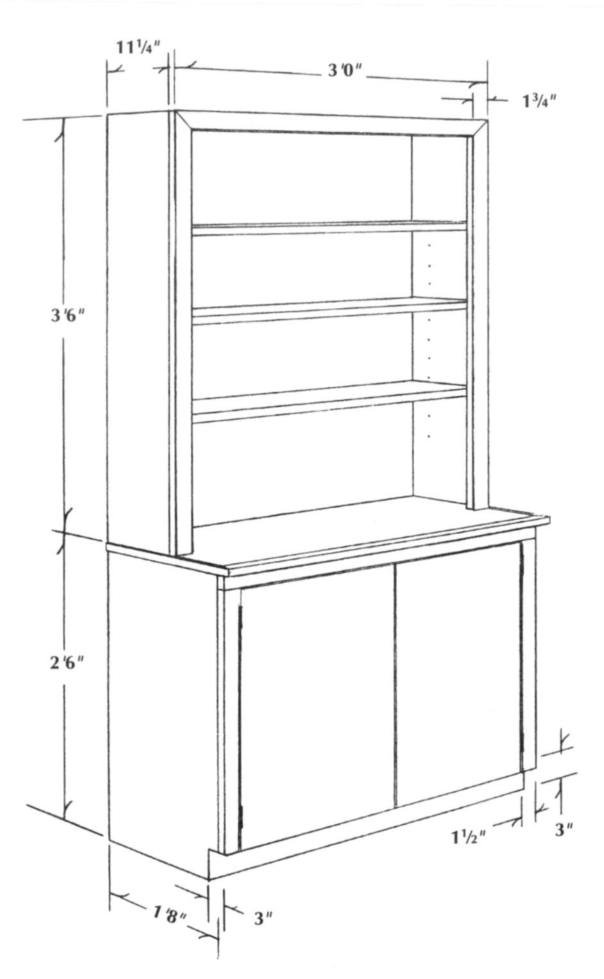

Figure 16-7. *The basic contemporary hutch with enhancing cabinetry.*

The Assembly Set-Up

Place a side on its back, with well-glued dado ready to receive its countertop; press the countertop down into the dado, properly located to meet the back panel and flush with the forward edge of the sides. Clamp a box or something else square to hold the countertop vertically while you tack a diagonal across. The dadoes should not be press-fit, but snug enough to hold upright all the shelves, countertop, lower shelf, and bottom. Have a batten handy long enough to tack to the parts and hold them perfectly vertical. Now wet the other set of dadoes with glue, and carefully lower that side so you can slip the parts into their respective slots (with help so you do not mess up the glue). Tap on a block lightly to press them gradually to the bottom of the dadoes. Tack a long diagonal across the shelf edges and another across the lower section, and turn the hutch onto suitable high and low blocking. Check with a level and straightedge from corner to corner of the back and/or sight on it to be sure no corners are askew.

Lay your doorskin or grained ⅛-inch plywood into the rabbet previously routed or sawed; if it does not drop in place, remove the diagonals and adjust things until it does. Of course your back is accurate!

Remove it and apply a good white carpenter's glue; replace it and nail on 4- or 5-inch centers with 3/4- to 1-inch headed wire nails. I advise you to have two or three bar or pipe clamps handy; it is not necessary to squeeze all the glue out of the joints, but rather to insure that the cabinet sides do not show bumps and hollows. Clamps are also advisable when fastening the top pieces during driving screws from above (and of course you may screw and plug from outside).

There is another simpler system: all parts screw-fastened and plugged over. Additionally, gluing would be a good idea if you can prevent messing up the surfaces to be finished natural. Tacking light scrap to locate shelves firmly would be a great help. But to sum up, dadoing is the easiest, fastest, and strongest way to install all the parts.

The photo of the original hutch shows the doors hinged directly on and overlapping the side, no stiles or risers, to reduce labor costs. As the simplest acceptable Danish style, I would want the doors hung flush on the inside of the trimmed side member, but having stiles and risers on sides and top, either half-lapped or mitered, would be superior. See a heavier version in the drawing, Figure 16-8.

When hanging the doors, see that the clearance all around is uniform, preferably not more than 1/16 inch (exaggerated in the drawings); spacing with bits of heavy cardboard works well. The original doors had touch latches, but ordinary knobs and magnetic latches are handy, too. Trim on the doors does not show up, but all edges were covered by glued and taped strips of Philippine mahogany.

Incidentally, the mahogany boards from which I ripped the trim were milled to 13/16 inch thick, so trim strips were finished 1/4 by 13/16 inch. I allowed the trim to overhang on the inside 1/16 inch; I used this in other pieces over the years. I do not know whether this milling is standard everywhere.

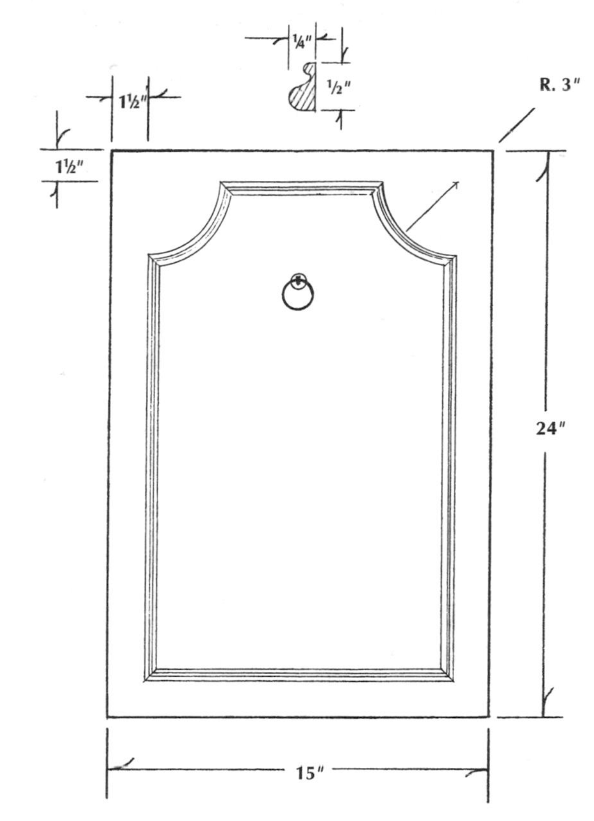

Figure 16-8. *Door molding pattern. Trim the dimensions for a 1/16-inch clearance.*

Finishing Basic or Contemporary Styles

While the hutch in the first two versions is quite plain, I cannot see saving a few bucks by going to fir plywood anywhere. Even for painted storage cabinets. You can never touch it with sandpaper until you have sealed it with Firzite or other brands made for that purpose, and then paint several coats, because that fir grain will leap out at you forever. Of course, you know it is never varnished or stained unless you insist on advertising it as a cheapy.

Philippine mahogany, hopefully available in lauan plywood or lumber, is a medium to dark reddish brown. Some plywood may come in a lighter shade, almost pinkish tan. They must be filled and stained, and a variety of filler-stains for this specific purpose are usually found in marine supply houses, rarely from neighborhood paint and hardware stores. They may try to sell you a "mahogany" stain that comes out purple. Philippine must take a thinned paste-filler that is *rubbed in*, then *rubbed off*! The filler must be seen in the grain, never on the surface; if you can scratch anything off with your nail, it needs more rubbing (across the grain, and if late, probably with a solvent). There are other finishing suggestions following the description of the Colonial Hutch.

As you may have deduced by now, ash plywood and solid lumber is my favorite combination, followed by oak as these strong grains add character. But your new piece may have to blend into your decor, so genuine mahogany plywood and lumber might be a candidate if price is acceptable. Be aware that all these, and especially birch, take stains beautifully. Birch itself can be a natural golden under ample coats of varnish, as well as many shades of tan and brown to blend in comfortably with other pieces in the room. I detest finishing, so make few claims of expertise. Please consult your nearby paint, lumber, marine supply, and hardware dealers before you start these projects. I would be surprised if they did not give more timely advice on materials and systems. But there is more to follow on finishing, a surprise, after the Colonial Hutch.

A Metamorphosis: A "Bogus Colonial" From a Modern Hutch

More than thirty years ago I made notes and thumbnail sketches about a colonial hutch I saw in a magazine. Since then I went through my second and third boatbuilding career, more than ten years or so operating my cabinet shop, writing and illustrating *Practical Yacht Joinery*, and designing the *Allegra* line of traditional sail yachts. I found recently that the proportions were approximately those of the modern hutch above but I cannot guarantee that the details are certifiably New England Colonial. Perhaps "imitation" would be a kinder, gentler word than "bogus"—or how about simply calling it an "interesting colonial"? In any event, I hope you will see it as a more challenging job for you, so long as it suits your decor.

The dimensions are from the Basic Hutch, except that the space above the countertop is fixed at 15 inches, no shelves being adjustable. The top overhangs and is trimmed with a molding, not set down into a rabbet, a quibbling change which I think may be a colonial characteristic. The curved "knee" addition to the sides extending out to the lower front is, I believe, another typical feature, and I have taken some pains to make it as graceful as possible. If your construction is from plywood, one piece per side, there would be no problem, of course. But if you choose to build from pine or hardwood lumber,it would be practical to join two pieces to make the 20 inch width, as indicated by the broken lines. I suggest that such a long joint should be stiffened by dowels as long as you can handle, grooved plentifully to release trapped air, perhaps 6 to 8 inches in length, spaced about 8 inches apart. Perhaps you can rely on the dadoed countertop and shelves to prevent warping, but if you are not positive about your lumber's curing, you might glue and screw hardwood supports under shelves and counter for stiffening.

I used few moldings in my shop, but I have drawn several sections from stock samples and numbers in my old files. I made my moldings now and then with a molding head and a selection of molding bit sets from Sears; you might check that out, but do not expect close to what I show here. See Chapter Three for basic molding construction. There should be a molding specialist near you.

The curious scalloped effect across the top is, I believe,a common colonial characteristic. I have spaced the verticals for stacking large plates and glassware. The attachment of the ¼ by 1-inch pieces backing the vertical moldings may not show clearly; they must be glued and bradded to the inside of the top and bottom faces, so the shelves must be notched accordingly. The moldings are the last to go on, glued and bradded.

I suggest that you assemble the entire face or frame, the joints dowelled or half-lapped, then simply glue and screw it to the box, plugged over, the top fastened into the edge of the upper face piece. During the assembly, clamp a diagonal to keep it square.

Of course, this hutch may be constructed from quality hardwood and plywood and trimmed and finished to match, but if you want a New England Colonial effect, consider pine lumber. You will find that #2 pine lumber is likely to offer large areas of clear wood with judicious cutting. Very *small* tight knots give the piece authenticity, or if that worries you, bore them out and glue in plugs, or graving pieces (rectangular patches). Character, character!

Ask your paint expert about the popular wash effects, but watch out for any filler or stain that raises the grain. There is no reason why this "antiquey" piece should not be built with hardwood lumber—ash, oak, birch, maple, butternut—you name it! Just so that it fits the decor.

A Super-Fast "Boiled Oil" Finish!

An age-old finish preferred by many experts is boiled linseed oil tediously rubbed in by the palm of your hand at least monthly for no less than a year. That woman for whom I made the two hutches was one such expert. But when I told her my technique, she sniffed, "Fiddlesticks and bosh! I have been using this process since you were being weaned?"

For those unfamiliar with it, here it is: Sand the

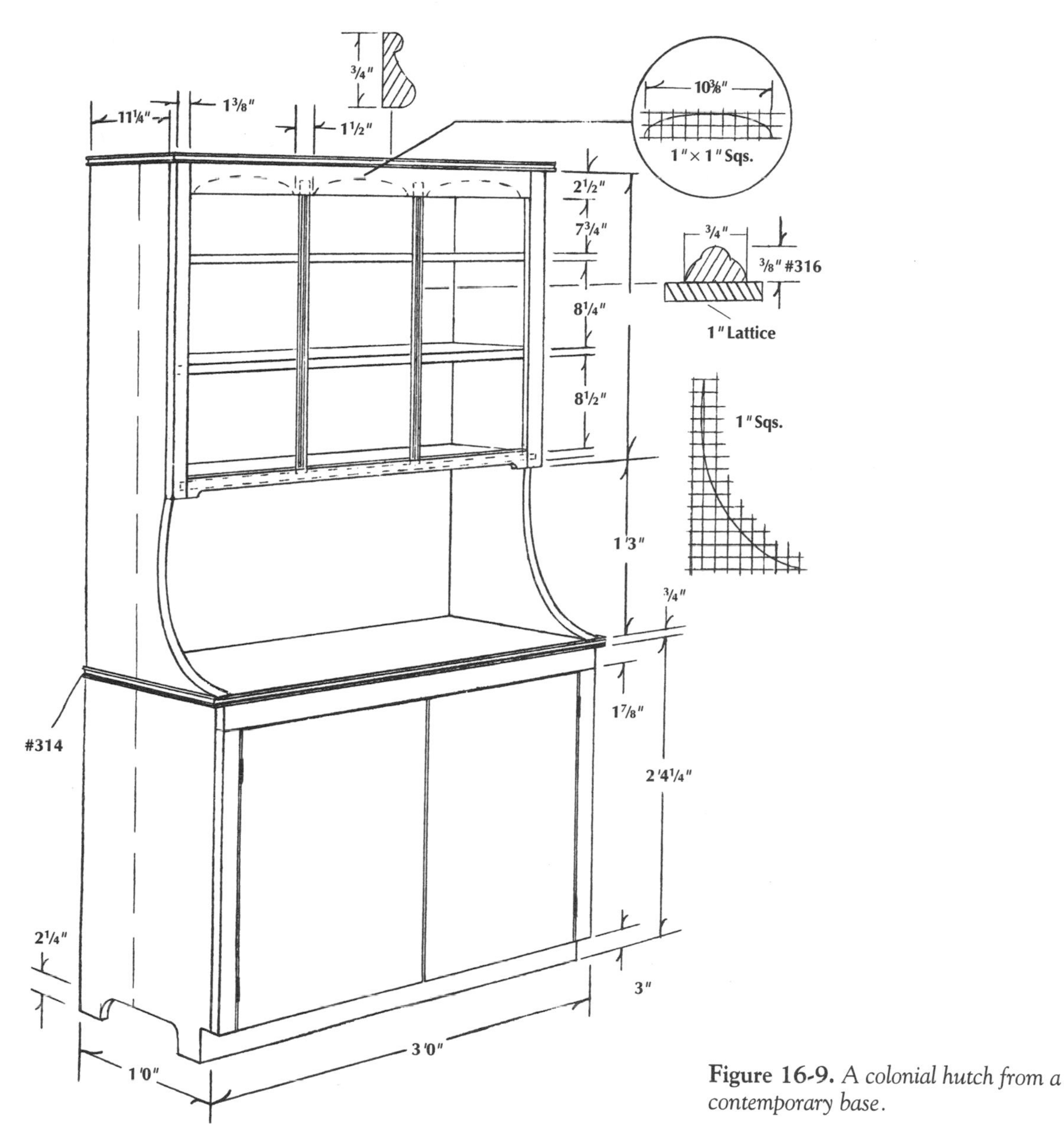

Figure 16-9. *A colonial hutch from a contemporary base.*

piece with #000 steel wool pads, dust with an old soft brush, then leave it in a room at 75 degrees for a day or so (and warmer is better). Mix up a two or three cup container of half-and-half white shellac and lacquer *thinner*. Wipe the surfaces well with a tacky rag soaked in shellac wrung out thoroughly, to remove every speck of dust. Apply very wet brushfuls as fast as possible. It's fast-drying so be ready to mix small cupfuls for the "holidays."

Hardware

I have shown only simple loose-pin brass hinges. I would pick antiqued copper, brass, or iron hinges and the same in drop-ring pulls, but I lack the skill to draw them. Look for a builder of furniture reproductions; he will have 16th to 18th century hardware or know where to get it. As for me, I would still go for today's touch latches, but you are the boss!

17 Build Your Own Trifle

Trifle is a delightful little sailboat you can build with only a nodding acquaintance with ordinary garden-variety tools, such as a crosscut handsaw, jack plane, and an egg-beater drill. I designed Trifle for an adult education industrial arts class on building dinghies suitable as yacht tenders, frostbiters, and cartoppers. The boat had to be cheap but durable, unsinkable, roomy but lightweight, and it had to be safe, lively, easy to handle, and, above all, easy to build.

Proved by hundreds of nonprofessionals and on dozens of small boats designs since 1965, the Stitch-'n-Tape system used in building Trifle requires no costly, finicky building jig, frames forms, mold, or strongback. The hull consists of just two side panels and two bottom panels of inexpensive ¼-inch exterior plywood (or fir marine plywood), plus bow and stern pieces (called transoms) of ½-inch exterior plywood. The basic structure is simple. You drill about fifty small holes around the edges where the panels for the bottom, sides, bow, and stern transom meet. Then you simply stitch the pieces together with copper wire, and there you have it. If you can tie your shoelaces, it's easy. This stage takes about a full day's work for you and an assistant, who can help hold the panels in place while you stitch. Of course, there remains the matter of gluing and finishing the boat.

A Nearly Exhaustive List of Materials

- 3 ¼-inch 4 × 8 sheets of ACX exterior, marine fir, lauan, or African mahogany plywood
- 1 ½-inch 4 × 8 sheet of plywood (matched to above options)
- 1 ¾-inch 4 × 4 half-sheet ACX plywood
- 21 pounds scrap lead
- 10 yards 44-inch 4- to 8-ounce fiberglass cloth
- 2 rolls (1½-inch and 3-inch) fiberglass tape
- 1–2 gallons epoxy resin (as needed to cover approximately 100 square feet.)
- 1 8-pound bag pulverized limestone flour (Cab-O-Sil)
- 2 gallons acetone
- 8 12-foot lengths hardwood (mahogany, teak, oak) dressed to ¾ inch × ¾-inch
- 18–20 board feet ¾ × 10-inch Sitka spruce (for mast; get longest lengths possible) or aluminum mast and hardware

Miscellaneous Hardware

chainplates, gooseneck, rigging wire (45 feet of ³⁄₃₂-inch stainless), end fittings, turnbuckles, bow and stern chocks, cleats, centerboard hoisting sheave (1½–2 inches), rudder gudgeons and pintles, miscellaneous braided nylon line (¼-inch and ⁵⁄₁₆ inch)

WHAT HOLDS TRIFLE TOGETHER?

Not copper wire! True, the wire ties *pull* the panels together, twist by twist, into accurate lines of the boat, but they are not a structural component. In fact, they are eventually nipped off and sanded before you sheath the hull. All the joints are bonded with a fillet you make from epoxy resin/limestone putty covered with fiberglass tape. This is how the hull gets its basic strength.

HOW BIG IS THIS JOB?

What follows is a sequence of operations. The details come later. I arranged things this way so you can make an informed decision early on whether to build Trifle from this book with raw materials, or from plans with full-size patterns, or from a kit, which you can order from me, 1425-A Longbranch St., Grover Beach, CA 93433. Trifle is a fun boat. No sailing dinghy in its class has ever been easier to put together or more versatile. But it does require work. How much you choose to do is up to you.

BUILDING SEQUENCE

1. Lay out sides and bottoms on ¼-inch plywood (Sheet 4), clamp the worst sides together; lay out bow and stern transoms on a single sheet of ½-inch plywood. Saw and dress to lines. You'll have to join 4-feet and 8-feet plywood hull panels with butt blocks using epoxy resin and strong thickener.
2. Drill holes, 3/32 inch or smaller, 14 inches apart, along all panel edges that will be joined. Roll and brush penetrating epoxy on inside.
3. Stitch bottom panels along centerline with copper wire by twisting pieces of wire loosely through matching holes; open panels like a book to reveal the shape of the boat's bottom.
4. Loosely stitch bow and stern pieces accurately in the bottom "vees." Stitch both sides to bottom and ends.
5. Place scrap seat and other spreaders to form sheer (Sheet 6); fasten temporarily with duct tape.
6. Bond all inside corners with epoxy resin and filler, plus tape. Let stand 24 hours.
7. Flip hull onto sawhorses or boxes. Nip off wire ties and set in; sand all corners to pencil-thickness radii.
8. Build centerboard trunk on bench; glass the inside surfaces. Bore centerboard for carriage bolt. Drill and saw centerboard slot 15/16 inch.
9. Assemble centerboard trunk/seat, bond to hull with fillets of epoxy putty and fiberglass tape; join the seat to sides with similar fillet or optional wood support (with epoxy adhesive) (Sheet 3 and Sheet 5).
10. Cover entire hull with 44-inch, 10-ounce (or 7-½-ounce) fiberglass cloth; wet through with epoxy resin. Allow 24 hours for curing; fine sand to make hull ready for paint.
11. Screw-fasten skeg; bond with resin, filler, tape, and fiberglass.
12. Construct sheer clamp or gunwale-seat, the last structural member in the hull (preferably hardwood to be finished bright later on).
13. Install bow and stern seats on face pieces, with openings for inserting flotation (plastic bags of foam "popcorn.")
14. Paint exterior and interior; stain and varnish gunwales and other trim. The hull is now complete.
15. Construct rudder and centerboard, both of ¾-inch plywood, streamlined, sheathed with fiberglass cloth. Cast lead insert finished to 20 pounds for centerboard thickness and fit in before fiberglassing.
16. Build solid mast sail track (Sheet 7).

TOOL POOR? NO PROBLEM!

Someone once said Trifle could be built with a dull Boy Scout hatchet. This is a slight exaggeration, but most of the basic tools can be found in almost every home. Essentially what one needs is a crosscut handsaw (sharp), or sabersaw, jack plane, eggbeater hand drill, screwdriver, sanding block, carpenter's square, a light hammer and nail set, a good ⅜ by ½-inch spruce or fir batten for fairing curves (you'll have to make this), and stuff for fiberglass work, including throwaway 3-inch bristle brushes, resin roller and pans, and several good 3- to 4-inch paint brushes.

It also helps to have some power tools, including a belt or disc sander (if you are experienced with these), a power saw with a fine blade, and a ¼-inch electric drill.

GETTING STARTED

To start everything, you have to transfer the lines of the panels from the drawings (Sheet 2) onto two sheets of ¼ 4 × 8 ACX fir or marine plywood. It is easiest to cut and finish matching pairs of panels simultaneously, bradded together, thus guaranteeing absolute duplication. Place the sheets so the good sides face each other, and hold them with about six small brads (outside the areas you will cut). Start to lay out the hull panels so that the "sheer" lines utilize the manufactured straight edge.

To transfer the dimensions to the plywood, drive 1-inch brads at the intersections of verticals and curves, plus a couple in between for fairing with a good batten. Mark the location of the verticals; these may be handy later on for locating the centerboard trunk and seat. Hold the batten against the brads with additional brads opposite (never drive nails through battens). *Sight along the batten*, making sure the batten touches the brads at all verticals. Mark the line with a sharp pencil.

Remember to *save* the lines when you cut. If you use a handsaw very close to the lines, make sure the strokes are perpendicular to the plywood, otherwise the saw might cut into the invisible line in the lower panel. Mark and saw the straight lines of the ½-inch bow and stern transoms. With the two bottom panels clamped together, plane and/or block-sand right to the lines, producing two identical panels. Do the same for the side panels.

HOW TO JOIN 8-FOOT PLYWOOD

First, study Sheet 4, which shows suggested arrangement of panels and other parts on ¼- and ½-inch plywood. Do not *think* of joining full plywood sheets! Only the side and bottom *panels* have to be butted. Use the manufactured ends as the location of the butt joints, since they are perfectly square. Only one side and bottom panel needs to be marked on the bad side, since the others will be sawed, dressed, and drilled together with the marked panels, good side to good side.

The ¼-inch butt-blocks are 8 inches wide (Sheet 4). These are cut off to lengths 2 inches shorter than the side and bottom dimensions, at angles parallel to chines and centerline and square under sheer. This provides clearance for gunwales and wire stitches along the centerline. For appearance, bevel butt block edges.

Mix a little batch of epoxy thickened with limestone or silica flour to a syrupy consistency. Drive 12 to 16½-inch copper tacks into the butt blocks so the points just show through; this is to prevent squeezing the adhesive out of the joint. Smear the plywood edges, too. Lay the parts on old plywood, with plastic wrap underneath to prevent adhesion to the under surface; drive tacks and place weights on top. Don't touch for 24 hours. Later sand off any points that protrude.

STITCHING SYSTEM

In 1968–69, I stitched up a Trifle with the help of my son. At the time, we drilled a couple of hundred holes and stitched the boat together with 50-pound nylon fishline. This wasn't a bad method, but it is faster, easier, and less expensive to use copper wire ties and drill fewer holes. It also makes for easier adjustments. Use 20 gauge wire, since finer wire tends to cut the plywood fibers. The twisted ties are easily nipped or chiseled off on the outside after bonding has cured.

Clamp panels outside to outside and drill 3/32-inch holes about ⅜ inch from the edges on 14-inch centers. Note: There are no holes along the sheer. Best to drill a pair of bottoms; lay the sides as close as possible on the matching bottom panel, and mark corresponding holes (Sheet 8). Drill the holes to match in both end pieces.

Begin stitching with the bottom panels, on the centerline. Clamp the panels so the good sides face each other and the edges line up exactly. Now you can proceed to stitch with copper wire.

Cut a few 5-inch lengths of wire. With your assistant steadying the bottom panels, put in a couple of ties and twist loosely. Do not twist too tight, since some play may be needed for later adjustment. Remember: the corners must meet perfectly. When all the ties are in place (plus an extra loose tie near each end corner), spread the panels like a book; they will form to the exact shape of the boat's bottom. But it is inside out; open the "book" again so the good sides are out. Check the corners. If misaligned, tap sharply, making the panels butt at the angle with no overlapping. Prop up this assembly. Now tighten all wire twists sufficiently to pull in but not damage panel fibers.

Now for the bow piece. Press it down into the open "vee" with the pointy end in dead center. Mark to matching holes and drill. If your work is accurate, twist in ties to bottom. Then repeat these same steps with the stern piece. Finally double those ties at the corners, just for insurance. Be sure that the outside surfaces and panel ends are flush.

Have your assistant support the sides from farther aft, and/or from the transom, while you insert loose ties. Another loose tie to the upper corners of bow and stern would stabilize everything further so you can drill the rest of the holes fairly accurately, then fasten. I have to mention here that it helps to install the opposite side panel in simultaneous stages. Perhaps light rope running from around one side and around the other is worth trying to hold things together as you twist the ties.

At this point, the hull is quite limp and only resembles a real boat. It's time to stiffen her.

TYING IT ALL TOGETHER

You must add temporary bracing to hold the shape of the boat until epoxy filler and fiberglass tape can be formed into all the joints and corners. The only member that is dimensioned in the plans is the 50-inch seat, which spreads the sides. Cut a scrap board to 4 feet 2 inches and press it down inside the hull to 13 inches above the centerline, its center to be 5 feet 7 inches forward of the inside of the transom at the same height. When both ends of the temporary support are equidistant below the sheer, tape securely with duct or packaging adhesive tape. Prepare to install four more temporary spreaders on 24-inch centers, springing the sheer into a lovely fair curve. With the spreaders duct-taped in place, you'll find that the hull is quite rigid. Build a pair of bucks or supports to support it under the bottom. To check that the hull does not twist, place a straightedge across at bow and stern, and lay a level at both locations. The boat must be absolutely level. If not, shift the bucks accordingly, or insert wedges to raise or lower a corner. Level fore and aft means nothing.

STRUCTURAL BONDING

Bonding means to join all hull parts with a fillet of putty and fiberglass tape. The putty consists of epoxy resin and mineral fillers or thickeners (no plastic spheres, balloons, wood flour, or other weaker material). A new, recommended thickener is pulverized (not ground) limestone flour (dolomite) that is as soft as talc. You can find it at nurseries in 8-pound bags for a nominal price. A better-known filler is Cab-O-Sil (silica), but it is more costly. Follow the resin manufacturer's directions for epoxy proportions.

Prepare for bonding the inside of the hull by marking ¾ inch and 1-½ inch from all joints so you can later lay the 1½-inch and 3-inch fiberglass tape neatly. Find a container lid (plastic or metal) the right diameter for forming the fillet, which should be pressed in, wide enough to clear the wires in all corners, chines, centerline, bow, and stern. Smooth and press the filler putty into the joints just enough to fill holes and gaps, extruding slightly outside the lines (this should be scraped off before it hardens). Clean up the excess putty inside beyond the marks and immediately lay the narrower tape in place, wetting it with unthickened epoxy, smoothing and adjusting it with a throwaway brush. Follow as soon as possible with the 3-inch tape, soaking it so no white spots (air pockets) are visible. Be neat! After the fillet has cured for 24 hours, fill and sand imperfections.

If you failed to slice off the extruded filler that squeezed out, you will now have to sand it off. Regardless, turn her over onto several strong cartons or a pair of low sawhorses. Nip or chisel off the wire ties, and set in what remains of them. Now sand all the joints and round them to the radius of a pencil. This is done so the fiberglass sheathing will cling crossing the corners instead of pulling away because of its inherent stiffness; glass resists making sharp bends. Be sure that the exterior is clean and smooth. Do not sand a radius on the top plywood edge, the sheer; the sharp corners must meet neatly with the four-piece gunwales to be installed later.

CENTERBOARD TRUNK

The new Trifle centerboard is constructed of ¾-inch plywood with a 20-pound lead insert, covered with fiberglass cloth and epoxy resin. This makes the total thickness of the centerboard 13⁄16 inch; the slot is cut 15⁄16-inch wide to allow clearance for sand or gravel. In addition, there must be fiberglass tape on each edge of the opening through the fillet and the plywood panels.

To cut the centerboard slot, rent a concrete drill 15⁄16 inch in diameter (remember that limestone or silica fillet is hard). Start drilling a series of holes along the centerline starting 55 inches from the bow vertical (taken at the sheer, not the camber). That allows 3 inches for the forward spacer in the trunk (see Sheet 3). Best to have someone sight on your drill to ensure it's perpendicular; and the more holes, the better, since they have to be connected with a rented metal cutting blade in your power saw, or a rented Sawzall. Block sand the edges.

To head off problems down the road, the centerboard trunk should be absolutely sealed. Sheath the inside plywood end to end. Make the end spacers ⅞ inch as shown in Sheet 3, but wrap each in epoxy/fiberglass cloth except for the outer edges and weight them down (plastic sheet between), to ensure bonding at the corners. Another way is to assemble the trunk while the fiberglass is wet, clamp all together, then drill and screw. Be sure the upper edges of the trunk are square to the sides; the lower curved edges must be spiled (fitted) to the centerline fillet. Allow the trunk to cure, then next day prop it accurately in its place.

Make a dry run of the centerboard trunk and seat (which you must make 50 inches wide and bevelled to fit the hull). The best way is to make a pattern of scrap strips clamped or tacked together to fit the slight curve of the hull side. If the ½-inch plywood or ¾-inch mahogany seat and the centerboard are accurate, all should go together nicely. Place a level on the side of the trunk and mark on the underside of the seat. Mark the seat's location on the hull side first to ensure it's located correctly (Sheet 3). On your bench, nail and epoxy glue both the seat and the trunk and form fillets to make one very strong unit.

The next step—a tricky one—involves making fillets around the trunk sitting on the hull centerline in order to bond the seat to the sides. First tap wedges up into the centerboard slot so the trunk is lined up properly and securely. Seat the trunk in a bead of epoxy putty. Form the fillet between the trunk and hull and tape around the joints. Remove wedges with care and clean the slot of any excess putty. Tape the seat in its correct location to hold the trunk vertically while curing.

The plans call for fillets supporting the seat at the hull, which means you'll have to turn the hull on one side, then the other, to apply the fillets.

SHEATHING THE HULL

Unless you have not taped the *exteriors* of the panel joints, use 44-inch, 10-ounce fiberglass cloth on the exterior. It will help protect the corners. You have the option of using fiberglass tape over the joints and 7-½-ounce cloth on the exterior of the hull—as some experts prefer—but if you are new to this process, I suggest you stick with the 10-ounce cloth.

Lay the 10-ounce cloth over the hull and let the excess material overlap the centerline, hang down, and lap across the bow and stern transom. Trim off the excess three or four inches below the sheer, but leave the doubled areas, since this will strengthen and protect the entire structure.

Follow the epoxy resin manufacturer's directions for the sheathing. You will need a couple of throwaway paint roller pans, a short-nap foam roller for epoxy resin, several 3-inch cheap bristle brushes, a coffee can or two and/or gallon cans. A gallon of acetone is necessary for cleaning brushes, rollers, pans, etc., immediately after use. I can't advise you on catalyzing because it differs with different brands. If you are nervous about a few voids in ACX plywood, pick up a rubber syringe for shooting resin 'way down there (and clean it immediately).

Stir the resin well, then roll it generously on the bare plywood, as you unroll the cloth. The brush and the roller will help you to slip the cloth around a bit, to straighten it. Make sure you roll and brush any white spots (trapped air) until the cloth is transparent everywhere and the surface of the weave is filled.

When the sheathing has cured for not less than 24 hours, sand the entire surface with #220 production paper to give the paint a little tooth to hang on to.

At this point, you need to make a decision about the skeg (Sheet 3). Although optional, it is necessary for towing and improves rowing. Its effects on sailing, if any, are minuscule. Make the skeg of mahogany, fir, or ¾-inch ACX plywood (use a remnant from the centerboard/rudder half sheet), 4 inches deep at the transom, tapering to feather out at the centerboard slot. Drill for six or more 1½ to 2 inch #10 stainless steel flathead wood screws on centerline. Bed the skeg in epoxy putty and form a taped fillet all around. Cover the skeg with 10-ounce fiberglass cloth and epoxy to head off rot. You may be able to sand out some of the laps in the cloth there and everywhere in the bottom and transoms, but it's a tedious job, and not essential.

ESSENTIAL COMBINATION SEATS/FLOTATION

The bow and stern seats are not easy. Both faces of ¼-inch ply are bonded to the sides and bottom, leveled at the heights in Sheet 3. Make tops from the remnants of ½-inch plywood, but rest them on faces with cleats for screw-fastening the tops (coat the inside surfaces with epoxy before bonding down the seats). Cut hand-holes 4 inches in diameter, large enough for stuffing in many plastic bags full of chopped foam or packing "popcorn" under the seats. Add a 1-inch drain hole over the centerline. These compartments should last for years if you coat them with epoxy inside.

THE GUNWALES

The last *structural* piece in the hull is the sheer clamp or gunwales. I refer to it as the gunwales/seat because it is more than 4-inches wide so that you can sit on it for balance when the wind is blowing great guns. But, most importantly, it serves to stiffen the topsides. Please note that drawing C on Sheet 3 and Sheet 5 show a 3- or 4-foot series of openings so sand can be cleaned out (3" spaces amidships"). Locate first so chainplates can be bolted 15 inches aft of the end of the centerboard slot.

Start with six blocks per side, ¾ by ¾ inch, 3 inches long. I use mahogany, teak, ash, oak, or any other hardwood. Screw from the outside and glue these (6 on each side) starting near the center; and fore and aft of these, continuous lengths to meet the transoms neatly. The other two gunwale pieces inside may be finish-nailed, all with epoxy glue. Break the inside corner slightly so varnish will last a little longer (sandpaper on a block will work for this). The outer piece, a rub strake, may also be nailed and epoxied, but use #6 stainless steel or bronze screws at the ends, plugged over. A small brass half-oval would be a nice touch—and prevent a lot of grief.

RUDDER AND CENTERBOARD

The rudder *can* be made from ½-inch plywood (Sheet 3) covered with 8-ounce fiberglass, but you'd be better off making it out of high quality (grade A or B) ¾-inch ACX or other moisture resistant plywood (see also Sheet 4). Both sections of the rudder must be streamlined, the trailing edge tapered to a rounded ¼ inch (the fiberglass can be bent over this edge, covered with Saran Wrap, and clamped between scrap until the epoxy has kicked off). The leading edge must be rounded, or better, elliptical. The tiller should be constructed of ¾- to 1-inch ash, oak, or mahogany, nicely rounded, about 3½ to 4 inches wide at the rudderhead, slotted to fit, with a through bolt; leave enough clearance for fiberglass sheathing. Be generous with the tiller, not less than three feet (but four feet can be reached from a more forward position). A ¼-inch brass or stainless steel bolt prevents splitting the tiller and permits you to swing it up or down to your comfort position.

Construct the centerboard of ¾-inch ACX plywood cut to the dimensions in Sheet 3. Ample stability is assured by a 20-pound, ¾-inch-thick lead insert with a 4⅝-inch radius. Avoid encroaching on the centerboard's streamlining. Saw out a casting mold from a ¾-inch piece of scrap. Sand the edges of the mold and nail it securely to another scrap board. Coat all with sodium silicate (called water glass at drug stores or foundry and patterns supply dealers). Scrounge or buy about 22 pounds of lead scrap (old tire weights are readily available). To melt the lead you can use an old pan on a barbeque or gas range—and be sure you are in a well-ventilated space. Rent a plumber's ladle and keep it hot in the molten metal. (You cannot pour from the pan, since the lead cools quickly when removed from the heat.) After the dross is skimmed off, the lead should be more than sufficient to fill the mold to the brim. After the lead is cool, break it out of the mold and press-fit it into the hole in the centerboard; epoxy glue it if not a tight fit, plane off flush both sides and fill spots with epoxy paste to make smooth and fair.

With the weighted centerboard complete, you need to build a hoisting system. In the drawing (Sheet 3), you will see the upper forward corner is cut off at 45 degrees to provide for two plates (bronze or stainless steel) to carry a bronze or plastic sheave ½ by 1½ to 2 inches, for a ⁵⁄₁₆-inch nylon pennant. The plates must be let into the sides of the ¾-inch board far enough for the sheave to clear amply, allowing for fiberglass sheathing under the plates. Dimension the plates ½ inch wider than the sheave so the pennant cannot jump out and jam. Best to fasten the plates with countersunk copper rivets filed flush.

The final job is to bore for the carriage bolt on which the centerboard swings. This is the simplest possible system; you could use a stainless steel pin, a snug ¼ or ⁵⁄₁₆ inch fit bored into the trunk side. The ⁵⁄₁₆-inch hole in the centerboard should be located at the center of the 2¼-inch radius of the forward lower corner (Sheet 3). Once this is done, measure carefully for the matching hole in the trunk sides. I find that the location 6 inches aft of the far forward end of the centerboard trunk will provide clearance for the sheave if you haul the boat out on a beach.

MAST

Plans buyers have discovered that the "improved" stainless steel sail track for small boats now costs $8.50 per foot, or $102. If you can afford that, go right into the mast construction (Sheet 7). If not, build the tricky all-wood sail track described below.

The plan shows two pieces of spruce or fir, ¾ × 3 inches tapered to 2 inches at the head. By all means, try to find full-length stock, 20 feet long (allowing for checking at the ends), 10 inches wide. Nesting the tapers should provide ample material for the 17-foot mast. The third piece starts with the same dimensions but it must be tapered in thickness from ½ inch at the foot to zero at the head, and be sandwiched between the other two. By using a chalkline or a long, stiff batten, you can mark a line to follow to rip the taper from 3 inches at the foot to 2 inches at the head, as per Sheet 7. This is better done with a hand power saw "free hand," unless you and a helper can use a table saw (also free hand). To taper the inside thickness, epoxy glue that piece (full thickness) to another. Let it cure for 24 hours lying on a flat surface or boxes placed carefully in line. The tapering by planing is now pure sweat! The assembly must be well-supported, of course. Glue the third piece, then dress neatly with plane or belt sander. Borrow or rent a router with a ½-inch rounding bit, and run it around all the corners.

A finishing touch is the groove in the head to take the halyard up to 5⁄16 inch or more. Use a rat-tail file or coarse sandpaper rolled on a ¼-inch rod or dowel. A careful saw-cut ¼-inch deep makes starting easy.

The boom is very simple. At most lumberyards you will find round fir sticks called dowels. Find a 6-footer of clear fir, 2 inches in diameter. Make a simple gooseneck of brass and a swiveling shoulder eyebolt. You can also find both of these in good marine supply shops.

You may finish the mast and wooden boom with clear varnish (four or five coats) or paint with Spar Tan or the color of your choice.

If you have rejected the $102 purchase of a stainless steel sail track, here is my wooden option. See the detail of the mast section in the upper right-hand corner of Sheet 7. This uses stock ¾-inch quarter round pine or fir, two lengths of 13 feet each. You will need the use of a table saw for a few minutes and a router set into a make-do table (Chapter Five) for less than an hour. You may have to rent a ⅜-inch corebox bit to form each half of the groove. The sectional view shows three steps. First, cut off cross-hatched area ⅛ inch to reduce thickness of the track, using the table saw with the rounded side against the fence, held securely with "feathers" (Chapter Four) clamped to fence and table, so no movement up or sideways is possible.

Second, see Chapters Three and Five for excellent little router-shapers. There is no other way to form the groove. Rent a bit, lock it firmly in the ¼-inch chuck. Secure a special fence shaped to guide the reduced quarter round on the table against another simple fence, or use "feathers" as on the table saw. Experiment with same-size scrap, moving the bit up in stages until you get the hang of it. Once fixed, don't move anything. Keep the material going on through; don't slow down or stop, and don't force it so that motor whine lowers its tone. The operations should take less than a minute per piece.

Third, set the saw blade to a height of 3⁄32-inch above the table to provide the space necessary for the sail cloth. Again use feathers for security; the part must be stable. And last, use sandpaper to knock off the sharp corners and rough surface from that cut; leave the sawn surface unsanded where it is glued to the mast and to the mating piece for better adhesion with epoxy glue.

Run a center line along the aft side of the mast. Brad and glue the pieces along the line and to each other, but wipe out glue that wells up inside the groove. Saw the angled end for the sail entry into the groove and sand it smooth. Furniture wax put into the groove with a small brush or Q-Tip will reduce friction. This track should be no heavier than the steel one, and will significantly stiffen any mast. And, of course, it saves a hundred bucks.

For those who prefer, extruded aluminum masts, with or without hardware, are available from me or your local marine store. Suitably sized masts are frequently available used.

RIGGING

Shrouds (side stays) and ¼-inch turnbuckles are attached to the hull before gunwales installation by means of stainless steel chainplates (check marine stores) or bronze pieces (you make), 3/32 by 3/4 by 6 inches, with 1/4-inch hole for turnbuckle pin (Sheet 1). Prepare to fasten outside or inside the hull with three #10 round-head bolts, at a 5½-degree angle to match the shroud, 15 inches aft of the mast, as shown. Glue a ½-inch-thick 3 × 9 plywood pad (Sheet 3) inside the hull. The two small jib-sheet leaders you buy are cast bronze, brass, or plastic. If in dire straits you can install brass shoulder eyebolts about 32 inches abaft the mast. But don't do anything permanent until you have proved the location by the set of the jib under sail. A small C-clamp will work for an hour or so, until you find the proper location for your jib-sheet lead.

The headstay turnbuckle goes to a shoulder eyebolt through the bow transom inside on a line from sheer to sheer. You may use another similar one for the jib tack leader and snap or shackle.

Get ideas for spar hardware— tangs (three), jib halyard block, cleats, gooseneck, etc.—from hardware catalogs, magazines, how-to books or from small sailboats in your area. If you are handy at all with metalwork, you should make your own.

SAILS

I've enclosed a sail plan kit (Sheet 1) not because I think you should make your own, but so you can show the plans to a professional sailmaker. I feel strongly that untrained people can no more make their own sails than they can build an engine for their cars. You know from my writings that I am all for saving a buck, but not on sails. Small imperfections can ruin the performance of a sail. Order from a sailmaker not too far away where you can get service in the future, if needed. Let him worry about weight, amount of roach, corner reinforcements, stitching, etc. If you don't have a local sailmaker, you can find some names in current sailing magazines in your library. You can also contact Sailrite Kits, 305 West Van Buren Street, Columbia City, IN, 46725. Now a full sail loft, Sailrite Kits has been making sails in kit form for more than twenty years. I'd estimate that sails for Trifle would cost about $350 or so, including battens and sail bag.

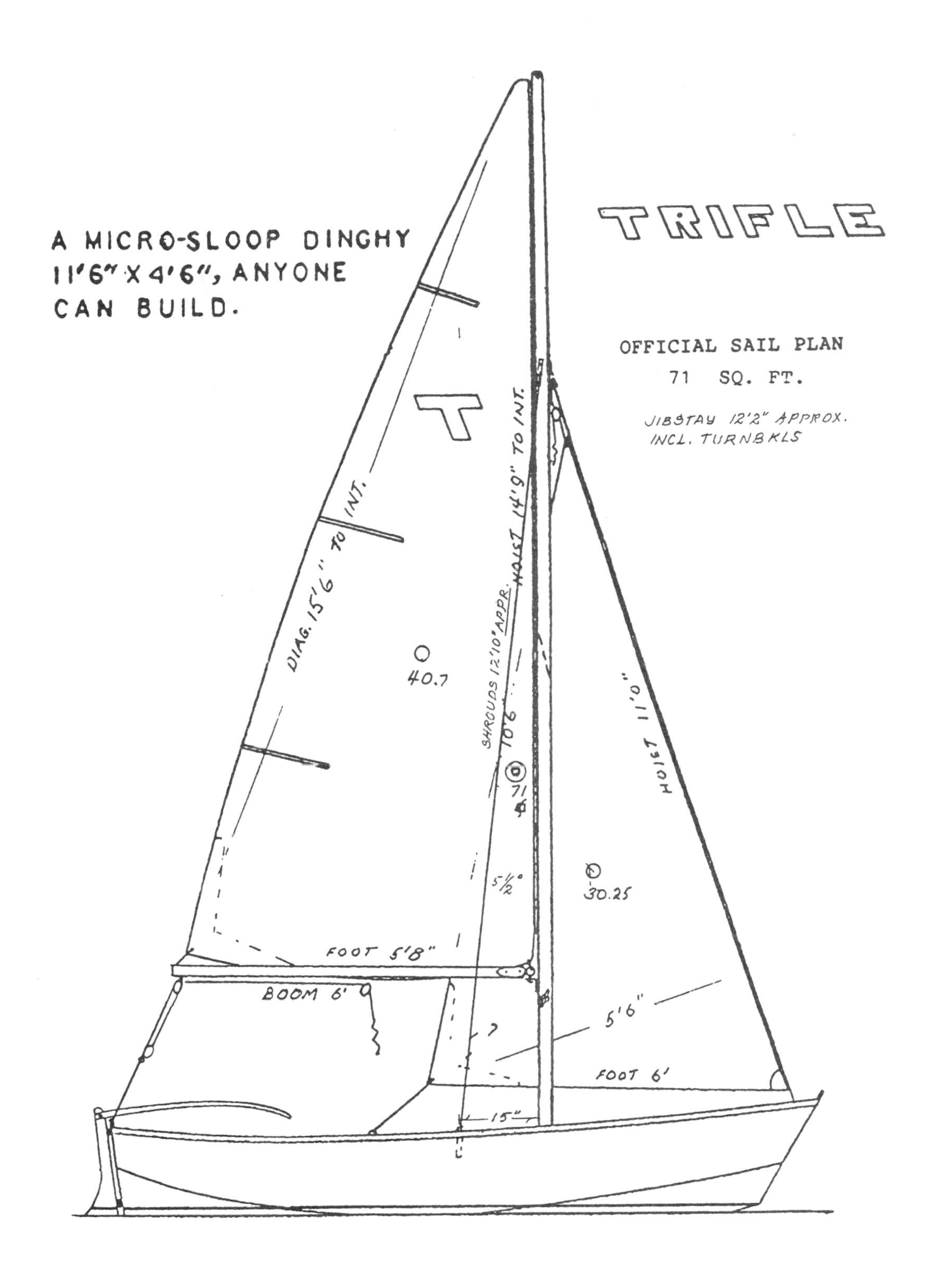
A MICRO-SLOOP DINGHY
11'6" X 4'6", ANYONE
CAN BUILD.
TRIFLE
OFFICIAL SAIL PLAN
71 SQ. FT.
JIBSTAY 12'2" APPROX.
INCL. TURNBKLS
DIAG. 15'6" TO INT.
HOIST 14'9" TO INT.
SHROUDS 12'10" APPR.
10'6"
40.7
71
HOIST 11'0"
5-1/2°
30.25
FOOT 5'8"
BOOM 6'
5'6"
7
FOOT 6'
15"

Sheet 1

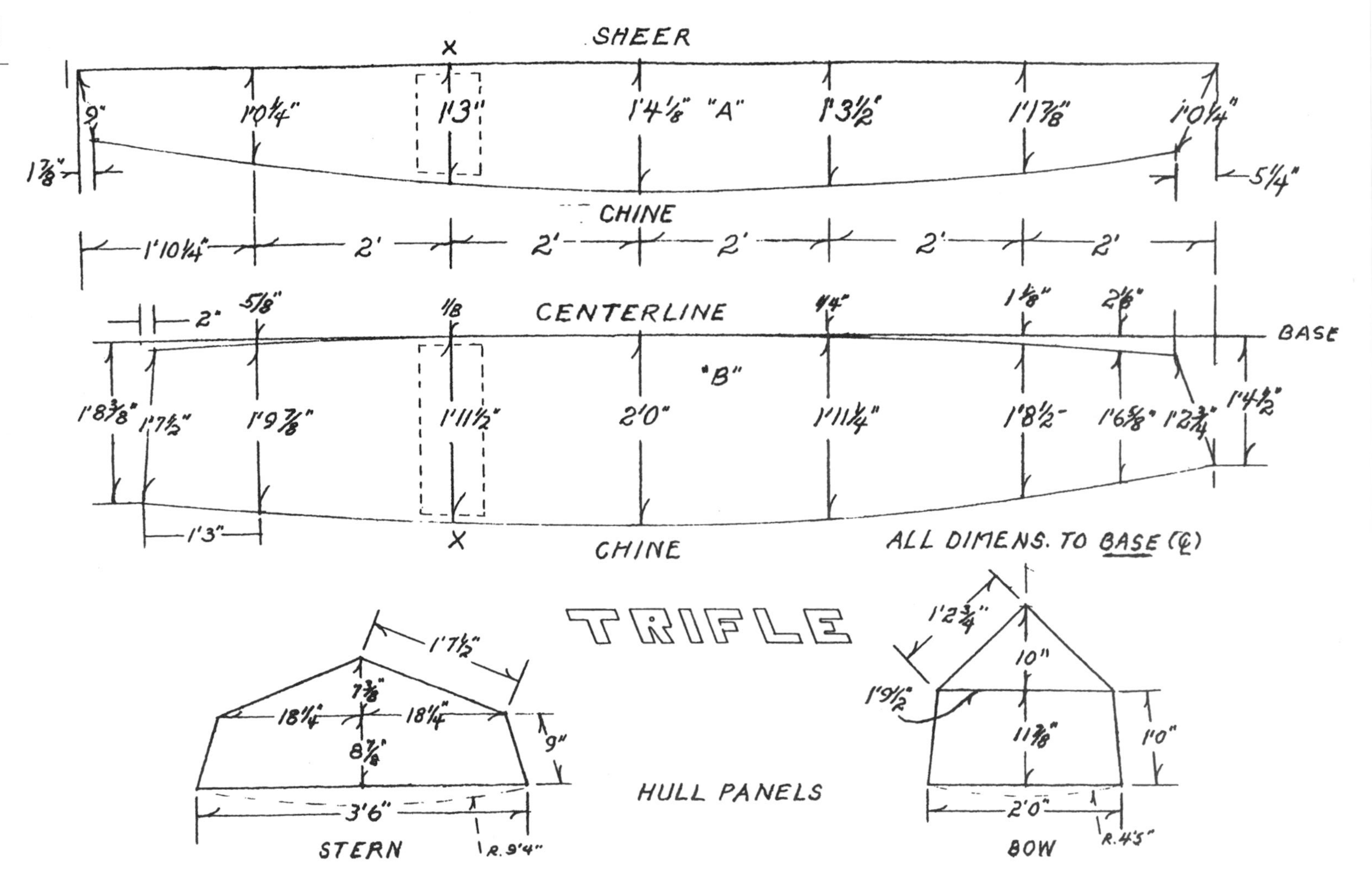
SHEER
CHINE
CENTERLINE
BASE
"A"
"B"
CHINE
ALL DIMENS. TO BASE (℄)
TRIFLE
HULL PANELS
STERN
BOW

Sheet 2

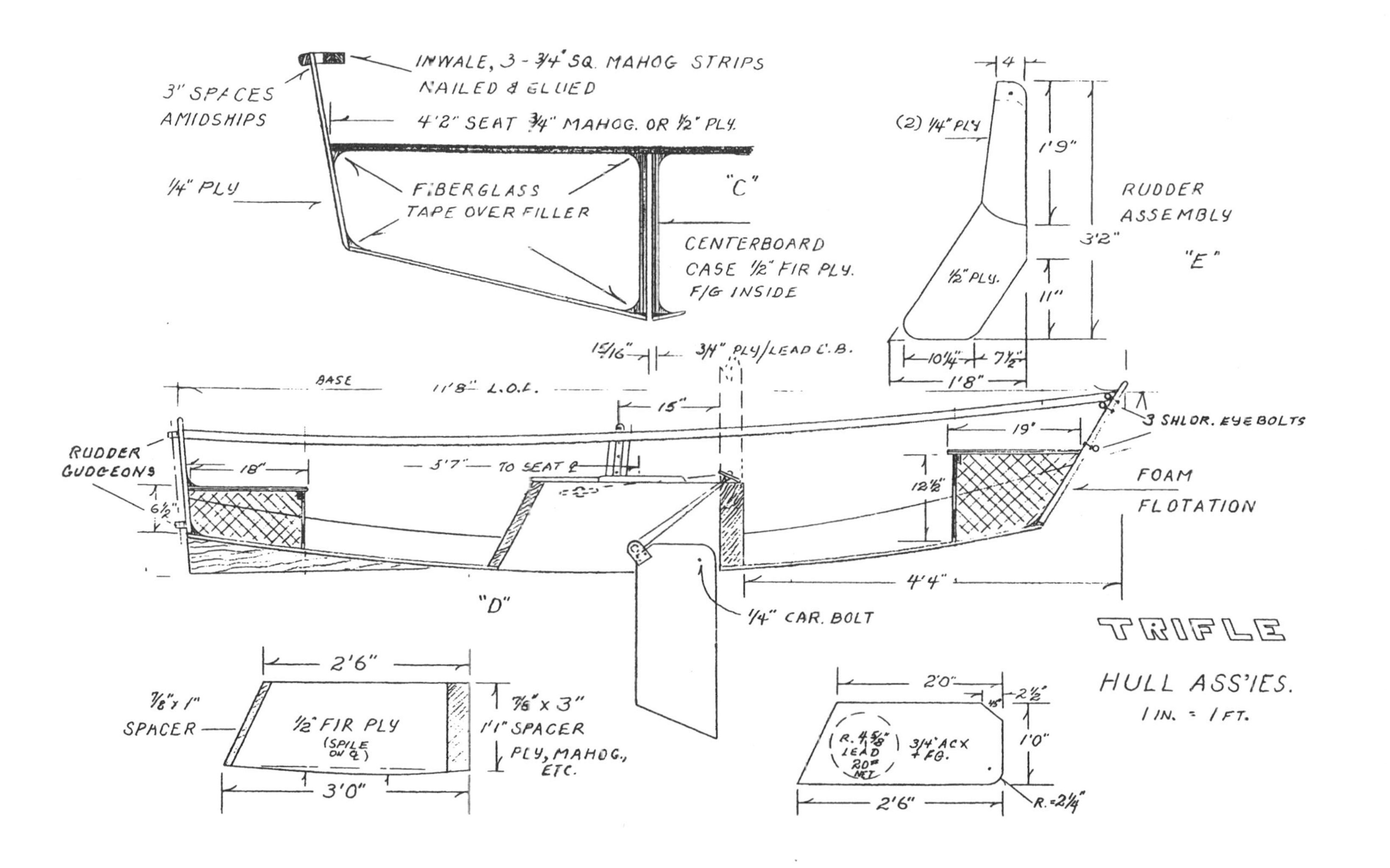
INWALE, 3 - 3/4" SQ. MAHOG STRIPS
NAILED & GLUED
3" SPACES
AMIDSHIPS
4'2" SEAT 3/4" MAHOG. OR 1/2" PLY.
1/4" PLY
FIBERGLASS
TAPE OVER FILLER
"C"
CENTERBOARD
CASE 1/2" FIR PLY.
F/G INSIDE
4
(2) 1/4" PLY
1'9"
RUDDER
ASSEMBLY
3'2"
"E"
1/2" PLY.
11"
10 1/4"
7 1/2"
1'8"
15/16"
3/4" PLY/LEAD C.B.
BASE
11'8" L.O.A.
15"
3 SHLDR. EYEBOLTS
19"
RUDDER
GUDGEONS
18"
3'7" TO SEAT
12 1/2"
FOAM
FLOTATION
6 1/2"
4'4"
"D"
1/4" CAR. BOLT
TRIFLE
2'6"
7/8" x 1"
SPACER
1/2" FIR PLY
(SPILE ON Q)
7/8" x 3"
1'1" SPACER
PLY, MAHOG.,
ETC.
3'0"
2'0"
2 1/2"
HULL ASS'IES.
1 IN. = 1 FT.
R. 4 5/8"
LEAD
20# NET
3/4" ACX
+ FG.
1'0"
2'6"
R.=2 1/4"

Sheet 3

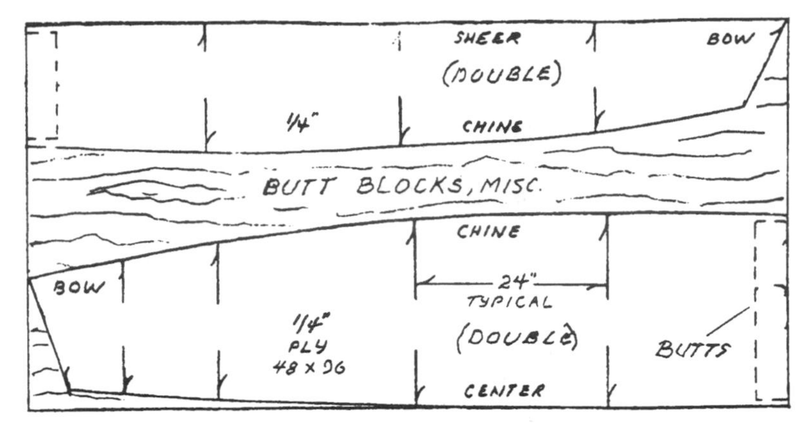

1. TWO 1/4-INCH SHEETS CLAMPED LIKE FACES TOGETHER SO BOTH CAN BE SAWED AND DRESSED SIMULTANEOUSLY

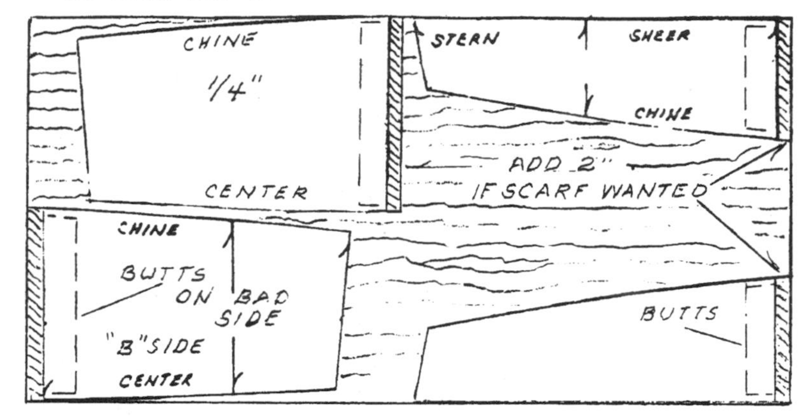

2. SINGLE SHEET REQUIRES SEPARATE SAWING, THEN LIKE FACES PLACED TOGETHER FOR PLANING AND FINISHING

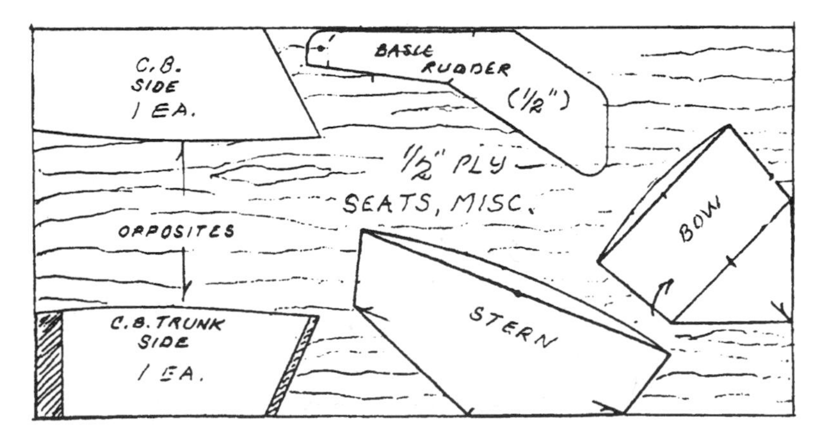

ARRANGEMENT OF PATTERNS ON PLYWOOD SHEETS SHOWING OPTIONS FOR BUTTED OR SCARFED JOINTS

Sheet 4

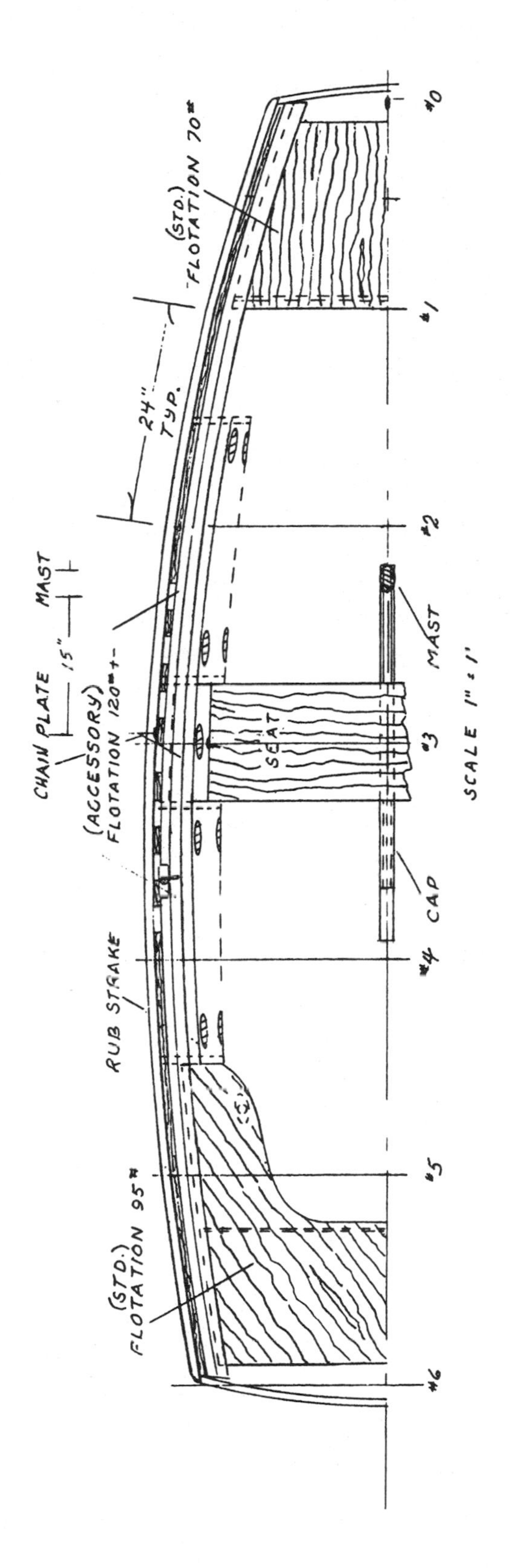

TRIFLE ARRANGEMENT PLAN:
FLOTATION, SEATS, GUNWALES, CENTERBOARD TRUNK

Sheet 5

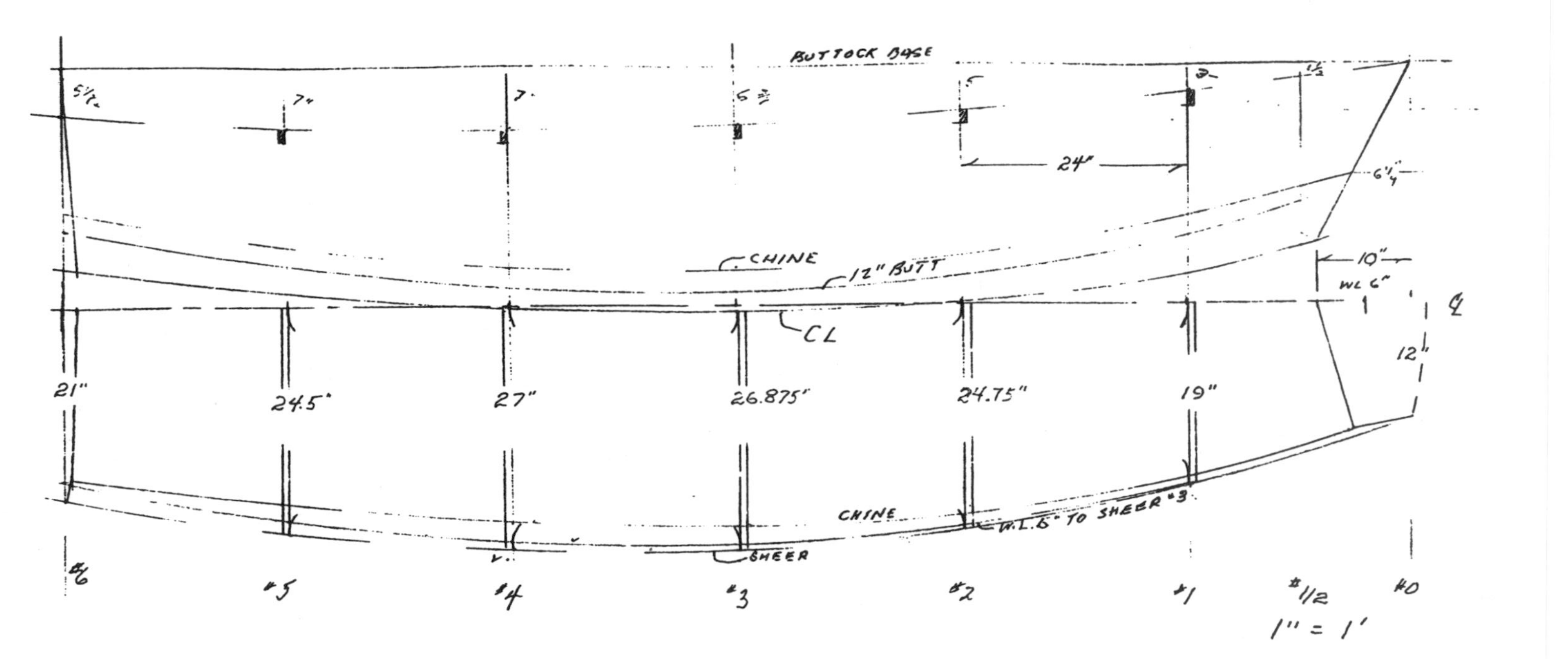

TRIFLE BEAM SPACERS

DIMENSIONS TO INSIDE OF 1/4" HULL
ON 24" centers starting from bow at sheer.

#0 24"   #1 38"   #2 49½'
#3 53 3/4"   #4 54"   #5 49"   #6 42"

Note: #0 and #6 are Transom widths

Suggest 1x2 shaped each end to meet inside of hull, taped temporarily in placed, each to be removed for forming fillets, installation of centerboard trunk, seats, etc.

TRIFLE LINES

AND BEAM SPACERS

Sheet 6

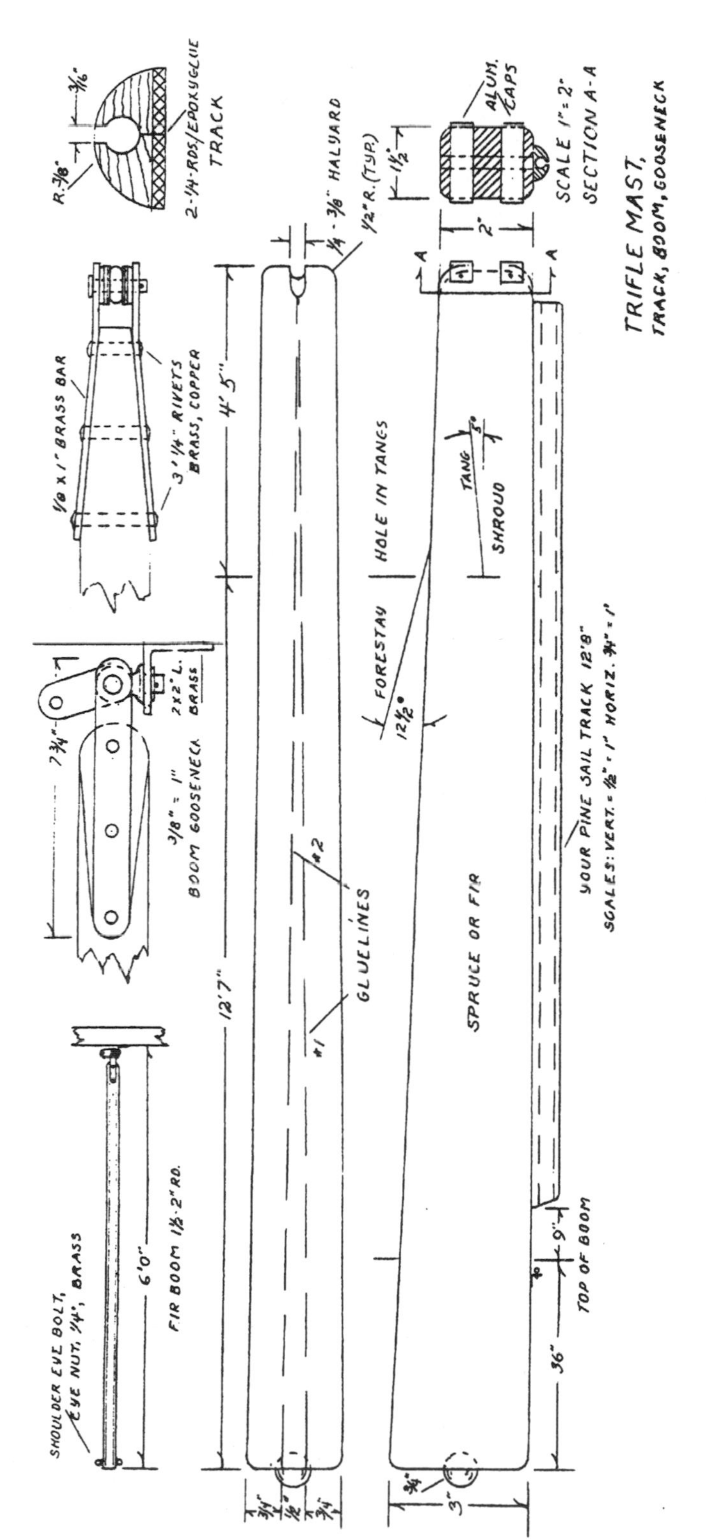
SHOULDER EYE BOLT, EYE NUT, 1/4", BRASS
6'0"
FIR BOOM 1 1/2-2" RD.
7 3/4"
3/8" = 1" BOOM GOOSENECK
2x2" L. BRASS
1/8 x 1" BRASS BAR
3 1/4" RIVETS BRASS, COPPER
R. 3/8"
3/16"
2-1/4-RDS/EPOXY GLUE TRACK
12'7"
4'5"
1/4 - 3/8" HALYARD
1/2" R. (TYP.)
#1
#2
GLUELINES
FORESTAY
12 1/2°
HOLE IN TANGS
SPRUCE OR FIR
TANG
5°
SHROUD
A
2"
1 1/2"
ALUM. CAPS
SCALE 1" = 2'
SECTION A-A
YOUR PINE SAIL TRACK 12'8"
SCALES: VERT. = 1/2" = 1' HORIZ. 3/4" = 1'
36"
9"
TOP OF BOOM
3"
TRIFLE MAST,
TRACK, BOOM, GOOSENECK

Sheet 7

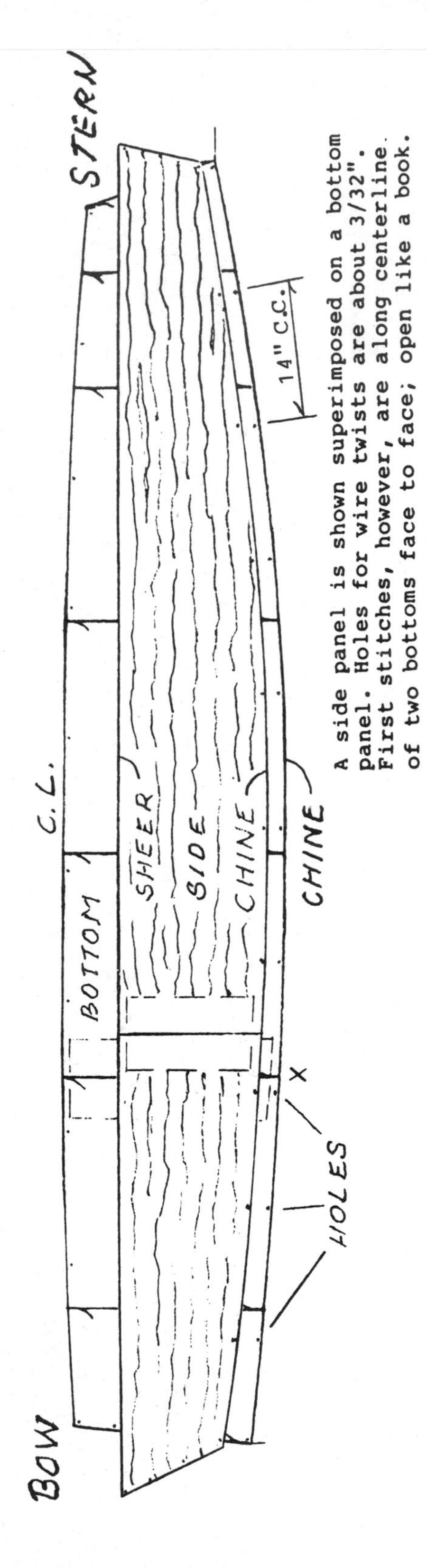

A side panel is shown superimposed on a bottom panel. Holes for wire twists are about 3/32". First stitches, however, are along centerline of two bottoms face to face; open like a book.

Sheet 8

Trifle under sail.

INDEX

T

V

W